Advanced Welding Technologies

Scrivener Publishing
100 Cummings Center, Suite 541J
Beverly, MA 01915-6106

Advances in Production Engineering

Scope: "Advances in Production Engineering" addresses recent developments and research issues related to the production engineering which necessitates the development of new materials and manufacturing processes in order to produce high-quality products at lower prices and in less time. This series aims to provide a scientific platform for researchers, practitioners, professionals, and academicians to discuss the most recent technological developments in metals, polymers, ceramics, composites, biomaterials, nanomaterials, special materials, metals, micro-forming, powder metallurgy, ceramics processing, non-traditional machining, high speed machining, micro and nanomachining, and laser processing. Tribological analysis, friction behavior, modelling, and optimization techniques in materials, machining, and manufacturing are also covered in the series.

Edited by: Dr. Sandip Kunar
Email: drsandipkunar@gmail.com

Publishers at Scrivener
Martin Scrivener (martin@scrivenerpublishing.com)
Phillip Carmical (pcarmical@scrivenerpublishing.com)

Advanced Welding Technologies

Edited by

Sandip Kunar
Department of Mechanical Engineering, Aditya University, Surampalem, India

and

Gurudas Mandal
Department of Metallurgical Engineering, Kazi Nazrul University, West Bengal, India

Scrivener Publishing

WILEY

This edition first published 2025 by John Wiley & Sons, Inc., 111 River Street, Hoboken, NJ 07030, USA and Scrivener Publishing LLC, 100 Cummings Center, Suite 541J, Beverly, MA 01915, USA

For more information about Scrivener publications please visit www.scrivenerpublishing.com.

Wiley Global Headquarters
111 River Street, Hoboken, NJ 07030, USA

For details of our global editorial offices, customer services, and more information about Wiley products visit us at www.wiley.com.

Library of Congress Cataloging-in-Publication Data

ISBN 9781394331895

Front cover image courtesy of Adobe Firefly
Cover design by Russell Richardson

Set in size of 11pt and Minion Pro by Manila Typesetting Company, Makati, Philippines

Printed in the USA

10 9 8 7 6 5 4 3 2 1

Contents

Preface

Welding has traditionally been considered more of a skill-based proficiency than a technological industrial method. The reliance on highly trained human operators, along with the high cost and low reproducibility of many welding processes, has contributed to this perception. As a result, welding is classified as a "Special Process" in international quality standards. Extensive research has now been conducted on the fundamentals of welding processes, and while it remains a complex, multidisciplinary subject, the basic concepts are well understood. Significant technological advancements are built on this solid scientific foundation.

This book covers a range of advanced welding technologies that can be leveraged by the production industry to boost productivity and socio-economic development. Additionally, it highlights current research in advanced welding technologies and offers insights into the latest advancements in the welding industry.

Chapter 1 explores the crucial role of advanced welding in sophisticated manufacturing, highlighting its potential for future innovation and its transformative impact on various welding industries. Chapter 2 details significant developments and recent progress in the use of advanced welding processes. Chapter 3 covers the key requirements for arc welding power supplies, advancements in power source technology, and the theories behind both traditional and complex power source designs, along with their benefits and challenges.

Chapter 4 examines the evolution of automation and robotics, highlighting their impact on efficiency, precision, and safety in various welding industries. Chapter 5 introduces the macroscopic modeling approach to the friction stir welding process, focusing on material flow and heat transfer analysis. Chapter 6 addresses the efficiency, energy yield, economic and social considerations in protecting people and property, as well as the environmental impact of atomic hydrogen welding. Chapter 7 discusses ultrasonic welding, an advanced technology for joining materials without melting, and explores its benefits, including the ability to join dissimilar metals and work near heat-sensitive components without causing damage.

Chapter 8 provides a thorough analysis of the explosion welding process, focusing on its metallurgical features, key process variables, and inherent constraints. Chapter 9 discusses the progress and applications of hybrid laser arc welding technology. Chapter 10 delves into the core concepts of ultrasonic welding, including the functions of critical components such as the ultrasonic generator, transducer, horn, and booster. The chapter also examines the impact of ultrasonic welding on various materials, including aluminum alloys, steels, titanium, magnesium, and fiber-reinforced composites, highlighting the benefits and challenges associated with each material.

Chapter 11 introduces a machine learning model using Kernel Ridge Regression to predict the current flow in Tungsten Inert Gas Welding, achieving a notably low root mean squared error (RMSE) of 2.10, outperforming other models in this field. Similarly, a Gaussian Process Classifier model is presented to predict void formation in Friction Stir Welding. Chapter 12 discusses the control and monitoring of real-life welding situations through machine learning-derived Integrated Master Models. Chapter 13 explores the potential of AI and ML in welding technologies to shape the future of smart manufacturing and define new standards in welding processes.

Chapter 14 highlights the IoT-based aspects of welding, emphasizing their role in driving revolutionary changes in smart manufacturing, along with significant improvements in productivity, quality control, and worker safety. Chapter 15 summarizes recent advancements, their impact on industry practices, and potential future projections, providing insight into the current state of welding quality assurance. Chapter 16 discusses the sustainability of welding operations, focusing on environmentally friendly methods that use minimal energy, maximize process efficiency, produce better results with fewer resources, and generate minimal waste during operation.

Chapter 17 examines the ductile strength, bending characteristics, hardness, and microstructural features of specimens produced through the friction stir welding process. Chapter 18 analyzes joint strength during microwave welding of polymer-based materials, focusing on different input process parameter combinations and optimizing these parameters. Chapter 19 explores the use of metaheuristic approaches to optimize key friction stir welding parameters, such as rotational speed, welding speed, and axial force. The study investigates how various factors, including pin shape, axial force, rotational speed, and welding speed, impact the tensile strength and elongation of welded joints.

Chapter 20 discusses the features of welding and laser beam arc welding combined in a single weld pool, known as the laser-arc hybrid welding process. It covers the basic principles, advantages, disadvantages, and applications of this technique. Chapter 21 highlights various green welding processes, emphasizing their advantages in terms of lower energy consumption, reduced emissions, and practical applications. Chapter 22 explores the multifaceted components of sustainability in different welding and processing methods, providing an in-depth analysis of energy conservation, material waste reduction, resource optimization, and process parameters, while underscoring the environmental and cost-efficiency benefits of diverse welding techniques.

Chapter 23 highlights a patent landscape analysis to explore new research directions and emerging trends in robotic welding sensing technology. Chapter 24 explains sensing systems, signal monitoring, supervised learning, adaptive robotic welding, and path planning. Chapter 25 provides an overview of wire arc additive manufacturing and its connection to welding, with a focus on the advancements this technology brings to automated systems, real-time monitoring, and sensing in advanced manufacturing. Chapter 26 discusses major developments and future trends in welding technology, emphasizing how smart manufacturing knowledge is being applied to sustain advanced materials, such as high-entropy alloys, in welded components.

Weldability and metallurgical research have played a major role in advancing welding techniques, enabling the joining of various materials, maintaining joint quality, and ensuring the integrity of welded joints. While further research is needed in materials science, especially for the successful joining of advanced materials, recent efforts have focused on improving our understanding of fundamental processes, equipment and consumable development, as well as automation and control.

The editors are grateful to everyone who has supported their work and research, and also wish to thank Martin Scrivener and Scrivener Publishing for their support and publication.

February 2025

Acknowledgements

I express my gratitude to my all authors, and colleagues in the welding industry worldwide, and my students, whose work is extensively featured in the pages that follow and who have taught me a great deal. I also thank them for their ongoing support, guidance, and encouragement. Finally, I want to express my gratitude to my family members for their patience during the lengthy gestation of this work.

Sandip Kunar

Part I

ADVANCED WELDING PROCESSES

1

Introduction to Advanced Welding Processes

Sandip Kunar[1]*, Jagadeesha T.[2], Chika Oliver Ujah[3], Norfazillah Talib[4], Gurudas Mandal[5], Akhilesh Kumar Singh[6] and Aezeden Mohamed[7]

[1]*Department of Mechanical Engineering, Aditya University, Surampalem, Andhra Pradesh, India*
[2]*Department of Mechanical Engineering, National Institute of Technology, Calicut, India*
[3]*Department of Mechanical and Industrial Engineering Technology, University of Johannesburg, Johannesburg, South Africa*
[4]*Department of Metallurgical Engineering, Kazi Nazrul University, Asansol, India*
[5]*Department of Manufacturing Engineering, Universiti Tun Hussein Onn Malaysia, Johor, Malaysia*
[6]*Department of Mechanical Engineering, Aditya College of Engineering and Technology, Surampalem, Andhra Pradesh, India*
[7]*Department of Mechanical Engineering, PNG University of Technology, Lae, Morobe Province, Papua New Guinea*

Abstract

The field of advanced manufacturing has explored a prominent revolution in the techniques of welding. This article reveals the development and importance of welding techniques in relation to state-of-the-art manufacturing practices. Among the cutting-edge technologies, promoting accuracy and efficiency in welding are friction stir, laser, electron beam, and ultrasonic welding. These procedures are essential in some advanced sectors, including aerospace, electronics, and medical device, because they provide benefits like decreased distortion, fewer heat-affected zones, and better weld quality. Combining additive manufacturing with welding opens new opportunities for creating complex structures and strengthening weak spots. Robotic welding systems have improved productivity and worker safety in production by producing uniform, high-quality welds. The qualities of several techniques are combined in hybrid welding procedures, which highlight flexibility

**Corresponding author*: sandip.sandip.kunar@gmail.com

Sandip Kunar and Gurudas Mandal (eds.) Advanced Welding Technologies, (3–24) © 2025 Scrivener Publishing LLC

and optimization for a range of applications. Although there are many advantages to welding in advanced production, difficulties still exist. Maintaining uniformity and quality throughout large-scale production is still an issue, especially when working with complex designs. Research into more environmental friendly and energy-efficient welding procedures is fueled by the pursuit of sustainability. In conclusion, companies have been able to attain improved precision, efficiency, and quality credits to the evolution of welding processes in tandem with advanced manufacturing techniques. This article clarifies the crucial role that welding plays in the context of sophisticated manufacturing, emphasizing both its potential for future innovation and its revolutionary effect on a variety of industries.

Keywords: Ultrasonic welding, robotic welding, friction stir welding, additive manufacturing, 3D printing

1.1 Introduction

A variety of cutting-edge methods are included in advanced manufacturing, which aims to maximize production processes' effectiveness, sustainability, and quality. It improves conventional production techniques by fusing cutting-edge technology including automation, robotics, additive manufacturing, and data analytics. With this strategy, industries may meet market expectations with more flexibility, less waste, and higher precision. Its disruptive impact benefits key industries like electronics, automobiles, and aerospace. Advanced manufacturing is still evolving, and research and development are shaping its uses [1–3]. Advanced manufacturing refers to a collection of state-of-the-art technologies that improve productivity, accuracy, and sustainability in manufacturing operations. In this regard, welding procedures have experienced notable changes that allow industry to push the limits of innovation. This article explores the diverse welding techniques that have surfaced in the field of advanced manufacturing emphasizing their advantages, uses, and contributions to the contemporary industrial environment. Modern industries are being driven by advanced manufacturing, which is transforming production methods, increasing sustainability, and boosting product quality. By optimizing manufacturing operations through the integration of cutting-edge technology and procedures, this transformative strategy helps companies maintain their competitiveness in the ever-evolving market landscape of today. Welding has been a crucial step in the production process joining materials to produce buildings and goods on a range of sizes [4–6]. But as sophisticated manufacturing methods have emerged, welding's function has changed dramatically. This article explains the critical role that welding plays in the context of advanced manufacturing highlighting the technology's revolutionary impact on a range of industries as well as its potential for future innovation.

1.2 Defining Advanced Manufacturing

A vast array of cutting-edge methods and tools that improve on conventional production processes are included in advanced manufacturing. To increase productivity, accuracy, and flexibility, it makes use of automation, robotics, additive manufacturing, data analytics, and other cutting-edge technologies. The main objective of advanced manufacturing is to produce goods of a higher caliber while cutting lead times and waste, better satisfying market, and consumer needs.

1.3 Technologies Enabling Sophisticated Manufacturing

1.3.1 Additive Manufacturing

Additive manufacturing, or 3D printing, is a fundamental component of advanced manufacturing. This method uses computerized designs to build objects layer by layer. It provides unmatched design freedom making it possible to create complex geometries and lightweight structures that were previously impossible to achieve with conventional techniques. Particularly benefiting from additive manufacturing's capacity to create complicated components with less material waste and quicker production cycles are the aerospace, healthcare, and automotive industries.

1.3.2 Robotics and Automation

Manufacturing has undergone a revolution for transforming to automation and robotics, which have improved production processes' accuracy, uniformity, and efficiency. An expanding number of functions, from material handling and packaging to assembly and quality control, are being performed by robots. In a variety of industries, collaborative robots, or COBOTS, augment human operators while increasing output and security.

1.3.3 Data Analytics and Industrial IoT

For welding processes to be optimized, data-driven insights are essential. Real-time data gathering and analysis are made possible by the Industrial Internet of Things (IIoT), which links machinery and devices. With the use of this data, decisions that improve operations and decrease downtime may be made, inefficiencies can be found, and maintenance requirements can be anticipated.

1.3.4 Benefits and Impact

The following numerous significant advantages of advanced welding methods have an impact on various industries:

- **Better Quality:** By integrating advanced welding technologies, human error is reduced, and product quality is raised, producing output that is more dependable and consistent.
- **Less Waste and Expense:** Advanced welding methods cut down on material waste, and labor expenses are decreased by automation, which takes the place of labor-intensive jobs.
- **Greater Customization:** Advanced welding methods can now offer a level of personalization that was previously unfeasible meeting the needs of each consumer.
- **Quicker Time-to-Market:** Advanced welding methods enable shorter manufacturing cycles, quick product development, and market entry.

1.3.5 Applications and Industries

Many different industries, as follows, are impacted by advanced welding methods:

- **Aerospace:** The development of lightweight, fuel-efficient components is made possible by advanced welding methods, which are essential for the aerospace industry's developments.
- **Automobile:** Production lines have been optimized, robotics and automation have changed the automobile industry, and the development of electric and driverless vehicles has been made possible.

- **Healthcare:** By creating specialized medical implants and gadgets, additive manufacturing improves surgical techniques and patient care.
- **Electronics:** The electronics industry depends heavily on precision and miniaturization, which means that sophisticated components must be produced using sophisticated advanced welding methods.

1.4 Welding Processes

1.4.1 Laser Welding

A concentrated laser beam is utilized in the high-precision welding process known as laser welding to produce a concentrated heat source. This method is well-known for producing complex and fine welds, which makes it a preferred option in sectors where accuracy is crucial. Because the laser beam is concentrated, it can be utilized to weld in tight spaces with less heat-affected zones. This makes it especially appropriate for uses involving sensitive materials, such electronics, medical equipment, and aerospace parts. Beyond accuracy, laser welding has many advantages. Less post-welding rework is required because distortion is reduced because of the localized heating. Furthermore, less wear and tear are placed on the equipment because there is no physical contact between the welding tool and the workpiece. Automation is also made possible by laser welding since robots can be trained to repeatedly produce incredibly accurate welds. With its unmatched accuracy and efficiency, laser welding has become a key technique in advanced production. By using a concentrated laser beam to produce high-intensity heat in a specific area, the materials fuse together quickly. Because laser welding can create complex and tiny welds, it is a crucial method in precision-driven industries like the fabrication of medical devices, electronics, and aircraft [7].

Reducing thermal distortion and maintaining material qualities by minimizing the heat-affected zone (HAZ) is one of the many impressive benefits of laser welding. This quality is especially important for maintaining the structural integrity of the finished product when working with delicate or heat-sensitive materials. The laser beam's focused quality also makes it easier to automate processes and weld in confined spaces, which improves production efficiency. Applications for laser welding can be found in fields that need both speed and accuracy. It is used in the aerospace industry to connect lightweight parts that are essential to aircraft efficiency.

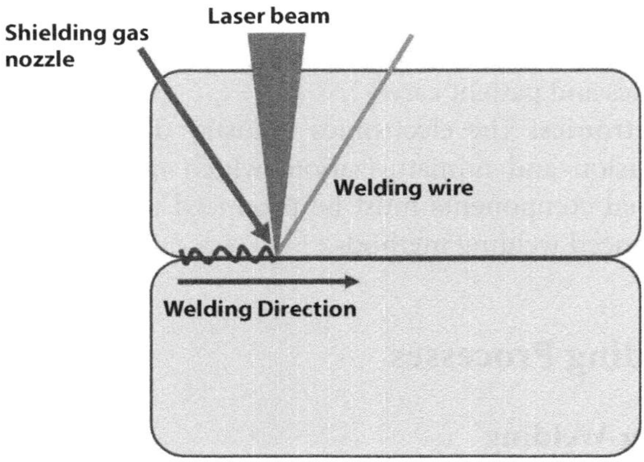

Figure 1.1 Laser welding.

The method's capacity to make complex connections with little harm to delicate components is advantageous to electronics. Laser welding is used in the production of medical devices to precisely combine tiny components while upholding strict quality and safety requirements [8]. Nonetheless, there are still issues with laser welding, namely, its sensitivity to surface conditions and material characteristics. Changes in the thickness and composition of the material might affect the welding process and lead to uneven weld quality. The goal of ongoing research is to better understand and control these elements to enhance process reliability. In summary, laser welding is a progressive technology in advanced manufacturing that offers the accuracy and productivity needed by contemporary businesses. Its expanding importance in industries that prioritize accuracy and innovation is attributed to its ability to produce precise welds, eliminate thermal distortion, and enable automation [9]. Figure 1.1 shows the picture of laser welding.

1.4.2 Electron Beam Welding

Another cutting-edge welding method that uses a concentrated energy beam in this example, a stream of fast electrons, is electron beam welding. The surrounding material is kept mostly intact since this method produces a deep, narrow weld with little heat dispersion. In industries like aerospace and automobile manufacture, where combining high-strength materials is essential, electron beam welding is very useful. Welding produces strong and dependable welds since there is no air interaction throughout the process. To avoid electrons being scattered by air molecules, electron beam welding necessitates a vacuum environment. Its application may be limited in some situations by this restriction. However, the method is a useful tool in advanced manufacturing since it can fuse disparate materials together and create deep welds with little deformation. In the field of advanced manufacturing, electron beam welding (EBW) has become well known for its accuracy and versatility in combining high-strength materials. Using a high-velocity electron stream, this method produces focused heat that allows for narrow, deep welds with little heat dispersion. Because of its special qualities, EBW is a priceless tool for sectors that depend heavily on strong, dependable welds like nuclear power generation, automotive, and aerospace [10].

The capacity of EBW to produce welds without the requirement for a filler material results in joints with fewer impurities and a lower danger of corrosion, which is one of its main advantages. In addition, the procedure is carried out in a vacuum or low-pressure setting to avoid air contamination that can erode the weld. Because of this, EBW is especially well-suited for applications involving materials like titanium alloys and refractory metals, which are difficult to weld using traditional techniques. The deep penetration capabilities of EBW are beneficial to sectors that need reliable and sturdy couplings. For parts like aviation engine turbine blades, EBW is utilized in aerospace to provide high-temperature performance and structural integrity. The capacity of EBW to combine incompatible materials with little distortion helps the automobile industry by enabling the creation of lightweight and fuel-efficient vehicles. EBW is not without difficulties, though. The technique is made more difficult and expensive by the vacuum requirement, which limits its use in some situations. Additionally, complex machinery and trained operators are needed for the accurate management of electron beams. These issues are still being researched to improve EBW's usability and efficiency in advanced manufacturing. To sum up, electron beam welding is an essential process in advanced manufacturing that allows high-strength materials to be precisely joined for sectors that need durable and dependable components. Its importance in industries where quality and strength are crucial is attributed to its deep penetration capabilities and minimum heat-affected zone [11].

1.4.3 Friction Stir Welding

In the solid-state welding technique known as friction stir welding (FSW), materials are joined without melting. Rather, heat is produced by friction as a rotating tool passes along the material-to-material joint. To create a solid weld, the softened material is then mechanically combined. For materials like aluminum alloys that are difficult to fuse using conventional fusion procedures, FSW is especially helpful. Decrease in thermal distortion and elimination of flaws associated to solidification are two of the main benefits of FSW. Because of this, it can be used in the automotive, aerospace, and shipbuilding industries. Because FSW produces less emissions and uses less energy than other welding techniques, it is also environmentally suitable. In the field of advanced manufacturing, friction stir welding (FSW) has become a suitable welding method transforming the solid-state joining of materials. Using a non-consumable rotating tool that rotates along the joint to create frictional heat, FSW produces plasticized material that is mechanically combined to form a solid weld. Numerous benefits make this novel technique an invaluable tool in a variety of industries, including shipbuilding and aerospace [12].

FSW's capacity to fuse materials that are typically difficult to combine using fusion techniques, like aluminum alloys and other high-strength materials, is one of its main advantages. By using this procedure, solidification-related problems are resolved, and porosity and other defects that are frequently seen in conventional fusion welding are avoided. Furthermore, by minimizing thermal distortion caused by the lack of a melting phase, complex components can be produced with greater accuracy. The applications of FSW are diverse throughout industry. Lightweight and strong components are essential for fuel efficiency in aircraft, and this technology produces them. The capacity of FSW to join thick plates without additional materials or post-welding procedures is beneficial to the shipbuilding industry. FSW is used by the automobile industry for structural parts improving vehicle performance and safety. Notwithstanding its benefits, FSW is not without difficulties. Precise control over parameters including traverse speed, rotational speed, and tool design is required for the process. These variables affect the weld's quality and need for knowledgeable operators. The goal of ongoing research is to further improve the efficacy and efficiency of FSW by adjusting its parameters for various materials and applications.

Conclusively, friction stir welding is a fundamental aspect of advanced manufacturing providing solid-state joining with enhanced characteristics above traditional fusion welding. Its minimum deformation properties and capacity to overcome the difficulties presented by hard-to-weld materials make it an essential tool in contemporary industries. FSW plays a significant role in the aviation sector. FSW has become recognized as a unique and important development in the field of welding technology. Weld strength, quality, heat generation, and material mixing are all greatly influenced by several parameters, some of which are as follows: rotation speed, feed rate, weld speed, pin length, pin diameter, tool angle, and shoulder diameter and profile. Investigating the process variables that affect the properties of welded products is the main goal of the current research [13].

1.4.4 Ultrasonic Welding

High-frequency vibrations are used in ultrasonic welding to generate heat at the interface between two materials fusing them together. When precise, quick welding and airtight sealing are required, this method is frequently applied to metals and polymers. Ultrasonic welding is more economical and cleaner because it does not use consumables or open flames.

Ultrasonic welding is useful in industries, like electronics, medical devices, and automobile manufacture, because it can fuse disparate materials together while preserving the integrity of small parts. Because of its efficiency and speed, the technique works well for large-scale production. With the use of high-frequency vibrations, ultrasonic welding has become an essential process in sophisticated production allowing for precise bonding. The two materials fuse together because of the localized heat created by this reaction at their interface. The benefits of ultrasonic welding, including its quick speed, low thermal distortion, and cleanliness, have made it an indispensable instrument for sectors requiring effective and superior connections.

The capacity of ultrasonic welding to produce hermetic and airtight seals is one of its main advantages; this is important for applications in sectors like electronics, medical devices, and automotive manufacture. This method works especially well for putting different materials together, such as metals and polymers, without the need for extra consumables or adhesives. Furthermore, ultrasonic welding is a clean and environmentally friendly method because it does not involve open flames or outside heat sources. Ultrasonic welding is used in many different industries. It guarantees dependable connections for fragile components in the electronic production process. Ultrasonic welding is used in the medical business to create precise and sterile medical devices. Its efficiency and speed are advantageous to automotive manufacturers, particularly for applications requiring high-volume production. There are restrictions with ultrasonic welding, though. Because the heat generation is localized, the technique works best for smaller components. For a strong bond, larger components may need more than one weld point. The growing use of fiber-reinforced thermoplastic (FRTP) in industries has led numerous researchers to concentrate on the welding of hybrid constructions made of both metal and

FRTP. One economical and effective way to combine metal and FRTP dissimilar materials is via ultrasonic welding. The current state of knowledge regarding metal/FRTP ultrasonic welding is compiled in this work from four perspectives as follows: mechanical properties, joining mechanism, welding process, and galvanic corrosion [14]. Solid-state ultrasonic spot welding was used to weld 2024Al alloy to SiC/2009Al composite in a dissimilar manner. It was possible to obtain sound spot welds that were flawless. During the strong high-frequency rubbing, it was observed that SiCp particles migrated from the composite side to the 2024Al side in tandem with the interfacial interlocking. This migration was observed to increase with increasing welding energy. Nearly continuous microhardness was observed across the ultrasonic spot-welded joints, both identical and different [15].

Furthermore, since certain materials could not react well to ultrasonic vibrations, the materials being welded must be compatible with the technique. Research is still being conducted to improve the parameters of ultrasonic welding and provide methods for a wider range of materials, which will increase the technology's usefulness in advanced manufacturing. In summary, ultrasonic welding is a crucial process in advanced manufacturing that offers dependable and effective bonding for sectors that need accuracy and speed. Its importance in contemporary production is highlighted by its capacity to produce airtight seals, operate with different materials, and reduce environmental effect.

1.4.5 Additive Manufacturing and Welding

The process of designing and creating items is being revolutionized by additive manufacturing, also referred to as 3D printing. Additive manufacturing (AM) can be combined with welding techniques, even though it is not a typical welding process, to build complex and useful objects. Larger and more intricate constructions can be created by occasionally using welding to attach 3D-printed components. With this hybrid method, a 3D-printed object's structural integrity can be improved by using welding to strengthen important sections. Furthermore, new design options may arise from the combination of welding and additive manufacturing enabling the construction of optimal, light-weight structures with little material waste. There is a steady rise in the application of welding processes in additive manufacturing [16]. Thin material layers are added to parts made with additive manufacturing (AM) using various welding techniques. Laser beam welding, gas metal arc welding (GMAW/MIG), gas tungsten arc welding (GTAW/TIG), plasma arc welding (PAW), friction stir welding (FSW), ultrasonic welding, and electron beam welding are a few of the most used metal AM welding techniques.

1.4.6 Robotic Welding and Automation

Automation is frequently associated with advanced manufacturing, and welding is no exception. Robotic welding technologies are now essential in businesses that demand consistent and high-quality welds. These systems can be programmed to execute complex welds with accuracy and swiftness minimizing human mistakes and enhancing production efficiency. Robotic welding is especially advantageous in businesses that need repetitive welding processes such as car manufacture and heavy machinery production. Additionally, it improves worker safety by minimizing their contact with perilous welding settings. With the continuous advancement in technology, collaborative robots (COBOTS) are being implemented to operate alongside human operators enhancing manufacturing processes. Robotic welding is a revolutionary technology in advanced manufacturing that provides exceptional accuracy and productivity by using automated welding methods. This method entails employing robotic systems that are equipped with welding equipment to execute complicated welds with exceptional precision. The use of robotics into welding processes has brought about a significant transformation in industries, as it has improved production efficiency, uniformity, and worker protection. Robotic welding offers a notable benefit in its capacity to carry out repetitive tasks with unwavering accuracy, hence minimizing human mistakes and fluctuations in the quality of welds. Robots can operate without interruption resulting in increased output and reduced production time, thereby improving efficiency in sectors including automotive, aerospace, and heavy machinery manufacture. Moreover, robotic welding systems have the capability to function in perilous surroundings, so protecting human workers from potential harm caused by extreme temperatures, noxious fumes, and other occupational hazards. Robotic welding enables the application of elaborate welding patterns and processes ensuring precise and consistent execution of complex welds. Robots possess a high degree of programming flexibility enabling manufacturers to swiftly adjust to evolving production needs and create tailored solutions for different welding tasks. Although there are benefits to adopting robotic welding, it necessitates investments in equipment, programming, and worker training. Achieving a smooth and efficient connection between the robotic system and the welding process is essential for achieving the best possible performance. Ongoing research and development efforts aim to improve the accessibility and effectiveness of robotic welding in advanced manufacturing settings. Robotic welding is a revolutionary technology in advanced manufacturing that combines automation and accuracy to consistently produce high-quality welds. The impact of AI on sectors ranges from automotive to aerospace enhancing productivity, quality, and worker safety. In the production process of welding, two parts are connected using filler material. The fusion process produces a pool of weld that, when cooled, turns into a fixed junction. There are many other kinds and classifications of welding. A versatile manipulator that can be controlled and programmed to carry out several tasks, and that can move tools, materials, or parts along different trajectories for use in industrial automation applications, is what is known as an industrial robot. Put differently, industrial robots may be programmed to perform a variety of tasks since they are multipurpose. Because of this, the primary factors to consider while designing industrial robots are degrees of freedom (number of axes), action speed, weight capacity, workspace area, and programming level [17].

1.4.7 Hybrid Welding Processes

Multiple welding techniques are used in hybrid welding procedures to maximize their strengths and minimize their shortcomings. For instance, laser–arc hybrid welding combines the deep penetration of an electric arc with the focused heat of a laser. This leads to increased weld quality and faster welding speeds, which makes it appropriate for use in the automotive and shipbuilding industries. By using these hybrid techniques, producers may optimize both speed and quality by customizing the welding process for certain materials and applications. They also demonstrate how welding techniques can be tailored to satisfy the various demands of sophisticated production. Hybrid welding technologies, which combine the best features of several welding techniques to provide greater outcomes, have become a potent and cutting-edge strategy in modern manufacturing. Hybrid welding delivers better weld quality, quicker production rates, and increased efficiency in a variety of applications by combining the advantages of several technologies. Laser–arc welding, which combines the concentrated heat of a laser beam with the deep penetration of an electric arc, is a well-known example of hybrid welding. This procedure has benefits for shipbuilding, automobile manufacturing, and other industries because it produces faster welding rates, less distortion, and better weld quality. In a similar vein, laser–gas metal arc welding (GMAW) combines the high deposition rates of GMAW with the accuracy of laser welding to facilitate the efficient construction of huge structures [18]. The limitations of separate approaches are addressed by hybrid welding, which also enables producers to customize the welding process for certain materials and applications. Hybrid welding can provide advantages including better control over the amount of heat input, decreased thermal distortion, and improved material properties by optimizing the combination of techniques. However, because hybrid welding requires precise control over several parameters and integrates multiple systems, it can be a difficult implementation. The main goals of this field's research are to increase the number of materials that can be used, develop automated solutions, and improve process parameters. To sum up, hybrid welding methods are a dynamic approach to modern manufacturing that maximize the benefits of different welding techniques to produce welds with improved accuracy, efficiency, and quality. By enabling more adaptable and efficient manufacturing processes, the ongoing research and development of hybrid welding techniques holds the potential to completely transform industries.

1.4.8 Atomic Hydrogen Welding

Early in the 20th century, the technique of atomic hydrogen welding was invented. As a shielding gas, hydrogen gas is used in this sort of arc welding. Between two tungsten electrodes in a hydrogen gas environment, an arc is created throughout the operation. With temperatures as high as 4,000°C, the hydrogen molecules produce a tremendous amount of heat as they split apart into atomic hydrogen. Metals, such as aluminum, stainless steel, and steel, may all be welded together thanks to this extreme heat. Regarding alternative welding techniques, atomic hydrogen welding provides several benefits. Deep penetration and rapid welding are the outcomes of its extremely high heat concentration production. Furthermore, ambient gases like nitrogen and oxygen cannot contaminate the weld because hydrogen gas acts as a pure and efficient shielding gas. Along with producing a very steady arc, the method produces welds that are of excellent quality and rarely spatter. But there are certain drawbacks and safety issues with atomic hydrogen welding. Careful handling is necessary to avoid burns or other damage due to the high temperatures involved. Hydrogen gas should also be utilized carefully to reduce the risk of explosion because it is extremely combustible. In businesses that require high-quality, high-speed welding, atomic hydrogen welding is still a crucial method despite these obstacles.

1.4.9 Magnetic Pulse Welding

A solid-state welding technique called magnetic pulse welding (MPW) joins two workpieces by applying magnetic forces. The welding process bears the closest resemblance to explosion welding. When solid-state welding was adopted by the car industry in the early 1970s, magnetic pulse welding was born. The main benefit of magnetic pulse welding is that it prevents the development of brittle intermetallic phases. As a result, metals that are incompatible can be fused together, something that fusion welding is unable to accomplish. Without the need of shielding gasses or welding supplies, magnetic pulse welding produces excellent welds in microseconds in both comparable and dissimilar metals [19].

The foundation of magnetic pulse welding is an extremely brief electromagnetic pulse (less than 100 μs) produced by rapidly discharging capacitors via low-inductance switches into a coil. An eddy current forms in one of the work pieces because of the high-density magnetic field produced by the pulsed current at a frequency and amplitude of 500 kA and 15 kHz. A strong magnetic pressure substantially over the material yield strength and repulsive Lorentz forces are produced accelerating the impact of one work piece onto the other at velocities of up to 500 m/s. The jet and high temperatures close to the collision zone cause strong plastic deformation, high shear strain, and oxide disruption during magnetic pulse welding. Because of the microstructure refinement, dislocation cells, slip bends, micro twins, and local recrystallization, this results in solid-state welding.

1.4.10 Explosion Welding

Explosion welding is a solid-state welding technique that joins overlapping metal sheets by rapidly accelerating one metal plate using chemical explosions. Depending on the material and technique, the joining components are positioned at an angle ranging from 1° to 15°, and they are primed with an explosive layer on top. The method preserves the material qualities of both materials by enabling the rapid joining of huge areas without melting any of them. Explosion welding is frequently used to apply a thin layer of a more durable or corrosion-resistant substance on aluminum or carbon steel plates. Numerous metals, including copper, brass, nickel, steel, aluminum, stainless steel, and aluminum alloys, can be welded using it. Explosion welding's main benefit is that it can create links between metal combinations that might not be possible to weld with conventional techniques. For instance, maintaining the mechanical, electrical, and corrosion-resistant qualities of both metals is necessary in the shipbuilding industry [20].

1.4.11 Plasma Arc Welding

Like gas tungsten arc welding (GTAW), plasma arc welding (PAW) is a type of arc welding. Between an electrode—typically, though not always, consisting of sintered tungsten and the workpiece, an electric arc is created. The primary distinction between PAW and GTAW is that the electrode in PAW is located inside the torch's body separating the plasma arc from the shielding gas envelope. The plasma is then driven through a fine-bore copper nozzle, which narrows the arc, and it emerges from the orifice with a temperature of at least 28,000°C and high velocities (nearing the speed of sound).

A gas can exist in an arc plasma state momentarily. Electric current flowing through the gas ionizes it, turning it into an electrical conductor. When an atom is ionized, it breaks down into electrons (−) and cations (+), and a mixture of ions, electrons, and extremely excited atoms are present in the system. Ionization can range from 1% to more than 100% (double and triple degrees of ionization are possible). These states arise when more electrons are extracted from their orbits. The electrical power used to produce arc plasma determines the energy of the plasma jet and, consequently, the temperature. A common plasma jet torch may reach temperatures up to 28,000°C, which is significantly higher than the 5,500°C (9,930°F) of a conventional electric welding arc. While all welding arcs are (partially) ionized plasmas, the arc plasma used in plasma arc welding is constrained [21].

1.4.12 High-Energy-Density Welding Processes

Welding methods with high energy density (HED) concentrate the energy required for welding into a very tiny region. This makes it possible for the work components to receive very little heat overall, which minimizes base metal degradation, residual stress, and deformation. Rapid welding speeds are possible. Laser beam welding and electron beam welding are the two primary methods that are recognized for achieving high energy densities. The three main parts of a typical laser welding machine are the lasing medium, the pumping source, and the lasing cavity. Machines that use electron beam welding can be classified as high, medium, or nonvacuum systems, and as high or low voltage. The machine's power and chamber size both affect how much the machine costs. Many devices, such as x–y tables and rotary/tilt positioners, provide a great deal of capability to maneuver the part inside the chamber [22].

1.4.13 Narrow-Gap Welding Techniques

For the construction of massive buildings and pressure vessels, narrow-gap welding is crucial because it reduces the amount of weld metal utilized and the distortion brought on by shrinkage. Both benefits, which set narrow-gap joint designs apart, come from the nearly parallel sides on the weld preparation, with the former increasing with plate thickness and the latter remaining constant. The authors propose that a joint aspect ratio (plate thickness to gap width) of five or greater is adequate even though there is no universal consensus on what constitutes a "narrow" gap [23]. Accordingly, the approximate minimum thickness for narrow-gap welds employing arc welding procedures to be economically advantageous would be 60 mm for a plate with a 12-mm gap. For a given thickness, other techniques like EBW or ESW would have various aspect ratios and, hence, different financial benefits. Three fundamental types of welding processes are used in narrow gap arc welding as follows: multi-pass, mono-pass, and single pass. Since the SAW process is the most widely used method in arc welding, di-pass and tri-pass (multi-pass) welds are the most prevalent because they reduce the likelihood of slag entrapment and sidewall fusion. Except for considerable thicknesses, bypass welding is not a real narrow gap procedure and is being replaced by di-pass processes. Several GMAW techniques involve the use of mono-pass welding. Both sidewalls are fused together using a single weld pass per layer. There are currently three basic systems available to guarantee sidewall fusion as follows: a straight single wire fed through an offset rotating torch head (NKK), a spirally wound double twist wire, and a sinusoidally kinked single wire. The argument put out for why SAW is preferred over GMAW in North America and Europe is that GMAW equipment is more complicated than SAW equipment, and sidewall fusion in mono-pass welding is more difficult to manage due to spatter adhering to sidewalls and gas nozzles. Nonetheless, there are strong arguments in favor of evaluating methods other than SAW for narrow gap operations. In a narrow groove, it is challenging to remove the slag from subsequent welds; specifically designed fluxes must be created for this purpose alone. Because fluxes are made of natural minerals, they have a high potential for oxygen and some residual elements that can prevent impurities from being removed from the weld metal.

1.4.14 AI and ML in Welding Technologies

Machine learning approaches have the potential to solve many welding processes' issues and increase their efficiency. Using machine learning algorithms has been shown to dramatically boost the accuracy and efficiency of welding processes. Artificial intelligence-trained industrial robots can solve a wide range of challenging manufacturing industry issues. Numerous welding procedures depend on human judgment when selecting ideal parameters, which makes them less effective and highly prone to human mistake. To lessen this reliance, neural networks that can produce consistent weld quality and increased efficiency are used in the training of robots and automated systems. Since visual inspection is essential to determine the quality of the weld, machine learning is also used to envision welding. Regression analysis is another tool that can be utilized with these methodologies to assess the causes of different health concerns [24].

1.4.15 IoT in Welding Industries

The Internet of Things (IoT), or the linking of smart objects to the global internet network, is one of the most intriguing new technological advances. Applying this kind of connection to consumer electronics opens a world of possibilities. The opportunities in industry are almost endless. For instance, IoT in welding might significantly increase weld quality, which would enhance the quality and dependability of complex systems used in manufacturing, aviation, and power generation. It has the potential to greatly simplify the reproduction of qualified welding parameters [25].

However, implementing IoT in welding is far from straightforward. Legacy technology and techniques still dominate the welding industry in many different sectors. Welding systems with the necessary networking technologies are not as common as welding systems without the technology to support Internet of Things. It has only been lately possible to gather and distribute data for cloud-based solutions. However, the benefits of IoT in welding are so great that, even in its infancy, projects that fully utilize the technology stand to reap significant advantages that will only compound as capacity grows.

1.4.16 Microwave Welding

A type of electromagnetic welding that uses 2.45-GHz radiation is called microwave welding. Most thermoplastics exposed to radiation at this frequency do not immediately undergo a temperature increase. For this method to function, a gasket or other substance that absorbs microwave energy must be placed at the joint interface. The surrounding thermoplastic is melted by the heat produced in this implant, and when it cools, a weld is created. When it comes to welding methods, microwave welding is a relatively new technology. Many potential benefits of the method include short heating times (due to volumetric heating, which is independent of thermal conduction), the ability to create complex and three-dimensional joint geometries, a decrease in equipment capital costs, especially when welding complex components, extremely low power consumption, and simplicity in disassembly. Faster heating occurs because of the higher frequency than in radio frequency welding. One of the drawbacks of microwave welding is that materials containing polar molecular groups can heat up quickly leading to material degradation if the heating rate and power level are not properly controlled. Another drawback is that a consumable microwave susceptible material is typically required at the joint interface requiring an additional step in the welding cycle and posing recycling challenges. The only conventional techniques for connecting polymers for industrial use are conduction heating, induction, or lasers. In large-scale applications, achieving efficiency and precision at the same time is challenging. In this research, a precise method of welding metallic macro-gaps together using microwave excitation is used to create localized surface plasmons [26].

1.5 Effect of Post Weld Heat Treatment

As cutting-edge methods and materials are revolutionizing the advanced manufacturing scene, it is critical to guarantee the best possible structural integrity and performance for manufactured components. Post Weld Heat Treatment (PWHT) is one method that is very important in accomplishing these objectives. PWHT is a carefully regulated thermal procedure that is used in conjunction with welding to reduce residual stresses and improve mechanical characteristics. This method helps to extend the overall dependability and durability of sophisticated manufactured goods. Complex geometries and the connecting of different materials have been made possible by advanced production processes like friction stir welding and additive manufacturing (3D printing). However, these approaches also create issues about residual tensions and microstructural modifications. These methods can leave residual stresses that weaken the mechanical integrity of welded components resulting in flaws like distortion and cracking. To overcome these obstacles and improve the caliber of sophisticated manufactured goods, PWHT becomes an essential procedure. PWHT's basic idea is to heat the welded component to a high temperature—usually between 500°C and 800°C for a predetermined amount of time, then allow it to cool under controlled conditions. The regulated thermal exposure improves mechanical characteristics and lowers stress concentrations within the component by relaxing residual tensions and promoting microstructure homogenization. Furthermore, PWHT lessens hydrogen-induced cracking by assisting in the release of hydrogen that becomes trapped during welding. The suitability of PWHT for advanced manufacturing depends on several variables, including the material being used, the welding process, and the final product's intended function. PWHT is essential, for example, in the aerospace industry, where strong, lightweight materials like titanium and nickel alloys are used to remove stress concentrations and avoid early failure in harsh environments. Like this, PWHT improves creep and corrosion resistance in the energy sector, where components are subjected to high temperatures and pressures prolonging the life of equipment. However, it is crucial to remember that PWHT benefits are subject to specific limitations. The performance of a component can be compromised by improper heat exposure during PWHT, which can lead to harmful grain growth and changes in mechanical properties. To attain the intended results, precise temperature monitoring and control of heating and cooling rates are therefore essential. The function that PWHT plays is changing in tandem with the advancement of modern manufacturing. Engineers and researchers are always coming up with new ways to enhance the PWHT process for certain materials and uses. To achieve real-time control and validation, this entails customizing heat treatment cycles, utilizing sophisticated simulations to forecast PWHT effects, and integrating *in situ* monitoring devices. The study demonstrated that precisely designed PWHT cycles significantly reduced residual stresses and enhanced the material's fatigue performance highlighting the crucial role PWHT plays in improving the mechanical behavior of sophisticated produced components. To sum up, Post Weld Heat Treatment (PWHT) continues to be a fundamental component of advanced manufacturing solving issues with residual stress and mechanical properties that result from cutting-edge production methods. Its capacity to reduce stress concentrations, boost mechanical qualities, and improve overall component performance confirms its place in maintaining the highest dependability and quality requirements in sophisticated produced goods. PWHT's versatility and efficacy will guarantee the durability and integrity of intricately constructed buildings even as technology pushes boundaries [27].

1.6 Challenges and Future Directions

Even though sophisticated manufacturing has many advantages, there are drawbacks. It takes significant financial outlays for personnel training and equipment to integrate new technology. Furthermore, it is imperative to guarantee cybersecurity in networked production environments. Regarding the future, investigations into the possibilities of advanced manufacturing persist. A few areas that are currently under development are more automation, better materials, and sustainable methods. The capabilities of advanced manufacturing will grow along with technology altering sectors and changing the face of production globally. Although there are many advantages to welding techniques in advanced production, difficulties still exist. It is still a worry to maintain quality and consistency during large-scale production, especially when working with complex designs or distorted materials. Furthermore, incorporating cutting-edge technology, like machine learning and artificial intelligence, into welding procedures may improve quality control and real-time monitoring. Furthermore, a crucial component of contemporary production is sustainability. Like other industrial processes, welding also produces emissions and uses energy. Research is being done to create welding methods that are more environmentally friendly and energy efficient, with an emphasis on lowering waste, energy usage, and hazardous byproducts.

1.7 Conclusion

In conclusion, in the context of sophisticated production, welding processes have significantly advanced. These procedures have made it possible for enterprises to produce goods more quickly, with more accuracy, and with better material usage. A variety of industrial needs can be met by the combination of laser welding, electron beam welding, friction stir welding, ultrasonic welding, and their integration with additive manufacturing. In addition, the industry's dedication to innovation is demonstrated by the cooperation between robotics and welding and the rise of hybrid welding techniques. Welding techniques will probably continue to be crucial in redefining advanced manufacturing in the future as technology advances.

References

1. Huang, Y., Leu, M.C., Mazumder, J., Donmez, A., Additive manufacturing: Current state, future potential, gaps and needs, and recommendations. *J. Manuf. Sci. Eng.*, 137, 1, 014001, 2015.
2. Gu, D., Meiners, W., Wissenbach, K., Poprawe, R., Laser additive manufacturing of metallic components: Materials, processes and mechanisms. *Int. Mater. Rev.*, 57, 3, 133–164, 2013.
3. Gibson, I., Rosen, D., Stucker, B., *Additive manufacturing technologies: Rapid prototyping to direct digital manufacturing*, Springer New York, New York, 2010.

4. Klein, R., *Laser welding of plastics: Materials, processes and industrial applications*, Wiley Publishing, Germany, 2011.
5. Xian, W., Yu, K., Han, F., Fang, L., He, D., *et al.*, Advanced manufacturing in industry 5.0: A survey of key enabling technologies and future trends. *IEEE Trans. Ind. Inf.*, 20, 2, 1055–1068, 2024.
6. Steen, W.M. and Mazumder, J., *Laser material processing*, Springer London, London, 2010.
7. Sun, G., Wang, Z., Lu, Y., Mingzhi, C., Kun, Y., *et al.*, Underwater laser welding/cladding for high-performance repair of marine metal materials: A review. *Chin. J. Mech. Eng.*, 35, 5, 1–19, 2022.
8. Mishra, R.S. and Ma, Z.Y., Friction stir welding and processing. *Mater. Sci. Eng.: R: Rep.*, 50, 1-2, 1–78, 2005.
9. Krishna Murthy, K.R., Akyel, F., Reisgen, U., Olschok, S., Mahendran, D., Modelling the Evolution of Phases during Laser Beam Welding of Stainless Steel with Low Transformation Temperature Combining Dilatometry Study and FEM. *J. Manuf. Mater. Process.*, 8, 2, 50, 2024.
10. Liu, X., Dong, Q., Wang, P., Chen, H., Review of electron beam welding technology in space environment. *Optik*, 225, 11, 165720, 2021.
11. Dutilleul, T., Widdison, R., Kyffin, W., Development of electron optical capabilities for manufacturing of large components by electron beam welding. *Weld. World*, 68, 819–828, 2024.
12. Sadeesh, P., Venkatesh Kannan, M., Rajkumar, V., *et al.*, Studies on Friction Stir Welding of AA 2024 and AA 6061 Dissimilar Metals. *Procedia Eng.*, 75, 145–149, 2014.
13. Alsalami, Z., Thiruramanathan, P., Krishnakumar, S., Kala Priyadarshini, G., Balmiki, V., Advancements in Friction Stir Welding: A Comprehensive Review of Process Variables and Emerging Developments. *E3S Web Conferences*, 491, 02026, 2024.
14. Liu, Z., Li, Y., Liu, Z., Yang, Y., Li, Y., Luo, Z., Ultrasonic welding of metal to fiber-reinforced thermoplastic composites: A review. *J. Manuf. Processes*, 85, 702–712, 2023.
15. Patel, V.K., Bhole, S.D., *et al.*, Ultrasonic spot welding of dissimilar 2024Al alloy and SiCp/2009Al composite. *J. Mater. Des. Appl.*, 233, 531–538, 2018.
16. Debroy, T., Wei, H.L., Zuback, J.S., Mukherjee, T., Elmer, J.W., Milewski, J.O., *et al.*, Additive manufacturing of metallic components – process, structure and properties. *Prog. Mater Sci.*, 92, 112–124, 2018.
17. Curiel, D., Veiga, F., Suarez, A., Villanueva, P., Advances in Robotic Welding for Metallic Materials: Application of Inspection, Modeling, Monitoring and Automation Techniques. *Metals*, 13, 4, 711, 2023.
18. Acherjee, B., Hybrid laser arc welding: State-of-art review. *Opt. Laser Technol.*, 99, 60–71, 2018.
19. Yu, H., Jiang, X., Zhang, M., An innovative coil for magnetic pulse welding of dissimilar sheet metals: Numerical simulation and experiments. *J. Mater. Process. Technol.*, 324, 118230, 2024.

20. Khaustov, S.V., Pai, V.V., Lysak, V.I., Kuz'min, S.V., Investigation of thermal processes in the gap during explosion welding. *Int. J. Heat Mass Transf.*, 209, 124166, 2023.
21. Wu, C.S., Wang, L., Ren, W.J., Zhang, X.Y., Plasma arc welding: Process, sensing, control and modelling. *J. Manuf. Processes*, 16, 74–85, 2014.
22. Babu, S.S., Reidenbach, F., David, S.A., Böllinghaus, T., Hoffmeister, H., Effect of high energy density welding processes on inclusion and microstructure formation in steel welds. *Sci. Technol. Weld. Joining*, 4, 63–73, 1999.
23. Meng, W., Chen, K., Pan, Q., Ye, K., Liu, Z., Ma, Q., Hu, L., Du, W., Yin, X., Narrow gap welding of 316L thick-section steel with cold metal transition pulsed arc swing. *Int. J. Adv. Manuf. Technol.*, 132, 4035–4053, 2024, https://doi.org/10.1007/s00170-024-13601-z.
24. Goka, S., Narayana, G.S., Divya Jyothi, G., Shaik, H.S., Moinuddin, S.Q., *AI and ML in Welding Technologies*, Wiley, Hoboken, USA, 2024, DOI:10.1002/9781394172948.
25. Sanjith, S., Balaji, N., Vijay Anand, L.D., Regis Anne, W., Shanmugapriya, P., Sheeba Rani, S., IoT enabled temperature sensing system of molten metal in welding. *Mater. Today Proc.*, 45, 2514–2517, 2021.
26. Hu, Y. and Hu, Y., Precise microwave welding induced by localized surface plasmon: A novel high-quality large-scale plastic joining method with low energy consumption. *J. Clean. Prod.*, 449, 141778, 2024.
27. Joshi, A., Gope, A., Gope, P.C., Effect of post-weld heat treatment on mechanical properties and fatigue crack growth behaviour of friction stir welded 7075-T651 Al alloy. *Theor. Appl. Fract. Mech.*, 123, 103714, 2023.

2

Developments and Trends in Welding Development

Sandip Kunar[1]*, Jagadeesha T.[2], Gurudas Mandal[3], Norfazillah Talib[4], Adduri S. S. M. Sitaramamurty[1], Akhilesh Kumar Singh[5] and K. Rakesh Varma[1]

[1]Department of Mechanical Engineering, Aditya University, Surampalem, India
[2]Department of Mechanical Engineering, National Institute of Technology, Calicut, India
[3]Department of Metallurgical Engineering, Kazi Nazrul University, Asansol, India
[4]Department of Manufacturing Engineering, Universiti Tun Hussein Onn Malaysia, Johor, Malaysia
[5]Department of Mechanical Engineering, Aditya College of Engineering and Technology, Surampalem, India

Abstract
Because of economic and societal factors, mechanization has become progressively prevalent, and more efficient tools and supplies have been developed. This has led to a need for welding process development. One substantial advancement in typical welding procedures is the deployment of improved control and monitoring systems. Nevertheless, both components can be significantly enhanced by making use of some of the cutting-edge process improvements. The history of some of the more significant developments as well as recent developments and patterns in the use of advanced welding processes are covered.

Keywords: Advanced process development trends, cost effectiveness, consumable trends, automation, advanced materials

2.1 Introduction

The main driving force behind the development of welding processes is the requirement to increase the overall cost effectiveness of joining activities in

**Corresponding author*: sandip.sandip.kunar@gmail.com

Sandip Kunar and Gurudas Mandal (eds.) Advanced Welding Technologies, (25–34) © 2025 Scrivener Publishing LLC

the production and fabrication sectors. However, there may be other factors that affect the need for supplementary procedures. The safety precautions of welding and the possible lack of qualified operators and specialists in every nation have emerged as substantial areas for investigation.

Most of the old welding procedures discussed are thought to be costly and hazardous; however, by utilizing some of the advanced process that are covered, both elements can be greatly improved. The sections discuss the history of some of the more important improvements as well as current advances and trends in the application of advanced processes.

2.1.1 Cost Effectiveness

All labor, material, power, and capital plant depreciation expenses added together equal the cost of manufacturing a welded junction. The cost of labor, which can make up 75–80% of the whole cost of welding operations in many typical welding techniques, greatly influences the overall cost of welding operations in western countries. A diagram representation of this can be found in Figure 2.1.

Weld cycle time is generally shortened, and labor costs are decreased with increasing deposition rate. The most recent advances in high deposition rate methods are explored. But if, for any reason, the superiority is compromised and greater maintenance is needed, the deposition rate can provide a false impression of cost effectiveness. Furthermore, when discussing autogenous techniques with high joint completion rates, such as laser and explosive welding, the term "deposition rate" is unsuitable.

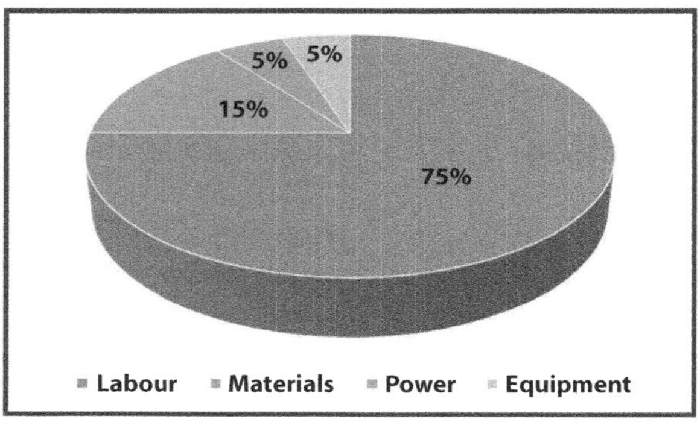

Figure 2.1 Welding costs.

It is obvious that the following more elements need to be considered for a more thorough evaluation of cost savings:

- Managing the quality of the joint,
- Joint design,
- Operational effectiveness, and
- Equipment and consumable cost.

• Managing the Quality of the Joint

The joint quality is frequently dependent on the optimization of numerous connected operational parameters in traditional welding procedures, in addition to the precise management of pre and post weld treatments. The operational parameters, which are described for every welding technique, are obtained from a mix of welding rules and trials to ensure reproducible joint quality [1]. A certifying authority may formally consent this welding technique and the operator for crucial structural joints. The process of developing new procedures takes a lot of time and money. Once a method is developed, implementing a new one may be too expensive.

The effectiveness of this control method also hinges on making sure that the established protocol is really followed in manufacturing; this calls for keeping an eye on the functionality of the machinery and making sure the operator follows the original protocol. The post-weld inspection and repairing of weld frequently result in additional expenses. Costs should be significantly impacted by the development of procedures that allow more effective control over the welding process. Improved overall process control is a result of the use of consumables that are more resilient, more repeatable machinery and procedures, automation, online supervising, and real-time control systems. Furthermore, there is a growing interest in utilizing parameter prediction and modeling approaches to determine the ideal welding parameters for a particular welding scenario.

• Joint Design

The cost of welding is significantly impacted by over-specifying the joint requirements; for a straightforward fillet weld, a mere 1-mm increase in the stipulated leg length can result in a 45% cost increase.

Selecting a particular joint design may automatically exclude using the most economical procedure; restricted access or intricate joint profiles, for instance, may restrict the process selection; it is crucial for the designer to be aware of the joining process's limitations to avoid these constraints. On the other hand, choosing the right procedure can shorten the time it takes to prepare for a joint and lower the associated expenditures [2].

• **Operational Effectiveness**
The operating efficiency of welding processes is generally known as the "operating factor." The ranges of operating factor are 16% to 20% for manual metal arc welding and 35% to 55% for gas metal arc welding. Enhancements in operating factor have a major effect on costs since they directly affect the labor element [3].

• **Equipment and Consumable Cost**
Welding may cause issues that need to be fixed once the procedure is completed. The most frequent issues of this kind are distortion and residual strains, but it is also important to consider metallurgical issues like grain growth and hydrogen-induced cold cracking, as well as aesthetic issues like damage to surface coatings and spatter deposits. The cost of fabrication will undoubtedly rise if extra mechanical or thermal procedures are required after welding; hence, it is desired to create processes that lessen this demand. The possibility of defects frequently results in the need for expensive non-destructive testing and post-weld inspections. Although new codes of practice permit the consequence of faults to be tied with the working conditions, rectification—if necessary—involves raising the value of work-in-progress, delays production, and is frequently labor intensive. Therefore, it is desirable and economical to identify possible quality issues early on [4].

2.2 Safety and Environmental Factors

Table 2.1 contains an inventory of some of the possible safety risks associated with welding. While local screening, ventilation, and protective

Table 2.1 Different instances of common welding safety hazards.

Hazard	Process
Radiation and UV	TIG, MIG, plasma, and laser
Electric shock	Arc and resistance electron beam
Noise	Friction and plasma
Toxic gases (e.g., ozone)	Arc methods
Ionizing radiation (x-rays)	Electron beam
Explosive gases	Oxy fuel processes and plasma cutting
Nontoxic gases	Most welding and cutting methods

clothes are typically sufficient to safeguard the operator, other workers in nearby regions may need extra protection. These actions could have an impact on the industrial operation's overall efficiency and be costly in and of itself. Therefore, it is preferable to have process innovations that enhance the working environment or take the operator out of more dangerous tasks [5].

2.3 Skill and Training Requirements

When the abovementioned procedural requirements need to be fulfilled, many of the conventional welding procedures need high levels of operator ability, which can comprise the expensive training programs. The more sophisticated machinery that has, regrettably, taken the place of the older processes in certain instances has reduced the total skill requirement somewhat, and the time required to set the parameters of the newer processes may have reduced the operating factor. This section will cover developments aimed at making equipment operation simpler. However, proper operator and supporting staff training are still necessary for the efficient use of even the most sophisticated methods and equipment. Usually, the expense of this training is soon retrieved in higher quality and productivity [6].

2.4 Areas for Development

The following benefits of improved welding techniques could make them justifiable:

- Higher deposition rate,
- Shorter cycle time,
- Better process management,
- Decreased repair rates,
- Shorter joint preparation times,
- Operator removal from hazardous areas; smaller welds,
- Less post-weld procedures,
- Enhanced operating factor,
- Decreased possible safety hazards, and
- Easier-to-use equipment configuration.

2.5 Process Application Trends

On a global scale, several noteworthy trends in the use of welding methods can be noted including the following:

- Modifications to consumable electrode for arc welding procedures,
- A rise in automation,
- A greater desire for novel techniques (such as laser welding), and
- The requisite to create cutting-edge materials.

2.5.1 Consumable Trends

In many developed nations, traditional metal arc welding (MMA) is being replaced by gas metal arc welding (GMAW) and flux cored arc welding (FCAW) techniques. According to the most recent data, the percentage of welding with MMAW electrodes in developed nations will settle at 10% to 20%, while slightly higher percentages are anticipated in emerging nations. Approximately 13% of the deposited weld metal has already seen a stabilization in the usage of submerged arc consumables.

This tendency highlights the significance of determining the whole cost of the welding process; in comparison to the solid wires used for GMAW, the flux-cored consumable is invariably more expensive to produce and frequently more than five times more costly to purchase. Despite the greater consumable costs, the weld's overall cost may be lowered due to factors such as better joint quality, more deposition rate, and higher operating factor. The overall cost will vary depending on the application. For instance, a multi-pass vertical butt weld for 20-mm-thick steel is more cost effective when rutile flux-cored wire is used.

2.5.2 Automation

Automation can be used more when single-shot procedures, like continuous processes like GMAW and resistance spot welding, are used. The introduction of robotics trends has been extensively covered in the media. Whereas the applications for robotic GMAW and GTAW systems are more varied, the automotive industry is home to several resistance welding robots. Furthermore, the adoption of robots in Japan has evidently surpassed both those of western Europe and the United States. Although there is less information available, non-robotic automation and simple mechanization are used in welding. However, in more recent techniques, like electron and laser beam welding, automation plays a crucial role in the system. Mainly for GMAW and FCAW processes, simple low-cost mechanization is a very economical technique of automation, and its application is anticipated to rise. More flexibility is possible with recently developed computer-numerical-controlled (CNC) systems, which carry out many of the tasks typically functioned by welding robots.

With the help of robot- or computer-controlled welding devices, it is now technically possible to incorporate welding in a modern manufacturing facility. Features, like automatic component recognition, automatic reporting of machine malfunction, online quality assurance, online data recording, and production statistics, can all be offered. Such integrated fabrication facilities have a high capital cost, but their overall productivity gains and end-product costs must be considered when evaluating their economic benefits [7].

2.5.3 New Processes

They indicate that the employment of novel joining methods, including diffusion bonding and laser welding, is growing. The utilization of these procedures has been limited in the past, but it is anticipated that this will change as more people become aware of the advantages of automation and the need for high-integrity joints in modern materials. When high-speed, distortion-free welding in ductile materials is needed, friction stir welding has been a tremendous advancement and is finding new uses. This process's application range will probably grow as it continues to be developed.

Industrial laser sales globally have been increasing at a rate of approximately 10% annually since 1988, when sales were anticipated to be approximately 3,000 units. This growth rate is anticipated to continue, with CO_2 laser sales rising at a rate of 13% annually and Nd:YAG sales growing at a rate of 7% annually. High-power (3–4 kW) devices are anticipated to drive the majority of the YAG growth. Approximately 20% of this rise is probably attributable to welding applications. Recently, a lot of research and development has gone into hybrid (laser/GMAW) welding, which finds use in the automotive and shipbuilding industries [8]. With output levels ranging from 3 to 20 kW, more effective diode and fiber lasers are now accessible, and extensive research is being done on their potential applications in surface engineering and welding. Recently, new methods have been introduced, such as magnetic impulse welding, and some parts of the automotive sector are already using them. The advancements in process control, consumables, and equipment have also increased the viability of many of these processes.

2.5.4 Advanced Materials

The necessity for using more advanced materials is developing as fabrications are subjected to harder service conditions and, often, environmental and economic concerns with higher strength-to-weight ratios. These characteristics can be seen in many other industries, for instance, the production of microcircuits, offshore structures, and aircraft. Examples of developments in advanced materials that have undergone thermomechanical treatment are fiber-strengthened composite materials ceramics, cermets, and polymers. The simplicity of creating strong connections between two parts of the same material or, more frequently, between interfaces that are different, such as ceramic-to-metal links, may determine how these materials are used. Although methods for high-yield-strength steels are currently accessible employing standard arc welding methods, promising outcomes have been obtained employing solid-phase bonding techniques for ceramic-to-metal junctions. Bonding techniques for these advanced materials are still in the development stage [9].

2.6 Summary

Due to social and economic causes, there has been a significant growth in the use of automation and the creation of more efficient equipment and consumables. This has created a demand for advancement in welding processes. A noteworthy development in traditional welding processes is the application of improved control and monitoring. Advanced capabilities and new problems in joining technology have been brought about by the advent of new methods and advanced materials.

References

1. Wahidi, S., II, Oterkus, S., Oterkus, E., Robotic welding techniques in marine structures and production processes: A systematic literature review. *Mar. Struct.*, 95, 103608, 2024.
2. Conde, J., Freire, F., Ljubinković, F., Vild, M., da Silva, L.S., Welded beam-to-column steel joints: Assessment of European design rules. *Thin Walled Struct.*, 199, 111844, 2024.

3. Vorobel, R., Student, O., Ivasenko, I., Maruschak, P., *et al.*, Development of a method for computer processing of fractographic images to assess the cohesion of inclusions to the matrix in the weld metal after its operational degradation and hydrogenation. *Materialia*, 34, 102074, 2024.
4. Loukas, C., Williams, V., Jones, R., *et al.*, A cost-function driven adaptive welding framework for multi-pass robotic welding. *J. Manuf. Processes*, 67, 545–561, 2021.
5. Okumus, D., Fariya, S., Tamer, S., *et al.*, The impact of fatigue on shipyard welding workers' occupational health and safety and performance. *Ocean Eng.*, 285, 115296, 2023.
6. Katheria, S.K., Kumar, D., Khan, T.A., Singh, M.K., Reality based skills development approach in welding technology: An overview. *Mater. Today Proc.*, 47, 7184–7188, 2021.
7. Gibson, B.T., Lammlein, D.H., Prater, T.J., *et al.*, Friction stir welding: Process, automation, and control. *J. Manuf. Processes*, 16, 56–73, 2014.
8. Su, R., Li, H., Chen, H., *et al.*, Stability analysis and porosity inhibition mechanism of oscillating laser-arc hybrid welding process for medium-thick plate TC4 titanium alloy. *Opt. Laser Technol.*, 174, 110569, 2024.
9. Motta, M.F., Santiago, K.G., Pessoa, E.F., *et al.*, Creating *in-situ* alloys by welding — new perspectives for advanced materials and applications. *J. Mater. Res. Technol.*, 9, 6950–6956, 2020.

3

Welding Power Source Technology

Sakthimurugan D.*, Thavasilingam K., Praveen Kumar A., Karthikeyan K. and Nagachandrika K.K.

Department of Mechanical Engineering, Easwari Engineering College, Chennai, India

Abstract
This chapter covers the basic requirements for an arc welding power supply, developments in welding power source technology, the principles of conventional and complex power source designs, and their advantages and disadvantages. Numerous recent advancements in arc welding have been made possible by improvements in the design of welding power sources, particularly the introduction of electronic control. Because they supply the electrical energy essential to fuse metals together in a strong and durable way, welding power sources are crucial. Technologies for welding power sources has improved with a focus on increases in economy, accuracy, and sustainability for the environment. The earliest welding power sources were transformer-based designs; the most advanced systems today use inverter technology. Welding capabilities have improved because of the increased control over parameters, such as voltage, current, and waveform shape, made available by these improvements. More specifically, inverter technology has made welding possible by enabling the development of compact, portable equipment that performs dependably in a range of settings. Furthermore, welding power sources have been further enhanced by the coupling of digital controls and microprocessors enabling adaptive welding procedures that optimize energy efficiency and reduce material waste. These smart devices can reduce human mistake and ensure optimal weld quality by mechanically modifying parameters depending on real-time data.

***Keywords*:** Pulsed welding, energy efficiency, digital control, rectifiers, inverters, gas metal arc welding (GMAW), magnetic pulse welding (MPW)

**Corresponding author*: skthids@gmail.com

Sandip Kunar and Gurudas Mandal (eds.) Advanced Welding Technologies, (35–46) © 2025 Scrivener Publishing LLC

3.1 Introduction

The welding power sources materials to fill gaps in arc welding gases used in sophisticated welding procedures gas metal arc welding (GMAW) measures with high energy density techniques for welding narrow gap process monitoring and control for welding operation robotics and automation welding [1]. The improvement of robot welding, semi-automated and automated welding, steel-framed building, and vehicle manufacturing exhibit that gas-shielded arc welding has emerged as the most widely used technique in recent years. This is due to constant growth and improvement that has been made to welding supplies, equipment, and welding power sources. Automation and improvements in welding efficiency are ideally suited for gas-shielded arc welding [2]. The material that acts as the interface between the two metals melts during the welding process, the materials are fused or mixed, and then the mixture solidifies to form a junction that is extremely strong and rigid. Filler metals and fluxes are two more essential components used in welding. Filler metals are used to supply the same material, that is, the base metals at the molten contact gap between them. Using fluxes, contaminants such as oxide forms in the weld pool are eliminated, and a layer of slag is formed above the molten pool to shield it from more oxidation [3].

3.2 Fundamentals of Welding Power Sources

3.2.1 Definition and Purpose

In the context of shipbuilding, electrical energy is most often used as an arc for different for electro-slag welding and as a barrier against molten slag during welding procedures. A power source competent of supplying electricity at high current and low voltage is essential for arc welding to form and continue a column of arc plasma between the work piece and the welding electrode. Direct current or alternating current may be output by the welding power sources. These power sources often have stable current or constant potential volt-ampere characteristics. There might be a pulsing mode on the output as well. The power sources are separated into groups according to their volt-ampere properties at rest. Sources of constant potential power, in contrast to constant-current power sources, often have virtually output voltage that is steady. Constant-current output is a general feature of constant-current power sources [4].

3.2.2 Basic Components

A welding transformer and a low-voltage series active power filter on power field effect transistors with a low channel resistance are used to create the source. It is possible that the transformer in use has prominent or normal leakage. The optional source has better technical attributes since it can form the V–I characteristic having the required rigidity without having to turn the taps on the welding transformer. The control technique makes sure that the welding current's DC component is eliminated. The welding machine, wire feeder system, control system, welding actuator, and remote control box are the parts of the system. The welding torch is situated atop the actuator so that it can track the actuator's rotational motion, which is generated by a circular track [5].

3.3 Key Parameters and Characteristics

The static output characteristic of the welding power source is a vital non-parametric parameter. An adaptable resistor is loaded into the source to recognize the parameter is shown in Figure 3.1. The correlation between the output voltage and output current in the case of a pulsed gas metal arc welding (P-GMAW) power source is reliant upon a number of factors. If these factors are not taken into account, the working point's variable slope may not be precisely assessed. To monitor the volt/amp characteristics of the power supply during the duration of the pulse and setting current, a separate worth for the arc parameters has been added to the welding

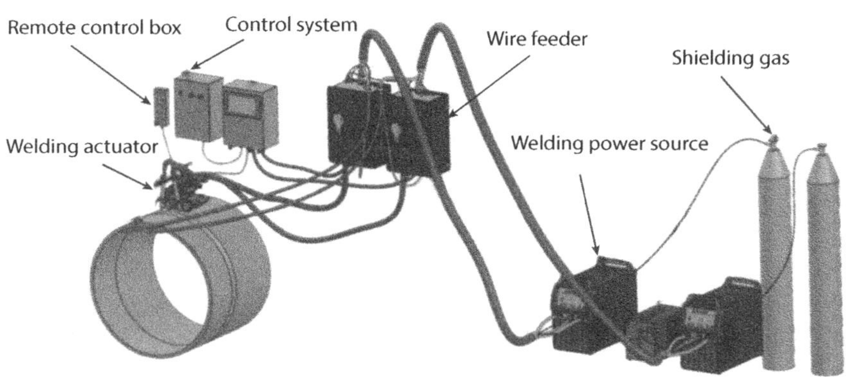

Figure 3.1 Elements of a power supply for welding (source: Yin, T *et al.* 2022) [6].

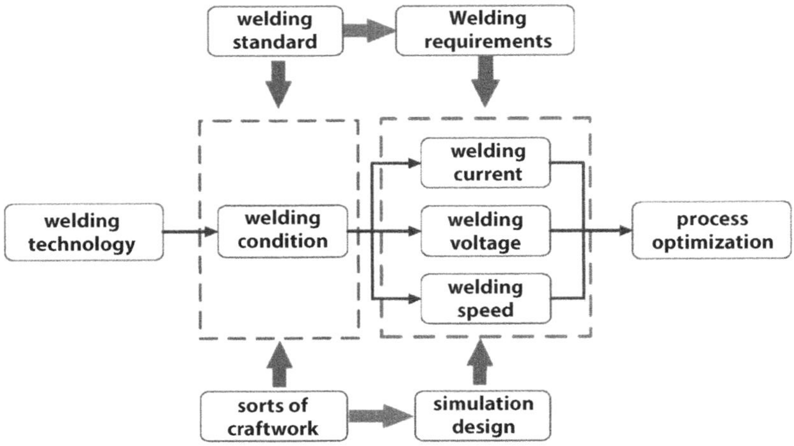

Figure 3.2 The welding process parameter [7] (source: Zhang, H *et al.* 2023) [8].

machine's synergistic control system is shown in Figure 3.2. The three most crucial parameters, welding voltage, welding speed, and welding current, are also particularly modeled in the section is shown in Figure 3.2.

3.4 Types of Welding Power Sources

3.4.1 Transformers

The most popular power source in recent years, the AC/DC transformer–rectifier, with a variety of control systems, has met most of the manufacturing needs. More developments in welding technology have resulted from the latest inverter-based GMAW power sources with enhanced features, whether it is in the area of practical efficiency or the welding of rather complex materials, joint configurations, or weld positions [9]. Alternating current welding has become more admired because of the advantages of using a transformer, which is simple to use and has no moving parts. It was necessary to build a welding transformer with a secondary voltage so high that the operator might be in danger when the only accessible welding electrodes were bare [10]. One feature of this specific circuit application is that 2 kV at 500 kHz is produced in a resonant circuit by means of the isolation transformer's magnetizing inductance. The transformer is subject to two loading conditions: high voltage at the secondary during open circuit, just before the arc strikes, and high secondary current typically greater than 100 A during the welding process itself is shown in Figure 3.3 [11].

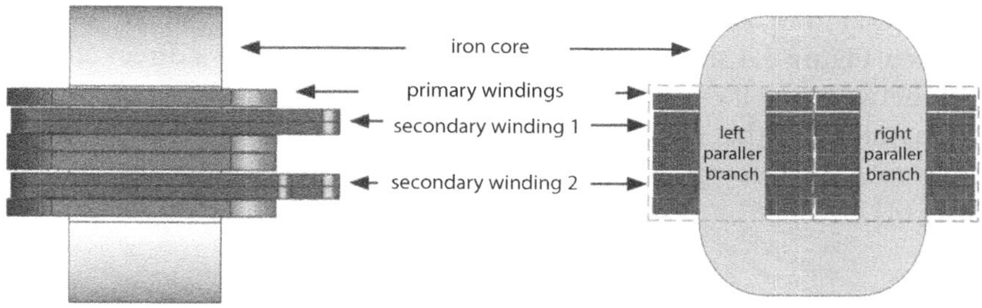

Figure 3.3 A typical design of a welding transformer (source: Habjan, G *et al.* 2019) [11].

3.4.2 Rectifiers

Rectifiers, which transform AC to DC, are essential parts of power electronics. Reversing essentials perform unidirectional switching, connecting transformer secondary windings to the load circuit at prearranged intervals, as part of their operating principle [12] since there are many various kinds of direct current loads, including batteries, front-end converter AC drives, DC motor drives, welding, and so on. AC power sources are more generally available; the power electrical circuits that are most frequently employed are probably AC to DC rectifiers [13].

3.4.3 Inverters

An inverter is a device that transforms DC electricity into AC power. The inverter operates on the basis of quickly switching the voltage and current of the power supply by utilizing the switching characteristics of semiconductor devices (e.g., field effect transistors or thyristors). This transforms

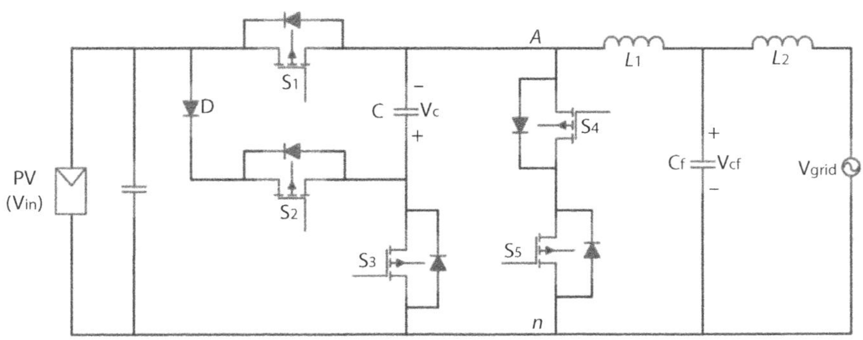

Figure 3.4 Transformerless inverter topology (source: Cardoso, J *et al.* 2022) [14].

DC power into alternating current with a subsequent voltage and frequency. Figure 3.4 shows the plan that takes into account a direct connection between the PV module's negative terminal and the AC grid's neutral; by maintaining this connection, the common mode voltage Vcm is kept constant, which lowers the probability that leakage currents would manifest [14].

3.4.4 Hybrid Power Sources

The welding robot, the CCD camera, the WSME-630, the self-designed K-TIG welding torch, the K-TIG welding power supply, and the YAG2400 W laser power source are all shown in Figure 3.5a schematic illustration of K-TIG laser welding systems. A single amendment method is used by WSME-630. The voltage used for arc welding is automatically changed in response to variations in the current used in arc welding. The robot welding platform's internal hybrid welding adjustment device may modify the parameters of the K-TIG hybrid welding method using laser in addition to managing the opening and closing of the laser and the K-TIG welding flame. The 45° angle twisted between the horizontal plane and the welding torch's axis is depicted in Figure 3.5b. The laser's axis and the horizontal plane make an angle of 85° [15].

3.5 Advanced Features and Technologies

3.5.1 Pulse Welding

As a welding power control system was enhanced, pulsating welding became escalating in popularity because it could use pulse current to

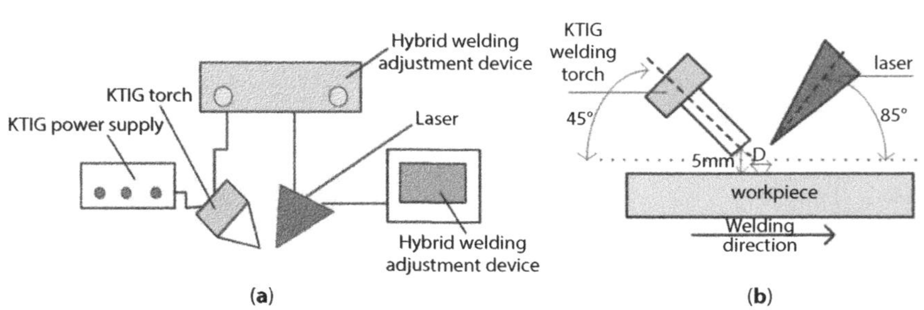

Figure 3.5 Hybrid welding power source (a) schematic diagram and (b) distribution of hybrid welding parameters (source: Zhang, H *et al.* 2022) [15].

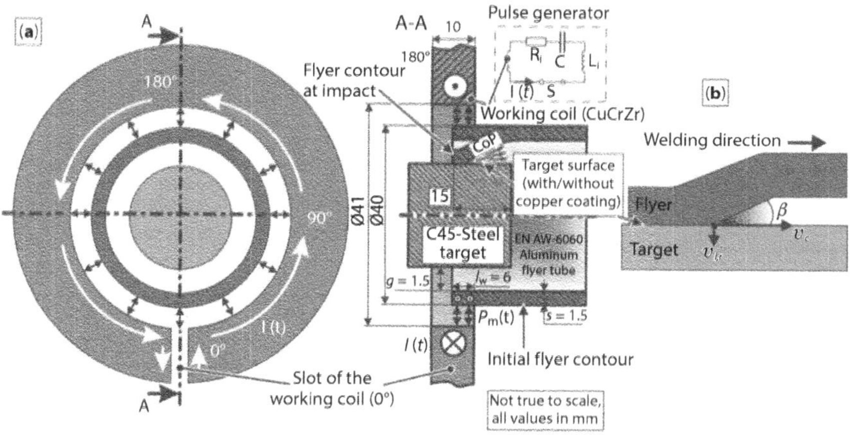

Figure 3.6 (a) Magnetic pulse welding setup and (b) the collision point C's velocity components (source: Bellmann, J *et al.* 2022) [18].

regulate the droplet transfer process [16]. Solid-state welding is skilled using magnetic pulse welding (MPW), which produces a real metallurgical link between two materials by impacting them at a high velocity is shown in Figure 3.6 (a) and (b). MPW is a magnificent tool for combining different metals, which is one of its many advantages. Due to the negligible heating and the enlargement of clean surfaces caused by the jet, there are no heat-affected zones, and the metal is not deteriorated. Additionally, this approach produces less brittle intermetallic complex production than other frequent welding processes. Even so, there has not been much of this process' applicability to industrial domains despite its many advantages [17].

3.5.2 Remote Monitoring and Control

To check the happening of weld faults during friction stir welding (FSW) manufacturing method, a unique cloud-based remote and real-time monitoring and control method has been created [19]. Large amounts of energy are consumed during processes like welding, which results in the production of important electrical data including current and voltage signals. To detect flaws in the production of welded parts or equipment, it is necessary to examine the quality of welding using such data as soon as feasible [20]. Figure 3.7 depicts the intelligent welding manufacturing system based on IoT-MAS created in this study. The intelligent welding production system known as IoT-MAS, which serves as the foundation for the work unit, is extracted using the intelligent welding manufacturing system model of each agent [21].

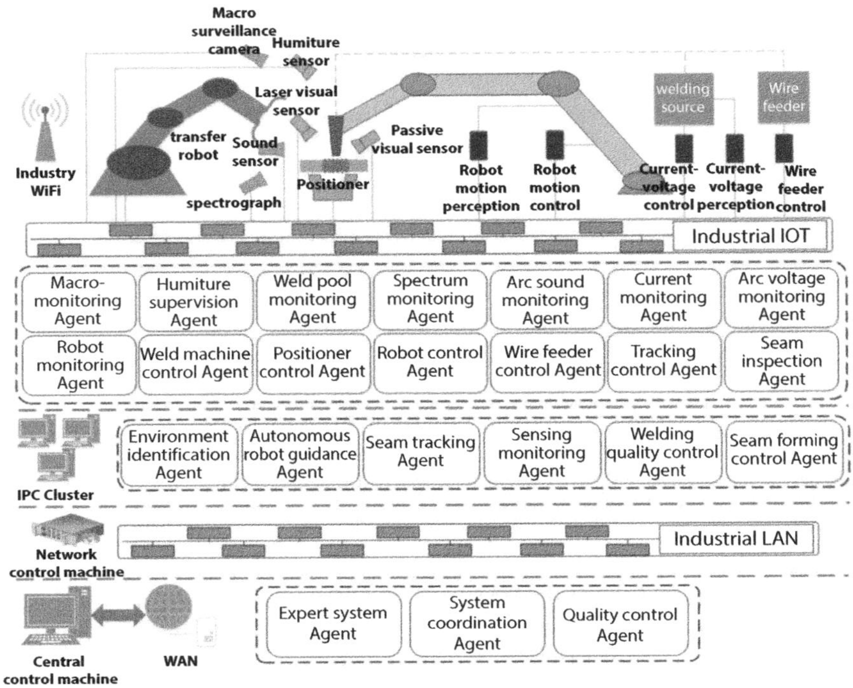

Figure 3.7 IoT-MAS-based model of systems and manufacturing for intelligent welding (source: Liu, Q *et al.* 2022) [21].

3.5.3 Energy Efficiency and Sustainability

A resonant converter with a valley-fill rectifier on the main side has been suggested as part of the topology of a welding power supply source. Resonant converters have negligible energy losses and a low number of power semiconductor switches. Using the original control technique will boost energy efficiency [22]. To supply the suitable welding currents and voltages from industrial supply mains (230/460 VAC 3 phi 60 Hz) (up to 32 V/up to 300 A), the power source development uses switch mode power supply techniques. The circuit design method that was preferred maximized welding performance, size/weight, consistency, costs, and energy efficiency using a series capacitor switch mode power circuit [23]. Sustainable manufacturing takes into account the energy and ecological effects as well as the economic implications of the product creation process. One of the key components in the manufacturing business is welding. The area of concern for the manufacturing industry nowadays is sustainability in welding [24].

3.6 Future Trends and Innovations

In terms of increased efficiency and control over an increased range of welding operations are involved, modern welding power sources have come a long way. Inverter-based systems are among the modern power sources with enhanced capabilities for welding difficult materials and joint types. Gas metal arc welding (GMAW) has greatly improved as a result of these advancements, with modern power sources offering higher functional efficiency and responsiveness to industrial demands [9]. These sources include advanced arc management features like intellectual arc management and progression repeat, which can improve the weld's quality, stabilize the process of welding, and reduce heat input especially in challenging locations [25]. Finally, the application of highly concentrated power sources, such as lasers, electron beams, and plasma, can allow the joining of complex materials and structures [26]. Since process technology, laser sources, and process indulgent have advanced, the field of laser beam welding has evolved dramatically. As a consequence of these advances, more accurate, efficient, and versatile welding power sources are being created for a range of applications [27].

3.7 Case Studies and Practical Applications

Applications of microplasma arc in surfacing, remelting, and welding procedures explain the uniqueness and scope of plasma welding and illustrate the impact of microplasma arc and its blazing conditions. In addition, the article showcases thin element microplasma welded joints and explores the use of microplasma arc in remelting and surface. It also highlights the benefits of plasma arc and its likely substitute for other welding power sources [28]. By utilizing the optional technique to make an alternating current source for arc welding, it is believable to update antiquated transformer energy sources by including a unit with an active power filter. Network voltage fluctuations might have less of an effect on the welding process's quality and that consumer qualities of the sources would be improved by enhancing the welding arc's stability and smoothly managing the welding current. It eliminates the need for ballast resistors to control the welding current and a block of capacitors to prevent a transformer from magnetizing during the TIG-AC process [5].

3.8 Conclusion

Power supplies for welding have altered considerably over time moving from obsolete transformer-based equipment to cutting-edge inverter-based systems. Improved control over welding conditions, portability, and enlarged energy economy are just a few benefits of using inverters. These developments have absolutely changed how welding is done in many different industries. They have decreased operating costs and their impact on the surroundings by using less energy, all while improving the quality, speed, and reliability of welding. Future welding power sources are probable to be centered on improving digital control integration, Internet of Things (IoT) connectivity for remote diagnostics and monitoring, and materials science developments to support new welding applications (e.g., dissimilar material joining for lightweight structures). There are still issues to be resolved, like achieving a cost-performance balance, maintaining reliability in a variety of ecological situation, and closing the skill gap by providing training in new technologies. Overall, the technology of welding power sources is spurring innovation in the building and industrial industries. In addition to guaranteeing product quality and pleasing changing consumer needs for sustainability and efficiency, it is essential to the development of industrial capacities.

References

1. Norrish, J., Advanced welding processes: technologies and process control. in: *Woodhead Publishing eBooks*, 2006. http://ci.nii.ac.jp/ncid/BA85287597.
2. Ueyama, T., Welding power sources. *Weld. Int.*, 24, 9, 699–705, 2010. https://doi.org/10.1080/09507111003655267.
3. Kumar, K., Kalita, H., Zindani, D., Davim, J.P., Welding, in: *Materials forming, machining and tribology*, pp. 65–81, 2019. https://doi.org/10.1007/978-3-030-21066-3_5.
4. Mandal, N.R., Fusion Welding Power Source, in: *Springer series on naval architecture, marine engineering, shipbuilding and shipping/Springer series on naval architecture, marine engineering, shipbuildung and shipping*, pp. 145–157, 2016, https://doi.org/10.1007/978-981-10-2955-4_13.
5. Burlaka, V. and Lavrova, E., Impoving energy characteristics of the welding power sources for TIG-AC welding. *Eastern-European J. Enterp. Technol.*, 5, 5, 101, 38–43, 2019. https://doi.org/10.15587/1729-4061.2019.180925.
6. Yin, T., Wang, J., Zhao, H., Zhou, L., Xue, Z., Wang, H., Research on Filling Strategy of Pipeline Multi-Layer Welding for Compound Narrow Gap Groove. *Materials*, 15, 17, 5967, 2022. https://doi.org/10.3390/ma15175967.

7. Makles, K., Winczek, J., Gucwa, M., Advanced Output Characteristics of Welding Power Source for Pulsed GMAW, in: *Lecture notes in mechanical engineering*, pp. 307–314, 2018. https://doi.org/10.1007/978-3-319-99353-9_33.
8. Zhang, H., Li, R., Yang, S., Zhan, L., Xiong, M., Wang, B., Zhang, J., Experimental and Simulation Study on Welding Characteristics and Parameters of Gas Metal Arc Welding for Q345qD Thick-Plate Steel. *Materials*, 16, 17, 5944, 2023. https://doi.org/10.3390/ma16175944.
9. Shome, M., Gas Metal Arc Welding – Simplified Perspective of Power Sources (Part I). *Indian Weld. J.*, 55, 2, 47, 2022. https://doi.org/10.22486/iwj.v55i2.212408.
10. Blankenbuehler, J.H. and Lester, R.V., Capacitor Stabilization of Arc-Welding Transformers. *Trans. Am. Inst. Electr. Eng.*, 65, 2, 54–55, 1946. https://doi.org/10.1109/t-aiee.1946.5059300.
11. Habjan, G. and Petrun, M., Impact of the Winding Arrangement on Efficiency of the Resistance Spot Welding Transformer. *Energies*, 12, 19, 3735, 2019. https://doi.org/10.3390/en12193735.
12. Herskind, C.C., Rectifier-Circuit Duty. *Trans. Am. Inst. Electr. Eng.*, 63, 3, 123–128, 1944. https://doi.org/10.1109/t-aiee.1944.5058906.
13. Agarwal, V., AC-DC converters (rectifiers), in: *Elsevier eBooks*, pp. 6–23, 2023. https://doi.org/10.1016/b978-0-12-821204-2.00156-2.
14. Cardoso, J., Orosco, R., Vazquez, N., López, H., Hernandez, C., Vaquero, J., High Efficiency Transformerless Photovoltaic DC/AC Converter with Common Mode Leakage Current Elimination: Analysis and Implementation. *Energies*, 15, 9, 3177, 2022. https://doi.org/10.3390/en15093177.
15. Zhang, H., Yu, J., Zhang, Z., Gao, J., Su, Z., Sun, Z., Li, Y., Arc Characteristics and Welding Process of Laser K-TIG Hybrid Welding. *Metals*, 12, 7, 1139, 2022. https://doi.org/10.3390/met12071139.
16. Lou, L.Y. and Gao, Y., Equipments of Pulsed Welding System and its Application. *Adv. Mater. Res.*, 989–994, 3314–3318, 2014. https://doi.org/10.4028/www.scientific.net/amr.989-994.3314.
17. Kang, B.Y., Review of magnetic pulse welding. *J. Weld. Joining*, 33, 1, 7–13, 2015. https://doi.org/10.5781/jwj.2015.33.1.7.
18. Bellmann, J., Roder, K., Zimmermann, M., Beyer, E., Kroll, L., Nestler, D., Influence of Copper Interlayers on the Magnetic Pulse Welding Process between Aluminum and Steel. *Metals*, 11, 6, 868, 2021. https://doi.org/10.3390/met11060868.
19. Mishra, D., Gupta, A., Raj, P., Kumar, A., Anwer, S., Pal, S.K., Chakravarty, D., Pal, S., Chakravarty, T., Pal, A., Misra, P., Misra, S., Real time monitoring and control of friction stir welding process using multiple sensors. *CIRP J. Manuf. Sci. Technol.*, 30, 1–11, 2020. https://doi.org/10.1016/j.cirpj.2020.03.004.
20. Mathias, S.G., Schmied, S., Grossmann, D., Monitoring of Discrete Electrical Signals from Welding Processes using Data Mining and IIoT Approaches. *2020 IEEE 32nd International Conference on Tools With Artificial Intelligence*, 2020. https://doi.org/10.1109/ictai50040.2020.00142.

21. Liu, Q., Chen, C., Chen, S., Key Technology of Intelligentized Welding Manufacturing and Systems Based on the Internet of Things and Multi-Agent. *J. Manuf. Mater. Process.*, 6, 6, 135, 2022. https://doi.org/10.3390/jmmp6060135.
22. Burlaka, V., Gulakov, S., Podnebenna, S., Kysliak, V., Plakhtii, O., Nerubatskyi, V., *Development of welding power supply source with higher efficiency*, Bulletin of the Azov State Technical University. Series: Technical Sciences, 43, 121–129, 2021. https://doi.org/10.32782/2225-6733.43.2021.15.
23. Thommes, J.M., *Highly efficient welding power supply*. Final report, May 1, 1978–September 30, 1980. Cyclomatic Industries, Inc., San Diego, CA (USA), 1980b. https://doi.org/10.2172/6743064.
24. Mehta, K.P., Sustainability in Welding and Processing, in: *Materials forming, machining and tribology*, pp. 125–145, 2018. https://doi.org/10.1007/978-3-030-03276-0_6.
25. Kolařík, L., Kolaříková, M., Kovanda, K., Pantůček, M., Vondrouš, P., Advanced Functions of a Modern Power Source for GMAW Welding of Steel. *Acta Polytechnica*, 52, 4. 83–88, 2012. https://doi.org/10.14311/1600.
26. Paton, B.E., Advanced Studies and Developments of the E.O. Paton Electric Welding Institute in Welding and Allied Technologies. *Biuletyn Instytutu Spawalnictwa W Gliwicach/Biuletyn Instytutu Spawalnictwa*, 2019, 1, 7–18, 2019. https://doi.org/10.17729/ebis.2019.1/1.
27. Martukanitz, R.P., A critical review of laser beam welding, in: *Proceedings of SPIE, the International Society for Optical Engineering/Proceedings of SPIE*, 2005, https://doi.org/10.1117/12.601655.
28. Golański, D., Chmielewski, T., Skowrońska, B., Rochalski, D., Advanced Applications of Microplasma Welding. *Biuletyn Instytutu Spawalnictwa W Gliwicach/Biuletyn Instytutu Spawalnictwa*, 2018, 5, 53–63, 2018. https://doi.org/10.17729/ebis.2018.5/5.

4

Welding Automation and Robotics

Jyoti Bhattacharjee and Subhasis Roy*

Department of Chemical Engineering, University of Calcutta, Kolkata, India

Abstract

Welding, a cornerstone of production, transforms automation and integrates robotics. This chapter examines the evolution of automation and robotics emphasizing their impact on efficiency, precision, and safety in various industries. Robots are very adaptable to multiple types of welding, such as resistance and arc welders. The function of sensors in robotic welding is explored, along with a detailed analysis of methodologies for extracting weld seam position and geometry features using various sensor types commonly used for seam tracking. We discuss the proposed welding reported task data model, which aims to rectify the deficiencies of existing robotic welders in aircraft manufacturing by incorporating precise geometry, dimension, and welding requirement information. This is achieved through a thorough requirement analysis that identifies and summarizes the relevant information necessary for a seamless welding process, including machine readability and compatibility considerations. This chapter also analyses robotic welding systems that produce structural steel components for bridges, skyscrapers, and industrial facilities with unprecedented efficiency and precision, shortening building timetables and decreasing on-site labor requirements. There is a highlight on the evolution and development of robotics by leveraging artificial intelligence (AI) algorithms in welding, including several advantages and disadvantages associated with using various robots for welders. Further discussion is held on using different sensors to compare their benefits and drawbacks.

Keywords: Welding automation, robotics, automation, sustainability, efficiency

**Corresponding author*: srchemengg@caluniv.ac.in; ORCID: 0000-0003-4197-535X
Jyoti Bhattacharjee: ORCID: https://orcid.org/0009-0006-6835-2899

Sandip Kunar and Gurudas Mandal (eds.) Advanced Welding Technologies, (47–72) © 2025 Scrivener Publishing LLC

4.1 Introduction

Welding automation and robotics represent a significant advancement in manufacturing bringing increased precision, efficiency, and safety to the welding process. The history of welding automation and robotics can be traced through several key developments over the past century [1].

4.1.1 Early Beginnings

Welding itself dates back to the early 20th century, with the invention of arc welding by Nikolay Slavyanov and C. L. Coffin in the late 1800s. However, it was not until the 1920s and 1930s that welding techniques began to be widely adopted in industrial settings, particularly in shipbuilding and construction [1, 2]. The advent of World War II further accelerated the development and adoption of welding technologies due to the need for rapid and reliable ship and aircraft production.

4.1.2 Post-War Innovations

The post-war period saw significant advancements in welding technology, including the development of gas–metal arc welding (GMAW) and submerged arc welding (SAW). These methods improved efficiency and quality laying the groundwork for further automation. In the 1950s, the first steps toward automated welding were taken with the introduction of basic mechanized systems that could perform repetitive welding tasks more consistently than human welders [2, 3].

4.1.3 Rise of Robotics in the 1960s and 1970s

The 1960s and 1970s were transformative decades for welding automation largely due to the rise of industrial robotics. The introduction of the Unimate robot by George Devol and Joseph Engelberger in 1961 marked the beginning of the robotics era in manufacturing. Unimate was initially used for tasks, such as die casting and spot welding, in the automotive industry. Companies, like Fanuc, ABB, and KUKA, emerged as leaders in the field, developing robots specifically designed for welding applications. These robots were equipped with sensors and vision systems that enabled them to adapt to variations in the welding environment and perform high-precision welds. In the 2000s and 2010s, welding automation and robotics continued to evolve with the integration of advanced technologies such as artificial intelligence (AI) and the Internet of Things (IoT). These technologies enhanced the capabilities of robotic welding systems allowing for real-time monitoring and adjustments to ensure optimal weld quality. Collaborative robots, or cobots, also emerged during this period, designed to work alongside human operators enhancing flexibility and safety in welding operations [1–3].

4.1.4 Modern Era

Today, welding automation and robotics are integral to many industries, including automotive, aerospace, and construction. Modern robotic welding systems are highly advanced incorporating machine learning algorithms that enable them to improve their performance over time. The trend toward industry has further driven the adoption of smart welding systems that are interconnected and capable of autonomous decision making [3, 4].

The future of welding automation and robotics looks promising, with ongoing research focused on further improving the precision, efficiency, and adaptability of these systems. Developments in areas, such as additive manufacturing (3D printing) and hybrid welding processes, are expected to open new possibilities for robotic welding applications, pushing the boundaries of what can be achieved in automated welding. In contrast, non-fusion welding, such as soldering and brazing, consists of heating the base metal without melting it and using a filler metal to create a strong bond upon cooling. Intelligent welding robots utilize vision sensors as central components [4, 5]. Using vision-sensing systems, researchers have developed various experimental platforms tailored to specific welding tasks. A schematic of a typical system framework for an intelligent welding robot is illustrated in Figure 4.1.

The art of welding boasts an ancient lineage. Evidence from the Bronze Age reveals the skill of Egyptians in creating welded gold boxes and iron tools. Throughout the Middle Ages, blacksmiths continued to refine their welding techniques for various iron implements. However, the 19th century marked a turning point with groundbreaking discoveries [5]. During this period, significant milestones included the discovery of acetylene gas and the development of a battery-powered device that could generate an electric arc between carbon electrodes. These breakthroughs paved the way for significant advancements in the late 19th century, particularly in carbon arc welding and metal electrode arc welding, fundamentally transforming the welding industry [5, 6].

Subsequently, the 20th century saw the emergence of various welding methods such as stud welding, submerged arc welding, Heliarc welding, and gas-shielded metal arc welding. Innovations, like CO_2 welding, electro-

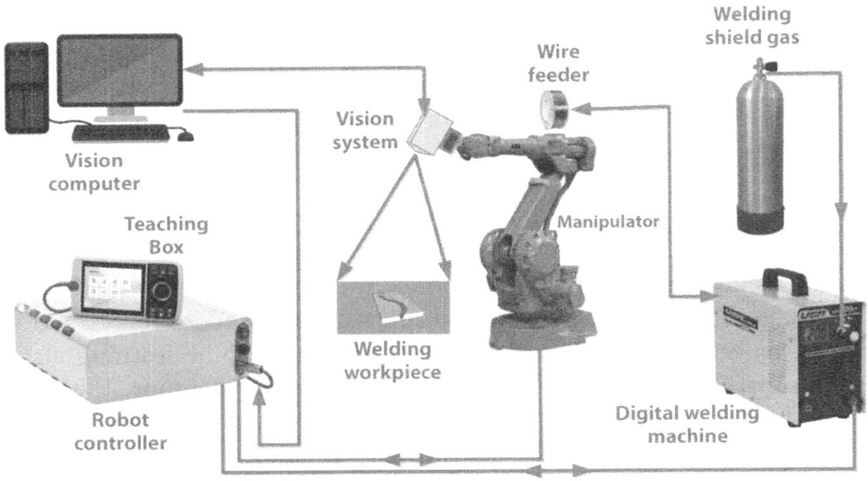

Figure 4.1 Exhibition of the welding process [4] (reproduced with permission).

slag welding, and plasma arc welding, continued to shape the welding industry [5–7]. In recent times, advancements in welding technology have led to the development of precision techniques, like friction welding and laser welding, which are now commonly used in automated manufacturing processes. These innovations represent a significant leap forward in the welding industry enabling precise and efficient welding processes. Over the last five decades, welding technology has undergone remarkable evolution. The advent of lasers has revolutionized automated welding, while substantial safety enhancements have made the practice safer through improved tools, equipment, and protective gear [6, 7].

4.1.5 Industrial Automation

Manufacturing organizations today contend with intense national and global competition necessitating enhanced performance. To meet these demands, many have adopted computer-integrated manufacturing (CIM) across a variety of production environments. CIM is designed to boost human productivity, enhance product quality, and optimize the use of financial resources, while also allowing for swift adaptation to market needs. By linking the information systems of various automation units within conventional manufacturing frameworks, CIM functions similarly to a closed-loop control system: product orders are received, and completed products are delivered [7, 8]. These robots automate a multitude of operations managing simple, repetitive tasks and improving the quality of manufactured goods. They are particularly advantageous in hazardous manufacturing processes, such as spray painting and welding, where they can significantly reduce health risks for human workers. Industrial robots have become indispensable in modern factories and are anticipated to become even more crucial in future manufacturing settings. Their versatility, intelligence, and ability to execute repetitive tasks reliably and economically fuel their demand. These tasks are performed with remarkable endurance, speed, and accuracy.

The following are some key points about industrial robots [8, 9]:

- Robots carry out specific tasks repeatedly with minimal deviation maintaining high precision through programmed routines that dictate direction, acceleration, speed, deceleration, and movement distances.
- Certain robots provide greater adaptability, adjusting to different object orientations or undertaking various tasks, often utilizing object recognition capabilities.
- Robots often integrate machine vision systems as their "eyes," connected to powerful computers or controllers for accurate guidance.
- Artificial intelligence plays an increasingly crucial role in enhancing the capabilities of modern industrial robots enabling them to adapt, learn, and optimize their operations.

This integration of robotics, machine vision, and AI contributes significantly to the efficiency and reliability of automated industrial processes across a wide range of sectors [10, 11].

Programmable transfer machines initially functioned using hydraulic actuators to move objects through short distances. These machines operated based on joint coordinates storing angles of various joints during a teaching phase and replicating them during operations [12–14].

4.2 Inspection of Welding Performance

To ensure a successful weld, it is paramount to begin by thoroughly assessing the metals that require joining, identifying the most suitable welding arc for the specific materials and their thicknesses, and meticulously setting the operational parameters for the chosen welding process. These critical parameters encompass current, voltage, welding speed, temperatures, gas flow, and torch angle [15, 16]. Any inaccuracies in defining or controlling these parameters can lead to a decrease in process efficiency, suboptimal quality outcomes, compromised mechanical properties of the weld, and the occurrence of defects that may necessitate costly rework resulting in significant economic repercussions [15–17].

Once the optimal welding technique and parameters have been determined, it is essential to anticipate and prepare for unforeseen occurrences during the welding process that could give rise to defects. Before initiating welding activities, it is advisable to calibrate both the robot and the entire welding system and establish mechanisms for evaluating weld quality and categorizing defects. These categorizations often rely on databases or leverage advanced technologies, such as neural networks and machine learning algorithms like decision trees, to detect defects accurately. For example, they can establish correlations between current and voltage signatures by employing radiographic image-processing techniques using adaptive neuro-fuzzy inference systems (ANFIS). Real-time defect classification during welding can be achieved through methodologies such as principal component analysis in conjunction with artificial neural networks (ANNs) or employing deep learning strategies that utilize digital radiography and optimized convolutional neural networks (CNNs) [16, 17].

The common welding defects and their detection methods include the following:

- Cracks are serious welding defects that occur when the applied loads or stresses exceed the material's strength. They can appear as "hot cracks" during the metal's solidification process or as "cold cracks" after the metal has cooled. Detection methods for cracks include eddy current testing, thermography using high-power infrared light sources, acoustic emission analysis with sensors, ultrasonic laser techniques, and real-time X-ray imaging systems [18, 19].
- Porosity arises from trapped gas in the molten metal leading to voids as the metal solidifies. This defect can result from improper material handling, welding processes, or contamination in the base metal. Factors contributing to porosity include issues with nozzles, guns, torches, electrodes, gas flow, improper shielding, and material cleanliness or moisture content. Ensuring compliance with conditions for material protection, cleanliness, and drying during welding is essential to prevent defects like porosity, as illustrated in Figure 4.2.

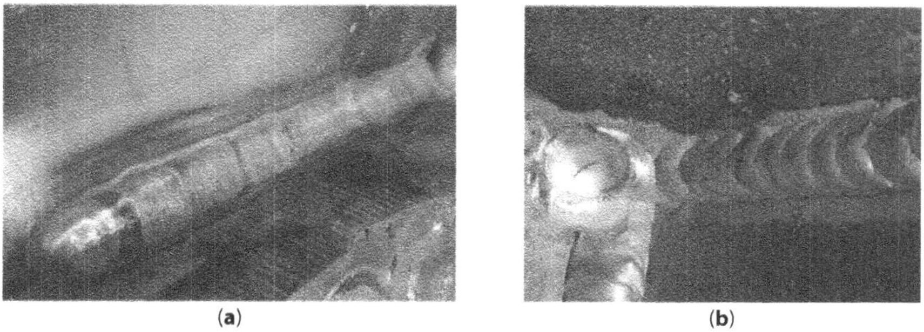

Figure 4.2 Illustration of two types of weld defects: (a) A longitudinal surface crack may occur in the weld metal when excessive clamping force is applied between plates of different thicknesses (e.g., 40 and 100 mm), and TIG welding is performed without sufficient preheating of the plates. (b) A transverse crater crack can develop in the weld metal due to an abrupt arc interruption creating tensile stresses across the weld surface [20] (reproduced with permission).

4.2.1 Welding Robots

The allure of robotics stems from its integration of precision-engineered devices, dependable power sources, and adaptable programming capabilities [21]. These standards enable unrestricted access to system capabilities, particularly crucial in research settings where resource accessibility is vital for implementing and testing novel concepts. For traditional robotics domains, such as industrial robot manipulators and mobile robots, which have benefited from decades of engineering refinement resulting in robust machines, open-source software is not always a prerequisite. However, in emerging fields like humanoid robotics, space exploration, and medical robotics, open-source software is pivotal in expediting development processes. It facilitates collaboration, accelerates innovation cycles, and lowers entry barriers for researchers and developers aiming to advance these specialized areas of robotics. Industrial robotic welding stands out as one of the most prevalent robotics applications on a global scale. Various products necessitate welding as a crucial component of their assembly processes, with the automotive industry serving as a prominent example due to the widespread utilization of spot welding and MIG/MAG welding in car body workshops. Moreover, smaller businesses catering to individual client needs increasingly produce small series or custom-made products emphasizing the demand for highly automated welding processes to meet client requirements with exceptional quality swiftly. Agile production plays a pivotal role in this context bolstered by adaptable manufacturing configurations [22, 23].

Despite the substantial interest in industrial robotic welding, it remains a multifaceted and unresolved technological endeavor. Welding poses challenges in terms of parameterization, monitoring, and effective control. Most welding techniques rely on empirical models derived from experience under specific conditions, with a limited comprehensive understanding of their effects on welding joints and surfaces [23, 24]. For instance, MIG/MAG welding often involves elevated temperatures in small areas leading to significant thermal expansion and contraction cycles that can impact mechanical properties and induce plastic deformation. Therefore, comprehending these transformations is essential to mitigate their adverse effects. Integrating robots for welding tasks is a complex area that has attracted considerable research and development initiatives. In the modern landscape, a diverse range of products necessitate welding for assembly. When a substantial portion of a product comprises welded connections, implementing automation becomes imperative to execute welding tasks efficiently. This transition enhances productivity and quality and reduces production expenses and labor requirements. However, the introduction of robots into welding setups introduces additional complexities. Robots frequently present challenges to everyday operators due to complexities in usability and programming, limited remote capabilities, and constrained programming environments within closed systems with restricted software interfaces [24–26].

4.3 Robotics in Welding

4.3.1 Motivations

Automated welding has transcended its origins in long production runs and large-scale applications propelled by advancements in digital technologies that seamlessly integrate welding equipment with robotics. This convergence has given rise to programmable cells tailored for smaller, more specialized tasks [27–29]. On the factory floor, the fusion of robotics and welding not only enhances quality and consistency but also accelerates production, reduces downtime, and ultimately boosts profitability. Recent innovations in welding automation have significantly expanded programming flexibility and digital connectivity among welding power sources, robotic arms, and positioning equipment making automation economically viable and productive across a wide range of manufacturing scales [29, 30].

Traditionally, automated welding was predominantly feasible only for extensive production runs due to the high costs associated with fixtures and specialized workstations. However, technological progress has reshaped this paradigm. Modern systems now offer user-friendly programming interfaces, adaptable work-holding solutions, and integrated cells that seamlessly combine weld control, robotic movement, and workpiece positioning. These advancements have democratized automation making it practical even for smaller batch sizes [31, 32]. Today, companies producing batches as modest as 50 to 100 weldments can reap substantial benefits from automation comparable to those manufacturing hundreds or thousands of identical parts. Moreover, larger enterprises can automate shorter runs or specialized welding tasks using pre-engineered systems that are easily programmed and adaptable to diverse production requirements.

Strategic collaborations among welding equipment manufacturers, robotics firms, and positioner providers have played a crucial role in making customized automation cells more accessible for small-lot applications. Previously, manufacturers often deemed automation impractical for their most complex parts due to the need for dedicated fixturing. However, with more flexible systems, the requirement for specialized fixturing has diminished allowing easier switching between different parts [33, 34]. The primary driver behind automation adoption is economic feasibility. Consistent quality represents another critical motivation for automation. These systems achieve superior and uniform weld quality by minimizing uncontrolled variables, thereby enhancing overall customer satisfaction. Flexible fixturing within contemporary robotic cells enables swift product transitions with minimal changeover times, further elevating productivity and ROI. The shortage of skilled welding operators further accelerates the shift toward automation. Manufacturers can reduce labor costs and dependency by delegating repetitive tasks to machines. In a robotic welding cell, operations are typically segregated into two zones: one where the operator loads and unloads materials and another where the robot performs welding operations simultaneously. This setup maximizes efficiency and operational throughput [33–36]. The integration and connectivity among system components distinguish modern welding automation from earlier practices. Advanced industrial communication networks facilitate continuous feedback loops between office computers and the factory floor enabling real-time process evaluation, internet connectivity for remote monitoring and control, and faster, more flexible digital communications over traditional analog controls. Equipment manufacturers increasingly form strategic partnerships to enhance synergy and effectiveness to develop integrated products that operate seamlessly as cohesive systems. This collaborative approach allows them to leverage global expertise while responding swiftly to local market demands ultimately delivering more efficient automated welding solutions than standalone work cells assembled independently by users [34, 35].

4.3.2 Disadvantages of Welding Robots

While robotic welders offer numerous benefits, they also come with several drawbacks. First, the high cost of robotic welders can be prohibitive for many businesses. Additionally, operating these machines requires trained personnel for programming and maintenance, and they can frequently break down or require repairs. Robotic arms have a limited range of motion, which may prevent them from reaching all necessary welding points necessitating human intervention to complete the job. Furthermore, if a workpiece is incorrectly positioned, the robot will weld in the pre-programmed locations potentially resulting in misplaced welds. Inaccurate programming can also lead to errors. Robots are typically capable of performing only a limited range of welding types, some of which may have longer cooling times or could weaken the metal if not used correctly [36].

4.3.3 Automated Welding

Robotic welding systems typically create a cubic workspace, employing rectilinear robots that move along X, Y, and Z axes with rotary wrist attachments. In contrast, articulated robots mimic human arm movements with rotating joints, offering a more flexible work arc. Establishing a robotic welding facility involves unique considerations compared to manual setups, such as managing start/stop functions, gas pre-flushing, electrode feeding, and nozzle cleaning. For approximately 15 years, robots have played a crucial role in welding automotive body assemblies and subassemblies. Automatic arc welding equipment is designed for high-duty cycles and must be compatible with advanced control systems [36–39].

The decision to automate hinges on the volume of items requiring welding to justify the investment. For example, rack-mounted automatic guns or torches are suitable for simple joints on accessible workpieces. These automatic guns can be fixed or mounted on curved tracks, facilitating tasks like welding pipes or attaching flat bases to cylindrical shapes through rotational manipulation of the workpiece. However, automating processes becomes challenging if parts require frequent adjustments, have varying joint positions or widths, or are scattered across different locations on each piece. In practical applications, tabletop-sized robots prove effective for welding one side of a rotating jig, providing access to both sides of the workpiece. Robots excel in repetitive tasks involving multi-axis welds or complex geometries. In manufacturing settings, they function either as components of production lines or as standalone units for batch production. Many companies transition gradually from manual to automated welding, often beginning with a standalone robotic cell [38].

Welding robots showcased at trade shows demonstrate their speed and flexibility through intricate maneuvers. Beyond demonstrations, modern robots handle various welding operations, such as arc welding, metal inert gas (MIG) welding, tungsten inert gas (TIG) welding, spot welding, and stick welding, highlighting their versatility and suitability for diverse industrial applications [38, 39].

4.4 Modeling of Welding Processes

In the earlier section focusing on inspection, the chapter delved into the primary defects and critical characteristics associated with welding. To mitigate these defects effectively, leveraging contemporary technological tools is essential. This includes utilizing diverse types of neural networks featuring modular software interfaces, which can analyze weld parameters through machine learning and deep learning techniques. Once the necessary parameter data is gathered, mathematical methodologies such as finite elements, regression analysis, algorithms, and Gaussian hypotheses will be applied to simulate and model the welding process. The following elements are defined for this purpose:

4.4.1 Neural Network

A neural network is an artificial intelligence tool designed to enable computers to process data similarly to the human brain. By employing deep learning techniques, neural networks utilize layers of nodes or neurons to model complex, non-linear relationships between input and output data [39]. These artificial neurons collaborate through mathematical computations to solve problems. Machine learning, another AI technique, empowers computers to learn from extensive datasets. Machine learning algorithms identify patterns within existing data and apply this knowledge to new data to make informed decisions. Human intervention is crucial in selecting relevant features for analysis. For instance, machine learning can establish correlations between arc sound and weld quality by considering parameters such as current, voltage, and travel speed. This process enables the classification of welds based on algorithms that derive insights from the data [39, 40].

4.4.2 Deep Learning

Deep learning involves inputting raw data into the software enabling it to learn and extract new features autonomously. This approach can analyze unstructured datasets and tackle complex issues. Applications of deep learning include the following:

- Analyzing the 3D surface of the weld pool to generate penetration depth maps for neural network classification [40–42].
- Detecting defects using an optimized convolutional neural network (CNN) with digital radiography.
- Predicting weld pool shape and depth based on interactions between the arc and electrodes.
- Categorizing defects in the weld bead.

To achieve precise modeling and defect identification in welds, gathering comprehensive data on the parameters for analysis is imperative. Subsequently, these data undergo analysis using diverse mathematical formulations [41, 42].

4.4.3 Finite Element Analysis (FE)

Finite element analysis is employed to scrutinize defects, such as cracks, porosity, and undercuts, through ultrasonic lasers or to assess thermo-mechanical behavior. For instance, it can compute Von Mises stresses near the weld bead and their distribution, as illustrated in Figure 4.3.

4.4.4 Regression Analysis

Regression analysis can forecast weld pool evolution through imaging, regulate pool width via current, and adjust bead geometry.

4.4.5 Algorithms

Algorithms are primarily developed for the following:

- Modeling Weld Pool Parameters: This includes imaging-metal flow on the surface [43–45].
- Bead Geometry and Metal Trans based modeling of current, velocity, pool width, pool size, and molten fer Modeling: Focuses on modeling bead geometry, metal transfer, and current and voltage.
- Controlling Robot and Tool Trajectories: Ensures precision by avoiding deviations and dimensional errors before welding [43, 44].
- Detecting Weld Defects: Utilizes image transformation, pattern recognition methods, sub-pixel algorithms, and temperature signal-based algorithms to identify weld defects.
- Classifying Defects and In-Line Fault Detection: Identifies defects, such as porosity, undercuts, lack of fusion, and spatter, based on frequency bands [45–48].
- Assessing Weld Quality: Uses optimization algorithms, plasma spectroscopy analysis techniques, and AI algorithms based on current and voltage parameters. It correlates weld

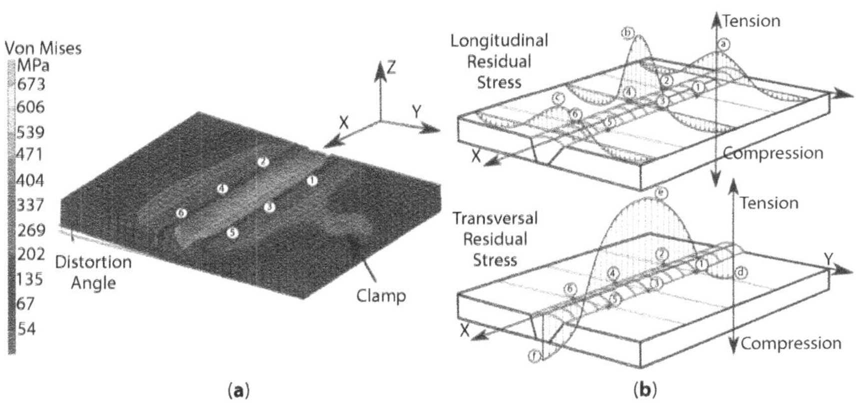

Figure 4.3 Illustration of two critical aspects: (a) Von Mises stresses derived from the finite element (FE) model of a butt joint with a single V-groove in the initial specimen studied and (b) theoretical representation of longitudinal and transverse stress distributions applicable to welding any butt joint with a single V-groove [48] (reproduced with permission).

quality with these parameters and conducts online quality inspections through voltage variance algorithms, as shown in Figure 4.3 (a, b).

4.5 Robotics in Different Welding Techniques

4.5.1 Spot Welding

Spot welding is a widely used welding process in manufacturing renowned for its efficiency and effectiveness in joining metal sheets together. It involves creating a series of overlapping welds or spots at specific points along the joint, where intense heat generated by electrical resistance fuses the metal surfaces. This process is particularly favored in automotive, aerospace, electronics, and appliance manufacturing industries, where rapid assembly of sheet metal components is crucial [49].

In automotive manufacturing, spot welding plays a pivotal role in assembling car bodies. For instance, panels and structural components made of steel or aluminum are often joined using spot welding due to their ability to quickly create strong, durable bonds. This method ensures the structural integrity of the vehicle while maintaining efficiency in production lines. Similarly, in aerospace, spot welding is utilized for constructing fuselage panels and other components where lightweight, yet sturdy, joints are essential for ensuring aircraft safety and performance.

Electronics manufacturing also heavily relies on spot welding, especially in the production of circuit boards and electrical contacts. Here, precision is key, and spot welding allows for the creation of reliable connections without damaging delicate electronic components. Appliances, like refrigerators and washing machines, also benefit from spot welding during their assembly process ensuring secure seams and joints that can withstand the rigors of daily use.

The process of spot welding involves specialized equipment known as spot welders, which typically consist of two copper alloy electrodes. These electrodes clamp the metal sheets together and pass an electric current through them for a precise duration generating heat at the contact points. The heat melts and fuses the metal surfaces forming a weld nugget that cools to create a solid joint. Modern spot welding machines are equipped with advanced controls to regulate current, pressure, and timing ensuring consistent weld quality and minimizing defects [49, 50].

Overall, spot welding remains a cornerstone of modern manufacturing due to its reliability, speed, and ability to create strong joints in various

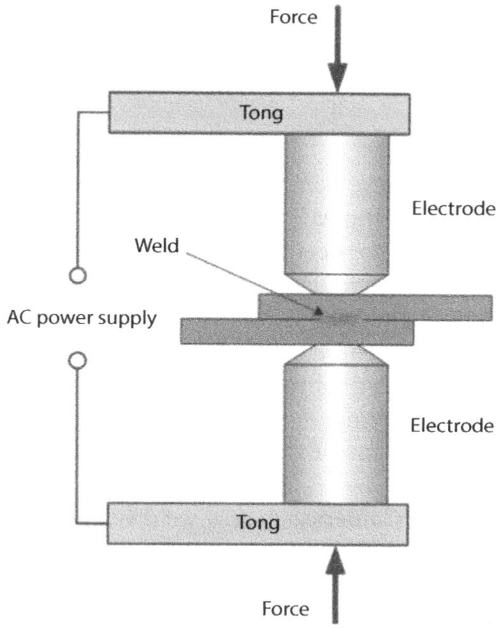

Figure 4.4 Process of spot welding [49] (reproduced with permission).

materials. The welding gun, mounted on the robot's end effector, is tailored to the process and comes in the following two forms: the cost-effective C-type and the X-type. These guns can be equipped with pneumatic actuators for consistent electrode force or hydraulic actuators for high-pressure applications in confined spaces. Recently, servo guns powered by servo motors have been developed offering more precise electrode force control than pneumatic guns as shown in Figure 4.4.

4.5.2 Bead Tracking, Weld Dimension, and Defect Monitoring and Control

This section focuses on the critical aspects of bead tracking, weld dimension control, and defect monitoring and control in robotic welding processes. Precise bead tracking is achieved through sensors, like laser, vision, or arc sensors, which detect and monitor the weld seam's position and shape. Weld dimension control involves dynamically adjusting welding parameters, such as wire feed rate and voltage based on sensor feedback and desired weld bead geometry. These capabilities enable robotic welding systems to autonomously and adaptively perform tasks in various industrial settings ensuring consistent quality and efficiency [50, 51].

Various sensors play essential roles in welding processes, including the following:

- Arc Sensors: These sensors adjust the torch-to-workpiece distance by utilizing variations in arc voltage or current signals eliminating the need for extra equipment in the welding area. They are particularly useful in confined spaces where flexibility is limited.
- Optical Sensors: These sensors monitor welding operations and relay process data to control models. Key sensor types include the following:
  - Vision Cameras: Evaluate process stability and weld quality.
  - Multi-view Cameras: Assess weld bead properties from multiple angles.
  - Speed Cameras: Monitor deposition height and molten pool temperature.
  - High-Speed Cameras: Provide detailed insights into physical welding phenomena.
  - CCD Cameras: Measure coating height accurately.
  - CMOS Cameras: Offer versatile imaging capabilities suitable for various welding applications.

Additionally:

- Spectroscopy: Enables detection of defects, such as porosity and imperfections, within the weld pool or bead.
- Infrared Sensors: Used for automatic defect identification, controlling bead geometry parameters and torch position, monitoring pool temperature, and evaluating weld pool and penetration depth [50–52].
- Sound Sensors:
  - Acoustic Sensors: Analyze acoustic signals through dynamic microphones to classify metal transfer modes.
  - Ultrasonic Sensors: Provide high-resolution data for *in situ* porosity monitoring. These sensor technologies collectively facilitate comprehensive monitoring and control of welding processes ensuring high-quality welds and operational efficiency in industrial settings. To maintain

electrode performance, an automatic tip dresser sharpens them after each cycle, typically completed in 1–2 s. Ensuring a well-maintained contact surface on the electrode is crucial for achieving high welding quality reducing production downtime and optimizing utility costs. A common challenge in welding automation involves managing the positioning of cables and hoses, which can restrict the movement of robotic systems [51–53]. To tackle this challenge, a rotating mechanism integrates compressed air, cooling water, electric current, and signals into a single swiveling unit. This innovation greatly boosts the effectiveness of robotic welding by optimizing reach to spot-welding zones. Installed directly onto the robot's end tool, it removes the necessity for external tubes or wires guaranteeing uniform, top-notch spot welds. Additionally, a spot weld timer controls crucial parameters such as welding time, current intensity, and the sequence and duration of different phases within the welding cycle. This control mechanism ensures precise execution of welding processes enhancing overall weld consistency and reliability [52, 54].

Laser back-reflected (LBR) and acoustic emission (AE) wavelet spectrograms were analyzed using an advanced deep-learning framework to identify distinctive, stable signatures associated with predefined quality-significant events. Statistical learning techniques were employed to discern variations within the datasets and develop a specialized function enabling data separation through non-linear transformations. This chapter fully leveraged state-of-the-art deep convolutional neural networks (CNNs) renowned for their adeptness in structuring complex data. Integration of LBR and AE signals with X-ray radiographic data facilitated the identification of patterns linked to different quality-significant events. Spectrograms derived from LBR and AE signals for each pattern were organized according to respective quality-significant events, as demonstrated in Figure 4.5 (b). Consistent with prior studies, 4,096 frequency bands were extracted from each spectrogram and fed into the CNN. Wavelet transformation was adapted to ensure uniform spectrogram size, crucial for CNN processing efficiency [51, 52].

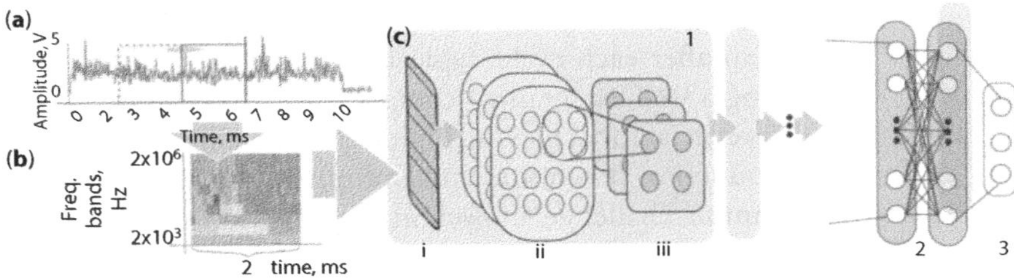

Figure 4.5 Analysis of the LBR signal involves three steps:

(A) LBR Signal Acquisition: Present the raw LBR signal obtained during the process [52].
(B) Wavelet Spectrogram Construction: Then converts the LBR signal into a visual representation using wavelet spectrograms [52].
(C) Convolutional Neural Network (CNN) Analysis: The data will be analyzed using CNN architecture. This CNN includes a block for automatic feature extraction, fully connected layers for processing the extracted features, and a softmax layer for classification purposes [52] (reproduced with permission).

4.6 Laser Welding and Laser–Arc Hybrid Welding

Laser welding and laser–arc hybrid welding techniques have been widely embraced in various industries due to their rapid speed, minimal distortion, and small heat-affected zones.

In monitoring systems, a neural network-based approach for high-power disk laser welding has been proposed [53–55]. This system utilizes signals from plume and spatter to detect welding defects in real time highlighting the potential of machine learning techniques in enhancing welding process monitoring and quality control.

Machine vision has also been extensively reviewed as a technology for welding monitoring [53, 54]. Discussions have centered on the challenges and opportunities associated with machine vision showcasing case studies that have improved welding inspection accuracy and efficiency, thereby advancing quality control practices.

Additionally, novel methods for detecting tight butt joint welds using optical flow and particle filtering of magneto-optical imaging data have been introduced [56, 57]. By analyzing magnetic field distributions during welding, innovative algorithms have been developed to identify weld seams in complex joint configurations demonstrating promise for enhancing inspection capabilities in challenging weld assessments. In additive manufacturing, Laser Metal Deposition (LMD) has been studied for its ability to create intricate geometries and reduce material waste compared to traditional methods [57, 58]. It provides valuable insights into the complex dynamics of welding and presents innovative strategies for optimizing welding quality and process efficiency [59].

4.7 Path/Motion Planning

Path planning is crucial for robotic welding guaranteeing accurate welds across diverse components. Researchers are exploring new methods, including the following:

- Dual-Robot Coordination: One robot holds the workpiece, while another wields the welding torch. Researchers developed a calibration method to synchronize the robots' coordinate systems to achieve this. Additionally, they implemented non-master/slave control schemes for motion planning. This allows for automated trajectory calculation based on the positions and timings of points on the workpiece [60, 61].
- Automated Path Layout for Thick Plate Welding: Automated path generation is essential for thick plates, particularly using multi-pass weaving techniques. Researchers created a user-defined path layout model for offline programming of multi-pass welds. This system allows users to define welding parameters, layer and pass details, and sequences. It then automatically generates torch positions, orientations, and weaving patterns for all passes. This research demonstrates effective multi-pass welding designs applicable to real-world production environment [61–63].
- Optimization Algorithms for Path Planning: Researchers have investigated algorithms like ant colony optimization to compute efficient welding paths. These algorithms have successfully optimized robot movements within production settings improving accuracy and efficiency.

These advancements highlight the progress in robotic welding technology, particularly in path planning strategies, multi-robot coordination, and automation of intricate welding processes. These innovations significantly enhance welding applications' precision, productivity, and quality control [60–64].

4.8 Challenges and Future Directions

While the benefits of welding automation and robotics are clear, several challenges remain as follows:

- High Initial Costs: The upfront investment in robotic systems can be significant, which may be a barrier for smaller enterprises.
- Skilled Workforce: There is a growing need for skilled workers who can design, program, and maintain these advanced systems.
- Complex Integration: Integrating robotic systems with existing processes and ensuring compatibility can be complex and require significant effort.

Despite these challenges, the future of welding automation and robotics is promising. Ongoing research and development are focused on making these technologies more accessible and user-friendly. Innovations, such as collaborative robots (COBOTS), which work alongside human operators, and advancements in AI-driven welding processes, are set to revolutionize the industry further [63, 64].

4.9 Conclusions

Integrating automation and robotics in welding processes marks a significant evolution in the manufacturing industry. These advancements enhance productivity and efficiency and ensure consistent quality and precision minimizing human error and workplace hazards. As technology progresses, the synergy between robotics and welding is set to revolutionize production lines fostering innovation and competitiveness. The future of welding lies in the seamless collaboration between human expertise and robotic precision paving the way for smarter, safer, and more sustainable manufacturing practices. Embracing these technologies will be crucial for industries that maintain a cutting edge in a rapidly changing global market. This adoption of welding automation and robotics is transforming the manufacturing landscape offering numerous advantages that drive efficiency, quality, and safety. As technology advances, these systems will become increasingly integral to modern manufacturing paving the way for more innovative and sustainable production methods. Embracing these technologies is crucial for staying competitive in the global market and meeting future demands.

References

1. Papakostas, N., Michalos, G., Makris, S., Zouzias, D., Chryssolouris, G., Industrial applications with cooperating robots for the flexible assembly. *Int. J. Comput. Integr. Manuf.*, 24, 650–660, 2001.
2. Bhattacharjee, J. and Roy, S., Density Functional Waltz: Molecular Theory of Spin Dynamics. *J. Phys. Opt. Sci. Sci. Res. Commun. Ltd.*, 1–4, 2024.
3. Chiroma, H., Abdullahi, U.A., Abdulhamid, S.M., Abdulsalam Alarood, A., Gabralla, L.A., Rana, N., Shuib, L., Targio Hashem, I.A., Gbenga, D.E., Abubakar, A.I. *et al.*, Progress on Artificial Neural Networks for Big Data Analytics: A Survey. *IEEE Access*, 7, 70535–70551, 2019.
4. Yang, L., Liu, Y., Peng, J., Advances techniques of the structured light sensing in intelligent welding robots: a review. *Int. J. Adv. Manuf. Technol.*, 110, 1027–1046, 2020.
5. Latombe, J.C., *Robot Motion Planning*, Springer, New York, NY, USA, 2012.
6. Chen, X., Stone, P., Sucar, L.E., Zant, T., *RoboCup 2012: Robot Soccer World Cup XVI*, Springer, Berlin/Heidelberg, Germany, 2013.
7. Yan, Z., Jouandeau, N., Cherif, A.A., A Survey and Analysis of Multi-Robot Coordination. *Int. J. Adv. Robot. Syst.*, 10, 1–18, 2013.
8. Gao, X., Mo, L., You, D., Li, Z., Tight Butt Joint Weld Detection Based on Optical Flow and Particle Filtering of Magneto-Optical Imaging. *Mech. Syst. Signal Process.*, 96, 16–30, 2017.
9. Ostra, T., Alonso, U., Veiga, F., Ortiz, M., Ramiro, P., Alberdi, A., Analysis of the Machining Process of Inconel 718 Parts Manufactured by Laser Metal Deposition. *Materials*, 12, 2159, 2019.
10. Şucan, I.A., Moll, M., Kavraki, L.E., The Open Motion Planning Library. *IEEE Robot. Autom. Mag.*, 4, 72–82, 2012.
11. Kuffner, J. and LaValle, S.M., RRT-connect: An efficient approach to single-query path planning, in: *Proceedings of 2000 IEEE International Conference on Robotics and Automation*, San Francisco, CA, USA, pp. 24–28, April 2000.

12. Şucan, I.A. and Kavraki, L.E., Kinodynamic motion planning by interior-exterior cell exploration, in: *Algorithmic Foundation of Robotics VIII*, Springer, Berlin/Heidelberg, Germany, 2009.
13. Saheed, B.A., Irina, L., Asmaa, K., Alexey, S., Effect of laser welding process parameters and filler metals on the weldability and the mechanical properties of AA7020 aluminum alloy. *J. Mater. Process. Technol.*, 2, 33, 2018.
14. Sandeep, S., Sachin, M., Research developments in laser welding—A review. *Int. J. Innov. Res. Sci. Technol.*, 3, 60–64, 2017.
15. Chavan, R., Ghatage, D.A., Bhosale, K.K., Review paper on laser welding machine. *Int. J. Appl. Sci. Eng.*, 5, 815–819, 2017.
16. Tanveer, M., Mohd, A.W., Faizan, A., Application of robotics in welding. *Int. J. Emerg. Manag.*, 7, 30–36, 2018.
17. Chen, X.Z. and Chen, S.B., The autonomous detection and guiding of start welding position for arc welding robot. *Ind. Robot*, 37, 70–78, 2010.
18. Zhang, Y.M., Kovacevic, R., Li, L., Adaptive control of full penetration gas tungsten arc welding. *IEEE Trans. Control Syst. Technol.*, 4, 394–403, 1996.
19. Curiel, D., Veiga, F., Suarez, A., Villanueva, P., Advances in Robotic Welding for Metallic Materials: Application of Inspection, Modeling, Monitoring and Automation Techniques. *Metals*, 13, 711, 2023. https://doi.org/10.3390/met13040711.
20. Wang, T., Gao, X.D., Katayama, S., Jin, X.L., Study of dynamic features of surface plasma in high-power disk laser welding. *Plasma Sci. Technol.*, 14, 245–251, 2012.
21. Bae, K.Y. and Park, J.H., A study on development of inductive sensor for automatic weld seam tracking. *J. Mater. Process. Technol.*, 176, 111–116, 2006.
22. Bhakta, D., Gayen, D., Bhattacharjee, J., Chatterjee, R., Dastidar, P., Roy, S., Repurposing Industrial and Agricultural Wastes as Sustainable Alternatives in Photovoltaic Cells and Dye Degradation, in: *ChemistrySelect*, vol. 9, no. 38, Wiley, 2024. https://doi.org/10.1002/slct.202403145.
23. Baek, D., Moon, H.S., Park, S.H., Development of an automatic orbital welding system with robust weaving width control and a seam-tracking function for narrow grooves. *Int. J. Adv. Manuf. Technol.*, 93, 767–777, 2017.
24. Du, R.Q., Xu, Y.L., Zhou, Z., Shu, J., Chen, S.B., Strong noise image processing for vision-based seam tracking in robotic gas metal arc welding. *Int. J. Adv. Manuf. Technol.*, 101, 2135–2149, 2019.
25. Zhu, Z.Y., Lin, T., Piao, Y.J., Chen, S.B., Recognition of the initial position of weld based on the image pattern match technology for welding robot. *Int. J. Adv. Manuf. Technol.*, 26, 784–788, 2005.
26. Chen, X.Z., Chen, S.B., Lin, T., Lei, Y.C., Practical method to locate the initial weld position using visual technology. *Int. J. Adv. Manuf. Technol.*, 30, 663–668, 2006.
27. Chen, H.Y., Liu, K., Xing, G.H., Dong, Y., Sun, H.X., Lin, W., A robust visual servo control system for narrow seam double head welding robot. *Int. J. Adv. Manuf. Technol.*, 71, 1849–1860, 2014.

28. Fang, Z.J., Xu, D., Tan, M., A vision-based self-tuning fuzzy controller for fillet weld seam tracking. *IEEE/ASME Trans. Mechatron.*, 16, 540–550, 2011.
29. Kiddee, P., Fang, Z.J., Tan, M., A real-time and robust feature detection method using hierarchical strategy and modified Kalman filter for thick plate seam tracking. *Int. J. Autom. Control*, 11, 428–446, 2017.
30. Xu, Y.L., Lv, N., Fang, G., Du, S.F., Zhao, W.J., Ye, Z., Chen, S.B., Welding seam tracking in robotic gas metal arc welding. *J. Mater. Process. Technol.*, 248, 18–30, 2017.
31. Fang, Z.J., Xu, D., Tan, M., Visual seam tracking system for butt weld of thin plate. *Int. J. Adv. Manuf. Technol.*, 49, 519–526, 2010.
32. Wang, W. and Fang, Z.J., Improving 2D camera calibration by LO-RANSAC. *Int. J. Inf. Electr. Eng.*, 7, 93–98, 2017.
33. Xu, G., Hao, Z.B., Li, X.T., Su, J., Liu, H.P., Zhang, X.Y., Calibration method of laser plane equation for vision measurement adopting objective function of uniform horizontal height of feature points. *Opt. Rev.*, 23, 33–39, 2016.
34. Kiddee, P., Fang, Z.J., Tan, M., A practical and intuitive calibration technique for cross-line structured light. *OPTIK*, 127, 9582–9602, 2016.
35. Bhattacharjee, J., Bhakta, D., Roy, S., Characteristics, Applications, and Limitations of Nanocomposites in Biosensing, in: *Advances in Chemical and Materials Engineering Book Series*, pp. 161–191, 2024, https://doi.org/10.4018/979-8-3693-1094-6.ch006.
36. Shin, H.H., Cakmak, S., Brion, O., Villeneuve, P., Turner, M.C., Goldberg, M., Jerrett, M., Chen, H., Crouse, D., Peters, P., *et al.*, Indirect adjustment for multiple missing variables applicable to environmental epidemiology. *Environ. Res.*, 134, 482–487, 2014.
37. Roman, R.C., Precup, R.E., David, R.C., Second order intelligent proportional-integral fuzzy control of twin rotor aerodynamic systems. *Procedia Comput. Sci.*, 139, 372–380, 2018.
38. Zhang, H.B., Liu, X.M., Ji, H.H., Hou, Z.S., Multi-agent-based data-driven distributed adaptive cooperative control in urban traffic signal timing. *Energies*, 12, 1402, 2019.
39. Saheed, B.A., Irina, L., Asmaa, K., Alexey, S., Effect of laser welding process parameters and filler metals on the weldability and the mechanical properties of AA7020 aluminium alloy. *J. Mater. Process. Technol.*, 2, 33, 2018.
40. Bhattacharjee, J. and Roy, S., A review on photocatalysis and nanocatalysts for advanced organic synthesis. *Hybrid Adv.*, 6, 100268, 2024, https://doi.org/10.1016/j.hybadv.2024.100268.
41. Rout, A., Deepak, B.B.V.L., Biswal, B.B., Advances in Weld Seam Tracking Techniques for Robotic Welding: A Review. *Robot. Comput. Integr. Manuf.*, 56, 12–37, 2019.
42. Tanveer, M., Mohd, A.W., Faizan, A., Application of robotics in welding. *Int. J. Emerg. Manag.*, 7, 30–36, 2018.

43. Kwak, Y.-M.; Doumanidis, C.C. Geometry Regulation of Material Deposition in Near-Net Shape Manufacturing by Thermally Scanned Welding. *J. Manuf. Process.*, 4, 28–41, 2002.
44. Hamzeh, R., Thomas, L., Polzer, J., Xu, X.W., Heinzel, H., A Sensor Based Monitoring System for Real-Time Quality Control: Semi-Automatic Arc Welding Case Study. *Procedia Manuf.*, 51, 201–206, 2020.
45. Bingul, Z., Cook, G.E., Strauss, A., Application of fuzzy logic to spatial thermal control in fusion welding. *IEEE Trans. Ind. Appl.*, 36, 1523–1530, 2000.
46. Bhattacharjee, J. and Roy, S., Synthesis of thermoelectric nanocomposites by incorporating reduced graphene oxide, in: *Archives of Advanced Engineering Science*, Bon View Publishing PTE, 2024. https://doi.org/10.47852/bonviewaaes42023514.
47. Lostado Lorza, R., Corral Bobadilla, M., Martínez Calvo, MA., Villanueva Roldán, P.M., Residual Stresses with Time-Independent Cyclic Plasticity in Finite Element Analysis of Welded Joints. *Metals*, 7, 136, 2017.
48. Warinsiriruk, E., Mahfudianto, F., Design the Feature Extraction for Real Time Inspection of Welding Quality, in: *Proceedings of the 2018 3rd Technology Innovation Management and Engineering Science International Conference (TIMES-iCON)*, Bangkok, Thailand, 12–14 December 2018.
49. Bhattacharjee, J. and Roy, S., Review on green resources and AI for biogenic solar power. *Energy Storage Convers.*, 2, 1, 2024. https://doi.org/10.59400/esc.v2i1.457.
50. Shaikh, A.A., Bhattacharjee, J., Datta, P., Roy, S., A comprehensive review of the oxidation states of molybdenum oxides and their diverse applications. *Sustainable Chem. Environ.*, 7, 100125, 2024. https://doi.org/10.1016/j.scenv.2024.100125.
51. Bhattacharjee, J. and Roy, S., Significance of Renewable Energy in Water Management and Irrigation, in: *Water Management in Developing Countries and Sustainable Development*, Water Resources Development and Management. Springer, Singapore, https://doi.org/10.1007/978-981-99-8639-2_12.
52. Shevchik, S., Le-Quang, T., Meylan, B., *et al.*, Supervised deep learning for real-time quality monitoring of laser welding with X-ray radiographic guidance. *Sci. Rep.*, 10, 3389, 2020. https://doi.org/10.1038/s41598-020-60294-x.
53. Wang, T., Gao, X.D., Katayama, S., Jin, X.L., Study of dynamic features of surface plasma in high-power disk laser welding. *Plasma Sci. Technol.*, 14, 245–251, 2012.
54. Hamzeh, R., Thomas, L., Polzer, J., Xu, X.W., Heinzel, H., A Sensor Based Monitoring System for Real-Time Quality Control: Semi-Automatic Arc Welding Case Study. *Procedia Manuf.*, 51, 201–206, 2020.
55. Feng, Y., Gao, X., Zhang, Y., Peng, C., Gui, X., Sun, Y., Xiao, X., Simulation and Experiment for Dynamics of Laser Welding Keyhole and Molten Pool at Different Penetration Status. *Int. J. Adv. Manuf. Technol.*, 112, 2301–2312, 2021.

56. Gao, X., Sun, Y., Katayama, S., Neural Network of Plume and Spatter for Monitoring High-Power Disk Laser Welding. *Int. J. Precis. Eng. Manuf. Green Technol.*, 1, 293–298, 2014.
57. Fan, X., Gao, X., Liu, G., Ma, N., Zhang, Y., Research and prospect of welding monitoring technology based on machine vision. *Int. J. Adv. Manuf. Technol.*, 115, 3365–3391, 2021.
58. Yang, Q., Zhang, P., Cheng, L., Min, Z., Chyu, M., To, A.C., Finite Element Modeling and Validation of Thermomechanical Behavior of Ti-6Al-4V in Directed Energy Deposition Additive Manufacturing. *Addit. Manuf.*, 12, 169–177, 2016.
59. Wang, Y., Xu, X., Zhao, Z., Deng, W., Han, J., Bai, L., Liang, X., Yao, J., Coordinated Monitoring and Control Method of Deposited Layer Width and Reinforcement in WAAM Process. *J. Manuf. Process.*, 71, 306–316, 2021.
60. Adolfsson, S., Bahrami, A., Bolmsjö, G., Claesson, I., On-Line Quality Monitoring in Short-Circuit Gas Metal Arc Welding. *Weld. J.*, 78, 59S–73S, 1999.
61. Zhang, Z., Chen, X., Chen, H., Zhong, J., Chen, S., Online Welding Quality Monitoring Based on Feature Extraction of Arc Voltage Signal. *Int. J. Adv. Manuf. Technol.*, 70, 1661–1671, 2014. Available online: https://link.springer.com/article/10.1007/s00170-013-5402-2.
62. Adolfsson, S., Bahrami, A., Bolmsjö, G., Claesson, I., On-Line Quality Monitoring in Short-Circuit Gas Metal Arc Welding. *Weld. J.*, 78, 59S–73S, 1999. [Google Scholar].
63. Lostado Lorza, R., Corral Bobadilla, M., Martínez Calvo, M.Á., Villanueva Roldán, P.M., Residual Stresses with Time-Independent Cyclic Plasticity in Finite Element Analysis of Welded Joints. *Metals*, 7, 136, 2017. [Google Scholar] [CrossRef] [Green Version].
64. Bhattacharya, S., Pal, K., Pal, S.K., Multi-Sensor Based Prediction of Metal Deposition in Pulsed Gas Metal Arc Welding Using Various Soft Computing. *Models. Appl. Soft Comput.*, 12, 498–505, 2012.

5

Material Flow and Heat Transfer Analysis in Friction Stir Welding

Sumanta Banerjee[1]* and Prokash C. Roy[2]

[1]*Department of Mechanical Engineering, Heritage Institute of Technology, Kolkata, India*
[2]*Department of Mechanical Engineering, Jadavpur University, Kolkata, India*

Abstract
In friction stir welding (FSW) process, basic understanding of thermal energy transport and material flow dynamics ensures creation of structurally sound, reliable, and virtually defect-free weld joints. Formulation and implementation of sophisticated mathematical models of FSW, aimed to compute the relevant process parameters, requires high-performance computing. This book chapter reviews the macroscopic modeling approach of the process, as reported in published literature, and interprets the governing transport equations and boundary conditions, which characterize the process. Based on the theories of plastic deformation, analytical models of heat transfer, and friction mechanisms at interfacial contacts, the dynamics of material transport and energy transfer are reviewed on the basis of published studies. The FSW process enables to join a variety of metallic materials (copper, steel, titanium) and composites. When compared with conventional FSW process, preheating applied to its hybrid variant enhances material plasticity and reduces plunging forces. To predict system behavior accurately, multi-scale modeling approaches have been highlighted, which are equipped to handle complex interactions among mechanical, thermal, and metallurgical phenomena. Results of analytical investigation of various process parameters of FSW must be validated with experiments to assess the potential for further improvements in process modeling.

***Keywords*:** Transport equations, process model, deformation, heat transfer, visco-plastic flow, friction

**Corresponding author:* sumanta.banerjee@heritageit.edu

Sandip Kunar and Gurudas Mandal (eds.) Advanced Welding Technologies, (73–106) © 2025 Scrivener Publishing LLC

5.1 Introduction and Background

In recent years, friction stir welding (FSW) has garnered considerable research interest. Although FSW was initially applied on metals, like aluminum and its alloys, this solid-state joining method is presently applied to relatively harder materials and plastics. Understanding the synergy between varied physical phenomena involved in the welding process holds practical importance and theoretical significance. Understanding the fundamental physics is facilitated through construction and comprehension of complex and sophisticated analytical models, which consider multiple parameters associated with the process. Therefore, this review work seeks to gather key physical concepts involved with energy and material transport during the FSW process. These include, namely, thermal energy transfer rates, volumetric rates of heat generation, and rates of material deformation in the plastic state.

The general configuration of the tool used in FSW is a solid cylinder terminating in a pin. The tool simultaneously rotates and traverses the contact surfaces of two substrates. The latter are rigidly clamped, and are supported by a backing plate. The top surface of the workpiece makes firm contact with the tool shoulder through the application of external loading, which generates considerable heat due to friction at the surfaces of contact. The tool shoulder, consequently, receives significantly more thermal flux than the pin surface. Significant material deformation and flow occurs around the tool due to plasticity. Flow of metal (in plastic state) occurs from the front to the trailing edge of the welding tool. Upon reaching the trailing edge, the plasticized material is forged into a permanent joint [1].

The transport of thermal energy during the FSW process can be visualized as a three-phase phenomenon. The dwelling period, or the first phase, allows a stationary rotating tool to preheat the material. This action results in significant heat generation, causing the temperature to rise sufficiently high, ahead of the tool, to facilitate tool traversal movement. This is followed by the time-dependent (transient) heating phase, or the second phase, which spans over the traversal of (rotating) welding tool. The final phase corresponds to the termination of the welding process, when heat transfer reaches a quasi-steady state.

The final phase, which follows the transient period, corresponds to the quasi-steady state. The system attains a spatially varying, yet time-invariant temperature field around the tool. Due to combined translational and (directional) rotational motions of the advancing tool, the relative velocities at symmetrically located points on the retreating side (RS) and advancing side (AS) differ. This results in asymmetry of heat generation and thermal flux transfer rates on the two sides of the weld joint. Consequently, the material flow is also markedly asymmetric, as confirmed in published studies [2].

As mentioned, the FSW process is a novel assembly of heat generation and plastic deformation phenomena resulting in complex metallurgical processes [3]. Heat is generated through combined modes of friction and plastic deformation, which alters microstructure and material properties. The kinetics of these changes is principally dependent on temperature and strain rates. The conduits of energy transport within the system evolves out of (material) microstructural changes. Coupled interactions between thermal, mechanical, and metallurgical phenomena requires thorough understanding of heat transfer and material flow physics. Metallographic analyses show that the center of stir zone (SZ) undergoes dynamic recrystallization. A region of recovered microstructure, at lower temperature levels and deformation rates, surrounds the stir zone. This feature dually increases the number of grain boundaries and the surface energy of grains [4, 5]. An increase in grain dislocation densities within deformed grain boundaries, which characterizes the thermo-mechanically affected zone (TMAZ), facilitates storage of a certain portion of plastic deformation energy. The heat-affected zone (HAZ) and the base material maintains a similar microstructure, as it is primarily influenced by temperature fields set up through conduction heat transfer modes.

In FSW, heat and mass transfer rates are primarily governed by welding parameters and material properties. The tool speed (transverse and/or rotational) is a primary parameter. Extrusion and forging of the material characterize the joining process, which involves high (plastic) strain rates [1]. These observations justify modeling of the mechanical behavior of workpiece material as either fully plastic, or visco-plastic, or viscous fluid continuum. Numerical modeling aims to unravel the kinetics of recrystallization during the FSW process through analysis of material flow and temperature fields. From review of open source literature, there are two modeling approaches for material flow [6–9]. The first approach, which solves the flow field parameters by fluid dynamics principles, models the plasticized material as a highly viscous fluid. The other approach uses the principles of plastic deformation and derives the velocity and acceleration fields through computation of the displacement field.

To predict the successive thermal episodes during welding, proper analysis of transport processes within the medium (workpiece) is necessary.

This enables tuning the thermo-physical properties which, in turn, improves the quality of weld joint. Heat transfer can be modeled as an initial boundary-value problem for both the workpiece and the tool. By combining experiments with numerical analyses, analytical solutions can be obtained using inverse methods. Heat generation (by friction) at the workpiece–tool shoulder interface is measured experimentally and used as primary data [10, 11]. As reported in scientific studies, at the tool–workpiece interface, the maximum temperature reaches up to approximately 80%–90% of the melting temperature of the welded material. The additional generation at the pin–workpiece interface, which accounts for approximately 20% of the total heat, has minimal effect on the overall thermal profile [12, 13].

Energy dissipated by friction gets irreversibly converted to heat; in FSW, the other heat generation source at the interfacial contacts (TMAZs) is plastic deformation [14, 15]. For ease of investigation, the entire tool–workpiece interface is segregated into tool pin–workpiece and tool shoulder–workpiece interfaces. At these interfaces, generation of heat is mainly due to friction, and contribution of plastic deformation is small in comparison. Additionally, the non-symmetrical heat flux distribution causes differences in temperature fields between the retreating and advancing sides (RS and AS). This requires formulation of complex transport models.

This review on FSW deals with published mathematical models that deal with material flow and heat transfer mechanisms. The objective is to elucidate interlinked underlying phenomena and present an overview of the entire process dynamics. The key issues are presented in dedicated sections. Section 5.2 commences with a basic description of the process in the context of transport phenomena, which govern transport of material and enthalpy. Section 5.3 deals with the basics of mathematical modeling. The basic principles underlying the conservation equations are discussed. The recipes of modeling the physical aspects of FSW process are outlined through the governing equations and boundary conditions. Analysis of heat transfer modes is dealt with in Section 5.4, while material flow due to combined thermo-mechanical strains is the subject of focus in Section 5.5. The heat generation models developed for stationary and moving sources of heat is reviewed next in sufficient detail (Section 5.6). Section 5.7 discusses the pertinent thermo-mechanical interactions. The chapter concludes with a brief overview of thermal, mechanical, and metallurgical considerations (Section 5.8). The final section provides future prospects and concluding remarks.

5.2 Material Flow and Heat Transfer: Review of Basic Principles

Factors like the tool geometry and its traversal kinematics, the applied torque, and the thermo-mechanical conditions at the tool-workpiece TMAZ-interfaces govern the modes of heat transfer, heat generation rates, as well as the flow rates of plasticized material. The latter are also regulated by factors such as tool–shoulder size, tool–surface angle, surface characteristics, and pin geometry. The phenomena of sliding and sticking determine the flow field of plasticized material, with heat generated by both these interactions and material flow driven by sticking alone. Typically, the tool-shoulder surface can be concave, or convex, or flat. Pin geometries can be tapered, inverse tapered, cylindrical, or triangular [16–18]. Experiments confirm that the geometrical shapes of tool shoulder influence the quality of welded joint. For example, a concave shoulder affects axial force and pressure distribution differently than a flat one [18]. Compared to its concave counterpart, a convex scroll shoulder provides a more stable FSW process with minimal flash, a consistent (nearly constant) plunge depth, and fewer defects [19].

The (tool) pin geometry also influences the quality of the joints, as it impacts the flow of material in plastic state. While the pin supports layer-by-layer material flow, the shoulder enables bulk material flow [16]. The tool-pin contour also exerts an influence; for example, a triangular pin causes greater rates of material flow than a cylindrical pin. In addition, the orientation and geometry of threads on the surface of the pin determines the axial force on the workpiece, in addition to causing material flow around the tool. Threads and flute profiles on the pin also affect axial and transverse forces. Due to (effectively) larger contact area, the rates of heat generation increase, and material flow is improved. Tool geometries are also important in this context. Compared to other shapes, the square-tool geometry favors more uniform particle distribution. Flat-faced tools undergo much larger wear than circular tools [19]. A conical shoulderless tool can reduce the (process) forces significantly. This makes it suitable for plates with variable thicknesses. As reported in experiments, increasing the angle between the pin surface and its axis causes more uniform temperature distribution (or reduced distortion) along the vertical direction.

Literature survey indicates that most experimental investigations (in FSW) are conducted using a vertical milling machine. As the FSW tools exert plunging and translational forces, test pieces are securely clamped to prevent movement. Thermocouples placed at strategic locations capture the time–temperature history. The average temperature difference between the retreating and advancing sides (RS and AS) can be recorded (approximately 13 K; [20]). Transient temperatures recorded from thermocouples at predetermined welding speeds, RPM, and average plunging force can be plotted for both the sides (refer to ref. [20] for more details). Compared to others, the thermocouple nearest to the joint line records the highest temperature. Therefore, experimental studies can estimate the magnitude and location of the peak temperature in defect-free welding. It is also observed that the temperature on AS exceeds that of RS at all thermocouple positions, even when measured at similar locations. The temperature gradients even out as the distance from the center line increases and eventually stabilizes owing to gradient-driven heat flux. This temperature difference, caused by spatially varying and asymmetric heat generation, coupled with flow of material in plastic state gives rise to an unsteady, non-uniform temperature field, which elevates the AS temperature over that of the RS temperature.

Compared to the FSW of similar materials, the same for dissimilar materials presents greater challenges. Difference in deformation behaviors, variations in thermo-physical properties, and potential formation of intermetallic compounds with adverse mechanical characteristics contribute to asymmetry in material flow and heat generation, and present hurdles in achieving defect-free welds. Property discontinuities at the butting surfaces also affect the material flow pattern. In line with a published study [20], one potential method for joining dissimilar metals (aluminum and copper) involves preheating the copper side to ~200°C using a localized heat source. However, the location and movement of the tool must be carefully controlled for optimized results.

For a butt joint configuration, a schematic of dissimilar material joining is shown in Figure 5.1. For this FSW method, preheating compensates for heat loss and enhances material flow. As a viable preheating option, plasma arc offers advantages of concentrated energy density, high arc stability, and low equipment cost. During welding, the copper and the aluminum sheets are, respectively, placed on the advancing and retreating sides. This is because copper, acting as bulk material, is relatively hard and difficult to transport to the AS [21]. Yet another crucial factor (in welding dissimilar metals) is pin-offset behavior. Published studies report improper mixing of aluminum and copper, which results from greater pin offset toward the copper side. This causes higher copper percentage in the welded joint leading to tunneling defects. Another disadvantageous feature (compromising weld strength) is the formation of brittle intermetallic compounds with decreased solubility. Large pin offsets increase chances of undesirable mixing of copper pieces into the aluminum matrix [22]. To address this

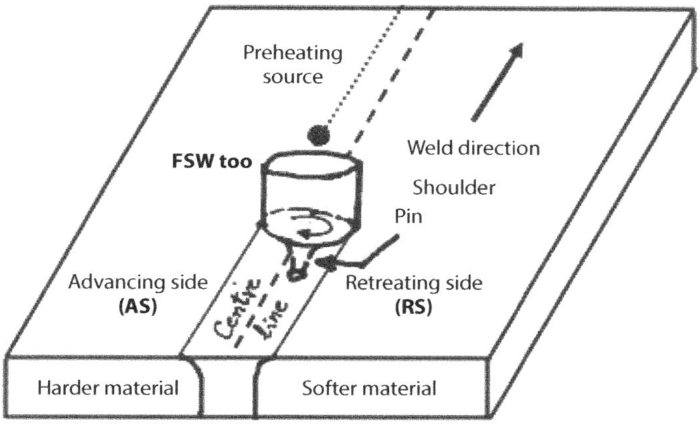

Figure 5.1 Schematic of joining dissimilar metals in butt joint by FSW.

issue, an optimized offset distance is maintained, and an additional heat source is employed to preheat the copper side to ~180°–250°C. In joining dissimilar materials, introducing plasma arc-assisted heating sets up favorable conditions to obtain metallurgically sound welds. For example, butt welding of pure Cu and commercial-grade Al alloy has been performed for 6-mm-thick plates, using plasma arcs. Use of DC (up to approximately 50 A) has been reported in open literature. The effects of plasma arc-assisted preheating have also been studied through plots [20]. For both AS and RS, these plots reveal the thermal history at various points from the weld center line, which depend on operating parameters. One prominent aspect is asymmetric heat generation, which enhances the temperature field within the AS. With preheating, the increase in heat input is reflected through the profiles of temperature plots. When compared with the plots without preheating, an increase in the area under the thermal profile is an obvious indicator of higher energy input. Preheating results in improved material properties of the welded joint, by altering the temperature profiles and favorably directing flow of heat and material. Among others, the mechanical strength is reduced around the weld zone through rise in temperature facilitating the FSW process.

Factors, like the motion of tool relative to workpiece, rotation of threaded pin attachment (which pushes the material in its vicinity downward), and (slug) flow of plasticized material, affect material transport around the tool through the AS and the RS zones. The motion (helical) of conical pins pulls the material upward at the rear and pushes the material downward at the front, thereby minimizing variation in properties across the cross-section of the workpiece [7]. For welding thick sheets, tapered tools are therefore preferred [23]. In case of cylindrical pins, axial force pushes down the material beneath the shoulder; near the periphery of the pin, the material flows upward. Consequently, as one recedes from the pin periphery, the trajectory of material particles curls upward near the pin's edge and then bends downward, forming a "hook." The pin and shoulder geometries influence the characteristics of this hook formation [6, 24].

Visualizing patterns for material flow in FSW includes tracer techniques or welding of dissimilar alloys. Material flow is rendered perceptible through spatial variations in (material) texture and movement of tracer particles [20]. One visualization technique involves using small steel balls as tracers [25]. The forward motion of the tool is suddenly stopped and simultaneously retracted, such that the rate of retraction allows sufficient time for the pin to unscrew itself. At the end of the welding process, material flow within the tool's threads can be studied by sectioning the keyhole of the pin. To track material flow, different marker materials (e.g., aluminum alloy [26–28], small steel shots [25, 29], copper foil [30], Al–W and Al–SiC composites [31], and tungsten wire [32]) have been used. This technique, through marking the terminal positions (initial and final) of the tracers, provides (plasticized) flow-tracking data. The flow is, thus, reconstructed.

Steel shots can also be indexed and tracked with X-ray imaging to observe their (secondary) positions along the welding path. It has been found that steel shots impregnated within AS move away to the retreating side (RS), while those implanted into the RS stay on the same side [33]. In other experimental studies [30, 34], tiny steel balls had been inserted within the (external) troughs (or grooves) of the threaded pin. After the process concluded, post-weld radiography tests showed minimal vertical drift of the (metal) spheres. In published studies on FS-welded lap joints, researchers inserted a copper foil at the faying surface of aluminum alloys dissimilar in composition. The material attached to the pin was observed to rotate with the tool's velocity. Significant degree of vertical mixing was observed in the weld's composition due to generation of a longitudinal vortex [35–38]. In several studies on FS-butt welding of similar and/or dissimilar aluminum alloys, examination of flow patterns have been reported. When softer aluminum alloy infuses into the harder alloy on the advancing side, distinct vortex-shaped flow patterns are observed. For alloys with similar hardness, these patterns are minimal. The (vortex) patterns are also minimal when a harder alloy infuses into a comparatively softer alloy [39].

It is well understood that material flow during FSW is an intricate process. Despite an upsurge in analytical, numerical, and experimental investigations, the transport processes are yet to be fully understood.

One viable approach of analysis is to consider material transport as an extrusion process. For instance, embedding a copper foil as tracer (approximately 0.5-mm thick), at various positions and orientations, presents a simplified method to analyze material flow within a butt joint (refer to ref. [20] for details). The procedure for tracking the tracer, by abruptly stopping the forward motion of welding tool, has been described. At the end of the welding process, the keyhole was sectioned to allow for material flow study. Radiography technique was also used to follow the path of the tracer material from the initial to the final position. This technique effectively distinguishes the copper marker from base material and does not significantly impact the analysis's effectiveness [39].

However, it must be acknowledged that marker tracing and direct/indirect measurements from experiments are challenging and difficult tasks, which impede characterization of material flow. Therefore, mathematical models are needed for analysis. Fully coupled, visco-plastic flow models, which incorporate thermal–mechanical phenomenological effects, can provide reasonably accurate results [9, 10, 40–42]. However, experimental validation of these models is generally difficult. Numerical analyses can be validated with experimental studies that make use of copper markers [10]. These studies show that tool pin shape and weld pitch significantly influence flow dynamics [34]. At smaller pitch values, for instance, the marker practically exhibits continuous flow.

The plasticized flow generates velocity field, fields of strain tensor, and also the strain rate fields. These metrics provide valuable insights into the process. Streamlines can potentially indicate rotational zones, as well as regions of recirculation flow around the tool pin [9, 10]. The strength and spread of the recirculation zones are primarily determined by viscosity (a material property) and heat flux intensities at the tool–material interface. Indeed, understanding the mechanism of heat transfer is crucial as material flow is dependent dually on the temperature and deformation fields. Welding defects are a direct consequence of insufficient material flow to desired sites. Creation of relatively stagnant zones potentially leads to defect formation [43]. In numerical studies on FSW of dissimilar materials, flow-field streamline plots indicate that mass transfer predominantly occurs on the RS, beyond the recirculating zone. Potential flow reversal takes place on the AS near the pin, which creates a relatively stagnant zone. In this way, prediction of weld defects can be rendered possible from computational studies.

Apart from the initial and final phases of FSW, the rate of volumetric heat generation during intermediate phases is nearly uniform. This is more so when the tool traverses at uniform speed and also rotates with uniform angular velocity allowing the system to attain a quasi-steady state. During this period, material properties and the weld profile remain time invariant. It becomes convenient to carry out dynamic analysis in a moving reference frame. The time variable is absorbed into a dimensionless velocity term, so that the governing equations do not have time as an explicit variable. The profiles for temperature and deformation rates then depend directly on the tool's (non-dimensional) velocity. In this way, the generic solution at a particular tool position can describe the entire process. The reader is referred to ref. [20] for a detailed discussion in this regard. However, the actual process can be simulated via transient analysis. Here, a fixed coordinate system is chosen, and the tool moves over successive time steps, such that the distance traveled over each time step is determined by the instantaneous velocity of the tool. Both the approaches are adopted in analyses and are therefore reviewed in the following subsections.

5.3 Foundations of Mathematical Modeling

Owing to its simplicity, heat transfer by conduction is preferred over other energy advection models in material transport analysis in FSW. Diffusion transport models are relatively easy to formulate, and the (numerical) solution schemes offer reduced computational time. However, these models can only predict temperature distributions and rates of cooling, and not transport of material. Solutions of transport equations (continuity, momentum, and enthalpy) enable us to solve the flow field due to plastic deformations, in addition to obtaining the temperature profiles during the process. In line with published studies, the analytical model for heat transfer has been cast in 3-D orthogonal Cartesian coordinate system and is applicable to both transient and steady-state processes. The following sections, through discussion of transport equations and boundary conditions, revisit the theoretical formulations for material flow and heat transfer analysis [20].

5.4 Analysis of Basic Modes of Thermal Energy Transfer

For stir welding, heat generation is controlled mainly by two mechanisms: friction and plastic deformation. Within the workpiece, heat transfer is governed by Fourier's heat conduction law. In transient numerical analysis, a nonlinear, 3-D transport equation is solved for $T \equiv T(x, y, z; t)$

(time-dependent temperature field). This equation is expressed in a moving Cartesian reference frame, with tool movement along the y direction:

$$\frac{\partial}{\partial x}\left(k_{WP,x}\frac{\partial T}{\partial x}\right)+\frac{\partial}{\partial y}\left(k_{WP,y}\frac{\partial T}{\partial y}\right)+\frac{\partial}{\partial z}\left(k_{WP,z}\frac{\partial T}{\partial z}\right)+\dot{Q}_G=\rho_{WP}C_{p,WP}\left(\frac{\partial T}{\partial t}-V_T\frac{\partial T}{\partial y}\right) \tag{5.1}$$

In Equation 5.1, ρ_{WP}, k_{WP}, and $C_{p,WP}$ denote the density, (direction dependent) thermal conductivity, and specific heat of workpiece (WP) material, respectively. Heat generation rate is $\dot{Q}_G$, and V_T is the tool speed. For generality, the WP material is assumed to be anisotropic. Consistent with the analysis presented in ref. [20], the thermal resistance at the plates' interface is neglected. For thermal modeling, the boundary condition is formulated by energy flux transport across the boundaries of solution domain. The Dirichlet-type initial condition is specified as follows:

$$T(x,y,z;t=0)=T_{in} \tag{5.2}$$

In Equation 5.2, T_{in} represents the initial (surrounding) temperature at t = 0. The schematic of energy interactions at the boundaries of the solution domain is depicted in Figure 5.2. The latter (i.e., domain of solution; white zone) is segregated from the tool and indicated by dotted patterns. For this finite domain, the sides and the top surfaces allow for convection

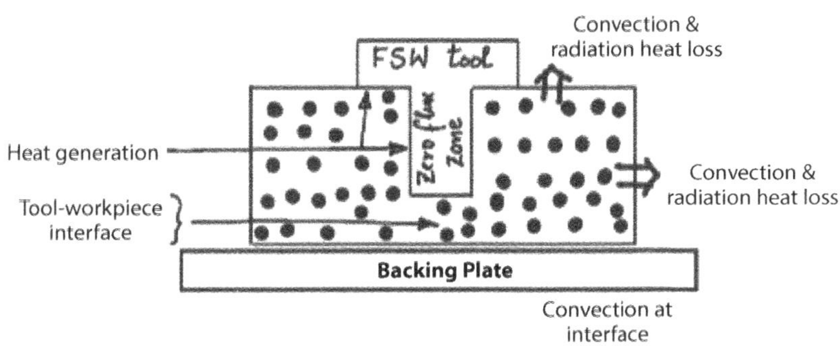

Figure 5.2 Schematic diagram representing the solution region (dot-filled domain) and boundary interactions of energy flux.

and radiation losses, while convection heat loss takes place at the bottom surface. Friction and plastic deformation occur at interfacial contact areas and acts as a secondary heat source.

Zero heat flux is assigned normal to the tool's surfaces meaning that heat transfer is neglected. The boundary condition is mathematically expressed as follows:

$$\underbrace{\epsilon\sigma(T_S^4 - T^4)}_{\text{radiation loss}} + h(T_S - T) - \dot{Q}_S = k_{WP}\frac{\partial T}{\partial n} \tag{5.3}$$

In Equation 5.3, n is the (direction) normal vector at the boundary, T_S is the substrate temperature, h is the heat convection coefficient, ϵ is the emissivity, $\dot{Q}_S$ denotes the surface heat flux, and σ denotes the Stefan–Boltzmann constant. The first term on the LHS of Equation 5.3 accounts for radiation loss. A backing plate, placed in direct contact with the bottom surface of the workpiece, acts as support. The conduction transfer (loss) is assumed uniform across the surface area of the backing plate. This simplification reduces the complexity of contact surface modeling and simply considers $T \equiv T(t)$ everywhere within the plate. From the bottom surface of the backing plate, the (diffused) heat through the plate matrix is lost due to convection (referred to as "contact conductance").

In steady-state analysis, a constant temperature field $T \equiv T(x, y, z)$ replaces transient considerations, which does not depend on the instantaneous position of the rotating tool. As the weld tool moves at a constant speed V_T in the y direction (transverse to interface), the moving coordinate reference frame (x, ζ, z) can be suitably selected:

$$y - V_T t = \zeta \tag{5.4}$$

In Equation 5.4, ζ represents the (instantaneous) distance between an arbitrary point and the heat source. Additionally, for any value of t, $V_T t$ is (instantaneous) the distance between the origin of the (fixed) coordinate system (x, y, z) and the heat source. In Equations 5.5 and 5.6, y and ζ are explicitly expressed as functions of time (=t) as follows:

$$\frac{\partial \zeta}{\partial t} = -V_T \tag{5.5}$$

$$\frac{\partial T}{\partial t} \equiv \left(\frac{\partial T}{\partial \zeta}\right)\left(\frac{\partial \zeta}{\partial t}\right) = -V_T \frac{\partial T}{\partial \zeta} \tag{5.6}$$

For the moving coordinate system, the transient heat conduction equation can be modeled as follows:

$$\frac{\partial}{\partial x}\left(k_{WP,x}\frac{\partial T}{\partial x}\right) + \frac{\partial}{\partial y}\left(k_{WP,\zeta}\frac{\partial T}{\partial \zeta}\right) + \frac{\partial}{\partial z}\left(k_{WP,z}\frac{\partial T}{\partial z}\right) + \dot{Q}_G = \rho_{WP} C_{p,WP} V_T \frac{\partial T}{\partial \zeta} \tag{5.7}$$

It is evident that the aforementioned equation (i.e., Equation 5.7), expressed in a transformed coordinate system, is not an explicit function of time (due to the absence of the $\frac{\partial T}{\partial t}$ term). Consequently, Equation 5.7 represents the heat conduction equation in quasi-steady state, where the temperature profile is implicitly time dependent and varies with the motion history of the tool. The thermal condition at the boundaries is analogous to that in Equation 5.3.

In Equation 5.3, one observes that the boundary condition involves the surface thermal flux term ($=\dot{Q}_S$). All contact surfaces effectively serve as heat sources. The volumetric heat flux ($\dot{Q}_G$) is included in Equation 5.1 and represents heat generation through viscous dissipation. The boundary interaction comprises three components: the tool shoulder–workpiece interface (site-1), the vertical (or inclined) interface of the pin and surrounding material (site-2), as well as the bottom interface of the pin and the substrate (site-3). Typically, for stainless steel, approximately 80% of heat generation happens at site-1, approximately 20% at site-2, while for site-3, the heat generation is quite small. On the overall, approximately 50% of the motive work is converted to heat, as experimentally estimated [8]. The incorporation of boundary thermal flux conditions is detailed in the following subsections.

5.5 Analysis of Material Flow in FSW

Material properties, process parameters, and modes of mechanical loading determine material flow dynamics in stir welding. As mentioned in open literature, the material in flow state can be modeled as viscous, or visco-plastic, or elastoplastic, or fully plastic continuum, taking into consideration appropriate rheologies in respective instances. In analyses pertaining to viscous or visco-plastic behavior, the flow-field parameters are derived from fluid dynamics principles. However, plastic deformation of material is modeled using principles of solid mechanics. In line with majority of published studies, which models the plasticized material as an incompressible, non-Newtonian, and visco-plastic continuum, this section reviews field analyses from fluid dynamics principles. The continuity and momentum equations are solved for an incompressible flow, within the domain of the solution. Additionally, energy equation is coupled or simultaneously solved for $T \equiv T(x, y, z; t)$ to estimate the magnitudes of temperature-dependent material properties, which effectively modify all transport equations for convective flow. The transient conservation equations for momentum and mass are expressed in Equations 5.8 and 5.9 [44] as follows:

$$\rho_{WP}\left[\frac{\partial \vec{V}_{lm}}{\partial t} + (\vec{V}_{lm} \cdot \nabla)\vec{V}_{lm}\right] = \nabla \cdot \bar{\bar{\sigma}}_S + \vec{F}_b \tag{5.8}$$

$$\nabla \cdot \vec{V}_{lm} = 0 \tag{5.9}$$

In Equations 5.8 and 5.9, t is the time variable, $\vec{V}_{lm} = (u\ v\ w)_{lm}$ denotes the liquid (material) velocity, $\nabla \equiv \begin{pmatrix} \partial / \partial x \\ \partial / \partial y \\ \partial / \partial z \end{pmatrix}$ is the (velocity) gradient operator, $\bar{\bar{\sigma}}_S$ is the stress tensor, $\rho_{lm} \equiv \rho_{WP}$ denotes the material density, while $\vec{F}_b$ is the (vector) body force. The stress tensor in Equation 5.8 can be expressed in Equation 5.10 below by the Stoke's law [45]:

$$\bar{\bar{\sigma}}_S = -P\bar{\bar{I}} + \mu\left[(\nabla\vec{V}) + (\nabla\vec{V})^T\right] \tag{5.10}$$

In Equation 5.10, μ denotes the material viscosity, P is the hydrostatic pressure, and $\bar{\bar{I}}$ stands for the symbol of identity matrix. The symbol $(\nabla\vec{V})$ denotes the gradient tensor of the velocity field and represents the strain rates of the fluid element. Using Equations 5.8 and 5.9, and including suitable modifications to account for movement of the tool, the momentum transport equations can be written as in Equation 5.11 as follows:

$$\rho_{WP}\left[(\vec{V}_{lm}\cdot\nabla)\vec{V}_{lm}+\frac{\partial\vec{V}_{lm}}{\partial t}\right]=\mu\nabla\cdot\left[(\nabla\vec{V})+(\nabla\vec{V})^{T}\right]+\nabla P+\rho_{WP}\vec{U}_{T}\cdot(\nabla\vec{V})+\vec{F}_{b} \quad (5.11)$$

In Equation 5.11, $\vec{U}_T$ denotes the weld tool velocity, equivalent to $\vec{V}_T$ (directional movement) along the y-axis. The continuity, momentum, and energy conservation equations are coupled, and must be solved simultaneously. In Cartesian coordinates, the equation, which governs diffusion heat transport (conduction) can be expressed as follows:

$$\nabla\cdot(k_{WP}\nabla T)+\dot{Q}_{G}=\rho_{WP}C_{p,WP}\left(\frac{\partial T}{\partial t}+\vec{U}_{T}\cdot\nabla T-\vec{V}\cdot\nabla T\right) \quad (5.12)$$

In Equation 5.12, $k_{WP}\equiv k_{lm}$ denotes the direction-dependent conductivity of the (orthotropic) solid. The term $\rho_{WP}C_{p,WP}\vec{V}\cdot\nabla T$ denotes the (transport) energy flux in convective flow of liquid metal. For a stationary heat source, Equations 5.11 and 5.12 may be modified by setting $|\vec{U}_T|=0$. This permits analysis with reference to (x,y,z) (fixed coordinate system). As also mentioned, the transient term is neglected in steady-state analysis, for which the solution is found with respect to a moving coordinate frame, with its origin attached to the (instantaneous) position of the tool.

Figure 5.3 illustrates, in top view, the schematic of the velocity boundary conditions applicable at the interface of the workpiece shoulder of a flat tool. On the workpiece, let us consider a point "A" at radius r, and angle θ relative to the direction of $\vec{V}_T$. The (absolute) rotational speed of the welding tool is ω_T.

On the boundary of the top surface of the tool shoulder, the velocity components are as follows:

$$v_{TS}=\omega_T r\sin\theta-V_T;\ u_{TS}=\omega_T r\cos\theta\ (R_{TP}\le r\le R_{TS}) \quad (5.13)$$

At the tool pin periphery, the velocity components are as follows:

$$v_{TP}\,|_{periphery}=\omega_T R_{TP}\sin\theta-V_T;\ u_{TP}\,|_{periphery}=\omega_T R_{TP}\cos\theta\ (r=R_{TP};\ 0\le h\le H_{TP}) \quad (5.14)$$

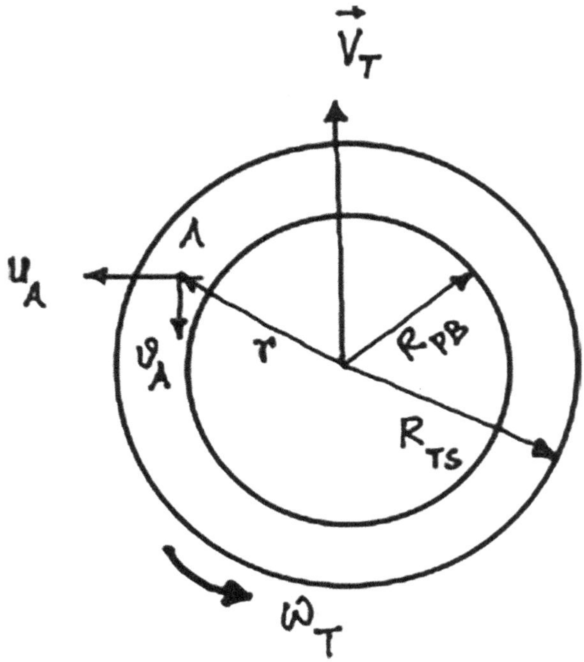

Figure 5.3 Boundary conditions for the calculation of velocity field (top view).

Boundary conditions for the velocity components at the bottom surface of the pin are as follows:

$$v_{TP}\big|_{bottom} = \omega_T r\sin\theta - V_T; u_{TP}\big|_{bottom} = \omega_T r\cos\theta \; (0 \le r \le R_{TP}) \quad (5.15)$$

In Equations 5.13–5.15, R_{TS}, R_{TP}, and H_{TP} represent the radius of the shoulder, radius, and height of the tool pin (for a cylindrical tool), respectively. For a stationary tool, the boundary conditions are adjusted (setting $|\vec{V}_T| = 0$).

From the above analytical formulations, it is clear that modeling of FSW must consider multiple parameters. The formulation becomes more complex as the material behavior model accounts for dependence of material properties on temperature or strain rate. As such, multiple approaches are mentioned in the reference literature for the prediction of material behavior. Considering non-Newtonian, visco-plastic flow modeling for the material transport, the effective viscosity can be calculated using Hart's simplified model [45]. As derived from experiments, this model suggests splitting of the bulk stress $(=\bar{\bar{\sigma}}_S)$ into plastic and viscous components as follows:

$$\bar{\bar{\sigma}}_S = \bar{\bar{\sigma}}_{plastic} + \bar{\bar{\sigma}}_{viscous} \tag{5.16}$$

In Equation 5.16, the viscous part $(=\bar{\bar{\sigma}}_{viscous})$ denotes frictional resistance to flow caused by dislocation; the plastic component $(=\bar{\bar{\sigma}}_{plastic})$ represents the internal resistance offered by the material to dislocation entanglement. Open-source literature deals with empirical correlations that relate these stresses to corresponding deformation rates and constitutive (material) parameters [3, 20, 45–47]. Equation 5.17 expresses the deformation rate tensor $\dot{\epsilon}_{mn}$ as follows [47]:

$$\dot{\epsilon}_{mn} = \frac{1}{2}\left(\frac{\partial u_m}{\partial x_n} + \frac{\partial u_n}{\partial x_m}\right); \quad m \neq n = 1(1)3 \tag{5.17}$$

Based on above-reviewed literature, the empirical functional dependences of viscosity, i.e. viscosity $f \equiv$ (strain rate, temperature), evaluated by experiments can be directly applied (see Equation 5.11). Typically, the relations report that viscosity $\downarrow \equiv f(\text{strain rate} \uparrow, \text{temperature} \uparrow)$, with the latter (i.e., rate of strain) being the predominant factor influencing viscosity.

The boundary conditions mentioned above do not consider the contact-state variable. For steady-state heat source analysis, the velocity boundary condition at the contact interface is hereby modified as follows:

$$v_{TS} = \delta_S \omega_T r \sin\theta; \; u_{TS} = \delta_S \omega_T r \cos\theta \; (R_{TP} \leq r \leq R_{TS}) \tag{5.18}$$

In Equation 5.18, r is the radial distance, and δ_S denotes the contact-state variable. It is a dimensionless parameter, which represents the degree of sticking. The remaining boundary conditions for the velocity are similarly derived from Equations 5.14 and 5.15.

Straightforward estimation of the extent of slip is challenging; therefore, the contact-state variable is determined using empirical relationships. The slip extent is inferred through statistical correlations, based on values that are experimentally measured at different relative velocities as follows [8]:

$$\delta_S = 0.6\left[1 - \exp\left\{-\delta_{S0}\left(\frac{r}{R_{TS}}\right)\left(\frac{\omega_T}{\omega_{T0}}\right)\right\}\right] + 0.2 \tag{5.19}$$

In Equation 5.19, r represents the radial distance from the tool axis, δ_{S0} is the (tunable) contact-state parameter, ω_T is the angular velocity of the tool, R_{TS} is the tool shoulder radius, and ω_{T0} is the normalizing rotational velocity typically chosen as the mean of the rotational speed range considered. For pure friction, one chooses $\delta_S = 0$. For FSW of aluminum 1xxx series, δ_{S0} is typically approximately 1.014, as suggested in ref. [8], with other parameters varying based on operating conditions. To quote a sample published result [20]: For tool-traverse speed of 1.63 mm/s, tool-rotation speed of 600 rpm, shoulder diameter measuring 50 mm, and the mean value of plunging force measured as 6.23 kN, the δ_S values are approximately 0.49 at the shoulder interface, 0.42 at the pin's side surface, and 0.41 at the pin's bottom surface. Hence, δ_S is nearly constant across all contact surfaces (≈0.44).

The boundary thermal flux for stationary analysis ($= q''_{steady}$) is depicted in Equation 5.20 as follows:

$$q''_{steady} = \varphi_{PF}[r\delta_S\omega_T\tau_{ys}(T) + (1-\delta_S)\omega_T r\tau_{fr}] \tag{5.20}$$

In Equation 5.20, $\tau_{ys}(T)$ represents the yield stress at the contact interface, which varies with temperature, and φ_{PF} denotes the power efficiency. This factor is a metric that measures the degree of conversion of mechanical power into heat. The frictional stress τ_{fr} (Equation 5.20) is related to the plunge pressure applied on the tool shoulder; the relation is as follows:

$$\tau_{fr} = \mu_{fr}P_{TS} \tag{5.21}$$

In Equation 5.21, P_{TS} represents the (plunge) pressure applied on the tool shoulder, and μ_{fr} is the friction coefficient. For specified ranges of tool speed, empirical correlations for the coefficient of friction are available for the cross-wedge rolling process [47, 48]. It may be assumed, for steady-state conditions, that the frictional (shear) stress balances the viscous shear stress at the contact areas as follows:

$$\tau_{fr} \approx \sqrt{\tau_X^2 + \tau_Y^2} \tag{5.22}$$

The flow along the z direction is neglected in Equation 5.22, and the viscous stresses τ_X and τ_Y are established at the boundary. This is due to the boundary conditions specified for the angular velocity (in the x–y plane). The power law viscosity model represents the constitutive behavior of the non-Newtonian viscous fluid used in the analysis. The shear stress $(\tau \equiv \tau(\dot{\gamma}, T))$ is dependent on the strain rate and the temperature as follows:

$$\tau = \eta_{eff}\dot{\gamma} \tag{5.23}$$

In Equation 5.23, $\eta_{eff} \equiv \eta_{eff}(\dot{\gamma}, T)$ is the effective viscosity, and $\dot{\gamma}$ is the strain rate. These two quantities are estimated as follows:

$$\eta_{eff,ref}(T) = \{\tau_{ref}(T)/\dot{\gamma}_{ref}\} = \tau_{ref}(T)/\dot{\varepsilon}_{ref} \tag{5.24}$$

$$\tau_{ref}(T) = \eta_{eff,ref}(T)\dot{\gamma}_{ref} \tag{5.25}$$

In Equations 5.24 and 5.25, $\dot{\varepsilon}_{ref}$ and $\tau_{ref}(T)$ are the reference values of the axial strain rate and shear stress, and computed from experimental observations.

5.6 Analysis of Heat Generation in FSW

Development of improved analytical models for FSW requires basic understanding of heat generation causes, mechanics of energy and mass transfer, and also their quantitative estimation to estimate process efficiency and predict weld quality. As outlined in preceding paragraphs, complex interactions between thermal and mechanical processes make process analysis a difficult job. The dissipation of mechanical energy into thermal energy at the workpiece–tool interface and plastic deformation within the TMAZ account for heat generation. Interfacial heat generation (rate) is the product of friction and sliding velocity. Deformation heat flux at interfaces is the product of (local) shear stress and (local) speed of workpiece adhering to the tool. A portion of deformation energy (in plastic state) is stored within the TMAZ as elongated grains and interstitial defects. Another portion of mechanical energy is simultaneously expended in the process of dynamic recrystallization and recovery. This section reviews various (published) macroscopic analytical heat generation models, including kinetics only at macroscopic length and time scales. The combined effects of adhesion and material deformation around the tool account for heat generation (friction + plastic deformation) around the vicinity of tool. The recipes for heat flux calculations are also reviewed allowing for the application of suitable numerical models (e.g., finite volume based) for analysis [20].

5.6.1 Analysis of Heat Generation from Stationary Heat Source

The heat produced by the tool traversal (along welding path) is much less than that generated by rotation. Interfacial contact conditions (e.g., sliding, sticking, or mix of both sliding and sticking conditions [17]) are of central importance in transport processes modeling. Under the tool shoulder, sliding friction can be described using μ_{fr} and interfacial pressure. The interfacial shear stress field, which determines the extent of sticking friction, is a function of the strain rate and temperature fields. Considering high rotational speeds (for the FSW tool) compared with traversal speeds, the (differential) heat generation rate for sliding friction $(= d\dot{Q}_{fr}^{slide})$ over an area element dA at the workpiece–tool interface is as follows [17]:

$$d\dot{Q}_{fr}^{slide} = (1 - \delta_S)\omega_T r \tau_C dA \tag{5.26}$$

In Equation 5.26, $\delta_S (< 1)$ represents the degree of sticking. The fraction $(1 - \delta_S)$ denotes the fraction of mechanical power dissipated as heat, into the workpiece, due to friction. The tool's angular velocity is ω_T, and τ_C is the contact stress. The elemental area dA is located at radial distance r from the axis. Contact parameter $(= \delta_S)$ varies linearly with (radial) distance $(\delta_S \sim r)$ from the axis of the pin. When the geometric characteristics of the deformation zone are specified, the parameter δ_S can be defined under the aforementioned assumptions as follows:

$$\delta_S = \left(1 - \frac{V_{WP}}{V_T}\right) \tag{5.27}$$

In Equation 5.27, V_T is the tangential (peripheral) velocity of the welding tool and V_{WP} the peripheral velocity of the workpiece. When no material adheres to the surface $(\delta_S = 1)$, all the heat generation is attributed to friction alone. On the other extreme, when $\delta_S = 0$, the entire heat produced is attributed to plastic deformation alone. Conforming to the tool's speed, 100% sticking condition is usually assumed at the pin–material interface.

For sliding friction, the contact stress is related to the applied pressure $(= P_a)$ as $\tau_C = P_a \mu_{fr}$. For sticking friction, the contact and yield stresses are equated $(\tau_C = \tau_Y)$. The (elemental) heat generated by shear deformation $(= d\dot{Q}_{fr}^{stick})$ is given by the following:

$$d\dot{Q}_{fr}^{stick} = \delta_S \omega_T r \tau_C dA \tag{5.28}$$

In terms of normal yield stress $(= \sigma_Y)$, τ_C is as follows:

$$\tau_C = \sigma_Y / \sqrt{3} \tag{5.29}$$

The influence of heat generation on plastic deformation $(= d\dot{Q}_{plastic})$ can be formulated as a differential quantity over volume dV, as in Equation 5.30 as follows:

$$d\dot{Q}_{plastic} = \beta \mu_{PM} \Phi dV \tag{5.30}$$

It must be noted that plastic deformation occurs away from the workpiece–tool interface. Besides, the phenomenon causes the material to stick onto the tool's surface. For the plasticized material, μ_{PM} is the (non-Newtonian) viscosity, and β is the fraction of deformation work dissipated, and (parameter) Φ is related to strain rates (velocity gradients) (Equation 5.30). In fact, the term Φ is calculated from the knowledge of velocity field (in line with ref. [49]) through Equation 5.31 as follows:

$$\Phi = 2\Sigma_{k=1}^{3}\left(\frac{\partial u_k}{\partial x_k}\right)^2 + \left(\frac{\partial u_1}{\partial x_2} + \frac{\partial u_2}{\partial x_1}\right)^2 + \left(\frac{\partial u_2}{\partial x_3} + \frac{\partial u_3}{\partial x_2}\right)^2 + \left(\frac{\partial u_1}{\partial x_3} + \frac{\partial u_3}{\partial x_1}\right)^2 \tag{5.31}$$

The proportion of mechanical energy spent on deformation work, which is converted to heat, ranges normally between 0.6 and 0.8 and increases at higher values of strain rates. Heat generation from friction losses (viscous dissipation) is relatively small and contributes approximately 5% of the overall generation rate [18, 50]. If the volumetric plastic deformation work (usually small) is neglected, the total heat transfer $(= \dot{Q}_{FSW})$ on the tool–workpiece interface is the sum of the contributions of sliding $(\dot{Q}_{fr}^{slide})$ and sticking $(\dot{Q}_{fr}^{stick})$ friction. The quantity is expressed in differential form below:

$$d\dot{Q}_{FSW} \equiv (d\dot{Q}_{fr}^{stick} + d\dot{Q}_{fr}^{slide}) = \{\delta_S \omega_T r \tau_C dA + (1-\delta_S)\omega_T r \tau_C dA\};$$
$$\Rightarrow d\dot{Q}_{FSW} = \delta_S \omega_T r(\sigma_Y/\sqrt{3})dA + (1-\delta_S)\omega_T r P_a \mu_{fr} dA \tag{5.32}$$

Thus, the extent of slip δ_S should be determined to accurately differentiate between frictional and deformation-induced heat generation (refer to Equation 5.19).

Expressions for (elemental) heat flux (such as Equation 5.32) can also be applied to tools with more complex shapes. Figure 5.4 illustrates a typical FSW tool with a truncated cone shape, lacking any threading, and featuring a shoulder with angular tilt. The geometric dimensions are clearly depicted in this diagram.

The thermal energy generated between a stationary workpiece and rotating conical (truncated)-shaped FSW tool, at their contact interface, is subdivided over the tool shoulder ($=\dot{Q}_{TS}$), tool pin's side surface ($=\dot{Q}_{PS}$), and the surface of the pin's tip ($=\dot{Q}_{PT}$), respectively. The analytical expression of $\dot{Q}_{TS}$ is as follows:

$$\dot{Q}_{TS} = \frac{2}{3}\pi\omega_T\tau_C(R_{TS}^3 - R_{PB}^3)(1+\tan\alpha_P);$$
$$\Rightarrow \dot{Q}_{TS} = \frac{2}{3}\pi\omega_T[\delta_S\tau_Y + \mu_{fr}P_{SW}(1-\delta_S)](R_{TS}^3 - R_{PB}^3)(1+\tan\alpha_P) \tag{5.33}$$

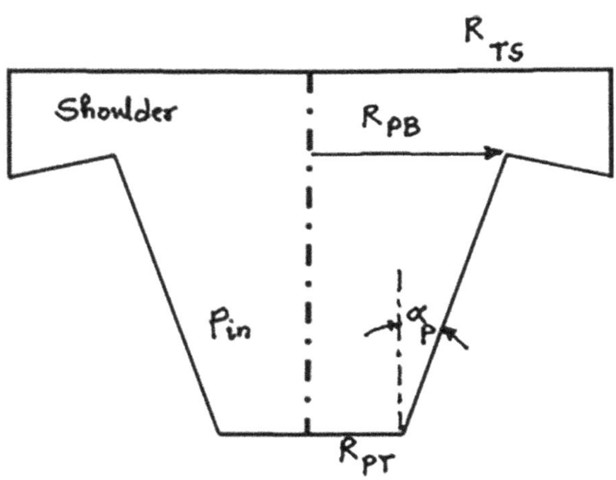

Figure 5.4 Geometrical parameters of an FSW tool of truncated conical shape.

In Equation 5.33, R_{TS}, R_{PB}, and R_{PT} denote the radius of the shoulder, the pin base, and the pin tip of the tool, respectively, measured from the tool axis. Let H_P denote the vertical distance (height) between the base and the tip of the pin. The pin angle α_P is therefore expressed as $\tan\alpha_P = \left(\frac{R_{PB} - R_{PT}}{H_P}\right)$. Also, the pressure (assumed uniform) across the tool shoulder–workpiece interface is denoted P_{SW}. Thermal energy generation rate over the pin's side surface $(= \dot{Q}_{PS})$ is given by Equation 5.34 below:

$$\dot{Q}_{PS} = \frac{2}{3}\pi\omega_T\tau_C\left(R_{PB}^3 - R_{PT}^3\right)\left(\frac{1+\tan\alpha_P}{\tan\alpha_P}\right);$$
$$\Rightarrow \dot{Q}_{PS} = \frac{2}{3}\pi\omega_T[\delta_S\tau_Y + (1-\delta_S)\mu_{fr}P_{PW}]\left(R_{PB}^3 - R_{PT}^3\right)\left(\frac{1+\tan\alpha_P}{\tan\alpha_P}\right) \quad (5.34)$$

In Equation 5.34, P_{PW} is the pressure (considered uniform) over the workpiece–pin lateral surface contact interface. The rate of (surface) heat generation over the pin tip contact area $(= \dot{Q}_{PT})$ is as follows:

$$\dot{Q}_{PT} = \frac{2}{3}\pi\omega_T\tau_C R_{PT}^3 = \frac{2}{3}\pi\omega_T[\delta_S\tau_Y + (1-\delta_S)\mu_{fr}P_{PT,W}]R_{PT}^3 \quad (5.35)$$

In Equation 5.35, $P_{PT,W}$ is the pressure (considered uniform) over the workpiece–pin tip interface area. At the contact interface, the total rate of heat generation $(= \dot{Q}_{FSW})$ equals the sum of all the components as follows:

$$\dot{Q}_{FSW} = \dot{Q}_{PT} + \dot{Q}_{PS} + \dot{Q}_{TS} \quad (5.36)$$

The above heat generation rate expression (Equation 5.36) is quite general in scope allowing scope for customization based on whether sticking or non-sticking welding conditions predominate, or if the tool is cylindrical or conical, and also if the shoulder surface of the tool is curved or flat. If one considers a flat shoulder surface of a cylindrical tool (CT), the components of the heat generation term are as follows:

$$\dot{Q}_{TS}\big|_{CT} = 2\pi\omega_T\tau_C\left(R_{TS}^3 - R_{PB}^3\right) \quad (5.37)$$

$$\dot{Q}_{PS}\big|_{CT} = 2\pi\omega_T\tau_C R_{PB}^2 H_P \tag{5.38}$$

$$\dot{Q}_{PT}\big|_{CT} = \frac{2}{3}\pi\omega_T\tau_C R_{PB}^3 \tag{5.39}$$

As previously discussed, the symbols R_{TS}, R_{PB}, and H_P denote the radius of the tool shoulder, radius of tool pin, and pin height, respectively. In general, the tool shoulder surface is the site of maximum heat generation. Therefore, estimation of appropriate (sliding or sticking) conditions is crucial for accurate prediction of heat generation. As confirmed by measurements [20], the plunging force (which the tool applies over the shoulder surface) sets up pressure of uniform intensity. For the remaining interfaces, sliding considerations (i.e., mutual rubbing of mating surfaces) alone provide reasonably good estimates (for heat generation). The proportion for the three surfaces depends primarily on tool geometry and contact conditions. Typically, as indicated by *in situ* measurements for a cylindrical tool, the shoulder generates approximately 85%, the pin side surface approximately 13%, and the pin bottom surface approximately 2% of the total heat. For FSW facilities, tool dimensions typically have a shoulder radius of 10 mm, pin radius of 3 mm, and tool shoulder height of 5.75 mm. While the pin generates nearly 15% of the heat, the bulk heat (approximately 85%) is generated from the tool shoulder surface when welding thin plates. For more details, the reader should refer to ref. [20].

The thermal energy, thus, generated gets distributed as boundary heat fluxes over the workpiece–tool interfaces. Figure 5.5 illustrates a schematic of thermal flux distribution on the weld tool surfaces. Over the tool shoulder surface, the heat flux magnitude linearly spreads out as follows:

$$q''(r) = r q''_{mean} \tag{5.40}$$

In Equation 5.40, q''_{mean} is the heat flux intensity per unit length, which varies in a linear fashion along r. Over the shoulder–surface of the tool, the law of variation of flux intensity is as follows:

$$q''_{TS}(r) = \frac{3}{2\pi}\left[\frac{r\dot{Q}_{TS}}{R_{TS}^3 - R_{PB}^3}\right] \tag{5.41}$$

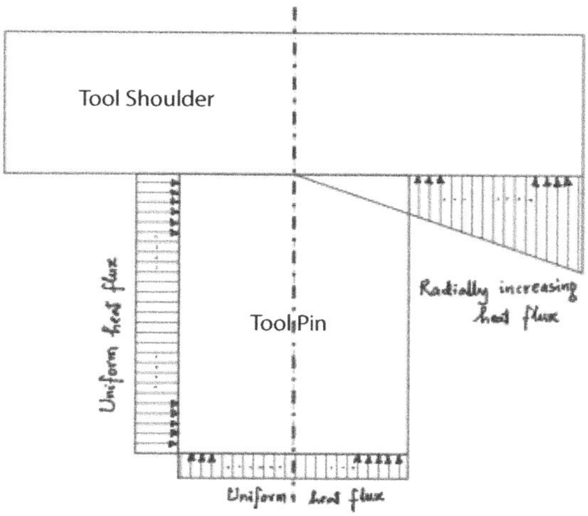

Figure 5.5 Heat flux distribution over the tool–workpiece interface for a cylindrical tool.

In Equation 5.41, r lies between R_{TS} and R_{PB}. Assuming uniform distribution of heat intensity, the expressions for flux densities on the pin–workpiece and pin–bottom interfaces are as follows:

$$q''_{PS} = \left[\frac{\dot{Q}_{PS}}{2\pi H_P R_{PB}} \right] \tag{5.42}$$

$$q''_{PT} = \left[\frac{\dot{Q}_{PT}}{\pi R_{PB}^2} \right] \tag{5.43}$$

It is important to recognize that the term $\dot{Q}_G$ in Equation 5.1 is the dissipation-induced heat generation rate due to viscosity, per unit volume, in the shear layer. As such, $\dot{Q}_G$ depends on viscosity, effective strain rate, and the fraction of motive mechanical power (for causing deformation) irreversibly dissipated. Due to localized material softening, friction decreases at high temperatures; this also affects $\dot{Q}_G$. The viscosity coefficient varies with the relative velocity of sliding, temperature, and surface contact conditions. The yield strength also decreases with temperature during plastic deformation, which reduces $\dot{Q}_G$. Last, it is observed that maximum heat generation occurs near the tool–workpiece interface, as strain rates rapidly diminish away from the tool [20].

5.6.2 Analysis of Heat Generation from Moving Heat Source

The (friction) heat generation rate mainly depends on the rotational speed and is not influenced much by the speed of the welding tool [40], as conformed by Equations 5.1 and 5.2. In analyses of situations where the FSW tool moves, steady-state analyses have been conducted in majority of published studies, and the heat generation rate is calculated by solving the transport equations in a moving reference frame considering the relative motion of the tool in relation to the workpiece. The heat transfer rate per unit area (or thermal flux), as a function of angle α_P, is specified by the following:

$$q''_{TS}(\alpha_P) = \eta_C[\delta_S \tau_C (r\omega_T - V_T \sin\alpha_P) + (1-\delta_S)\mu_{fr} P_{SW}]; \quad h = H_P, R_{TS} \le r \le R_{PB} \tag{5.44}$$

$$q''_{PS}(\alpha_P) = \eta_C\left[\delta_S \tau_C (R_{PB}\omega_T - V_T \sin\alpha_P) + (1-\delta_S)\mu_{fr} P_{PW}\right]; \quad r = R_{PB}, 0 \le h \le H_P \tag{5.45}$$

$$q''_{PT}(\alpha_P) = \eta_C[\delta_S \tau_C (r\omega_T - V_T \sin\alpha_P) + (1-\delta_S)\mu_{fr} P_{PT,W}]; \quad h = 0, 0 \le r \le R_{PT} \tag{5.46}$$

In Equations 5.44–5.46, η_C represents the efficiency factor indicating the extent of mechanical power dissipation into heat. The heat flux intensity is computed locally for all points of interfaces. Over the entire solution domain (refer to Figure 5.2), the recipes for surface flux calculations are specified by Equations 5.44–5.46. The combined effects of the tool's curvilinear traversal motion and axial rotation result in asymmetrical fields of surface heat flux and temperature. Of the total heat produced over the interfacial contact area of the tool shoulder and workpiece, a portion is absorbed by the tool material $(=\dot{Q}_T)$, while the remainder portion is transferred to the workpiece $(=\dot{Q}_{WP})$. Employing 1-D, steady-state heat flow analysis, the heat flux ratio across the interface of dissimilar metal joints is thus obtained as follows:

$$\eta_C \equiv (\dot{Q}_{WP}/\dot{Q}_T) = \sqrt{(\rho C_p k)_{WP}/(\rho C_p k)_T} \tag{5.47}$$

The analysis is detailed in ref. [51]. It shows that, out of the total generated heat, the fraction received by the workpiece is (η_C/η_C+1).

5.7 Overview of Thermo-Mechanical Process in FSW

In analysis, the temperature field obtained from the solution of the energy transport equation serves as input for determining the thermo-mechanical stresses induced within the tool (structural analysis). Typically, static mechanical analysis is preferred over dynamic analysis, and inertial effects are ignored. This is because the weld tool employs constant rotational and longitudinal speeds. Therefore, the equilibrium equations (to determine unknown forces and moments) and the constitutive equations (to determine internal stresses and strain rates) are employed in mechanical analysis. The equations of equilibrium are as follows:

$$\sigma_{ij,j}+\rho_{WP}F_{bi}=0;\ \sigma_{ij}=\sigma_{ji}(i,j=1,2,3)\ldots\ \ldots\ \ldots\ \ldots\ \ldots \tag{5.48}$$

In Equation 5.48, σ_{ij} represents the stress tensor. The symbol $\sigma_{ij,j}$ denotes the spatial derivative of σ_{ij} along the jth direction. More specifically, the symbol $\sigma_{ij,j}$ denotes the quantity $\partial\sigma_{ij}/\partial x_j$. The quantity F_{bi} is the ith component of body force. Using Lagrangian formulation for a thermo-elastoplastic model, the constitutive equations are expressed in differential forms of stress–strain interactions. As shown in Equation 5.49, the total infinitesimal strain increment $(=d\varepsilon_{ij})$ is the sum of incremental thermal $(=d\varepsilon_{ij}^{Thermal})$, elastic $(=d\varepsilon_{ij}^{Elastic})$, and plastic $(d\varepsilon_{ij}^{Plastic})$ strains as follows:

$$d\varepsilon_{ij}=d\varepsilon_{ij}^{Elastic}+d\varepsilon_{ij}^{Plastic}+d\varepsilon_{ij}^{Thermal} \tag{5.49}$$

As elaborated in ref. [52], if the material behavior is elastoplastic, the constitutive stress–strain relations can be mathematically modeled to develop the elastoplastic matrix, plastic stiffness matrix, elastic stiffness matrix, thermal stiffness matrix, increments of stress and strain, and changes in the temperature field. Appropriate hardening mechanisms (linear or multi-linear hardening, which may be either stationary or time dependent) can be considered for refinement of the elastoplastic model. The strain rates can be better-predicted by considering dislocation hardening or including more complex phenomena like plasticity, creep, or other microstructural changes. The von Mises yield criterion is then applied as follows:

$$\sigma_{avg}^2 = \frac{1}{2}(\sigma_1 - \sigma_2)^2 + (\sigma_2 - \sigma_3)^2 + (\sigma_3 - \sigma_1)^2 \tag{5.50}$$

In Equation 5.50, the principal stresses are symbolized as σ_1, σ_2, and σ_3, while σ_{avg} is average 1-D stress. The structural analysis can be coupled, or decoupled, from thermal analysis. An integrated thermo-mechanical modeling approach is recommended for FSW, as attested by published studies, for the underlying coupled mechanisms, which link heat generation, thermal flux, and the stress field. Although de-coupled analysis may be carried out for small displacement fields, coupled (thermo-mechanical) analysis, using techniques like the Arbitrary Lagrangian Eulerian (ALE) method, is more appropriate for large deformation processes. The force balance is modeled by considering parameters such as density, damping coefficient, stiffness coefficient, forcing function (body force), and system response (displacement vector). For a discussion on a comprehensive Finite Element Method (FEM)-based numerical modeling scheme, the reader is referred to ref. [20].

5.8 Conclusion: State-of-the Art and Future Directions in FSW

To conclude, multiple methods designed for FSW modeling to solve the temperature and flow fields have been surveyed. Implicit or explicit computational schemes in FEM or finite volume methods (FVMs) are equipped to tackle the coupling between flow and heat transfer in the FSW process. One of the primary steps in successful modeling is identification of the solution domains over which the temperature field and the (material) deformation fields are computed. Iterative coupled approaches are necessary as material properties, which influence the flow field, are temperature dependent, while the velocity field itself determines energy transport. In all numerical solution schemes, solution progresses to the next (load and/or time) iteration step so long as the (absolute) difference between the values of all variables in two consecutive iterations exceeds specified values of error. The solution converges when the predefined accuracy levels are attained. In the finite volume-based approach, suitable algorithms

(SIMPLE, SIMPLER) are equipped to handle coupling of temperature and velocity fields [53].

One of the key challenges, as apparent from literature review, is suitable prediction of material behavior as consequences of work hardening, material relaxation, and creep. Likewise, influence of microscopic phenomena, like phase transformation, dislocation density, and grain properties, must be reflected in macroscopic material behavior. One of the principal hurdles lies in the lack of comprehensive data for materials to model the pertinent physical phenomena accurately. Models have been developed and validated, which describe visco-plastic material deformation and consider a broad range of allied phenomena [44]. More sophisticated models provide recipes for measuring work hardening by considering metrics like strain tensor (in plastic flow and dislocation density) [50–52]. With the aid of appropriate modeling, the FSW system provides information on strain rate, and temperature fields can be gathered in the plastically deformed zone. Local distortions and/or residual stresses, and also mass fractions of individual components (metallographic analysis) in critical zones can be assessed.

Figure 5.6 shows the connectivity between various processes in FSW. It is clear from Figure 5.6 that there is an intricate connectivity between the various field parameters. For example, to calculate the field $T \equiv T(x, y, z; t)$,

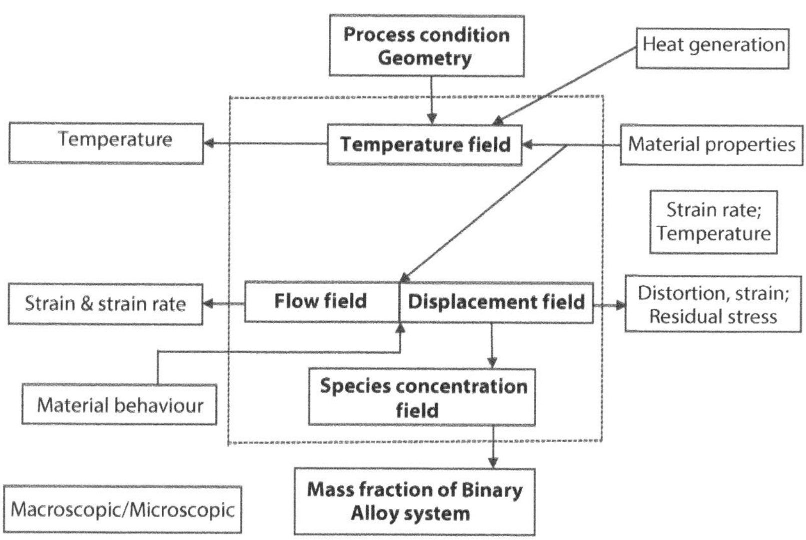

Figure 5.6 Flow chart schematic of multi-field interactions in friction stir welding process.

an appropriate model for heat generation is helpful. Suitable charts, which contain information on temperature-dependent material properties, and physically realistic models for friction and other interfacial contact conditions are also required.

Material properties depend on strain rates, which in turn affects material flow. The energy equation also relies on the advection of material. These considerations require a coupled approach for modeling. More so, a multiscale and multi-physics modeling approach is warranted, as solutions of displacement and/or stress fields require knowledge of the temperature field, thermal-induced strains, and material behavior at macroscopic and microscopic scales. Likewise, the solution of species transport equation, necessary to assess weld properties, requires knowledge of temperature-dependent material behavior. All of these leaves ample scope of future work in the field of FSW technology.

References

1. Nicholas, E. D. and Thomas, W. M., A review of friction processes for aerospace applications. *Int. J. Mater. Prod. Technol.*, 13, 1–2, 45–55, 1998.
2. Dawes, C.J. and Thomas, W.M., Friction stir process welds aluminum alloys. *Weld. J.*, 75, 3, 41–45, 1996.
3. Cho, J.H., Boyce, D.E., Dawson, P.R., Modeling strain hardening and texture evolution in friction stir welding of stainless steel. *Mater. Sci. Eng., A*, 398, 146–163, 2005.
4. Su, J.Q., Nelson, T.W., Mishra, R., Mahoney, M., Microstructural investigation of friction stir welded 7050-T651 aluminium. *Acta Mater.*, 51, 3, 713–729, 2003.
5. Frigaard, O., Grong, O., Midling, O.T., A process model for friction stir welding of age hardening aluminium alloys. *Metall. Mater. Trans. A*, 32, 1189–1200, 2001.
6. Schmidt, H. and Hatte, J., A local model for the thermo-mechanical conditions in friction stir welding. *Model. Simul. Mater. Sci. Eng.*, 13, 77–93, 2005.
7. Buffa, G., Hua, J., Shivpuri, R., *et al.*, Design of the friction stir welding tool using the continuum based FEM model. *Mater. Sci. Eng., A*, 419, 381–388, 2006.
8. Nandan, R., Roy, G.G., Lienert, T., *et al.*, Threedimensional heat and material flow during friction stir welding of mild steel. *Acta Mater.*, 55, 883–895, 2007.
9. Nandan, R., Roy, G.G., DebRoy, T., Numerical simulation of three-dimensional heat transfer and plastic flow during friction stir welding. *Metall. Mater. Trans. A*, 37, 4, 1247–1259, 2006.

10. Chao, Y.J., Qi, X., Tang, W., Heat transfer in friction stir welding: Experimental and numerical studies. *ASME Trans.*, 125, 138–145, 2003.
11. Chao, Y.J. and Qi, X., Thermal and thermo-mechanical modeling of friction stir welding of aluminum alloy 6061-T6. *J. Mater. Process. Manuf. Sci.*, 7, 215–233, 1998.
12. Tang, W., Guo, X., McClure, J.C., *et al.*, Heat input and temperature distribution in friction stir welding. *J. Mater. Process. Manuf. Sci.*, 7, 163–172, 1998.
13. Colegrove, P., Pinter, M., Graham, D., *et al.*, Three dimensional flow and thermal modeling of the friction stir welding process, in: *Second International Symposium on Friction Stir Welding*, Gothenburg, Sweden, 26–28 June, p. 26, 2000.
14. Askari, A., Silling, S., London, B., *et al.*, Modeling and analysis of friction stir welding processes, in: *Symposium on Friction Stir Welding and Processing*, Warrendale, PA, pp. 43–54, 2001.
15. Fonda, R.W. and Lambrakos, S.G., Analysis of friction stir welding using an inverse-problem approach. *Sci. Technol. Weld. Join*, 7, 3, 177–181, 2002.
16. Nandan, R., DebRoy, T., Bhadeshia, H.K.D.H., Recent advances in friction stir welding: process, weldment structure and properties. *Prog. Mater Sci.*, 53, 980–1023, 2008.
17. Schmidt, H., Hattel, J., Wert, J., An analytical model for the heat generation in friction stir welding. *Model. Simul. Mater. Sci. Eng.*, 12, 143–157, 2004.
18. Colegrove, P.A. and Shercliff, H.R., Development of Trivex friction stir welding tool. Part 1: Two-dimensional flow modelling and experimental validation. *Sci. Technol. Weld. Join*, 9, 345–351, 2004.
19. Elangovan, K. and Balasubramanian, V., Influences of pin profile and rotational speed of the tool on the formation of friction stir processing zone in AA2219 aluminium alloy. *Mater. Sci. Eng., A*, 459, 7–18, 2007.
20. Besharati-Givi, M. and Asadi, P., *Advances in Friction-Stir Welding and Processing*, Elsevier Science, Netherlands, 2014.
21. Xue, P., Ni, D.R., Wang, D., *et al.*, Effect of friction stir welding parameters on the microstructure and mechanical properties of the dissimilar Al–Cu joints. *Mater. Sci. Eng., A*, 528, 4683–4689, 2011.
22. Agarwal, S.P., Nageswaran, P., Arivazhagan, N., *et al.*, Development of friction stir welded butt joints of a 6063 aluminium alloy and pure copper, in: *International Conference on Advanced Research in Mechanical Engineering*, pp. 46–50, 2012.
23. Zhao, Y.H., Lin, S.B., Qu, F., *et al.*, Influence of pin geometry on material flow in friction stir welding process. *Mater. Sci. Technol.*, 22, 45–50, 2006.
24. Khandkar, M.Z.H. and Khan, J.A., Thermal modeling of overlap friction stir welding for Al-alloys. *J. Mater. Process. Manuf. Sci.*, 10, 91–105, 2001.
25. Colligan, K., Material flow behavior during friction stir welding of aluminum. *Weld. J. Res. Suppl.*, 78, 7, 229–237, 1999.

26. Reynolds, A.P., Seidel, T.U., Simonsen, M., Finite element simulation of material flow in friction stir welding, in: *First International Symposium on Friction Stir Welding*, Thousand Oaks, CA, 14-16 June, 1999.
27. Reynolds, A.P., Visualization of material flow in autogenously friction stir welds. *Sci. Technol. Weld. Joi.*, 5, 120–124, 2000.
28. Seidel, T.U. and Reynolds, A.P., Visualization of the material flow in AA2195 friction stir welds using a marker insert technique. *Metall. Mater. Trans.*, 32A, 2879–2884, 2001.
29. Colligan, K., Material flow behaviour during friction stir welding of aluminium, in: *First International Symposium on Friction Stir Welding*, Thousand Oaks, CA, 14-16, June, 1999.
30. Guerra, M., McClure, J.C., Murr, L.E., *et al.*, Material flow during friction stir welding, in: *Symposium on Friction Stir Welding and Processing*, Warrendale, PA, pp. 25–34, 2001.
31. London, B., Mahoney, M., Bingel, W., *et al.*, Material flow in friction stir welding monitored with Al-SiC and Al-W composite markers, in: *Symposium on Friction Stir Welding and Processing II*, Warrendale, PA, pp. 3–10, 2003.
32. Midling, O.T., Material flow behaviour and microstructural integrity of friction stir butt weldments, in: *Fourth International Conference on Aluminum Alloys*, Georgia Institute of Technology, Atlanta, GA, pp. 451–458, 1994.
33. Pourahmad, P. and Abbasi, M., Materials flow and phase transformation in friction stir welding of Al 6013/Mg. *T. Nonferr. Metal. Soc.*, 23, 1253–1261, 2013.
34. Guerra, M., Schmidt, C., McClure, J.C., *et al.*, Flow patterns during friction stir welding. *Mater. Charact.*, 49, 95–101, 2003.
35. Kazi, S.H. and Murr, L.E., Complex flow phenomena associated with friction stir welding of aluminium alloys, in: *Symposium on Friction Stir Welding and Processing*, Warrendale, PA, pp. 139–150, 2001.
36. Murr, L.E., Trillo, E.A., Li, Y., *et al.*, Solid-state flow associated with the friction stir welding of dissimilar metals, in: *Fluid Flow Phenomena in Metals Processing*, Warrendale, PA, pp. 31–40, 1999.
37. Li, Y., Murr, L.E., McClure, J.C., Flow visualization and residual microstructures associated with the friction stir welding of 2024 aluminium to 6061 aluminium. *Mater. Sci. Eng.*, 271A, 213–223, 1999.
38. Ouyang, J.H. and Kovacevic, R., Material flow and microstructure in the friction stir butt welds of the same and dissimilar aluminium alloys. *J. Mater. Eng. Perform.*, 11, 51–63, 2002.
39. Schmidt, H.B., Dickerson, T.L., Hattel, J.H., Material flow in butt friction stir welds in AA2024-T3. *Acta Mater.*, 54, 1199–1209, 2006.
40. Schmidt, H.B. and Hattel, J.H., Thermal modeling of friction stir welding. *Scr. Mater.*, 58, 332–337, 2008.
41. Santiago, D., Urquiza, S., Lombera, G., *et al.*, 3-D modeling of material flow and temperature in friction stir welding. *Soldag. Insp. São Paulo*, 14, 3, 248–256, 2009.

42. Santiago, D., Urquiza, S., Lombera, G., Numerical modeling of joints welded by friction stir welding process. *Mater. Res.*, 7, 4, 569–574, 2004.
43. Seidel, T.U. and Reynolds, A.P., Two-dimensional friction stir welding process model based on fluid mechanics. *Sci. Technol. Weld. Join*, 8, 175–183, 2003.
44. Bag, S. and De, A., *Computational Models for GTA and Laser Welding Processes*, Lap Lambert Academic Publishing, Germany, 2013.
45. Hart, E.W., Constitutive relations for the non-elastic deformation of metals. *J. Eng. Mater. Technol.*, 98, 193–202, 1976.
46. Fisher, E.S., Temperature dependence of the elastic moduli in alpha uranium single crystals. Part IV: (298° to 923°K). *J. Nucl. Mater.*, 18, 1, 39–54, 1966.
47. Kong, H.S. and Ashby, M.F., Friction-heating maps and their applications. *MRS Bull.*, 16, 4148, 1991.
48. Zienkiewich, O.C. and Cormeau, I.C., Visco-plasticity-plasticity and creep in elastic solids-a unified numerical solution approach. *Int. J. Numer. Method. Eng.*, 8, 821–845, 1974.
49. Simar, A., Lecomte-Beckers, J., Pardoen, T., *et al.*, Effect of boundary conditions and heat source distribution on temperature distribution in friction stir welding. *Sci. Technol. Weld. Join*, 11, 170–171, 2006.
50. Bastier, A., Maitournam, M.H., van Dang, K., *et al.*, Steady state thermomechanical modelling of friction stir welding. *Sci. Technol. Weld. Join*, 11, 278–288, 2006.
51. Lienert, T.J., Stellwag, W.L., Grimmett, B.B., *et al.*, Friction stir welding studies on mild steel. *Weld. J. Res. Suppl.*, 82, 1–9, 2003.
52. Gupta, O.P., *Finite and Boundary Element Methods in Engineering*, Oxford and IBH Publications, New Delhi, 2002.
53. Patankar, S.V., *Numerical Heat Transfer and Fluid Flow*, Mc-Graw-Hill, New York, 1988.

6

Atomic Hydrogen Welding

Chika Oliver Ujah[1,2*], Rodolphe N'Dedji Sodokin[2] and Daramy Vandi von Kallon[1]

[1]Department of Mechanical and Industrial Engineering Technology, University of Johannesburg, Johannesburg, South Africa
[2]Africa Centre of Excellence for Sustainable Power and Energy Development (ACE-PED), University of Nigeria, Nsukka, Enugu State, Nigeria

Abstract

Welding methods have been developed over several decades. These techniques focused on efficiency, energy yield, economic and social considerations with regard to the protection of people and property, as well as consideration of the environmental impact of the technique used. One of the techniques was atomic hydrogen welding. This was one of the most promising technologies, thanks to the use of hydrogen, which is a clean source of energy, and its ability to reach very high temperatures during welding. This ability made it well suited to some types of welding and very thin materials. However, the high cost of installation and certain risks associated with its application have limited its use for many people. The use of specific equipment, the constant checking of electrical equipment before use, and the fact that the technique is only suitable for particular uses, have not really helped its commercial utilization, as was intended when Langmuir introduced it in 1926. Even in 2020, this technology was considered obsolete by the standardization body, which withdrew its ISO 4063 registration granted in 1990 in 1998. Knowing that this method has undeniable advantages for the environment and in terms of efficiency thanks to the use of hydrogen and the possibility of reaching high temperatures, it is imperative to find ways of improving on the technique. The search for clean biomass solutions to produce hydrogen is one of the most reliable solutions for reducing the cost of hydrogen production and reducing the environmental impact. In addition, reliable hydrogen storage solutions need to be developed to reduce the risk of explosion. Finally, research needs to be carried out

**Corresponding author*: omega.ujah@gmail.com

Sandip Kunar and Gurudas Mandal (eds.) Advanced Welding Technologies, (107–126) © 2025 Scrivener Publishing LLC

into making the operating temperature variable, so that the technique can be used in a wide range of applications.

Keywords: Hydrogen welding, arc welding, high temperature, metal melt pool, gas cylinder

6.1 Introduction

Welding techniques have multiplied over the last few decades. The choice of one technique over another is not a haphazard one. It is based on efficiency, energy yield, installation cost, maintenance rate, and, above all, its impact on the environment. Too high hydrogen content in welded steel causes a decrease in the plasticity of the metal and is a reason for the formation of welding failures and irregularities, including pore formation and thermal fissures [1]. Consequently, many of the challenges associated with hydrogen levels in compounds have been an area of concern for a number of years [2]. There are numerous techniques used in determining the quantity of H_2 in a welded section. They share the use of different working fluids immersing the weld specimens and collecting hydrogen in the closed chamber above the fluid [3]. Every hydrogen present in a weld can be separated into the following two types: residual hydrogen, which remains permanently lodged in the steel and cannot be released at room temperature, and diffusible hydrogen, which dissolves in the steel in its atomic or ionic state and is released at room temperature [4]. The primary cause of the discrepancy in outcomes between the glycerin and mercury procedures is the different solubility of hydrogen in the working liquid. The influence of this effect can be forecast by utilizing pertinent correlations between the outcomes of the mercury and glycerin procedures [5, 6].

6.2 Welding Techniques

There are various welding techniques available joining components together or repairing worn out cracked components. In this sub-section, the existing welding techniques will be discussed.

Gas Metal Arc Welding (GMAW/MIG): This kind of welding shields the weld from oxidation and degradation by using a shielding gas. It is the welding technique most frequently used in the business. GMAW, sometimes referred to as metal inert gas (MIG) welding, is a semi-automated welding process that uses an inert gas to shield the arc and an intact wire electrode [4, 7]. The GMAW consists of the following functional principles: (a) The generated arc welding consists of an electrically charged flame, which creates an arc between the wire from the welding cylinders and the workpiece (Figure 6.1). (b) A protective gas is supplied from one of the cylinders, and the major function is to protect the welding arc from atmospheric gases, like oxygen or nitrogen, which oxidizes the welding and brings about defects to the welded part. The shielding gas could be an inert gas (He or Ar). (c) As the arc is created by igniting a combination of oxygen gas from one cylinder and fuel gas from the welding cylinder, the arc melts a welding wire and the workpiece to create a molten pool of metal, which will eventually cool and thicken to a very hard bond.

Gas metal arc welding's quick welding completion time is one of its advantages [8]. It is one of the fastest welding techniques. The protective gas, which shields the arc, helps in ensuring very clean welds devoid of porosity. This technique does not require a highly skilled worker, and it is useful in welding a variety of materials like Al, stainless steel, mild steel, metal composites, etc. The major challenges encountered in this welding technique is the high cost of the equipment and difficulty in controlling the melt pool of the weld [9].

Gas Tungsten Arc Welding (GTAW): GTAW, as shown in Figure 6.2, sometimes referred to as tungsten inert gas (TIG), uses a disposable tungsten electrode. Ferrous and non-ferrous metals can be welded with it [10].

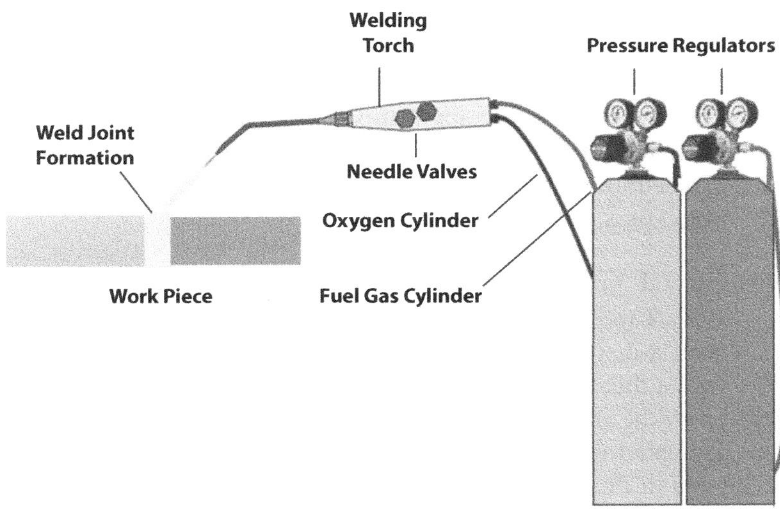

Figure 6.1 Gas metal arc welding setup.

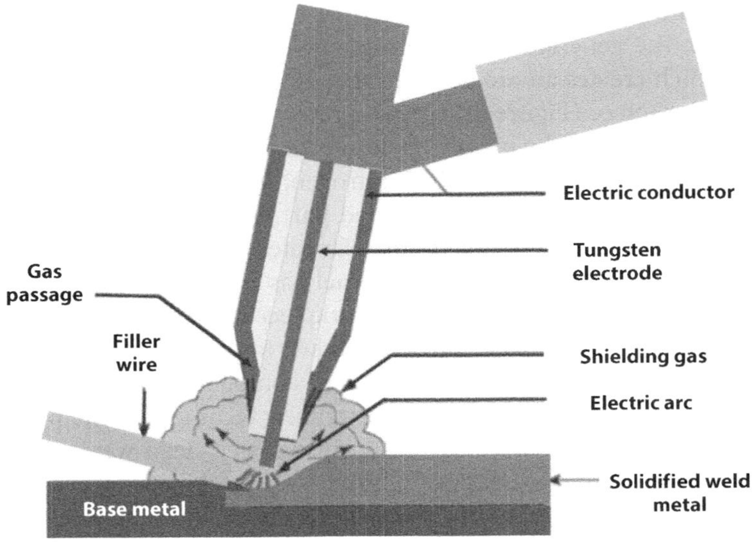

Figure 6.2 Gas tungsten arc welding setup.

An electric arc is created between the workpiece and the tungsten electrode during GTAW. To create the welding pool, a portion of the workpiece is melted by the electric arc. To screen the weld from ambient contamination, inert gas (helium or argon) is used. The advantages of the process are manifold. It gives welds of high quality and improved mechanical properties [11]. It is useful in welding a variety of materials like carbon steel, aluminum, stainless steel, etc. There is considerable control over the weld pool and heat generation. It generates very low residual stress. It is useful for some precise industrial applications. Concurrently, it is slower than many other welding techniques [12]. A highly skilled operator is needed in operating the equipment. It is restricted to thin materials of not more than 6.35-mm thickness.

Shielded Metal Arc Welding (SMAW): This welding technique is also known as "stick welding." It is a well-known arc welding technique, which makes use of a disposable electrode, just like GTAW, but its electrode is covered with a flux to shield the welding arc and molten metal from contamination by the atmospheric gases [13]. As shown in Figure 6.3, an electric arc is created in principle between the workpiece and the electrode that is coated in flux. The arc melts the metal and forms a weld pool. The function of the flux that covers the electrode is to melt and form a protective slag and protect the weld from atmospheric gas contamination.

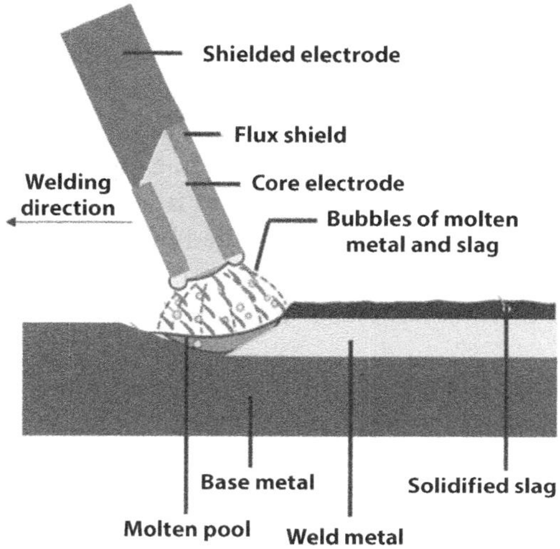

Figure 6.3 Shielded metal arc welding.

This welding technique can use AC or DC power source. The advantages of this technique is that it can weld thick workpieces of up to 25.4-mm thickness [14]. It can weld in a windy atmosphere, and another advantage is that the equipment is affordable, relatively. Again, it is a technique that is easy to learn and operate, and does not require a sophisticated skilled operator. Concurrently, there are some challenges that are inherent in this process. The slag created by the shielding flux needs to be removed after welding to make the work neat. The quality of the weld is not as high as that produced by GTAW [15]. There is little or no control on the penetration or shape of the weld. It endangers the operator's health as it produces splashes and smokes. This technique is usually applied in ship building, automotive repairs, and for heavy fabrications and welding of pipelines. Examples of electrodes used in SMAW include cellulosic and rutile electrodes.

Electroslag Welding (ESW): This type of welding joins metal by melting it with the help of electricity. ESW is a popular welding technique because it gives weld of high quality, with high productivity. The distortion of the weld is very minimal. In the process, a slag layer is created at the weld joint to shield the weld from being contaminated by the environmental gases [16]. The slag equally makes the welding arc stable. The welding current is passed through the shielding slag, which induces the slag's melting to create an arc. The arc then melts the metal to be welded, while the slag shields the weld. The molten metal cools off, solidifies, and forms a strengthened bond. The slag is usually removed after the weld's solidification. This type of welding can handle a workpiece of 5,000 × 6,000-mm cross-section [17]. It is worth to note that this technique can weld dissimilar metals. It is applied in ship building, welding heavy machinery, offshore works, and large steel structures [18]. Its advantages include high quality of weld, low distortion and scanty pores in the weld, and it gives a very strong bond. However, there are some limitations exhibited by this technique. The weld pool cannot be controlled easily. The technique needs highly skilled operators, and the equipment is somewhat complex and costly [19].

Plasma Arc Welding (PAW): This is a welding technique that makes use of high-speed plasma arc to weld metals. This is a popular method because it welds at high precision, with very low distortion and high quality weld [20]. The technique uses an electric current to ionize nitrogen or argon gas and creates plasma. Through a nozzle, the created plasma is passed to generate a high-speed arc, which melts and welds the workpiece. One of the characteristics of PAW is its ability to attain an elevated temperature of approximately 30,000°C making it capable of welding almost all metals. It can weld all types of metals like Al, steel, all kinds including stainless steel. Its weaknesses include the following: high operating cost as only specialized operators and equipment are involved, it can only weld medium thick plates [21]. It is applied in aerospace, medical, automotive, and food processing industries for welding of critical and high-performance products that require high precision, for example, medical implants, exhaust systems, and engine components [22].

Laser Beam Welding (LBW): This is another welding method that makes use of high-powered laser beam in melting and joining metals. In principle, a laser source like CO_2 or Nd:YAG generates a beam [23]. The beam is channeled unto a small aperture to generate a high-powered beam. The high-intensity beam melts the metal to create a weld pool, which forms a strong bond when cooled and solidified. It gives a weld of good precision, with high welding speed of up to 100 m/s [24]. It exhibits a low heat-affected zone because its heat input is low, and so, distortion by the weld is very low. Slag after weld is low, and the weld possesses low porosity. This method gives a high-quality finish. It can weld a lot of metal types like Al, Ti, steel of all kinds, etc. [25]. Concurrently, the equipment is costly and limited to thin materials of 10-mm thickness. LBW can cause eye injury and may bring about fire outbreak. It is used in precision-demanding applications like medical implants and surgical blades, aerospace engine components, and electronics components.

Atomic Hydrogen Welding (AHW): This is a welding technique that makes use of atomic hydrogen to melt and join metals. It is popular as it gives several beneficial services. The atomic H_2 is generated from electrolysis of H_2O [26]. The H_2 gas is thereafter broken into atomic hydrogen with an electric arc. Then, the atomic hydrogen creates an arc welding electrode and the metal being welded. The arc then generates a weld pool that gets solidified to create a strengthened bond. The AHW arc can get up to temperatures of 3,400°C to 4,000°C; hence, it is usually applied for high-melting metals [27]. It possesses high energy density, and so, it generates low weld pool with a minute heat-affected zone. It welds at high speed with low distortion of the workpiece. It gives a high-quality weld with low porosity and minimal slag. However, the process is challenged by high cost of generating H_2, high operating cost, and high cost of employing specialized operators. It welds only thin materials. Its application is in aerospace for welding the engine parts, energy industry for welding of power-generating components, medical devices like implants, and so on.

6.3 Overview of Atomic Hydrogen Welding

As an arc welding method, atomic hydrogen welding (AHW) melts metal using an arc. A hydrogen gas and two tungsten electrodes combine to form the arc. Irving Langmuir's studies on atomic hydrogen were the source of this welding technique's invention [28, 29]. The hydrogen molecules split into hydrogen atoms and then rejoin because of the electric arc utilized creating a significant quantity of heat that can reach temperatures of up to 3,400°C to 4,000°C. This welding procedure is one of the three hottest flames utilized in welding because of its ability to reach such high temperatures. The maximum temperature of an acetylene torch, which is far hotter than what the average welder is used to, is 3,300°C. This is not the case with atomic hydrogen welding [27]. Atomic welding technique is shown in Figure 6.4.

The year 1926 witnessed the invention of a device capable of producing a temperature of over 3,700°C called Atomic Hydrogen Blowtorch [8]. This device, which produces a temperature of 3,700°C, can melt tungsten, diamonds, carbon nanotubes, etc. [30, 31]. Nobel P.O. of the General Electric Company created automatic welding in 1920. A bare electrode wire that was linked to a direct current was employed in this method. General Electric rebuilt worn-out crane wheels and motor shafts using automated welding. Rear axle housings were another application for it in the auto sector. The benefits of heavy-coated welding rods over light-coated rods were hotly debated in the 1920s. Heavy-coated electrodes were invented in 1927 by A.O. Smith Company's Langstroth and Wunder. Then, in 1929, the public began

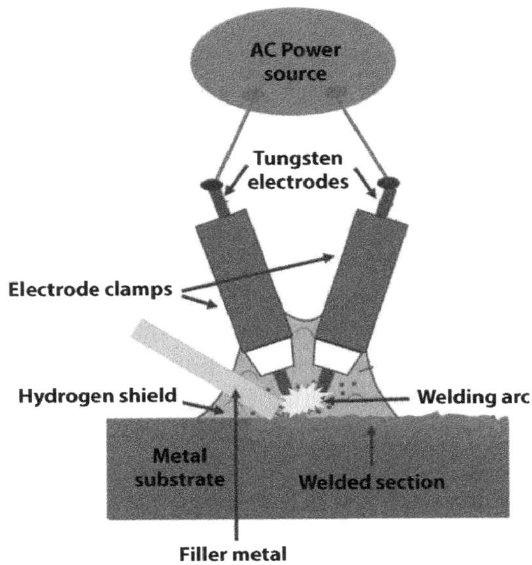

Figure 6.4 Atomic hydrogen welding.

to purchase extruded electrode rods made by Lincoln Electric Company. Coated electrodes were widely used by 1930. With the introduction of welding regulations that demanded improved welding quality, the selection of coated electrodes expanded [32].

Since the 1920s, a great deal of research has been conducted on the topic of externally applied gases sheltering the arc and weld area. It was observed that brittle and sometimes porous welds resulted from the contamination of molten weld with atmospheric oxygen and nitrogen [33]. By using gas shielding technique, researchers were able to solve the porosity and brittleness of welds. Using hydrogen as the welding atmosphere, Alexander and Langmuir worked in chambers. They used two electrodes. Subsequently, they switched to tungsten electrodes from carbon electrodes. In the torch's arc, the hydrogen was changed into atomic hydrogen. After that, it was blasted out of the arc, igniting an extremely hot atomic hydrogen flame that later transformed into a molecule form and emitted heat. This evolved into the method of atomic hydrogen welding. It was used in specialized welding applications in the 1930s and 1940s, and then tool steel welding. Similar research was done by H.M. Hobart and P.K. Devers using atmospheres of argon and helium. Their 1926 patent applications demonstrated arc welding with a gas supply surrounding the arc, which was a precursor to the gas tungsten arc welding method. A wire-like electrode material is fed through the tip. This served as the model for the gas metal arc welding technique [9]. In the second edition in 1990, the ISO 4063, the method was still listed as No. 149. However, it disappeared in the subsequent edition in 1998.

6.4 Operating Principle of Atomic Hydrogen Welding

The welding equipment consists of a special torch with two tungsten electrodes adjusted to ensure a stable arc (Figure 6.5). These tungsten electrodes are surrounded by ring-shaped nozzles, which facilitate the supply of hydrogen gas from gas cylinders. As far as the power source is concerned, alternating current is advantageous over direct current because it distributes the heat evenly between the two electrodes. A transformer with an open circuit voltage of roughly 300 V is required to start and maintain the arc. Within the atomic hydrogen welding facility, the workpieces are thoroughly cleaned to remove contaminants, such as dirt, oxides, and impurities, to ensure a flawless weld. Once the hydrogen gas supply and the welding current have been activated, an electric arc is created by bringing the two tungsten electrodes into contact for a short time and then immediately separating them by a tiny distance of 1.5 mm. As a result, the electric arc between these two electrodes remains intact [27]. As a jet of hydrogen gas passes through the arc, it dissociates into atomic hydrogen. In the endothermic reaction, there is an absorption of a significant portion of the heat that the arc produces.

$$H_2 \rightleftharpoons 2H \qquad \Delta H = 422\ \text{kJ} \tag{6.1}$$

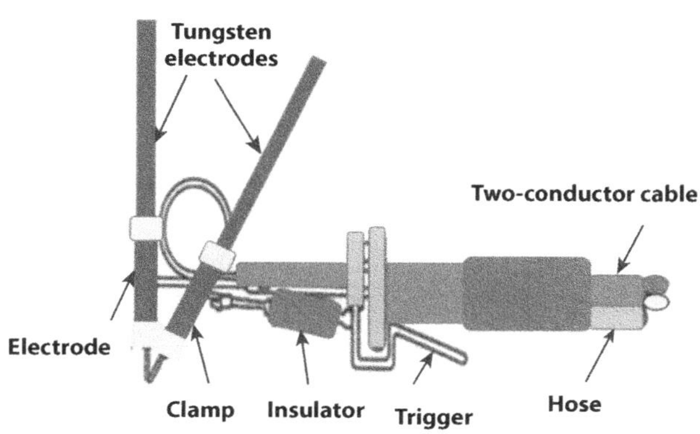

Figure 6.5 Working principle of atomic hydrogen welding.

Tungsten, the hardest metal of all, can be melted and welded using this process thanks to the tremendous heat produced by atomic hydrogen during the welding process. As a gas shield, the melted W keeps the metal from being welded, exposed to contaminants like carbon, oxygen, and nitrogen, which can lead to many weld failures [34]. Flux is not required to be used by the welder. The arc and the components being fused are kept apart during atomic welding. Normally diatomic, hydrogen gas transforms into an atomic state at such high temperatures absorbing a significant portion of the arc's heat. When hydrogen comes into contact with a cooler surface, such as the components being welded together, it quickly loses the heat it has stored and recombines into its diatomic state. The amount of energy generated during atomic hydrogen welding can be readily changed by varying the distance between the arc and the components being welded. Shielded metal arc welding is replacing shielded arc welding despite the former's advantages due to the latter's use of less expensive inert gasses.

6.4.1 Application Area of Atomic Hydrogen Welding

Applied in Joining of Thick Sections: Atomic hydrogen welding is effective in welding of thick sections (Figure 6.6) because it has high heat input as it generates a temperature of approximately 10,000°C, which permits a high heat input making it suitable for welding thick sections [35]. The thick section that AHW can weld is consequent upon its ability to penetrate deeply into the material. It equally has high welding speed.

AHW is used in industries where strong, ductile, and high-quality welds are of essence: This welding type produces strong and ductile weld as the presence of hydrogen gas/atom prevents the weld from being contaminated by carbon, Ni, and O_2 in the atmosphere. The presence of hydrogen equally prevents oxidation of the weld, which causes most weld defects like porosity or distortion [4]. A remnant of hydrogen molecule readily

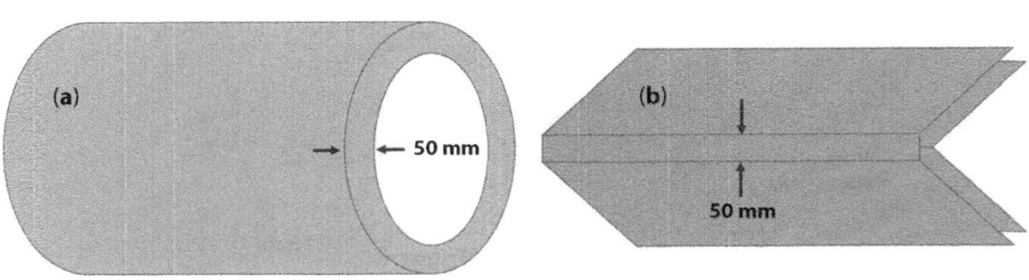

Figure 6.6 Thick metal sections. (a) Metal pipe. (b) Metal angle.

combines with oxygen in the surroundings to produce water, which evaporates easily as steam.

Applied in Welding Thin Structures: Ordinary gas welding cannot weld thin structures (Figure 6.7) because the heat input is not easily controlled and can result in burn-off of the thin plate. But in AHW, the arc is concentrated permitting for easy control of heat input [36]. Another characteristic that permits AHW for welding thin plates is its high heating and cooling rate, which inhibits high buildup of heat on the thin workpiece. The welding technique is fast, which prevents heat build-up, distortion, and burning of thin plates.

Applied in Welding of Refractory Materials: Those materials that are resistant to heat, chemical, and maintain both rigidity and strength at elevated temperature tend to resist most common welding techniques. But AHW arc, with its high heat input and penetration level can conveniently weld refractory plates and structures. The AHW arc, which can generate a very high temperature, can melt the strongest of metals like W, diamond, etc. [37]. (See Figure 6.8 for exploits of AHW.)

Applied in Plasma Joining and Cutting: A mixture of atomic hydrogen and other gases like Ar, Ni, and He are used in plasma welding and cutting [38]. The mixture of gases is used to shield the weld and prevent it from oxidation, distortion, and porosity. Plasma welding and cutting have a dual purpose. There is a plasma generating gas to maintain the arc, and a small amount of auxiliary gas mixed with the plasma generating gas. The most commonly used gases are argon, helium, nitrogen, and hydrogen [34].

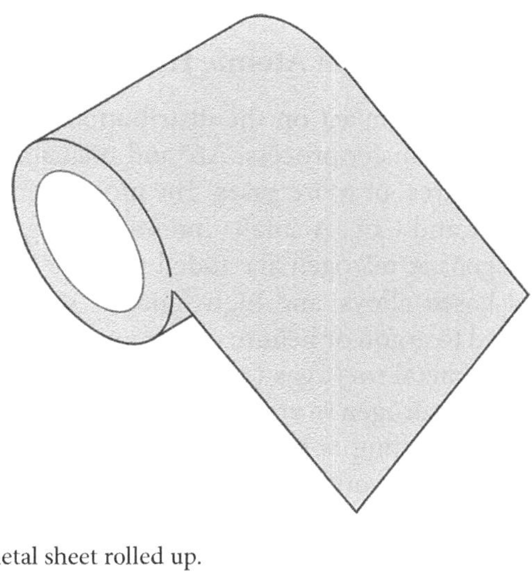

Figure 6.7 Thin metal sheet rolled up.

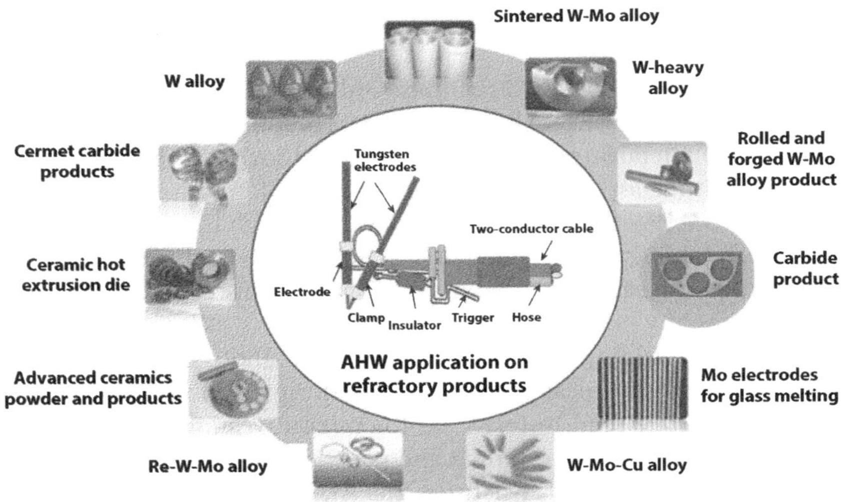

Figure 6.8 Application of AHW for refractory metal products.

Researchers found that plasma cutting and welding would be enhanced by the addition of hydrogen to the gas mixture. It is utilized at argon/nitrogen mixing ratios of 5% to 25%. Utilizing hydrogen raises the arc voltage enabling the use of higher power levels. The plasma arc is much more limited as a result of hydrogen's strong thermal conductivity, which raises the energy concentration. As a result, welds with greater penetration are produced, and the kerf width is reduced during cutting.

6.4.2 The Characteristics of Atomic Hydrogen Welding

The shield gases have an impact on the distribution of heat, the form of the arc, and the metal transfer process. Arc and melt shielding now use a combination of two, three, or more gases. The most often utilized gases in various mixes are Ar and CO_2. A small amount of oxygen and, occasionally, helium, hydrogen, or nitrogen are added to these two gases. To weld aluminum, nickel-based alloys, and high stainless steel alloys, hydrogen has lately been added to argon or helium. An argon/helium/hydrogen mixture can be used for metal inert gas (MIG) and tungsten inert gas (TIG) welding. The effect of hydrogen in argon on the amount of molten material during stainless steel welding is being studied by the Welding Institute. High enthalpy and thermal conductivity are characteristics of hydrogen at temperatures between 3,000 and 5,000 K (temperatures attained in the arc).

When hydrogen is added to argon, the arc's voltage drop and arc power both rise [38]. Deeper penetration is achieved because of the strong thermal conductivity of hydrogen, which causes the arc to shorten and the energy concentration to rise. Figure 6.9 displays two macro-sections of the fusion zone acquired during TIG welding at constant welding current and with various shielding materials. Because of the various shielding materials, the arc voltage fluctuates.

It has been noted that the evaluation of TIG and MIG welding processes with the addition of hydrogen is made easier by the welding process' melting efficiency. The following equation gives the melting efficiency of TIG and MIG welding [34]:

$$n = \frac{MEt}{\int_0^t IU\,dt} \times 100\% \tag{6.2}$$

where I is the current, V is the voltage, and M (g/s) is the quantity of base and filler metal that melt in a unit of time. E (J/g) is the theoretical energy needed to melt 1 g of austenitic stainless steel.

Of theoretical energy, 891 J is needed to melt 1 g of stainless steel. The results for TIG welding's melting efficiency in relation to argon's hydrogen content are displayed in Figure 6.10. As the hydrogen content rises, so does the melting efficiency. Higher currents also result in higher melting efficiencies [34]. Figure 6.10 illustrates how the base metal and filler material melted in argon during MIG welding are affected by the hydrogen content of the argon.

When hydrogen is added to argon, the amount of molten material is increased in the range of 11% to 30%.

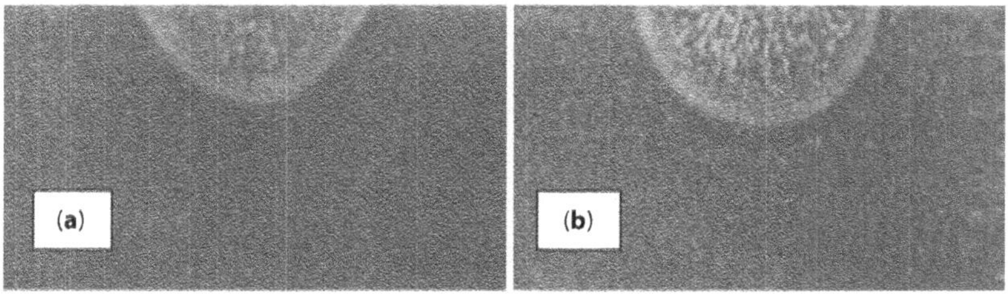

Figure 6.9 Cross-sections of welding arc with the same current intensity but with different gas shield compositions. (a) 100Ar, (b) 90Ar + $10H_2$ [34].

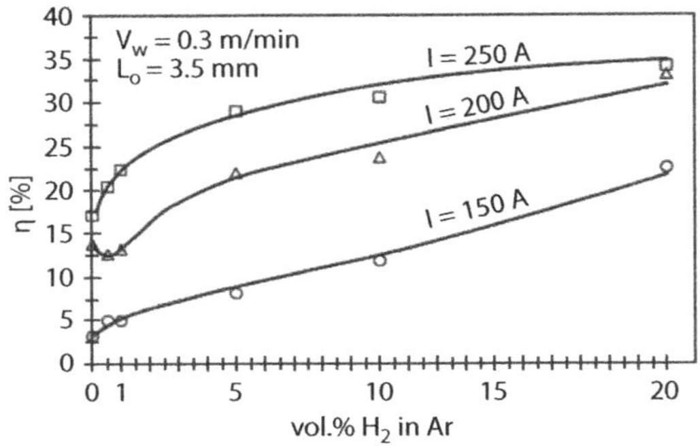

Figure 6.10 Effect of hydrogen content on argon protection of weld and welding current intensity on melting efficiency in TIG [34].

6.5 Prospects and Limitations of Atomic Hydrogen Welding

6.5.1 Prospects of Atomic Hydrogen Welding

Welding Faster—An outstanding advantage of the atomic hydrogen welding process is that it can be carried out at a much higher speed than the other welding methods. This is stimulated by the high temperature it can generate, which implies that the metal melts faster allowing a faster welding time [34]. The faster the welding, the higher the efficiency as shown in Figure 6.11.

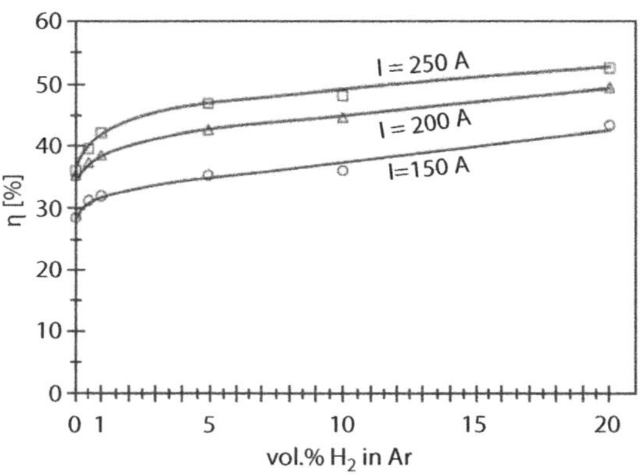

Figure 6.11 Effect of the hydrogen content in the shielding gas on the welding performance of M.I.G. welding of austenitic stainless steel [34].

Cleaner Weld Seams—The presence of hydrogen and shielding gases (argon, helium, etc.) prevent the formation of metal oxide during the AHW process, which makes it produce cleaner welds than other methods that are not shielded by protective gases [38]. This means that it is less likely to become contaminated, and the weld will be stronger.

Less Heat Input Is Required—Atomic hydrogen welding requires less heat input as little heat will give high heat value. This is because the dissociation of hydrogen gas into atomic hydrogen in the weld area will herald much higher temperature than other welding processes due to higher calorific value of H_2.

Increased Effectiveness—Compared with other forms of welding, such as traditional arc welding, atomic hydrogen welding is much more efficient [38]. This is because less heat is required to weld the metal together, as the atomic hydrogen welds at a much higher temperature than the metal hydrogen itself.

Relatively Thick Sections Can Be Effectively Welded Using Atomic Hydrogen Arc Welding—Atomic hydrogen welding provides excellent penetration to the welded workpiece. As the process uses a combination of hydrogen gas and an electric arc, it is able to penetrate to a greater depth than other welding techniques [34]. As a result, it is ideal for the creation of strong welds in thick materials or for joining of two pieces of thick metal.

6.5.2 Limitations of Atomic Hydrogen Welding

A major drawback to atomic hydrogen welding is that it requires special equipment and consumables, which are capital intensive. This includes the welding machine itself, as well as electrodes, electrode carriers, H_2 gas, and cooling systems for safe operation. When compared with welding techniques, atomic hydrogen welding tends to be more expensive [39].

The use of a large volume of hydrogen gas during atomic hydrogen welding can lead to explosive reactions in the welding area. Once the volume required for work is exceeded, the excess volume is very flammable and can cause fire explosion, which will result to laboratory accident and its consequences [40]. Therefore, adequate precautions must be taken to work with or use precisely what is needed for the AHW.

Specialized Operator Requirements—There is also a need for specialized equipment operator for atomic hydrogen welding, which is not readily available. The high temperatures needed for welding must be able to be produced by the welding torches, and managing such high temperatures demands a high level of ability [41]. In addition, atomic hydrogen welding can only be performed in a vacuum chamber, which further reduces the practicability of atomic hydrogen welding.

Requires High Temperatures—The need for extremely high temperatures to be effective is another disadvantage of atomic welding. Temperatures between 5,000°C and 10,000°C are required to weld metals using atomic hydrogen welding. The high temperatures are risky to the operator, people within the vicinity, and the environment [42].

Limited Uses—One of its main drawbacks is its limited applicability. Only metals with a high melting point, such as tungsten, molybdenum, and tantalum, can be effectively welded using atomic hydrogen welding technique. This feature makes it rarely useful for household and generalized welding technique. Most cheaply available metals, like Al, Cu, and Ti [43, 44], for developing commonly used products in homes, hospitals, offices, churches, and industries, cannot be easily welded with AHW.

6.6 Further Research in Atomic Hydrogen Welding

For the development of atomic hydrogen welding to an enviable height, the following studies are encouraged:

Efforts must be made to improve hydrogen production techniques in such a way that cost of hydrogen will be affordable. Research should be done to lower the price of producing hydrogen and the product's total cost. The use of biological hydrogen from biomass would be of great utility for both environmental and economic reasons.

Research must be carried out on the development of thermal resistant welding equipment, which can withstand the high operating temperatures of AHW without sudden degradation and failure. Hence, more durable materials must be developed to reduce the long-term cost of AHW.

New personal protection techniques need to be developed to reduce accidents and the risk of explosions due to the high flammability of dihydrogen. Moreover, more advanced studies should be conducted on safety measures to be adopted to reduce welding accidents.

6.7 Conclusion and Recommendation

Atomic hydrogen welding is an advantageous welding technique with a number of limitations. Due to the intense heat generated during the AHW welding process, operators must follow safety guidelines and take preventative measures to avoid the inherent risks of the work. The use of hydrogen for both oxy-fuel and arc welding is increasing. The introduction of hydrogen into the welding flame has been facilitated by the affordability of hydrogen production from the electrolysis of water and the safety level of the equipment. While acetylene still dominates welding, the benefits of hydrogen are currently being exploited mainly for brazing and cutting. The use of biological hydrogen from biomass through thermochemical techniques, in particular gasification, will be a great opportunity and will allow atomic hydrogen welding technology to take the lead as it becomes the cleanest and most environmentally friendly, not forgetting the economic implications. AHW gives clean welds, precise welds, strong welds, applied on thick and thin structures, used in the welding of refractory materials, and for cutting very hard materials. However, it is challenged by the high cost of hydrogen and the welding equipment, high temperature of operation, and dearth of skilled operators. These limitations should be worked on to make the AHW technique take its deserved position in the industry.

Acknowledgment

The authors wish to appreciate the financial support of the University of Johannesburg and NRF, South Africa, for offering a Postdoctoral Research Fellowship Grant to the corresponding author of this work.

References

1. R. Kurji, R. and Coniglio, N., Towards the establishment of weldability test standards for hydrogen-assisted cold cracking. *Int. J. Adv. Manuf. Technol.*, 77, 1581–1597, 2015.
2. Kühn, S., Unterumsberger, F., T. Suter, T., Pohl, M., Neue Methoden zur Analyse von diffusiblem Wasserstoff in hochfesten Stählen. *Mater. Test.*, 55, 9, 648–652, 2013.
3. Fydrych, D. and Łabanowski, J., Estudio experimental de procesos de soldadura con alto contenido en hidrógeno. *Rev. de Metalurgia*, 51, 4, e055–e055, 2015.
4. Padhy, G.K. and Komizo, Y.I., Diffusible hydrogen in steel weldments: A status review. *Trans. JWRI*, 42, 1, 39–62, 2013.
5. Pokhodnya, I., *Metallurgy of Arc Welding: Interaction metal with gases*, Naukova Dumka, Kyiv, Ukraine, 2004.
6. Kotecki, D., Hydrogen reconsidered. *Welding J.*, 71, 8, 35–43, 1992.
7. Singh, R., *Arc welding processes handbook*, John Wiley & Sons, New Jersey, United States, 2021.

8. Cheng, Y., Yu, R., Zhou, Q., Chen, H., Yuan, W., Zhang, Y., Real-time sensing of gas metal arc welding process–A literature review and analysis. *J. Manuf. Processes*, 70, 452–469, 2021.
9. Suzuki, R., State of the art of process control of molten droplet and pool in gas metal arc welding. *Weld. Int.*, 26, 3, 178–186, 2012.
10. Mvola, B., Kah, P., Martikainen, J., Suoranta, R., State-of-the-art of advanced gas metal arc welding processes: Dissimilar metal welding. *Proc. Inst. Mech. Eng., Part B: J. Eng. Manuf.*, 229, 10, 1694–1710, 2015.
11. Kumar, S. and Shahi, A., Effect of heat input on the microstructure and mechanical properties of gas tungsten arc welded AISI 304 stainless steel joints. *Mater. Des.*, 32, 6, 3617–3623, 2011.
12. Stenbacka, N., On arc efficiency in gas tungsten arc welding. *Soldagem Inspeção*, 18, 380–390, 2013.
13. Sidhu, G.S. and Chatha, S.S., Role of shielded metal arc welding consumables on pipe weld joint. *Int. J. Emerging Technol. Advanced Eng.*, 2, 12, 746–750, 2012.
14. Dias, J.M.S., dos Santos Paes, L.E., Santos, A.G.M., Borba, T.M.D., da Paixão Carvalho, L., Vilarinho, L.O., A methodology for multipass gas metal arc welding of shipbuilding steel plates with different thicknesses. *Int. J. Adv. Manuf. Technol.*, 127, 1, 751–773, 2023.
15. Thakur, A., Gebrelibanos, H., Gabrey, T., Arc welding process selection through a quality and costs. *Int. J. Curr. Eng. Technol.*, 9, 3, 383–394, 2019.
16. Alvaraes, C.P., Jorge, J.C., de Souza, L.F., Araujo, L.S., Mendes, M.C., Farneze, H.N., Microstructure and corrosion properties of single layer Inconel 625 weld cladding obtained by the electroslag welding process. *J. Mater. Res. Technol.*, 9, 6, 16146–16158, 2020.
17. L. Medovar, L. and Stovpchenko, G., Electroslag welding and related technologies, in: *Welding of Metallic Materials*, Elsevier, pp. 73–111, 2023.
18. Harada, Y., Iyama, J., Matsumoto, Y., Oki, K., Electroslag Welding Applications for Steel Building Construction in Japan: A State-of-the-Art Review. *Eng. J.*, 60, 2, 93–110, 2023.
19. Medovar B.I., B., II and Boyko, G.A., *Electroslag technology*, Springer Science & Business Media, Tiergartenstrasse 17, Heidelberg, 2013.
20. Zhang, Q.L., Yang, C.L., Lin, S.B., Fan, C.L., Horizontal welding of aluminium alloys by soft plasma arc. *Proc. Inst. Mech. Eng., Part B: J. Eng. Manuf.*, 228, 11, 1481–1490, 2014.
21. Zhou, W., Jia, C., Zhou, F., Wu, C., Dynamic evolution of keyhole and weld pool throughout the thickness during keyhole plasma arc welding. *J. Mater. Process. Technol.*, 322, 118206, 2023.
22. A. Ghosh, A., T. Bhattachraya, T., N. Deshpande, N., Basak, S., Microstructural and mechanical properties analysis of plasma ARC welded 316L austenitic stainless steel joints without addition of filler material. *Int. J. Trend Sci. Res. Dev.*, 3, 3, 849–865, 2019.

23. Kashaev, N., Ventzke, V., Fomichev, V., Fomin, F., Riekehr, S., Effect of Nd: YAG laser beam welding on weld morphology and mechanical properties of Ti–6Al–4V butt joints and T-joints. *Opt. Lasers Eng.*, 86, 172–180, 2016.
24. Biswas, A., Banerjee, N., Sen, A., Maity, S., Applications of Laser Beam Welding in Automotive Sector-A Review, in: *Advances in Additive Manufacturing and Metal Joining: Proceedings of AIMTDR 2021*, pp. 43–57, 2023.
25. Çalıgülü, U., Calıgulu, U., Dıkbas, H., Taskın, M., Microstructural characteristic of dissimilar welded components (AISI 430 ferritic-AISI 304 austenitic stainless steels) by CO2 laser beam welding (LBW). *Gazi Univ. J. Sci.*, 25, 1, 35–51, 2012.
26. J. Rossmeisl, J., A. Logadottir, A., Nørskov, J.K., Electrolysis of water on (oxidized) metal surfaces. *Chem. Phys.*, 319, 1-3, 178–184, 2005.
27. Thorat, S., *Atomic Hydrogen Welding*, 2024, LearnMech.com, Karad, Pune, Maharastra, Available: https://learnmech.com/atomic-hydrogen-welding-working-advantages-and-application/.
28. Miller, D.P., The political economy of discovery stories: the case of Dr Irving Langmuir and General Electric. *Ann. Sci.*, 68, 1, 27–60, 2011.
29. Suits, C.G., *Langmuir, the Man and the Scientist: With Contributions in Memoriam Including a Complete Bibliography of His Works*, Elsevier, Amsterdam, Netherland, 2013.
30. Ujah, C.O., Kallon, D.V.V., Aigbodion, V.S., Overview of Electricity Transmission Conductors: Challenges and Remedies. *Materials*, 15, 22, 8094, 2022.
31. Ujah, C.O., Kallon, D.V.V., Aigbodion, V.S., Tribological Properties of CNTs-Reinforced Nano Composite Materials. *Lubricants*, 11, 3, 95, 2023.
32. Greitmann, I.I.M.J., Welding through the Ages. *Am. Weld. Soc.*, 2013.
33. Singh, R., Singh, S., Singh, G., Singh, A., A Review of Porosity Formation and Recommendations on the Avoidance of Porosity in TIG Welding of Steels. *Int. J. Of Mech. Eng. Inf. Technol.*, 2, 8, 700–712, 2014.
34. M. Suban, M., Tušek, J., Uran, M., Use of hydrogen in welding engineering in former times and today. *J. Mater. Process. Technol.*, 119, 1-3, 193–198, 2001.
35. Khoshnaw, F., Krivtsun, I., Korzhyk, V., Arc welding methods, in: *Welding of Metallic Materials*, Elsevier, pp. 37–71, 2023.
36. Nevasmaa, P., Prevention of weld metal hydrogen cracking in high-strength multipass welds. *Weld. World*, 48, 2–18, 2004.
37. Olson, D.L., Mishra, B., Wenman, D., Welding, brazing and joining of refractory metals and alloys. *Miner. Procesing Extr. Metall. Rev.*, 22, 1, 1–23, 2001.
38. Tusek, J. and Suban, M., Experimental research of the effect of hydrogen in argon as a shielding gas in arc welding of high-alloy stainless steel. *Int. J. Hydrogen Energy*, 25, 4, 369–376, 2000.
39. Mirza, R. and Gee, R., Effects of shielding gases on weld diffusible hydrogen contents using cored wires. *Sci. Technol. Weld. Joining*, 4, 2, 104–111, 1999.

40. Oliver, U.C., An Integrative Review of Laboratory Accidents among Metallurgical Technicians: Types, Causes, Effects and Prevention Strategies. *Int. J. Home Economics Hospitality Allied Research Rev.*, 2, 2, 129–139, 2023.
41. Messler Jr., R.W., *Principles of welding: processes, physics, chemistry, and metallurgy*, John Wiley & Sons, Hoboken, New Jersey, US, 2008.
42. Jay, O., Capon, A., Berry, P., Broderick, C., de Dear, R., Havenith, G., Honda, Y., Kovats, R.S., Ma, W., Malik, A., Morris, N.B., Reducing the health effects of hot weather and heat extremes: from personal cooling strategies to green cities. *Lancet*, 398, 10301, 709–724, 2021.
43. Ujah, C., Popoola, A., Popoola, O., Afolabi, A., Uyor, U., Mechanical and Oxidation Characteristics of Ti20-Al16-V16-Fe16-Ni16-Cr16 High-Entropy Alloy Developed via Spark Plasma Sintering for High-Temperature/Strength Applications. *J. Mater. Eng. Perform.*, 32, 1, 18–28, 2023.
44. Ujah, C.O., Popoola, A., Popoola, O., Uyor, U., Analysis of the Microstructure and Tribology of Ti36-Al16-V16-Fe16-Cr16 HEA Developed with SPS for Engineering Applications. *JOM*, 74, 11, 4239–4249, 2022.

7

Modeling and Designing of Ultrasonic Welding

Jagadeesha T.[1]* and Sandip Kunar[2]

[1]*Department of Mechanical Engineering, National Institute of Technology, Calicut, India*
[2]*Department of Mechanical Engineering, Aditya University, Surampalem, Kakinada, AP, India*

Abstract
This chapter covers ultrasonic welding, an advanced technology for joining materials without melting. It discusses the process' benefits, including its ability to join dissimilar metals and work near heat-sensitive components without damage. The limitations, such as its suitability for nonferrous metals and lightweight components, are also addressed. Detailed descriptions of ultrasonic welding process and equipment are provided, including power systems, transducers, and coupling systems. Design considerations for ultrasonic horns, which amplify and transmit ultrasonic vibrations, are explained focusing on factors like material compatibility and resonance frequency. Process parameters, like power, clamping load, and welding duration, are discussed emphasizing their impact on weld quality and cycle time. The versatility of ultrasonic welding, spanning from electronics to automotive industries, is highlighted showcasing its broad applicability. Despite its advantages, challenges, like specialized equipment requirements and limitations on part size, are acknowledged. Resonance issues and material adherence to weld tips are noted as potential complications. Ultrasonic welding's significance in modern manufacturing is underscored offering efficient and reliable material joining while preserving properties. The chapter provides insights valuable for researchers, engineers, and practitioners in materials joining technology.

Keywords: Coalescence, transducer systems, coupling system, ultrasonic horns, resonance frequency, clamping load, amplitude amplification

**Corresponding author*: jagdishsg@nitc.ac.in

Sandip Kunar and Gurudas Mandal (eds.) Advanced Welding Technologies, (127–154) © 2025 Scrivener Publishing LLC

7.1 Introduction

Ultrasonic welding achieves coalescence between a diverse array of metals, both similar and dissimilar, through the localized use of mechanical oscillations of high frequency in the order of 15–25 kHz. These short-cycle time vibrations help in creating a welding without melting of parent material. Ultrasonic welding stands out as a cutting-edge advancement in joining technology, characterized by the fusion of high-frequency vibrations and pressure to swiftly and securely unite two materials. Unlike conventional heat-based welding methods, this process operates without generating significant heat offering a host of advantages. Notably, it enables the bonding of metals with vastly different melting points circumventing issues like the formation of brittle alloys at weld junctions. Ultrasonic welding is also useful in joining semiconductor materials and electronic circuits where high heat energy cannot be used due to circuit damage and heat sensitivity of electronic materials. What is more, ultrasonic welds are consumable-free, unlike traditional techniques that rely on solder or filler, and they consume substantially less energy. Despite its benefits, ultrasonic welding does have limitations. Ultrasonic joining is mostly used to weld plastic materials and nonferrous materials. It is not recommended for high thick and heavy materials as there is no melting in this process. This requirement stems from the considerable energy needed to vibrate heavier components at the requisite frequency effectively restricting the process to small-scale components and wires.

The ultrasonic welded joints are known for high strength compared to adhesive joints and other mechanical joints. There is no concentration of stress, and it is not a labor-intensive process. High-skilled workers, like those in traditional welding processes, are not required in ultrasonic welding. There is no melting in ultrasonic welding process, and it is a solid phase welding process. It is commonly used in thin sheet metal and cannot be used for heavy weight material joining [1]. Ultrasonic welding is also commonly used in welding various plastic materials. Plastic weldment requires complex dual vibration systems with a provision to high amplification and frequency ranging from 50 to 1,000 kHz [2]. Thermoplastics can be welded very fast and are cost effectively using ultrasonic welding processes. There is also increasing demand for bonding thermoplastic material and metal, and ultrasonic welding is a promising choice [3]. With the advent of electrical vehicles, there is more demand for unconventional welding processes such as ultrasonic welding. Dissimilar materials, like copper and aluminum can be joined effectively using USW processes [4]. Various dissimilar material combination of polycarbonate, polymethyl methacrylate, polylactic acid, and polyformaldehyde can be effectively welded using third-phase USM processes. High production rates and high efficiency joints are possible using third-phase USM processes [5]. Carbon epoxy joints with high lap joint efficacy of up to 95% is possible with coupling layers in USM. Without coupling layers, poor joint efficiency is observed [6]. Composite prepreg can be welded using USM, but the resin flow at the interface and localized plastic deformation at the interface of the welding is challenging in welding prepregs [7]. Amplitude and frequency are the two important process parameters in USW. A little vibration at the tool tip or the weld tip can be achieved through proper design of a sonotrode. Modal frequency, axial displacement, von Mises stresses, and material selection are important in developing a sustainable USW [8]. Lightweight materials like aluminum and magnesium alloys can be joined using USM. However, vibration of a sonotrode must be carefully chosen to avoid the weldment damage. Shear stress at the faying surface, weld geometry, and spacing of weldments are important consideration in welding combination of lightweight alloying materials [9]. Proper process parameters and design of the sonotrode is essential for effective weldment in USW. Optimum heat transfer routes, dwell time, weld time, and compression of weld pool are important in getting good spot welds [10]. Copper-to-copper sheet of 1-mm thickness can be welded using USM. Duration of weld time, applied axial force, amplitude, frequency, surface roughness, and other metallurgical properties affect the quality of welded joints. Good surface finish gives better mechanical strength of the joints. Welding energy and amplitude of the vibration are to be considered carefully for effective weldment of copper-to-copper material [11]. Tool-to-workpiece area contact ratio is one of the important factors to be accounted in USW, which affects the tool tip wear and tool life [12].

Aluminum sheet, copper and brass sheets USW can be modeled using second-order equations to predict weld strength and spot weld shear strengths. Mesoscopic models must account for flow of polymer, and quality and quantity of adhesion to accurately analyze the final welded joints [13]. Design, modeling, and fabrication of an ultrasonic unit can be challenging and involves design of transducers, sonotrode, power supply systems, clamping systems, validation and testing of models developed numerically, and electrical and mechanical characterization of the whole setup [14].

7.2 Description of Process

Originating in the 1950s, ultrasonic welding leverages high-frequency vibrations to accomplish material joining. Workpieces are pressed together under moderate static force, while oscillating shear stresses of ultrasonic frequency are applied parallel to their interface typically for a brief duration. The process begins with a power supply converting low-frequency electrical energy into high-frequency impulses. These impulses are then transformed into mechanical vibration using a suitable transducer. Function of a coupling system is to act as bridge to amplify, modify, and transmit mechanical vibration of low amplitude, but high frequency, to weldment interface with axial force applied as shown in Figure 7.1.

Clamping forces generate consistent and symmetrical stress that spreads through the material from the clamping point. When lateral welding force is exerted, stress distribution within the material moves to one side of the clamp point, delineating one segment of the vibration cycle. In the following segment of the cycle, lateral forces alter direction causing stress distribution to transfer to the opposite side of the clamping point. This rapid alternation of subsurface stresses, occurring tens of thousands of

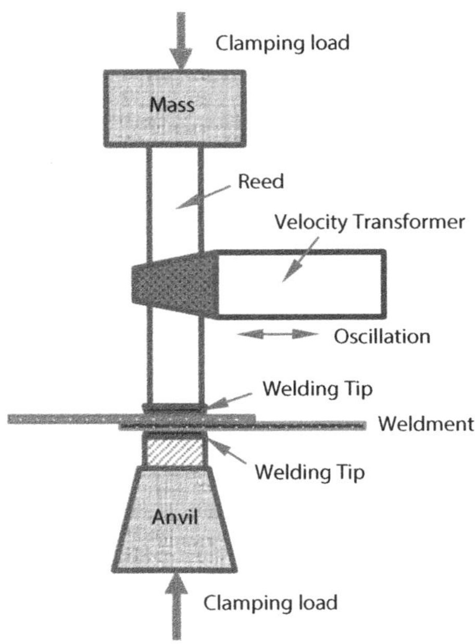

Figure 7.1 Ultrasonic welding processes.

times per second, leads to the dispersion of surface oxides and films, the smoothing of surface irregularities via plastic deformation, and ultimately, inter-atomic bonding through atomic diffusion. Remarkably, the welding interface experiences a maximum temperature that is only 35%–50% of the material's melting point. As there is no melting at the weld interface, metals can be joined without the need for fluxes or shielding gases eliminating the risk of undesirable porosity or gas formation.

Due to these capabilities, ultrasonic metal welding (USW) has been experiencing rapid expansion, particularly within the electronics and electric motor sectors, where it has become indispensable for wire and sheet welding applications. Its efficacy has prompted its adoption in even more demanding industries, such as aerospace and automotive, where it is now employed for critical structural components.

7.3 Description of Equipment

Figure 7.2 depicts a schematic layout of USW. The primary function of an ultrasonic welding machine is to produce and convey high-frequency mechanical motion to a clamped workpiece. This task is achieved through the following subsystems of a USW machine.

1. Power Systems
2. Transducer Systems
3. Coupling system

(1) Power Systems: The ultrasonic welding power supply plays a crucial role in the process by converting a 50-Hz electrical signal into a much higher frequency signal. This conversion is facilitated through the utilization of a frequency generator and amplifier. Frequency generators employed for different welding applications can operate within a wide range of frequencies spanning from 15,000 to 85,000 Hz. Generally, higher frequencies are preferred for lower power, light-duty tasks.

In the second stage of the power supply, amplifier circuits come into play, tasked with amplifying the high-frequency signal to a level suitable for the specific job at hand. This amplification process varies depending on the requirements ranging from just a few watts for light-duty tasks, like wire welding, to power levels as high as 15,000–17000 W for more demanding applications such as sheet and plate welding.

(2) Transducer Systems: The transducer serves a critical function in the ultrasonic welding process by converting the electrical energy from the power supply into mechanical energy, which is then utilized for welding [6]. There are two primary types of transducers employed in ultrasonic metal welding (USW), each chosen based on the specific application. They are as follows:

Magnetostrictive Transducer: This type of transducer harnesses the property of magnetostriction. Magnetostrictive materials exhibit changes in length in response to varying magnetic flux density. In the construction of this transducer, a heavy gauge wire is wound around a laminated stack comprised of nickel sheets forming an electromagnet. Whenever a pulse of current is applied to the wire, it induces a magnetic response within the stack causing a momentary change of approximately 0.04 to 0.006 mm in the overall length of the stack. Magnetostrictive transducers are typically utilized for applications requiring lower powers. While they are characterized as low-efficiency devices, they are exceptionally rugged and have a long lifespan.

The Piezoelectric Transducer operates on the principle that piezoelectric materials undergo slight changes in length when subjected to direct current passing through them. These transducers typically consist of a stack of ceramic disks often composed of lead zirconate titanate. They find extensive use in high-power applications. However, due to the inherent brittleness of the ceramic material, piezoelectric transducers are considerably more fragile compared to their magnetostrictive counterparts. Both types of transducers necessitate air or water cooling to dissipate the waste heat produced during their oscillations.

(3) Coupling System: The coupling system serves the crucial role of mechanically transmitting high-frequency vibrations from the transducer to the workpiece. Two primary methods are employed for transmitting energy to the weld as follows: the wedge-reed system and the lateral drive system.

The wedge-reed system, suitable exclusively for spot welding, comprises a transducer that propels a wedge-shaped coupler connected to a vertical reed (as depicted in Figure 7.1). When the transducer vibrates the reed, nodes and antinodes emerge along its length. The reed's length is carefully chosen so that at a specific operating frequency, a node (a point of maximum displacement) aligns with the welding tip. In cases demanding exceptionally high power, multiple transducers can be linked to the reed at appropriate nodal points.

To ensure durability, the reed and coupler must be crafted from materials with excellent fatigue resistance; titanium, Monel, or stainless steel are commonly utilized for this purpose. Both the tip and anvil contribute to the clamping force. The anvil's role is to secure the bottom metal piece stationary against the lateral forces induced by the vibrations.

Wedge–reed systems deliver substantial clamping force, albeit with low amplitude, rendering them suitable for high-power applications such as large-area bonding, heavy-section welding, and stranded wire welding [15]. In contrast, the lateral drive system involves a horizontally mounted transducer connected to a coupling horn and tip (as illustrated in Figure 7.2). This configuration generates tip motion parallel to the weld interface. Clamping force results from the pivoting motion of the ultrasonic assembly. This system accommodates both low-power spot welds and continuous seam welding.

In the lateral drive system, energy initially traverses through a tapered coupler or horn. This horn is meticulously crafted with a diminishing cross-sectional area facilitating an amplified amplitude at its output end. Essentially, the horn functions as a mechanical step-up transformer in relation to the motion of the weld tip.

The energy transmitted through the horn is then coupled into the welding tip, typically crafted from materials like tool steel or nickel alloys. Notably, the lateral-drive system is not confined solely to spot welds; with the appropriate welding tip, it can seamlessly execute high-speed continuous lap seam welds. This feat is achieved through the utilization of a rotating roller tip. Due to its characteristic low clamping force and substantial mechanical amplitude, the lateral drive system is predominantly employed in applications necessitating thin materials.

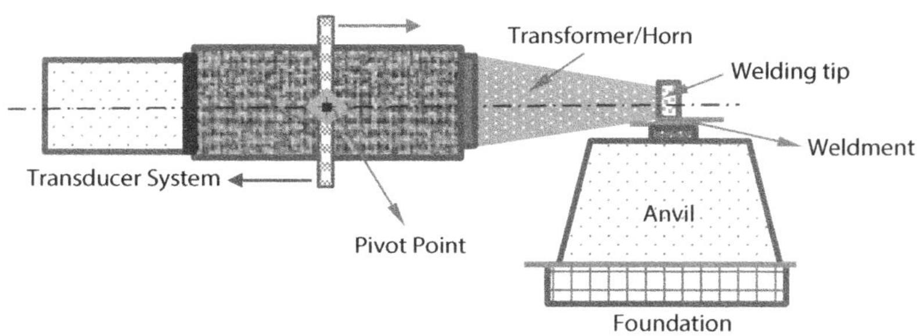

Figure 7.2 Lateral drive USW unit.

7.4 Design of Ultrasonic Horns (Velocity Transformers) for Welding Applications

Ultrasonic horns used for ultrasonic welding are specialized components designed to efficiently transmit and amplify ultrasonic vibrations for the purpose of joining materials through localized melting and bonding. Here is how their design considerations differ:

- ***Material Compatibility***: The material selection for ultrasonic welding horns depends on factors, such as the materials being welded, operating environment, and desired performance characteristics. Common materials include titanium, aluminum, and steel chosen for their high strength and fatigue resistance.
- ***Geometry and Shape***: Ultrasonic welding horns are designed with specific geometries optimized for the welding process. The shape often includes features, such as a flat or contoured working surface, depending on the shape of the parts being welded and the desired weld joint configuration.
- ***Resonance Frequency***: Like other ultrasonic horns, welding horns are designed to resonate at the desired ultrasonic frequency, typically ranging from 15 to 70 kHz. The horn's dimensions and geometry are carefully calculated to match the frequency output of the ultrasonic transducer.
- ***Amplitude Amplification***: The primary function of an ultrasonic welding horn is to amplify the amplitude of ultrasonic vibrations generated by the transducer. This amplification is achieved through the horn's mechanical advantage, which may involve tapering toward the tip or utilizing other geometric features.
- ***Energy Distribution***: Welding horns are designed to distribute ultrasonic energy evenly across the weld interface to ensure consistent bonding and minimize defects. The horn's geometry and surface finish play crucial roles in controlling energy distribution.
- ***Horn Tip Configuration***: The tip of an ultrasonic welding horn is carefully designed to optimize energy transfer and focus ultrasonic vibrations at the weld interface. Depending on the application, the tip may be flat, pointed, or contoured to match the shape of the weld area.

- ***Tooling Integration***: Ultrasonic welding horns are often integrated with tooling fixtures or sonotrode adapters that hold the parts in place during the welding process. The design of these attachments should complement the horn's vibrations and facilitate efficient welding.
- ***Cooling and Ventilation***: In high-power ultrasonic welding applications, cooling and ventilation systems may be incorporated into the horn design to dissipate heat generated during operation and maintain temperature stability.
- ***Durability and Longevity***: Welding horns are subjected to repetitive mechanical stresses during operation, so their design must prioritize durability and fatigue resistance. Finite element analysis (FEA) is often used to optimize the horn's geometry and ensure structural integrity.
- ***Process Optimization***: Horn design for ultrasonic welding involves iterative optimization to achieve the desired weld quality, cycle time, and process efficiency. Prototyping and testing are essential for validating the performance of the horn design under real-world welding conditions.

Designing effective ultrasonic welding horns requires a deep understanding of ultrasonic welding principles, materials science, mechanical engineering, and process optimization techniques. Collaboration between engineers, welders, and material scientists is often necessary to develop welding horn designs that meet the specific requirements of each welding application.

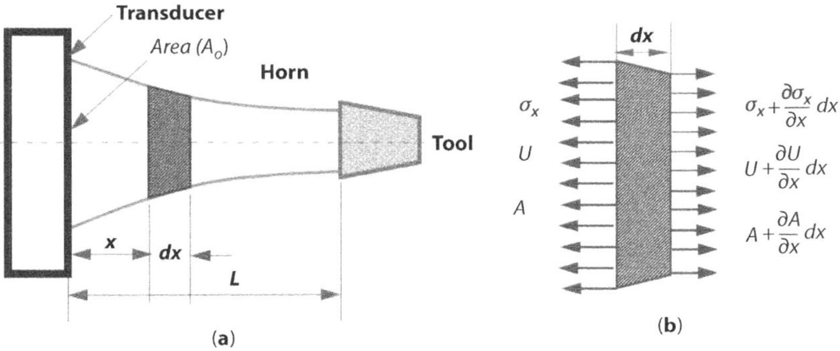

Figure 7.3 (a) Transducer unit. (b) Force balance on the differential element dx.

Consider a taper transducer as shown in Figure 7.3a. Consider an element of dx length at a distance of x from the base transducer. The force balances on differential element dx are shown in Figure 7.3b.

$$A = \text{Cross section area of the } U \text{ horn}$$

$$U = \text{Displacement in x direction}$$

$$\sigma_x = \text{normal stress in x direction}$$

From Hook's law, we can write the following:

$$\textit{Strain in the element in x direction} = \varepsilon_x = \frac{[U + (\partial U / \partial x)dx] - U}{dx} = \frac{\partial U}{\partial x} = \frac{\sigma_x}{E}$$

where U = Displacement of section at x.

The element of dx length is strained as the displacement of its two faces are not the same.

$$\sigma_x = \text{normal stress in x direction on the section at x}$$

$$\sigma_x = \mathrm{E}\frac{\partial U}{\partial x}$$

Unbalanced force in the X direction will accelerate the element.

The unbalanced force on the strip is as follows:

$$F_U = \left(\sigma_x + \frac{\partial \sigma_x}{\partial x}dx\right)\left(A + \frac{\partial S}{\partial x}dx\right) - \sigma_x . A \tag{7.1}$$

Using the second law of Newton, we can write unbalanced force as follows:

$$F_U = mass \times acceleration = \rho.\ A.\ dx. \frac{\partial^2 U}{\partial t^2} \tag{7.2}$$

Equating Equations 7.1 and 7.2, we get

$$\left(\sigma_x + \frac{\partial \sigma_x}{\partial x} dx\right)\left(A + \frac{\partial A}{\partial x} dx\right) - \sigma_x .\ A = \rho .\ A .\ dx .\ \frac{\partial^2 U}{\partial t^2}$$

$$\sigma_x A + A\frac{\partial \sigma_x}{\partial x} dx + \sigma_x \frac{\partial A}{\partial x} dx + \left(\frac{\partial \sigma_x}{\partial x} dx\right)\left(\frac{\partial S}{\partial x} dx\right) - \sigma_x .\ A = \rho .\ A .\ dx .\ \frac{\partial^2 U}{\partial t^2}$$

$$\sigma_x \cancel{A} + A\frac{\partial \sigma_x}{\partial x} d\cancel{x} + \sigma_x \frac{\partial A}{\partial x} d\cancel{x} + \left(\frac{\partial \sigma_x}{\partial x} \cancel{d}x\right)\left(\frac{\partial S}{\partial \cancel{x}} dx\right) - \sigma_{\cancel{x}} .\ A = \rho .\ A .\ d\cancel{x} .\ \frac{\partial^2 U}{\partial t^2}$$

Simplifying we get,

$$A\frac{\partial \sigma_x}{\partial x} + \sigma_x \frac{\partial A}{\partial x} = \rho .\ A .\ \frac{\partial^2 U}{\partial t^2} \tag{7.3}$$

$$\text{Also, we know that } \sigma_x = \mathrm{E}\frac{\partial U}{\partial x} \tag{7.4}$$

Differentiating, we get

$$\frac{\partial \sigma_x}{\partial x} = \mathrm{E}\frac{\partial^2 U}{\partial x^2} \tag{7.5}$$

Using Equations 7.4 and 7.5 in Equation 7.3, we get,

$$A\left(\mathrm{E}\frac{\partial^2 U}{\partial x^2}\right) + \left(\mathrm{E}\frac{\partial U}{\partial x}\right)\frac{\partial A}{\partial x} = \rho .\ A .\ \frac{\partial^2 U}{\partial t^2}$$

$$A .\ E\frac{\partial^2 U}{\partial x^2} + E\frac{\partial U}{\partial x} .\ \frac{\partial A}{\partial x} = \rho .\ A .\ \frac{\partial^2 U}{\partial t^2} \tag{7.6}$$

Assume displacement as simple harmonic motion with angular frequency ω

$$U = A\cos(\omega t) + B\sin(\omega t)$$

$$Velocity = \dot{U} = \frac{\partial U}{\partial t} = -A\omega\sin(\omega t) + B\omega\cos(\omega t)$$

$$Acceleration = \ddot{U} = \frac{\partial^2 U}{\partial t^2} = -A\omega^2 cos\,(\omega t) - B\omega^2 \sin\,(\omega t)$$

$$Acceleration = \ddot{U} = -\omega^2 (A\ cos\,(\omega t) + B\sin\,(\omega t)) = -\omega^2 U \quad (7.7)$$

Using Equation 7.7 in Equation 7.6, we get

$$A.\,E\frac{\partial^2 U}{\partial x^2} + E\frac{\partial U}{\partial x}.\frac{\partial A}{\partial x} = \rho.\,A.\,(-\omega^2 U)$$

$$\frac{\partial^2 U}{\partial x^2} + \frac{1}{A}\frac{\partial U}{\partial x}.\frac{\partial A}{\partial x} + \frac{\rho}{E}\omega^2 U = 0$$

$$\frac{\partial^2 U}{\partial x^2} + \frac{1}{A}\frac{\partial U}{\partial x}.\frac{\partial A}{\partial x} + \frac{\omega^2}{c^2}U = 0 \quad (7.8)$$

where

$$c^2 = \frac{E}{\rho} \quad or \quad c = \sqrt{\frac{E}{\rho}} \quad (7.9)$$

where c is the velocity of sound in the material of the horn.

Case 1: Constant Uniform Cross Section
For a uniform cylindrical rod

$$\frac{\partial A}{\partial x} = 0$$

$$\frac{\partial^2 U}{\partial x^2} + \frac{1}{A}\frac{\partial U}{\partial x}.\cancel{\frac{\partial A}{\partial x}} + \frac{\omega^2}{c^2}U = 0$$

$$\frac{\partial^2 U}{\partial x^2} + \frac{\omega^2}{c^2}U = 0 \quad (7.10)$$

This is a second-order differential equation, and the solution is

$$U = A\cos\left(\frac{\omega}{c}x\right) + B\sin\left(\frac{\omega}{c}x\right) \tag{7.11}$$

Putting the first boundary condition of the Free–free end

$$when\ x = 0\ then\ \frac{\partial U}{\partial x} = 0 \tag{7.12}$$

Differentiating Equation 7.11 and

$$\frac{\partial U}{\partial x} = -\frac{A\omega}{c}\sin\left(\frac{\omega}{c}x\right) + \frac{B\omega}{c}\cos\left(\frac{\omega}{c}x\right) \tag{7.13}$$

Using Equation 7.12 in Equation 7.13, we get

$$-\frac{A\omega}{c}\text{Sin}\left(\frac{\omega}{c}\times 0\right) + \frac{B\omega}{c}\cos\left(\frac{\omega}{c}\times 0\right) = 0$$

$$-\frac{A\omega}{c}\text{Sin}\left(\frac{\omega}{c}\times 0\right) + \frac{B\omega}{c}\cos\left(\frac{\omega}{c}\times 0\right) = 0$$

$$therefore\ B = 0$$

$$U = A\cos\left(\frac{\omega}{c}x\right) \tag{7.14}$$

There is no amplification, but simply, the rod will vibrate.
Putting the second boundary condition of the Free–free end

$$when\ x = L,\ then\ \frac{\partial U}{\partial x} = 0$$

$$-\frac{A\omega}{c}\text{Sin}\left(\frac{\omega}{c}L\right) + \frac{B\omega}{c}\cancel{\cos}\left(\frac{\omega}{c}L\right) = 0 \quad (since\ B = 0)$$

$$-\frac{A\omega}{c}\text{Sin}\left(\frac{\omega}{c}L\right) = 0$$

Since A cannot be zero,

$$\text{Sin}\left(\frac{\omega}{c}L\right) = 0$$

$$\frac{\omega L}{c} = n\pi$$

$$L = \frac{n\pi C}{\omega} = \frac{n\pi C}{2\pi f} = \frac{nC}{2f}$$

$$also\ \lambda = \frac{C}{f}$$

$$L = \frac{n\lambda}{2} \tag{7.15}$$

To obtain peak vibration at the opposing ends of the rod, it is optimal for the length to correspond to half the wavelength denoted as $\frac{\lambda}{2}$. This configuration ensures equal amplitudes at both ends resulting in a gain factor of amplitude amplification set at 1. In ultrasonic applications, amplification of vibration amplitude from the transducer end to the desired level at the tool end is essential [15]. Hence, a horn is fashioned from a tapered rod adhering to a specific area variation law. The first fundamental mode of vibration of a uniform bar is shown in Figure 7.4.

Case 2: Constant Uniform Cross Section

In case of a horn with exponentially decaying cross sectional area, the differential Equation 7.8 can be solved by assuming

$$A = A_0 . e^{-\propto x} \tag{7.16}$$

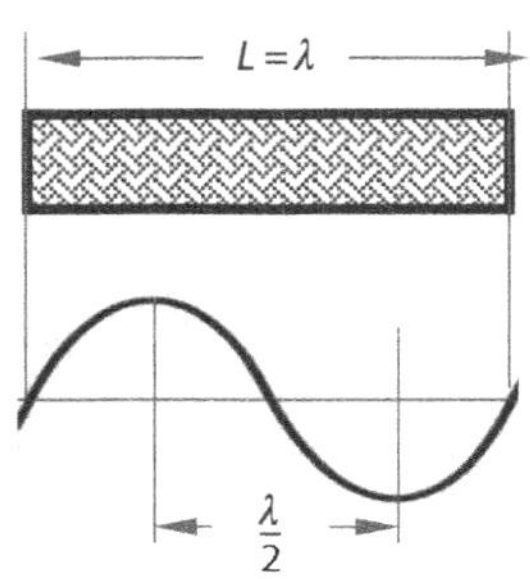

Figure 7.4 First fundamental mode of vibration of a uniform bar.

$$\frac{\partial A}{\partial x} = A_0(-\alpha)e^{-\propto x} \tag{7.17}$$

Using Equations 7.16 and 7.17 in Equation 7.8, we get,

$$\frac{\partial^2 U}{\partial x^2} + \frac{1}{A}\frac{\partial U}{\partial x}\frac{\partial A}{\partial x} + \frac{\omega^2}{c^2}U = 0$$

$$\frac{\partial^2 U}{\partial x^2} + \left(\frac{1}{A_0 e^{-\propto x}}\right)\frac{\partial U}{\partial x}(A_0(-\alpha)e^{-\propto x}) + \frac{\omega^2}{c^2}U = 0$$

$$\frac{\partial^2 U}{\partial x^2} - \propto \frac{\partial U}{\partial x} + \frac{\omega^2}{c^2}U = 0 \tag{7.18}$$

This is a linear differential equation of second order, and the solution can be written as

$$U = e^{sx} \tag{7.19}$$

$$\frac{\partial U}{\partial x} = se^{sx} \tag{7.20}$$

$$\frac{\partial^2 U}{\partial x^2} = s^2 e^{sx} \tag{7.21}$$

Using Equations 7.19, 7.20, and 7.21 in Equation 7.18

$$\frac{\partial^2 U}{\partial x^2} - \propto \frac{\partial U}{\partial x} + \frac{\omega^2}{c^2}U = 0$$

$$s^2 e^{sx} - \propto se^{sx} + \frac{\omega^2}{c^2}e^{sx} = 0$$

$$s^2 - \propto s + \frac{\omega^2}{c^2} = 0$$

$$s_{1,2} = \frac{\propto \pm \sqrt{\left(\propto^2 - 4\left(\frac{\omega}{c}\right)^2\right)}}{2}$$

$$s_{1,2} = \frac{\propto}{2} \pm \sqrt{\frac{\propto^2}{4} - \frac{\omega^2}{c^2}} \tag{7.22}$$

We will have vibration only when roots are complex (analogous to underdamped vibration as against overdamped or critically damped).

$$s_{1,2} = \frac{\propto}{2} \pm \sqrt{\frac{\omega^2}{c^2}\left(\frac{\propto^2}{4}\frac{c^2}{\omega^2} - 1\right)}$$

$$s_{1,2} = \frac{\propto}{2} \pm i\left(\frac{\omega}{c}\right)\sqrt{1 - \frac{\propto^2}{4}\frac{c^2}{\omega^2}}$$

$$U = Ae^{s_1 x} + Be^{s_2 x}$$

$$U = Ae^{\left\{\frac{\propto}{2} + i\left(\frac{\omega}{c}\right)\sqrt{1-\frac{\propto^2}{4}\frac{c^2}{\omega^2}}\right\}x} + Be^{\left\{\frac{\propto}{2} - i\left(\frac{\omega}{c}\right)\sqrt{1-\frac{\propto^2}{4}\frac{c^2}{\omega^2}}\right\}x}$$

$$U = Ae^{\left\{\frac{\propto}{2}\right\}x} e^{\left\{i\left(\frac{\omega}{c}\right)\sqrt{1-\frac{\propto^2}{4}\frac{c^2}{\omega^2}}\right\}x} + Be^{\left\{\frac{\propto}{2}\right\}x} e^{\left\{-i\left(\frac{\omega}{c}\right)\sqrt{1-\frac{\propto^2}{4}\frac{c^2}{\omega^2}}\right\}x}$$

$$U = e^{\left\{\frac{\propto}{2}\right\}x}\left[Ae^{\left\{i\left(\frac{\omega}{c}\right)\sqrt{1-\frac{\propto^2}{4}\frac{c^2}{\omega^2}}\right\}x} + Be^{\left\{-i\left(\frac{\omega}{c}\right)\sqrt{1-\frac{\propto^2}{4}\frac{c^2}{\omega^2}}\right\}x}\right]$$

Using the relation $e^{\pm ia} = \cos a + i\sin a$,

$$U = e^{\left\{\frac{\propto}{2}\right\}x}\left\{\begin{matrix}\left[A\cos\left(\frac{\omega x}{c}\sqrt{1-\frac{\propto^2}{4}\frac{c^2}{\omega^2}}\right) + iA\sin\left(\frac{\omega x}{c}\sqrt{1-\frac{\propto^2}{4}\frac{c^2}{\omega^2}}\right)\right] \\ +\left[B\cos\left(\frac{\omega x}{c}\sqrt{1-\frac{\propto^2}{4}\frac{c^2}{\omega^2}}\right) + iB\sin\left(\frac{\omega x}{c}\sqrt{1-\frac{\propto^2}{4}\frac{c^2}{\omega^2}}\right)\right]\end{matrix}\right\}$$

$$U = e^{\left\{\frac{\propto}{2}\right\}x}\left[(A+B)\cos\left(\frac{\omega x}{c}\sqrt{1-\frac{\propto^2}{4}\frac{c^2}{\omega^2}}\right) + i(A-B)\sin\left(\frac{\omega x}{c}\sqrt{1-\frac{\propto^2}{4}\frac{c^2}{\omega^2}}\right)\right]$$

$$U = e^{\left\{\frac{\propto}{2}\right\}x}[P\cos kx + Q\sin kx]$$

$$\text{where } P = (A+B) \text{ and } Q = i(A-B) \text{ and } k = \left(\frac{\omega}{c}\sqrt{1-\frac{\propto^2}{4}\frac{c^2}{\omega^2}}\right)$$

$$U = e^{\left\{\frac{\propto}{2}\right\}x}[P\cos(kx) + Q\sin(kx)] \tag{7.23}$$

Here, amplitude is proportional to $e^{\left\{\frac{\propto}{2}\right\}x}$. It is clear from the expression that the amplitude is rising exponentially with the length—this means that there will be an increase in the amplitude down the length of the horn.

Putting the first boundary condition of the Free–free end,

$$when\ x = 0\ then\ \frac{\partial U}{\partial x} = 0$$

$$U = e^{\left\{\frac{\propto}{2}\right\}x}[Pcos(kx) + Q\sin(kx)]$$

$$\frac{\partial U}{\partial x} = e^{\left\{\frac{\propto}{2}\right\}x}[-Pk\ sin(kx) + kQ\cos(kx)] + \left\{[Pcos(kx) + Q\sin(kx)]e^{\left\{\frac{\propto}{2}\right\}x}\frac{\propto}{2}\right\}$$

$$e^{\left\{\frac{\propto}{2}\right\}x}[-Pk\ sin(kx) + kQ\cos(kx)] + \left\{[Pcos(kx) + Q\sin(kx)]e^{\left\{\frac{\propto}{2}\right\}x}\frac{\propto}{2}\right\} = 0$$

$$e^{\left\{\frac{\propto}{2}\right\}0}\left[-Pk\ sin(k\times 0) + kQ\cos(k\times 0)\right] + \left\{[Pcos(k\times 0) + Q\sin(k\times 0)]e^{\left\{\frac{\propto}{2}\right\}0}\frac{\propto}{2}\right\} = 0 \tag{7.24}$$

$$\begin{aligned} kQ + P\frac{\propto}{2} &= 0 \\ kQ &= -P\frac{\propto}{2} \end{aligned} \tag{7.25}$$

Using Equation 7.25 in Equation 7.24, we get

$$\frac{\partial U}{\partial x} = e^{\left\{\frac{\propto}{2}\right\}x}\left[-Pk\ sin(kx) - P\frac{\propto}{2}\cos(kx)\right] + \left\{\left[Pcos(kx) - \frac{P\propto}{2k}\sin(kx)\right]\frac{\propto}{2}e^{\left\{\frac{\propto}{2}\right\}x}\right\} \tag{7.26}$$

$$\frac{\partial U}{\partial x} = e^{\left\{\frac{\propto}{2}\right\}x}\left[-Pk\ sin(kx) - P\frac{\propto}{2}\cos(kx)\right] + \left\{\left[P\frac{\propto}{2}cos(kx) - \frac{P\propto}{2k}\frac{\propto}{2}\sin(kx)\right]e^{\left\{\frac{\propto}{2}\right\}x}\right\} \tag{7.27}$$

$$\frac{\partial U}{\partial x} = e^{\left\{\frac{\propto}{2}\right\}x}\left[-Pk\, sin(kx) - \cancel{P\frac{\propto}{2}\cos(kx)} + \cancel{P\frac{\propto}{2}cos(kx)} - \frac{P\propto^2}{4k}\sin(kx)\right]$$

$$\frac{\partial U}{\partial x} = e^{\left\{\frac{\propto}{2}\right\}x}\left(-Pk - \frac{P\propto^2}{4k}\right)\sin(kx) \tag{7.28}$$

Putting the second boundary condition of the Free–free end in Equation 7.28

$$when\ x = L\ then\ \frac{\partial U}{\partial x} = 0$$

$$\frac{\partial U}{\partial x} = e^{\left\{\frac{\propto}{2}\right\}L}\left(-Pk - \frac{P\propto^2}{4k}\right)\sin(kL) = 0 \tag{7.29}$$

Equation 7.29 implies that

$$\sin(kL) = 0$$
$$kL = n\pi$$
$$n = 1\ for\ sine\ wave$$
$$kL = \pi$$

Using $k = \left(\frac{\omega}{c}\sqrt{1 - \frac{\propto^2}{4}\frac{c^2}{\omega^2}}\right)$ that we have used earlier

$$\left(\frac{\omega}{c}\sqrt{1 - \frac{\propto^2}{4}\frac{c^2}{\omega^2}}\right)L = \pi$$

$$\sqrt{1 - \frac{\propto^2}{4}\frac{c^2}{\omega^2}} = \frac{\pi c}{\omega L}$$

$$1 - \frac{\propto^2}{4}\frac{c^2}{\omega^2} = \left(\frac{\pi c}{\omega L}\right)^2 \tag{7.30}$$

We know that

$$A = A_0 . e^{-\propto x}$$

$$e^{\propto x} = \frac{A_0}{A}$$

Taking log on both sides

$$\ln\left(\frac{A_0}{A}\right) = \propto x$$

At $x = L$ (*boundary condition*)

$$\ln\left(\frac{A_0}{A}\right) = \propto L$$

$$\ln(N) = \propto L$$

$$\propto = \frac{\ln(N)}{L} \tag{7.31}$$

Using Equation 7.31 in Equation 7.30, we get,

$$1 - \left(\frac{\ln(N)c}{2\omega L}\right)^2 = \left(\frac{c\pi}{\omega L}\right)^2$$

$$\left(\frac{\ln(N)c}{2\omega L}\right)^2 + \left(\frac{c\pi}{\omega L}\right)^2 = 1$$

$$\left(\frac{\ln(N)c}{2(2\pi f)L}\right)^2 + \left(\frac{c\pi}{(2\pi f)L}\right)^2 = 1$$

$$\left(\frac{\ln(N)c}{4\pi f}\right)^2 \frac{1}{L^2} + \left(\frac{c}{2f}\right)^2 \frac{1}{L^2} = 1$$

$$\left(\frac{\ln(N)}{2\pi}\right)^2 \frac{1}{L^2}\left(\frac{c}{2f}\right)^2 + \left(\frac{c}{2f}\right)^2 \frac{1}{L^2} = 1$$

$$\left(\frac{\ln(N)c}{4\pi f}\right)^2 + \left(\frac{c}{2f}\right)^2 = L^2$$

$$L = \frac{c}{2f}\sqrt{1 + \left(\frac{\ln(N)c}{4\pi f}\right)^2 \left(\frac{2f}{c}\right)^2}$$

$$L = \frac{c}{2f}\sqrt{1 + \left(\frac{\ln(N)}{2\pi}\right)^2}$$

$$L = \textit{length of half wave horn} \tag{7.32}$$

Worked Examples:

Problem 1: There is a horn with a circular section, and its cross-section area is going exponentially from the transducer end to the tool end. Find out the frequency below which horn will not amplify.

So, the variation in cross section is

$$A = A_o e^{-\alpha x}$$

$$\frac{4\omega^2}{c^2} - \alpha^2 > 0$$

$$\frac{2\omega}{c} > \alpha$$

$$f_c = \frac{\propto \times c}{4\pi}$$

Below this, the horn will not amplify. In fact, the threshold frequency is kept even higher than this; say, it is never less than 1.5 times this value.

Problem 2: There is a horn, which works at 20 kHz, and its length is to be half the wavelength of the vibration. At the transducer end, the horn has a cross section of 8 cm × 8 cm, and from this cross section, two of the sides exponentially reduce to 4 mm. The other two sides remain at 8 cm only. If the velocity of sound in the medium is 4,800 m/s, find what should be the length of the horn.

$$\ln\left(\frac{A_o}{A_1}\right) = \ln(N) = In\left(\frac{80 \times 80}{80 \times 4}\right) = \ln(20) = 2.995732$$

$$L = \frac{c}{2f}\sqrt{1 + \left(\frac{\ln(N)}{2\pi}\right)^2}$$

$$L = \frac{c}{2f}\sqrt{1 + \frac{2.99573^2}{(2\pi)^2}} \approx 133mm$$

$$L = \frac{4.8\times10^3}{2\times20\times10^3}\sqrt{1+\frac{2.99573^2}{(2\pi)^2}} \approx 133mm$$

To check for the frequency,

$$\alpha = \frac{\ln(N)}{L} = \frac{2.995732}{0.133} = 22.524$$

Hence,

$$\frac{2\omega}{c} = \frac{4\pi\times20\times10^3}{4.8\times10^3} = 52.359\,,$$

which is much greater than the critical frequency.

The critical frequency is

$$f_c = \frac{\propto\times c}{4\pi} = \frac{22.524\times4.8\times10^3}{4\pi} = 8603.6Hz \approx 8.6kHz$$

20 kHz is more than double of that. Hence, the design is safe.

7.5 Process Parameters

The establishment of process parameters hinges on several factors as follows: workpiece material thickness, ambient material temperature, surface condition, and material properties. Consequently, any alterations to the alloy, surface finish, or weld geometry necessitate corresponding adjustments to these parameters. Within ultrasonic welding (USW), the adjustable parameters encompass power, clamping load, welding duration, and frequency.

- **Power:** The power required for ultrasonic spot welding varies widely ranging from a few watts for delicate wire welding to as much as 14,000–16,000 W for robust selection welding. Seam welds, on the other hand, typically operate within power levels ranging from 150 to 600 W. Increasing power augments the weld thickness capability while concurrently reducing cycle time.

- **Clamping Load:** The magnitude of clamping force in ultrasonic welds spans from under 200 g to several kilograms. As the contact area of the weld zone expands, so must the clamping force. For instance, wire welding applications necessitate light forces, whereas plate welding demands more substantial forces [12]. Insufficient force may lead to tip sticking, while excessive force unnecessarily escalates power requirements.
- **Welding Duration:** Ultrasonic spot welding typically transpires within 0.03 to 1.2 s per weld cycle. Prolonged welding times signal inadequate power usage. Inadequate power, resulting in lengthened cycles, heightens the risk of tip sticking, intensifies workpiece heating, and elevates the likelihood of surface marring at the welding tip location. Hence, maximizing power levels is desirable to minimize welding cycle time.

In ultrasonic seam welding, the parameter of time is governed by traverse rate. The traverse rate dictates the energy delivered per unit length of the weld, thereby influencing weld penetration. For instance, an aluminum foil that is 0.03-mm thick can be welded at a rate of 4 to 5 m/min, whereas a 0.15-mm-thick aluminum requires a slower rate of 0.3 to 0.5 m/min.

7.6 Process Capability

Due to the non-thermal nature of ultrasonic welding, a broad spectrum of metals and metal combinations can be effectively joined. Materials suitable for ultrasonic welding include aluminum, copper, gold, silver, platinum, their alloys, and various steels. However, certain refractory and high-strength metals, like molybdenum, tungsten, zirconium, tantalum, columbium, titanium, and nickel, pose challenges limiting their thickness to thin foils and reducing welding tip lifespan [13].

The welding thickness is constrained by the thickness of the top-joint member. Multiple layers of foil can be spot or seam welded depending on layer thickness and machine power. Remarkably, up to 30 layers of foil have been successfully welded in production applications. Aluminum sheets up to 2.5-cm thick can be welded, while refractories are typically limited to a maximum thickness of 0.4 to 1.2 mm. Ultrasonic welding can handle wires ranging from 0.03 to 0.8 mm in diameter, with the minimum reliable sheet metal thickness for welding being 0.003 to 0.005 mm.

When executing multiple spot welds on a single workpiece, there is a risk of fracturing previously made welds if the workpiece resonates with the machine's operating frequency. Every workpiece possesses natural resonant frequencies typically distinct from the operating system's frequencies. Resonance issues, if encountered, can be mitigated by adjusting workpiece dimensions, altering clamping locations or methods, or adding temporary weights to various spots on the workpiece.

Ultrasonically welded parts exhibit a rough surface where the welding tip and anvil contact the part. Compared to resistance welding, ultrasonic welds offer greater ductility since no melting occurs. Additionally, ultrasonic welding presents cost and quantity advantages due to simplified surface preparation requirements, lower energy consumption, and the ability to weld workpieces with vastly different compositions, properties, or thicknesses.

7.7 Applications of USW

a) Traditionally employed in the electronic, electrical, and defense industries, ultrasonic welding finds diverse applications. For instance, electronic appliances, such as switches, sensors, and data storage keys, are commonly crafted using this method.
b) In the medical field, ultrasonic welding plays a pivotal role in manufacturing various parts including filters, catheters, medical garments, and masks.
c) It is utilized for welding wires and containers for microcircuits, as well as for attaching wires and contacts to motors and switches.
d) Ultrasonic welding contributes to the hermetic sealing of volatile substances and the fabrication of nuclear fuel elements.
e) Continuous seam welding is employed to assemble components of corrugated heat exchangers using ultrasonic welding techniques.
f) Packaging applications benefit from ultrasonic welding enabling the fabrication of blister packs, pouches, tubes, storage containers, and carton spouts.

g) In the realm of energy storage, ultrasonic welding is instrumental in connecting anodes and cathodes, as well as interconnecting cells of lithium-ion batteries.
h) Threaded inserts, grub screws, or other components can be embedded in thermoplastics using ultrasonic welding methods [15].
i) Ultrasonic welding facilitates the melting and forming of a plastic stud to retain or lock another material, often dissimilar, in place.
j) This technique is adept at joining two thermoplastic parts without pre-formed holes or energy directors making it suitable for large or intricately shaped parts, as well as thermoformed or blow-molded parts lacking a joint.
k) Within the automotive sector, ultrasonic welding finds extensive use in fabricating headlamp parts, dashboards, buttons and switches, fuel filters, fluid vessels, seat-belt locks, electronic key fobs, lamp assemblies, and air ducts.

7.8 Advantages of USW

a) Surface preparation and cleanliness play a minor role.
b) Arc, gases, and filler metals do not generate defects.
c) Glass can be welded.
d) Minimal surface deformation results from USW.
e) Welding is feasible for very thin metals and plastics.
f) Dissimilar metals with significantly different melting points can be joined.
g) USW can join thin and thick sections.
h) Low-temperature requirements preserve material characteristics in the weld zone.
i) High-conductivity materials, like gold, silver, copper, and aluminum, are easily welded.
j) Various material combinations can be welded.
k) Power consumption is low.
l) Workpiece surfaces experience minimal deformation.
m) High-quality welds are achieved.
n) The process can be integrated into automated production lines.
o) Moderate operator skill suffices.

7.9 Disadvantages of USW

a) The capital investment required for ultrasonic machines is considerably higher compared to resistance welding.
b) Ultrasonic welding is limited to small and thin parts primarily for joining thin materials like sheet foil and wire. It is most effective with softer materials, such as aluminum, up to a maximum thickness of 3 mm and, for harder materials, up to 1 mm.
c) There is a tendency for the material being welded to adhere to the weld tip and anvil.
d) Resonance issues may lead to damage to parts.
e) Reciprocating loads from ultrasonic vibration may cause fatigue in workpieces and equipment components.

7.10 Conclusion

In the realm of materials joining, ultrasonic welding stands as a beacon of innovation, heralding a departure from conventional methods by accomplishing coalescence without the need for melting. As elucidated throughout this chapter, the ultrasonic welding process is a symphony of high-frequency vibratory energy and precise pressure application meticulously orchestrated to unite materials swiftly and securely. Its benefits are manifold ranging from the ability to join dissimilar metals with vastly different melting points to its compatibility with heat-sensitive components, a feat unattainable by traditional welding methods.

However, amid its myriad advantages, ultrasonic welding is not without its limitations. Its application is primarily suited for nonferrous metals and relatively lightweight components owing to the considerable energy required for vibrating heavier materials at ultrasonic frequencies. Despite these constraints, the technology's versatility shines through, finding applications across a spectrum of industries, from electronics to aerospace, where its efficiency and reliability are paramount.

The heart of ultrasonic welding lies in its equipment, meticulously designed to harness and channel high-frequency mechanical motion into the weld zone. From power systems converting electrical energy into high-frequency impulses to transducers transforming this energy into mechanical vibrations, each component plays a crucial role in the welding process. Coupling systems, ultrasonic horns, and welding tips further refine and amplify these vibrations ensuring uniform stress distribution and optimal energy transfer at the weld interface.

Designing ultrasonic horns, or velocity transformers, requires a delicate balance of material compatibility, resonance frequency, and amplitude amplification. These components are tailored to transmit and amplify ultrasonic vibrations efficiently facilitating localized melting and bonding of materials. The intricacies of horn design underscore the interdisciplinary nature of ultrasonic welding requiring collaboration between materials scientists, mechanical engineers, and welding experts.

Process parameters, such as power, clamping load, and welding duration, are meticulously adjusted to optimize weld quality and cycle time. The establishment of these parameters hinges on factors like workpiece material thickness, surface condition, and ambient temperature highlighting the intricacies involved in achieving consistent and reliable welds.

Ultrasonic welding's broad applicability is evidenced by its diverse range of applications, spanning from electronics to automotive and beyond. Whether it is joining wires in electronic appliances or fabricating structural components in aerospace, ultrasonic welding offers a cost-effective and efficient solution for material joining.

In conclusion, ultrasonic welding represents a paradigm shift in materials joining technology offering a potent blend of efficiency, reliability, and versatility. While challenges and limitations persist, the technology's continued advancement and widespread adoption underscore its significance in modern manufacturing. As researchers, engineers, and practitioners continue to refine and innovate, the future of ultrasonic welding holds promise for further breakthroughs and advancements in materials joining technology

References

1. Abbas, Z., Zhao, L., Deng, J., Wang, S., Wanlu, H., Advances in ultrasonic welding of lightweight alloys: a review. *High Temp. Mater. Processes (London)*, 42, 1–25, 2023.
2. Tsujino, J., Recent developments of ultrasonic welding, IEEE Ultrasonics Symposium. *Proc. An Int. Symposium*, Seattle, WA, USA, 2, 1051–1060, 1995.
3. Bhudolia, S.K., Gohel, G., Leong, K.F., Islam, A., Advances in ultrasonic welding of thermoplastic composites: a review. *Materials*, 13, 6, 1284, 2020.
4. Leon, M. and Hyung-Seop, S., Review of the advancements in aluminum and copper ultrasonic welding in electric vehicles and superconductor applications. *J. Mater. Process. Technol.*, 307, 117691, 2022.

5. Qiu, J., Zhang, G., Sakai, E., Liu, W., Zang, L., Thermal welding by the third phase between polymers: a review for ultrasonic weld technology developments. *Polymers*, 12, 759–765, 2020.
6. Gohel, G., Bhudolia, S., Kantipudi, J., Leong, K.F., Barsotti, R., Ultrasonic welding of novel Carbon/Elium with carbon/epoxy composites. *Compos. Commun.*, 22, 100463, 2020.
7. Senders, F., Van Beurden, M., Palardy, G., Villegas, I.F., Zero-flow: a novel approach to continuous ultrasonic welding of CF/PPS thermoplastic composite plates. *Adv. Manuf. Polym. Compos. Sci.*, 2, 3–4, 83–92, 2016.
8. Mughal, K.H., Ahmad, N., Hayat, N., Qureshi, M.A.M., Bugvi, S.A., Jamil, M.F., Khan, M.A., Design and performance evaluation of a novel ultrasonic welding sonotrode for Langevin transducer using finite element approach. *Int. J. Ind. Eng.Theory Appl. Pract.*, 30, 4, 971–985, 2023.
9. Huang, H., Chen, J., Feng, Z., Sun, X., Modal analysis of ultrasonic spot welding for lightweight metals joining. *Metals*, 13, 1735, 2023.
10. Zhao, T., Zhao, Q., *et al.*, Enhancing weld attributes in ultrasonic spot welding of carbon fibre-reinforced thermoplastic composites: Effect of sonotrode configurations and process control. *Compos. B: Eng.*, 211, 108648, 2021.
11. Nunes, R., Faes, K., Meester, S., De Waele, W., Kubit, A., Influence of welding parameters and surface preparation on thin copper–copper sheets welded by ultrasonic welding process. *Int. J. Adv. Manuf. Technol.*, 123, 1–16, 2022.
12. Chandra, N. and Rahman, M., Effect of machining parameter in ultrasonic vibration cutting. *Int. J. Mach. Tools Manuf.*, 48, 9, 965–974, 2007.
13. Chavhan, G.R. and Raipure, A.S., Recent advances in ultrasonic welding of dissimilar materials: a review. *Multi. J.*, 11, 2, 15–23, 2022.
14. Milewski, A., Kluk, P., Kardyś, W., Kogut, P., Modelling and designing of ultrasonic welding systems. *Arch. Acoust.*, 40, 1, 93–99, 2015.
15. Jagadeesha, T., *Unconventional Machining Processes*, I K International Publication House Pvt. Ltd, New Delhi, 2016.

8

Microstructural Control and Heat Treatment Effects in Explosion Welding

Sonika Chauhan*, Nisha Rani, Deepa Singh and Neeraj Bisht

Mechanical Engineering Department, College of Technology, Govind Ballabh Pant University of Agriculture and Technology, Pantnagar, Uttarakhand, India

Abstract

Explosion welding stands as a formidable solid-state welding technique that unites materials through high-velocity oblique collisions induced by powerful explosions. It was accidentally discovered by ordnance specialists during World War I. Later the method was formally reported in 1944 by Carl. Explosion welding can fuse different combinations of dissimilar and similar metals, which make it useful in a wide range of industries, including communications, electronics, and aerospace. The main advantage of explosion welding is its ability to rapidly fuse vast surface areas and produce strong metallurgical bonds by localized plastic deformation. Unlike conventional approaches, explosive welding works well for welding dissimilar metals because it removes oxide coatings from surfaces and forms high-pressure interatomic bonds. Plate separation, surface preparation, explosion load and detonation velocity are important factors that affect this process. Though it has many benefits, it has problems related to high-strength, difficult-to-weld metals because of the brittle intermetallic compound production tendency. Heat treatment helps improve the characteristics of materials that are fused together by lowering their hardness and brittleness. Recent research has focused on investigating the variation in microstructure at the weld interfaces and examining how annealing influences the growth of intermetallic compounds. This chapter offers a thorough analysis of the explosion welding process examining its metallurgical features, important process variables and intrinsic constraints.

Keywords: Explosion welding, micro-structure, heat treatment, dissimilar metal welding

**Corresponding author*: sonika.chauhan.me@gmail.com

Sandip Kunar and Gurudas Mandal (eds.) Advanced Welding Technologies, (155–176) © 2025 Scrivener Publishing LLC

8.1 Introduction

Explosion welding is distinct from conventional welding processes because of a greater range of possible metal combinations. Explosion welding is also characterized by its ability to join large surface areas rapidly, hence, creating strong bonds at the metallurgical interface through local plastic deformation. Explosion welding is simply a solid-state welding procedure where the components to be welded are accelerated into an oblique collision at a high velocity using high explosions. Ordnance specialists first discovered the method accidentally as they were unaware of welding procedure [1]. Carl found that two thin brass discs used to position explosive charge were welded together, and the first study on explosive welding was published in 1944 [2]. Subsequently, Philipchuk [3] reported that part of the aluminum U-channel was welded with steel die throughout an explosively formed process. From these unusual origins, the explosion welding technique has evolved to the point that many applications now acknowledge it as a useful joining method. By using explosion welding procedures, over 260 possible combinations of dissimilar and similar metals have been welded till date [4]. This process is being employed in many different industries, such as the electronics and communication sectors [5] and the construction of massive metal clads [6]. The titanium–steel transition joint created explosively for the Apollo spacecraft stands as one of the most renowned examples of an explosion welding. Another common application is the explosion coating used on sections of sandwich coinage currently in circulation in the United States. Explosion welding has been the focus of several notable studies [1, 4, 7–12]. Due to its remarkably brief processing time, explosion welding produces bonding interfaces with distinct microstructural characteristics. This chapter offers a thorough analysis of the explosion welding process examining its metallurgical features, important process variables, and intrinsic constraints. It also explores the microstructural properties at the welding interface and presents underwater explosive welding as an advanced technique for working with challenging materials. Explosion welding is among the best methods to join two metal plates with non-identical or comparable metallurgical characteristics that cannot be bonded by any other traditional manufacturing process. This method of solid-state welding creates a welded surface by carefully setting off an explosion. With this technique, a large surface area may be joined since the high energy density of the explosion may be distributed over the welding area. The metal plates are joined at an interior site that causes significant local plastic deformation at the contact where metallurgical bonding naturally occurs strengthening the bond even more than that of the parent metals [13]. The high-velocity oblique collision, which aids in jetting process, forms a connection. The process of jetting is crucial in joining metal plates. It cleans up non-metallic films, like oxide, from the top surfaces of both metals that obstruct the formation of metallurgical bonds and facilitates a seamless welding process. The bonding mechanism of an explosion welding operates on the principle of momentum conservation [14]. To form interatomic bonds, sufficient pressure must be applied over a sustained period. The duration available for bonding is dictated by the collision point velocity (v_c). Key process parameters influencing the outcome includes surface preparation, explosive loading, plate separation and detonation velocity (v_d) [15]. Optimal adjustment of these variables enables precise control over the bond quality and strength of the welded joint. Both the collision velocity (v_c) and plate velocity (v_p) should remain below the speed of sound [16]. Selecting a flyer plate with the lowest density and tensile yield strength is crucial [17]. It is also recommended to avoid explosives with detonation velocities exceeding 120% of the metal's sonic velocity [18]. The pressure impact at the collision point ranges from 0.5 to 6 GPa [19]. It was also found that the standoff distance has a noticeable impact on bond strength; however, the thickness of the explosive charge has a comparatively minor effect [20]. Due to the ultra-rapid nature of collision during explosion welding, diffusion is negligible. Explosion welding finds extensive application in manufacturing of metal-laminated composite materials composed of dissimilar metals [21]. The work by Carl in 1944 acknowledged the potential of producing metallic welds using explosively propelled plates [22], and in 1962, the first patent in this field marked a significant milestone in commercialization and practical implementation of explosion welding techniques [23].

8.1.1 Types of Explosion Welding

There are two types of explosion welding processes distinguished by the relative positioning of the parent plate and flyer plate. The two configurations used in explosion welding are oblique and parallel. Figure 8.1(a) illustrates the oblique configuration, which is used when the plates are thin and tiny. Figure 8.1(b) illustrates the parallel layout that is employed for larger plates. To get the best bonding possible, this approach involves careful cleaning and gentle polishing of the plates before beginning the welding process. The flyer plate is positioned above the base plate at a specific distance, or stand-off distance, to sustain the flyer plate's weight and survive the explosion force, while the base plate stays grounded. A buffer sheet is

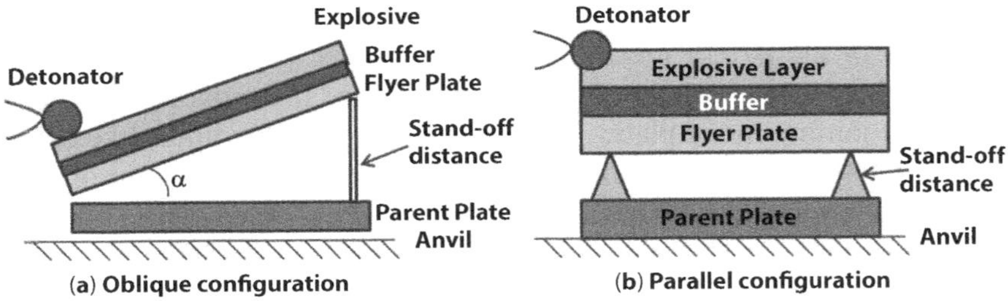

Figure 8.1 Different configurations of explosion welding.

placed on the upper surface of the flyer plate to protect it from the explosion's shockwave. The explosive charge is positioned above the flyer plate inside a box-shaped structure to efficiently direct and focus the explosive energy [24].

8.2 Factors Affecting Explosion Welding

There are several critical factors like stand-off distance, formation of jet, detonation velocity, etc., that affect the weld properties and quality. The quality and strength of the welded joint can be effectively controlled by regulating these parameters.

8.2.1 Stand-Off Distance

It is one of the essential factors because it facilitates metal acceleration and allows the weld to achieve the desired angle. For thin components (up to 6.5 mm or 0.26 in.), the stand-off distance is twice the flyer plate thickness; for heavier components (up to 13 mm or 0.5 in.), the stand-off distance is equal to the thickness of the flyer plate [25]. An increase in the stand-off distance was the reason behind the bond's decreased melted area fraction [26]. The stand-off distance has a direct bearing on the shear strength (Rs) and peel strength (Ro). As the stand-off distance increases, the bond's melt area percentage decreases. The steel begins to heavily harden after the 2-h mark is exceeded between the plates [26].

8.2.2 Formation of Jet

The development of the jet is determined by the collision angle, which is a phenomenon that happens during the weld formation process. In principle, the impact point's velocity remains subsonic causing jetting to occur. In practice, however, the magnitude of pressure should be high enough to surpass the material's dynamic elastic limit for welding to take place. The jet formation is symmetric if it originates from both plate jets and asymmetric if it originates from just one [15]. Higher yield strength of flyer plate causes low amplitude wave, which prevents the plate from deforming properly. When the explosion velocity is higher, the jetting materials become confined and result in a shortage of binding area [27].

8.2.3 Detonation Velocity

The detonation velocity of the explosion should ideally remain within approximately 120% of the material's sonic velocity [22]. Because of harmful effects of shock rarefaction, explosions with a broad range of low detonation velocities are more suitable [20]. It is recommended to employ explosions with detonation velocity ranging from 2,000 to 6,000 m/s. Higher detonation velocity has the potential to result in a comparatively elevated impact angle β, which in turn generates significant turbulence at the interface [28]. The explosion type is indicated by the detonation velocity, which is directly proportional to the explosion density and is derived empirically for nitroguanidine explosion [29].

8.2.4 Angle of Collision (β)

The angle of collision is the measure of jet formation. The change in the collision angle directly influences the transition from wavy to smooth interface morphology at the welded front [30]. This transition depends on the stand-off distance typically ranging from 5 to 250 and the controlled detonation velocity guidelines [23].

Through proper adjustment of variables such as the flyer plate's collision velocity, collision angle and horizontal point collision velocity, researchers have investigated the controlling parameters in explosion welding. As the detonation pressure impacted the base plate at a specified collision angle (β), the flyer plate accelerated at a high velocity (v_p) as shown in Figure 8.2 [31]. The flyer plate velocity (v_p) depends on the horizontal point collision velocity (v_c) and collision angle (β) as per Equation 8.1 [12].

$$v_p = 2v_c \sin\left(\frac{\beta}{2}\right) \tag{8.1}$$

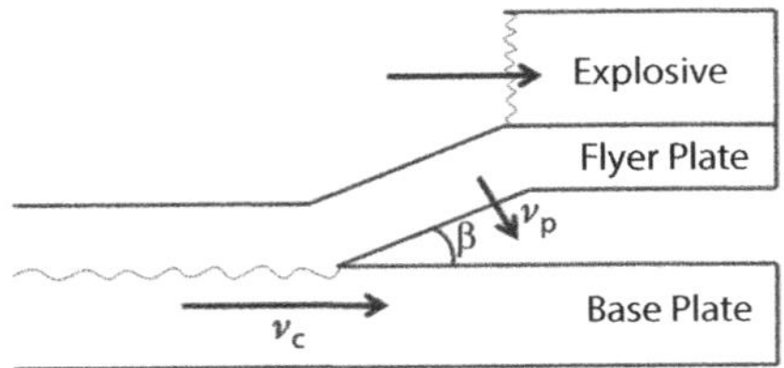

Figure 8.2 Schematic diagram of explosion welding.

In the context of explosive welding, the minimal bending angle correlates closely with the collision angle (β), which is nearly proportional to the velocity of the flyer plate (v_p). Crucial to successful welding are these fundamental elements: the collision angle (β) and the horizontal collision point velocity (v_c). This combination defines the "explosive welding window," illustrated in Figure 8.3 [32, 33]. Parameters, such as metal jet formation, maximal impact velocity, critical impact pressure, and the transition from wavy to smooth interfaces are determined within this window. Figure 8.3 shows the impact of collision velocity and dynamic collision angle on the weldability in explosive welding processes [34, 35]. The lower limit of explosive welding, as indicated by the material hardness curve [36], is defined by Curve 1. Curve 2 establishes the upper limits of explosive welding condition, considering factors like maximum welding energy,

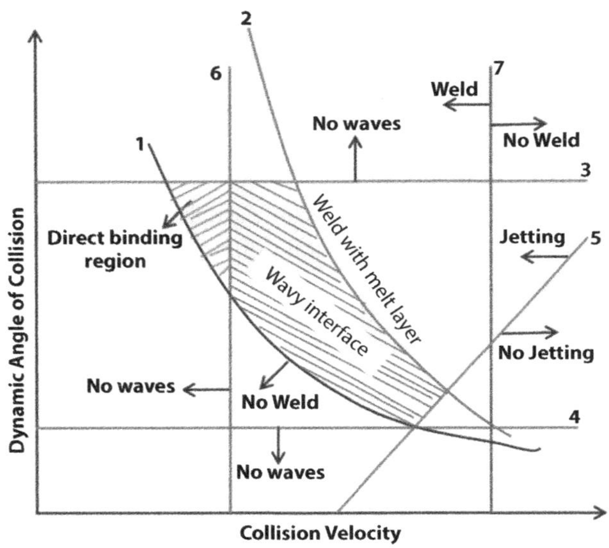

Figure 8.3 Explosive welding window.

excessive melting and solidification times, and interface reflection waves [32]. Curves 3 and 4 show the conditions of jet formation detailing the interface jet conditions at both ends of the spectrum [37]. Curves 5 and 6 represent the velocity at the collision location where wavy patterns form. Curve 7 shows the maximum velocity at the material impact point, typically 1.2–1.5 times the velocity of its sound propagation [38].

The weldability window in explosive metal welding is shown by the enclosed areas in Figure 8.3. The red enclosed area signifies the region of wavy interface, while the blue area signifies direct bonding region. Bonding strength in directly bonded dissimilar metals is mentioned to be superior by some researchers, while some acknowledge the strength of explosively welded joints [39]. Although explosive welding window resolves many challenges encountered in welding dissimilar metals, metals with high strength, hardness, susceptibility to oxidation, or significant differences in physical and chemical properties still face difficulties in determining optimal explosion welding parameters [40]. The smooth-wavy boundary is a sign of "laminar–turbulent transition" seen in fluid mechanics and can be compared to Reynolds number. Equation 8.2 presents the Reynolds number (R_e) for explosion welding [41, 42], where v_t represents the transition velocity, H_{VF} and H_{VB} denotes the Vickers hardness of flyer and parent plates, respectively, and ρ_F and ρ_B are the densities of metals involved. This equation effectively captures the transition when the Reynolds number R_e is set to 10.6.

$$R_e = \frac{(\rho_F + \rho_B)V_t^2}{2(H_{VF} + H_{VB})} \tag{8.2}$$

8.3 Microstructures at the Junction of Materials Fused by Explosion Welding

Ferrous and titanium alloys are generally metallurgically incompatible in the conventional fusion welding techniques due to the formation of intermetallic compound, such as FeTi, Fe_2Ti, and other secondary phases at the joint interface, resulting in embrittlement [43]. Explosion welding is a solution to such cases and allows bonding for two or more similar as well as dissimilar metals [44, 45]. Explosion welding is distinct from other traditional welding techniques because it can fuse metals and alloys with different mechanical properties, melting points, and coefficients of thermal expansion [46–50]. The joining of dissimilar metals, like titanium and steel in many critical applications, including petrochemical, nuclear fusion, and aerospace sectors has attracted a lot of researchers toward explosion welding [51, 52]. The joining interfaces during explosion welding have distinctive periodic wave-like pattern in the microstructure imparting anisotropic mechanical properties to the welded parts. Therefore, it is important to understand the effects of microstructural anisotropy with an emphasis on the mechanical characteristics of joining. This knowledge is essential for developing applications in the automobile sector. When explosion welding is used, the high velocity of the base and flying plates as a result of the blast causes high temperature and pressure zones to form at the joining interface. This collision triggers a series of additional interactions and metallurgical phenomena resulting in the formation of wave-like joining interfaces [53]. The physical and mechanical properties like microstructure, micro-hardness. and chemical constitution of the interfaces of the explosion welding materials have been explored by many researchers. For instance, the features of microstructure for an explosion welding interface bonded between Ti and steel have been discussed by Song *et al.* [54]. They also discovered that intermetallic compounds, which tend to disintegrate significantly under shear or bending stresses, are typically brittle and play a crucial role in determining material's overall mechanical characteristics. The TA2 (pure Ti alloy) and Q235 (mild steel) explosion welding composite plate has applications in aerospace, petrochemical industry, machinery industry, electronics industry, and nuclear industry [44, 45, 52, 55, 56]. There are various degrees of melting layers in the micro areas of the explosion welding interface between titanium and steel. This high-brittleness melting zone often leads to cracking of the welded joints. When carrying out the explosion welding process, the flyer plate was TA2, while the base plate was Q235. Finally, TA2/Q235 composites for explosion welding were made by employing the parallel placement method as presented in Figures 8.4(a), (b). In another investigation, Zhou *et al.* [57] conducted explosion welding of pure Ti alloy (TA2) and mild steel (Q235). Powder explosion was used to propel the flyer plate and impact the base plate at an angle (β). The collision angle β is determined using Equation 8.3 [58].

$$\beta = \left(\sqrt{\frac{k+1}{k+1}} - 1 \right) \times \frac{\pi}{2} \times \frac{r}{r + 2.71 + 0.184 t_e / S} \tag{8.3}$$

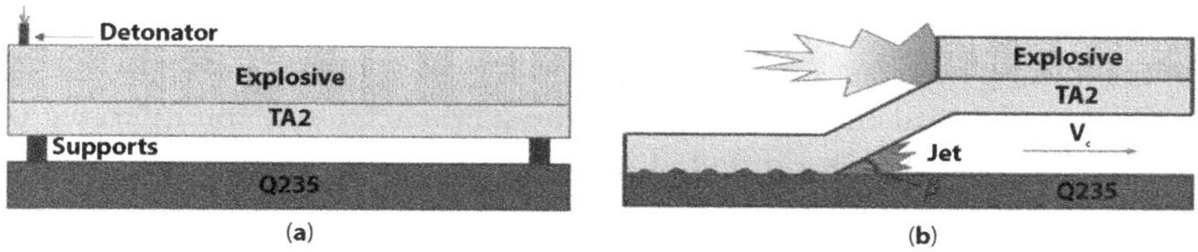

Figure 8.4 TA2/Q235 explosion welding: (a) setup of apparatus and (b) overview of welding process [59].

where t_e = the explosion thickness, k = a constant with values ranging between 1.81 and 2.6 depending on t_e, r = the loading ratio with a value of 1.5, and s = the stand-off distance between the flyer plate and the base plate.

The morphology and microstructure analysis of TA2/Q235 welding interfaces was conducted using optical microscopy (OM). Metallographic samples underwent polishing and etching processes (5% nitric acid + 95% ethyl alcohol) prior to microstructure analysis with a LEICA DMI 3000 M OM. In addition, mechanical characterization of nanocomposites was carried out post-shear testing on fracture surfaces using a ZEISS Gemini 300 field emission scanning electron microscope (SEM) fitted with energy dispersive spectroscopy (EDS) to examine microstructural alterations and the distribution of chemical composition. According to the study, a wavy pattern with increasing thickness in the direction of detonation was seen at the welding interface. Consequently, shear properties of interfaces were found to be dependent on this detonation direction. The longitudinal section (A zone) provided insights into the stripping surface morphology, while the transverse section (B zone), as depicted in Figure 8.5 [59], allowed for metallographic examination of the joining interface's microstructure. Moreover, samples with an "S" shape at 0°, 45°, and 90° to the

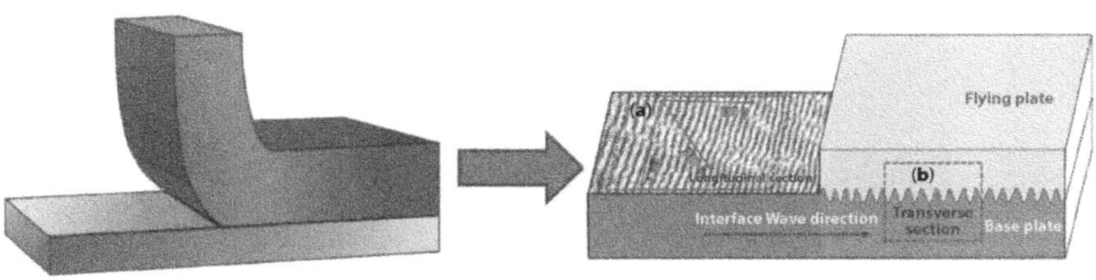

Figure 8.5 The microstructural anisotropy of two zones of TA2/Q235 explosion-welded interfaces [59].

detonation direction were prepared to investigate the anisotropic behavior of the joining interfaces in zone A.

The mechanical microstructure of the base metal and the interface of TA2/Q235 welding are presented in Figure 8.6. The bonding interface is periodic with wave form with a maximum height of bonding interface equal to 300 μm and width of the wave form equal to 800 μm. Bataev *et al.* [60] claimed that this wavy interface occurs due to shift in pressure dissemination at the point of collision arising from self-excited oscillations inside the bond region. In addition, the grains on the Q235 side were enormously stretched within the detonation direction, and this suggested that the material at the interface was severely deformed throughout the explosion welding. In the dynamics of hydromechanics within the melting pool, intermetallic compounds formed during the initial stages of melting are transported to the wavefront through a vortex mass flow mechanism. The temperature gradient (ΔT) is the main cause of the structural inhomogeneity and the wave detonation leading to the intermixing of the liquid and partially solid states of the material along with the intricate severe plastic deformations in the joint zone. Moreover, some defects, like microcracks, voids, and intermetallic compound, mostly develop at the center of the bonding interface. More importantly, a number of intermetallic phases exist especially between dissimilar metals [61]. This is due to the extreme temperatures and pressures generated at the collision point when the flyer plate strikes the base plate resulting in rapid thermal escalation and the

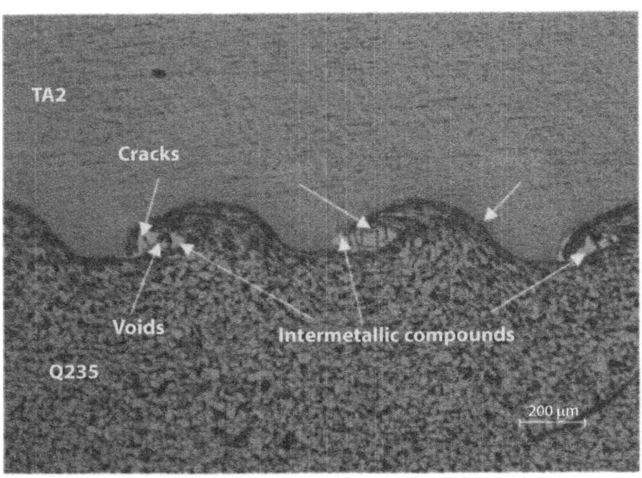

Figure 8.6 Welding interface microstructure [60].

creation of molten zones at the bonding interface. Bataev *et al.* [60] also noted that the explosion welding can potentially achieve a heating rate of 109 K/s, which is similar to the observations of Crossland [62] and Kacar [63]. However, Gloc [64] reported that due to inadequate dissipation of heat, voids and intermetallic compounds are usually found at the welding interface. This results in crack initiation because of poor toughness and presence of internal stresses at welding interface.

8.4 Effect of Heat Treatment on Weld Interfaces

Heat treatment is an important method to enhance the mechanical properties of joints welded by explosion welding. Annealing at various temperatures improves their performance and durability. Honarpisheh *et al.* [65] claims that annealing results in thickening of intermetallic compounds and decreases hardness across the weld interface. Bataev [66] observed that more deformed plates initially produced Al_3Ti intermetallic compounds at a high growth rate. The layers were less distinct over time, but the micro-cracks revealed brittle fractures within IMC. Furthermore, Foadian *et al.* [67] also employed explosion welding to join the Al sheets to CP–Ti and also investigated on the growth of intermetallic compound after annealing for 1–260 h at 630°C. Similar intermetallic compounds were found to form there in the study of Bataev [66], which were $TiAl_3$. To substantiate the observations, Gibbs standard free energy and interface energy calculations were conducted, which confirmed the stability of the $TiAl_3$ phase.

To elucidate the characteristics of grain distribution and corresponding evolution mechanism, a schematic diagram illustrates the microstructural evolution (Figure 8.7). Plastic deformation induced by explosion welding results in an elongated pearlite and ferrite grains along the detonation direction near the interface, shown in Figure 8.7(a). These elongated grains on the steel side transform into ferrite within a temperature range of 723 to 923 K, as illustrated in Figure 8.7(b). The ferrite grains continue to grow without undergoing phase transformation in the α-Fe and γ-Fe two-phase field (Figure 8.7(c)). Moreover, carbon elements concentrate in the mixing zone leading to the formation of intermetallic compounds. At 1,173 K, both ferrite and pearlite emerge near the interface, and the grain size was observed to be increasing with temperature, as depicted in Figure 8.7(d). Carbon elements also accumulated at the interface resulting in the formation of a thin NbC layer, whose thickness escalates with annealing temperature. Figure 8.7(e) shows the appearance of Nb2C carbide at 1,273 K between the NbC layer and the Nb side.

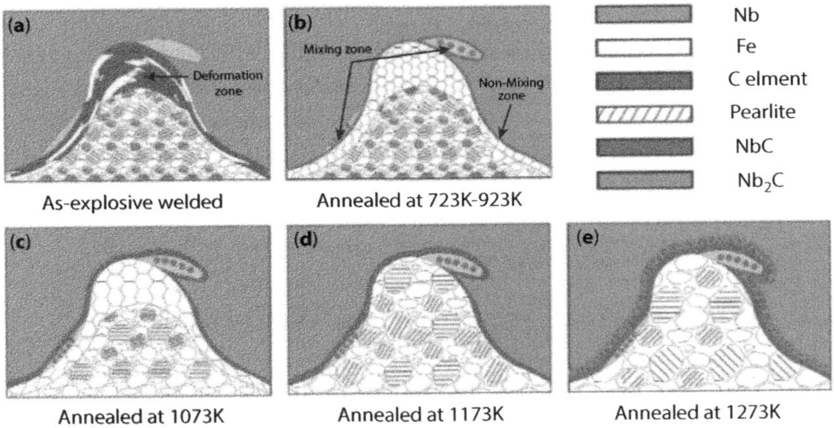

Figure 8.7 Microstructural evolution: (a) elongated pearlite and Fe on steel side; (b) recrystallized Fe and concentration of C-elements at interface; (c) growth of Fe grains and presence of NbC near interface; (d) normalizing structure and presence of NbC; (e) presence of NbC and Nb_2C [68].

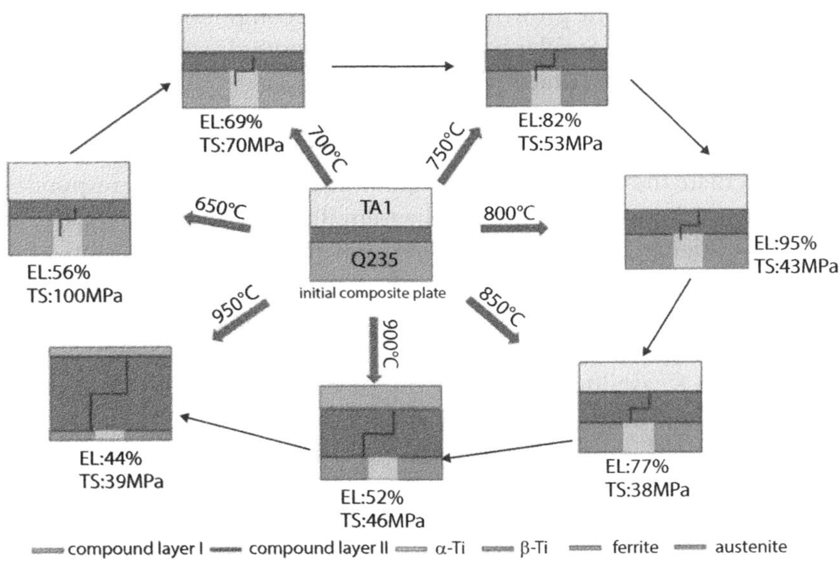

Figure 8.8 Variation in mechanical properties of TA1/Q235 composite plate with temperature [68].

Wang *et al.* [69] found that the thickness of the interfacial layer significantly affects both elongation and tensile strength. Figure 8.8 shows tensile specimens tested at various temperatures. Elongation increased with rising temperatures up to 800°C, while tensile strength decreased with temperature till 900°C. Therefore, it was concluded that, the mechanical properties are remarkably dependent upon the thickness of the interfacial layer, which increases as the temperature increases showing the critical impact of thermal conditions on material performance.

8.5 Underwater Explosion Welding

Underwater explosion welding (UWEW) process creates metallurgical bonds that are incredibly strong because of explosion-driven water pressure [70]. In UWEW, a high explosion is created at an inclined angle to reduce horizontal impact point velocity of plates, a critical factor in the success of the welding process [71]. Underwater explosion welding is particularly suited for welding materials that are difficult to weld because of less thickness of the material, high degree of hardness of surface, and fragile quality or low formability. These characteristics create immense difficulties in welding by standard conventional explosion welding technique. Air explosion welding produces significant force and heat, which thin foils cannot bear and, hence, leads to the destruction of thin foils. Underwater explosion welding solves this problem by utilizing water pressure as the method of pressure transmission, which is important in providing pressure at the appropriate locations. Water being 800 times denser than air is much less incompressible compared to air. This property prevents the explosion forces from producing high temperature on the water as at the same time the water can follow the direction the foil material gives it and, hence, not tamper it. Moreover, shock waves in water are stronger compared to shock waves that travel in the air because the peak pressure of underwater shock waves is much higher. This is possible since the thin plate can reach high velocities over small distances, thereby making it possible to produce the high energy required to weld. The presence of water above and below the welding specimen minimizes distortion and ensures integrity of composite laminate [72]. Currently, underwater explosion welding has successfully joined dissimilar foil materials such as ceramic and aluminum foils [73], amorphous and stainless steel foils, metallic glass and stainless steel foils [74], tungsten foil and copper plates [75], magnesium foil and steel, and NiTi shape memory alloy foil and copper [72]. These combinations have traditionally posed significant challenges for engineers highlighting the pioneering nature of underwater explosion welding in overcoming these complexities.

Studies done by Hokamoto *et al.* [73] explain the theoretical performance and the actual experiences of using UWEW, especially for thin plates over large areas. This technique involves underwater firing of shockwaves likely from an explosion to fuse thin metal plates together through the use of an insignificantly small amount of explosions making it efficient and precise. Mori *et al.* [74] carried out some welding tests where thin magnesium plates were weld onto different substrates such as aluminum (Al), titanium (Ti), and mild steel (MS) plates. According to the study, formation of smooth interfaces occurs in the combination of magnesium–titanium and magnesium–mild steel indicating minimum kinetic energy dissipation prior to interface creation. The welding of relatively thick metallic glass was largely focused, while the formation of metallic glass films through welding remains unexplored. Underwater explosion welding, as shown in Figure 8.9, was used by Hokamoto *et al.* [76] to fuse

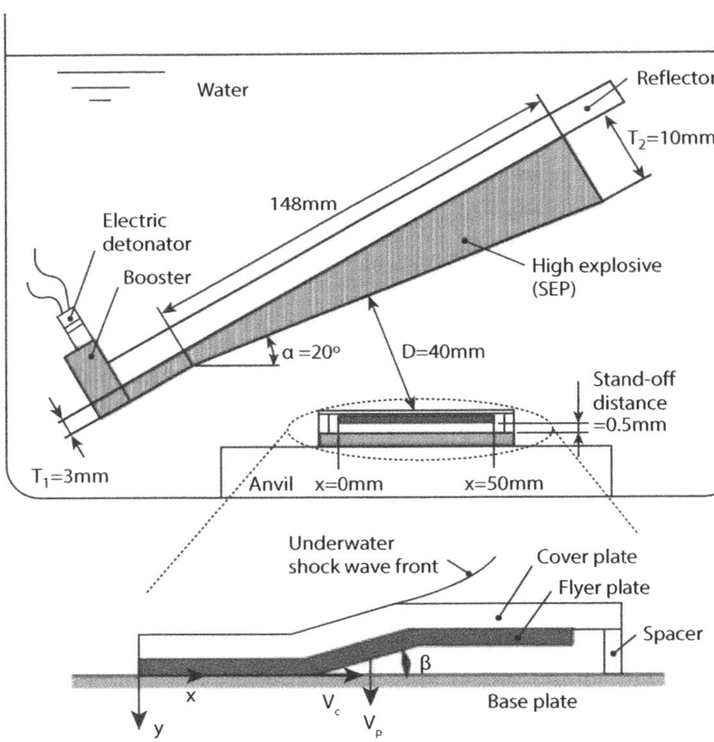

Figure 8.9 Setup for underwater explosion welding [74].

non-equilibrium thin metallic glass foils using a variable thickness explosion assembly. Through the successful interfacial welding of thin foils, the weld site displayed unique wave-like patterns. Hanliang *et al.* [77] successfully fused composite Zr-based bulk metallic glass (BMG) with pure aluminum in an innovative investigation demonstrating the adaptability and effectiveness of advanced welding technique. The results highlighted the efficiency of submerged explosion welding in creating BMG/Al composite-based zinc. Zr-based BMG and Al plates were flawlessly welded using this method resulting in a significant transition layer that is around 7-μm thick and free of visible flaws. Notably, a hardness enhancement was observed near the joining interface of Zr-based BMG and Al. In detailed investigations, scanning electron microscopy (SEM) and energy dispersive spectroscopy (EDS) revealed localized controlled melting within specific regions of metal jet. This targeted melting phenomenon gives rise to different metallurgical characteristics that significantly influence mechanical properties and overall integrity of welded joint. These findings underscore the critical importance of meticulous process control in achieving desired welding outcomes. These findings underscore the critical importance of meticulous process control in achieving desired welding outcomes.

8.6 Applications of Explosion Welding

Owing to the outstanding versatility and capacity to create strong metallurgical bonds between dissimilar and similar metals, explosion welding is a commonly employed technique in many different industries. The materials that were difficult to be welded by other welding techniques can be easily welded by explosion welding resulting in strong, long-lasting, durable joints that satisfy the specifications of modern engineering and manufacturing applications. Explosion welding is important in the following applications:

1. **Marine Engineering:** Steel and aluminum alloys are widely used in shipbuilding and offshore structures because of their exceptional corrosion resistance and structural strength. Explosion welding is useful in joining these dissimilar metals wherever needed in the challenging marine conditions [78].
2. **Cylindrical Components:** Explosion welding also plays an important role in the structural components where high mechanical strength and reliability are needed like

concentric cylinders, tube-to-tube plates, and tubular transition joints, which are critical for pressure vessels and pipelines [79].

3. **Heat Exchangers:** Explosion welding is an important technology in improving thermal effectiveness and reliability in heat exchangers, including sheet and clad tube sheets in various industries such as power generation, chemical processing, HVAC systems, etc. [80].
4. **High-Strength Welds:** Explosion welding develops high-strength joints, which are required in complex engineering projects of defense and aerospace applications [81].
5. **Feedwater Heater Tubes:** Energy efficiency and operational reliability of feedwater heater tubes in power plants increases when the joints are made using explosion welding technique [24].
6. **Dissimilar Metal Joining:** Allows challenging connections between steel and aluminum overcoming the limitations of standard welding and improving structural integrity in the construction of infrastructure, industrial machines, and automobiles [82].
7. **Similar Metal Bonds:** Explosion welding is also used to create strong joints between metals that are similar to one another, such as aluminum and aluminum, which are important for lightweight structural parts used in aerospace, automotive, and marine sectors [30].
8. **Lightweight Materials:** Explosion welding technique is also helpful in the production of lightweight and high-strength metal sheets. High strength-to-weight ratio is important for achieving high performance and fuel efficiency in the aerospace and automotive industries [83].

Explosion welding's versatility and effectiveness in creating durable bonds across these diverse applications underscore its critical role in modern industrial processes. It addresses specific challenges related to material compatibility, mechanical strength, and durability making it a preferred choice in various high-performance and demanding applications.

8.7 Conclusions

Explosion welding is a formidable solid-state welding technique that unites materials through high-velocity oblique collisions induced by powerful explosions. It is widely used in nuclear, petrochemical, automotive, aerospace, defense, marine engineering, and pressure vessel industries. Its advantages include the ability to join dissimilar metals quickly, rapid bonding with full coverage and low heat impact, and strong bond formation at the interface through high-pressure plastic deformation. However, there are some challenges, such as the need for optimizing the process parameters of explosion wielding, like detonation velocity and collision angle, as well as ensuring microstructural control to manage intermetallic compound formation and uniform bonding across large areas, alongside critical surface preparation to avoid contaminants affecting bonding or causing metallurgical defects. Though there are a number of challenges in explosion welding, it has proved to be important for various applications because it can create strong metallurgical bonds without melting. As research will progress to refine and optimize the process parameters; explosion welding will evolve as a valuable solution for demanding engineering applications across various sectors.

References

1. Bahrani, A.S., Black, T.J., Crossland, B., The mechanics of wave formation in explosion welding. *Proc. R. Soc. Lond. Ser. A-Math. Phys. Sci.*, *296*, 1445, 123–136, 1967.
2. Carl, L.R., Brass welds, made by detonation impulse. *Met. Prog.*, 46, 102–103, 1944.
3. Philipchuk, V., Creative Manufacturing Seminar, in: *Paper SP65-100. Amer. Soc. Tool Manuf. Eng*, New York, 1965.
4. Linse, V.D., Wittman, R.H., Carlson, R.J., Explosion Bonding. Battelle Memorial Institute. *Def. Met. Inf. Center*, 225, 1–9, 1967.
5. Carpenter, S.H. and Witman, R.H., University of Denver Research Institute Report on Theory & Application of Explosion Welding, vol. 67 pp. 11, Ezra, AA, 1973.
6. Edwards, J.L., Cranston, B.H., Krauss, G., Miniature Explosion Bonding With a Primary Explosion, in: *Metallurgical Effects at High Strain Rates*, Boston, MA: Springer US, pp. 559–569, 1973.
7. Rinehart, J.S. and Pearson, J., *Explosion working of metals*, The Macmillan Company, New York; A Pergamon Press Book, 1963.

8. Nolan, M.C., Gadberry, H.M., Loser, J.B., Sneegas, E.C., *High-Velocity: A Survey. NASA SP-5062*, p. 5062, NASA Special Publication, Washington, D.C., 1967.
9. Crossland, B. and Williams, J.D., Explosion welding. *Metallur. Rev.*, *15*, 1, 79–100, 1970.
10. Carpenter, S.H. and Wittman, R.H., Explosion Welding: A review. *Univ. Denver Res. Inst. Rep. Theory Appl. Explos. Weld.* 67, 941–959, 1972.
11. Ezra, A.A., *Principles and Practice of Explosion Metal-working*, p. 173, Industrial Newspapers Limited, London, 1973.
12. Crossland., B., *Explosion Welding and Its Application*, Oxford University Press, Oxford, UK, 1982.
13. Crossland, B., Review of the present state-of-the-art in explosion welding. *Met. Technol.*, *3*, 1, 8–20, 1976.
14. Patterson, R.A., Fundamentals of explosion welding, vol. 6, pp. 160–164, ASM International, 1993, https://doi.org/10.31399/asm.hb.v06.9781627081733.
15. Mousavi, A.A. and Al-Hassani, S.T.S., Numerical and experimental studies of the mechanism of the wavy interface formations in explosion/impact welding. *J. Mechan. Phys. Sol.*, *53*, 11, 2501–2528, 2005.
16. Lysak, V.I. and Kuzmin, S.V., Lower boundary in metal explosion welding. Evolution of ideas. *J. Mater. Proc. Technol.*, *212*, 1, 150–156, 2012.
17. Karolczuk, A., Kluger, K., Derda, S., Prażmowski, M., Paul, H., Influence of impact velocity on the residual stress, tensile strength and structural properties of an explosively welded composite plate. *Materials*, *13*, 12, 2686, 2020.
18. Philipchuk, V. and Bois, F.L.R., *Explosion welding*, US Patent 3024526, 1962.
19. Crossland, B., Development of exlosive welding and its application in engineering. *Met. Mater.*, *5*, 12, 402–413, 1971.
20. Patterson, R.A., Explosion bonding, in: *Welding: Theory and practice*, 1990.
21. Carpenter, S., Wittman, R.H., Carlson, R.J., Relationships of explosion welding parameters to material properties and geometry factors, in: *Proc first int-conf of the center for high energy forming*, University of Denver, p. 124, 1967.
22. Merriman, C., The fundamentals of explosion welding. *Weld. J.*, *85*, 7, 27–29, 2006.
23. Akbari-Mousavi, S.A.A., Barrett, L.M., Al-Hassani, S.T.S., Explosion welding of metal plates. *J. Mater. Process. Techno.*, *202*, 1–3, 224–239, 2008.
24. Sherpa, B.B., Pal, D.K.B., Batra, U., Mishra, G.C., Singh, B.B., Study of the explosion welding process and applications, in: *Advances in applied physical and chemical sciences-a sustainable approach*, pp. 33–39, 2014.
25. Wolf, H., The Energy Balance of Plane and Cylindrical Explosion Cladding, in: *Proc. Int. Symposium on Intense Dynamic Loading and its Effects*, Pergamon Press, 1988.
26. Prażmowski, M. and Paul, H., The effect of stand-off distance on the structure and properties of zirconium-carbon steel bimetal produced by explosion welding. *Arch. Metall. Mater.*, *57*, 1201–1210, 2012.

27. Raghu, N., Rai, S.K., Kumar, A., Jayakumar, T., Kasiviswanathan, K.V., Raj, B., Characterization Explosive Weld Interfaces, 205, 20–22, 2004.
28. Findik, F., Yilmaz, R., Somyurek, T., The effect of heat treatment on the microstructure and microhardness of the explosion welding. *Sci. Res. Essays*, 6, 19, 4141–4151, 8 September, 2011.
29. Groschopp, J., Heyne, V., Hofmann, B., Explosively clad titanium steel composite. *Weld. Int.*, *1*, 9, 879–883, 1987.
30. Grignon, F., Benson, D., Vecchio, K.S., Meyers, M.A., Explosion welding of aluminum to aluminum: analysis, computations and experiments. *Int. J. Impac. Eng.*, *30*, 10, 1333–1351, 2004.
31. Hokamoto, K., *Explosion, Shock-Wave and High-Strain-Rate Phenomena of Advanced Materials*, Elsevier, Netherlands, 2021, https://www.google.co.in/books/edition/Explosion_Shock_Wave_and_High_Strain_Rat/5n0eEAAAQBAJ?hl=en&gbpv=0.
32. Wittman, R.H., The influence of collision parameters of the strength and microstructure of an explosion welded aluminium alloy, in: *Proc. 2nd Int. Sym. on Use of an Explosion Energy in Manufacturing Metallic Materials*, pp. 153–168, 1973.
33. Deribas, A.A., Simonov, V.A., Zakcharenko, I.D., Investigation of explosion welding parameters for arbitrary combinations of metals and alloys, in: *Proc. 5th int. Conf. On high energy rate fabrication*, vol. 4, University of Denver Denver, CO, pp. 1–4, 1975.
34. Baoxiang, R., Gang, T., Peng, W., Changxing, D., Study on weldability window and interface morphology of steel tube and tungsten alloy rod welded by explosion welding. *Int. J. Refrac. Met. Hard Mater.*, *84*, 105005, 2019.
35. Saravanan, S. and Raghukandan, K., Weldability windows for explosion cladding of dissimilar metals. *Adv. Mater. Res.*, *445*, 729–734, 2012.
36. Carvalho, G.H. S. F. L., Galvao, I., Mendes, R., Leal, R.M., Loureiro, A., Weldability of aluminium-copper in explosion welding. *Int. J. Adv. Manufac. Technol.*, *103*, 3211–3221, 2019.
37. Yu, Y., Xu, P., Li, K., Zhao, J., Dai, Y., Ma, H., Yang, M., Study on modalities and microdefects of Al-Cu bimetallic tube by underwater explosion cladding. *Arch. Civ. Mech. Eng.*, *19*, 4, 1390–1398, 2019.
38. Linse, V.D. and Lalwaney, N.S., Explosion welding. *JOM*, *36*, 5, 62–65, 1984.
39. Changgen, S., Yefa, T., Jin, H., Peng, L., Behavior and mechanism of the bonding interface of 0Cr18Ni9/16MnR by explosion welding. *China Weld.*, *19*, 4, 59–62, 2010.
40. Wu, X., Shi, C., Feng, K., Gao, L., Li, W., Qian, K., Experimental and numerical approach to titanium-aluminum explosion welding. *Mater. Res. Expr.*, *8*, 9, 096503, 2021.
41. Cowan, G.R. and Holtzman, A.H., Flow configurations in colliding plates: explosion bonding. *J. Appl. Phys.*, *34*, 4, 928–939, 1963.

42. Cowan, G.R., Bergmann, O.R., Holtzman, A.H., Mechanism of bond zone wave formation in explosion-clad metals. *Metall. Mater. Trans. B*, *2*, 3145–3155, 1971.
43. Prasanthi, T.N., C Sudha, R., Saroja, S., Explosion cladding and post-weld heat treatment of mild steel and titanium. *Mater. Des.*, *93*, 180–193, 2016.
44. Gulenc, B., Investigation of interface properties and weldability of aluminum and copper plates by explosion welding method. *Mater. Des.*, *29*, 1, 275–278, 2008.
45. Findik, F., Recent developments in explosion welding. *Mater. Des.*, *32*, 3, 1081–1093, 2011.
46. Kundu, S., Ghosh, M., Laik, A., Bhanumurthy, K., Kale, G.B., Chatterjee, S., Diffusion bonding of commercially pure titanium to 304 stainless steel using copper interlayer. *Mater. Sci. Eng. A*, *407*, 1-2, 154–160, 2005.
47. Gao, X.L., Zhang, L.J., Liu, J., Zhang, J.X., Porosity and microstructure in pulsed Nd: YAG laser welded Ti6Al4V sheet. *J. Mater. Proc. Technol.*, *214*, 7, 1316–1325, 2014.
48. Geng, H., Mao, J., Zhang, X., Li, G., Cui, J., Strain rate sensitivity of Al-Fe magnetic pulse welds. *J. Mater. Proc. Technol.*, *262*, 1–10, 2018.
49. Liu, J.G., Cai, W.C., Liu, L., Han, J.T., Liu, J., Investigation of interfacial structure and mechanical properties of titanium clad steel sheets prepared by a brazing-rolling process. *Mater. Sci. Eng.: A*, *703*, 386–398, 2017.
50. Zeng, X.Y., Wang, Y.X., Li, X.Q., Li, X.J., Zhao, T.J., Effects of gaseous media on interfacial microstructure and mechanical properties of titanium/steel explosion welded composite plate. *Fusion Eng. Des.*, *148*, 111292, 2019.
51. Kundu, S. and Chatterjee, S., Diffusion bonding between commercially pure titanium and micro-duplex stainless steel. *Mater. Sci. Eng.: A*, *480*, 1–2, 316–322, 2008.
52. Hao, X., Dong, H., Xia, Y., Li, P., Microstructure and mechanical properties of laser welded TC4 titanium alloy/304 stainless steel joint with (CoCrFeNi) 100-xCux high-entropy alloy interlayer. *J. Alloys Compd.*, *803*, 649–657, 2019.
53. Zhou, B.B., Zhou, C.Y., Chang, L., Yu, X.C., Ye, C., Zhang, B.J., Investigation on fatigue crack growth behavior of Zr702/TA2/Q345R explosion welding composite plate with a through-wall crack. *Compos. Struct.*, *236*, 111845, 2020.
54. Song, J., Kostka, A., Veehmayer, M., Raabe, D., Hierarchical microstructure of explosion joints: Example of titanium to steel cladding. *Mater. Sci. Eng.: A*, *528*, 6, 2641–2647, 2011.
55. Borchers, C., Lenz, M., Deutges, M., Klein, H., Gärtner, F., Hammerschmidt, M., Kreye, H., Microstructure and mechanical properties of medium-carbon steel bonded on low-carbon steel by explosion welding. *Mater. Des.*, *89*, 369–376, 2016.

56. Xia, Y., Dong, H., Hao, X., Li, P., Li, S., Vacuum brazing of Ti6Al4V alloy to 316L stainless steel using a Ti-Cu-based amorphous filler metal. *J. Mater. Process. Technol.*, *269*, 35–44, 2019.
57. Zhou, Q., Lu, H., Lan, X., Guo, Y., Liu, K., Huang, G., Jia, B., Chen, P., Anisotropy in microstructure and shear properties of TA2/Q235 explosion welding interfaces. *J. Mater. Res. Technol.*, 25, 6472–6491, 2023.
58. Yang, M., Ma, H., Shen, Z., Huang, Z., Tian, Q., Tian, J., Dissimilar material welding of tantalum foil and Q235 steel plate using improved explosion welding technique. *Mater. Des.*, *186*, 108348, 2020.
59. Zhou, Q., Liu, R., Ran, C., Fan, K., Xie, J., Chen, P., Effect of microstructure on mechanical properties of titanium-steel explosion welding interface. *Mater. Sci. Eng. A*, *830*, 142260, 2022.
60. Bataev, I.A., Lazurenko, D.V., Tanaka, S., Hokamoto, K., Bataev, A.A., Guo, Y., Jorge Jr., A.M., High cooling rates and metastable phases at the interfaces of explosively welded materials. *Actamater.*, *135*, 277–289, 2017.
61. Chu, Q., Zhang, M., Li, J., Yan, C., Experimental and numerical investigation of microstructure and mechanical behavior of titanium/steel interfaces prepared by explosion welding. *Mater. Sci. Eng.: A*, *689*, 323–331, 2017.
62. Crossland, B., *Explosion Welding of Metals and Its Application*, Clarendon Press, Oxford, UK, 1982.
63. Kacar, R. and Acarer, M., An investigation on the explosion cladding of 316L stainless steel-din-P355GH steel. *J. Mater. Process. Technol.*, *152*, 1, 91–96, 2004.
64. Gloc, M., Wachowski, M., Plocinski, T., Kurzydlowski, K.J., Microstructural and microanalysis investigations of bond titanium grade1/low alloy steel st52-3N obtained by explosion welding. *J. Alloy Compd.*, *671*, 446–451, 2016.
65. Honarpisheh, M., Asemabadi, M., Sedighi, M., Investigation of annealing treatment on the interfacial properties of explosion-welded Al/Cu/Al multilayer. *Mater. Des.*, *37*, 122–127, 2012.
66. Bataev, I.A., Bataev, A.A., Mali, V.I., Pavliukova, D.V., Structural and mechanical properties of metallic–intermetallic laminate composites produced by explosion welding and annealing. *Mater. Des.*, *35*, 225–234, 2012.
67. Foadian, F., Soltanieh, M., Adeli, M., Etminanbakhsh, M., A study on the formation of intermetallics during the heat treatment of explosively welded Al-Ti multilayers. *Metallur. Mater. Trans. A*, *45*, 1823–1832, 2014.
68. Wang, H., Li, X., Wang, Y., Wang, J., Microstructural evolution and interfacial properties of explosively welded Nb/steel composite plate during post-heat treatment. *J. Mater. Res. Technol.*, *25*, 7376–7388, 2023.
69. Wang, C., Han, J., Zhang, C., Su, Y., Zhang, S., Zhang, Z., Wang, T., Temperature dependence of interface characterization and mechanical properties of TA1/Q235 composite fabricated by explosion welding. *Adv. Eng. Mater.*, *25*, 11, 2201759, 2023.

70. Hokamoto, K., Ujimoto, Y., Fujita, M., Basic characteristics of the explosion welding technique using underwater shock wave and its possibilities. *Mater. Trans.*, *45*, 9, 2897–2901, 2004.
71. Mori, A., Tamaru, K., Hokamoto, K., Fujita, M., Underwater explosion welding, discussion based on weldable window, in: *AIP Conference Proceedings*, vol. 845, American Institute of Physics, pp. 1543–1546, 2006.
72. Sun, W., Li, X.J., Hokamoto, K., Numerical simulation of underwater explosion welding process, in: *Materials Science Forum*, vol. 767, pp. 120–125, Trans Tech Publications Ltd., 2014.
73. Hokamoto, K., Fujita, M., Shimokawa, H., Okugawa, H., A new method for explosion welding of Al/ZrO2 joint using regulated underwater shock wave. *J. Mater. Process. Technol.*, *85*, 1–3, 175–179, 1999.
74. Hokamoto, K., Nakata, K., Mori, A., Tsuda, S., Tsumura, T., Inoue, A., Dissimilar material welding of rapidly solidified foil and stainless steel plate using underwater explosion welding technique. *J. Alloys Compd.*, *472*, 1–2, 507–511, 2009.
75. Manikandan, P., Lee, J.O., Mizumachi, K., Mori, A., Raghukandan, K., Hokamoto, K., Underwater explosion welding of thin tungsten foils and copper. *J. Nucl. Mater.*, *418*, 1–3, 281–285, 2011.
76. Mori, A., Hokamoto, K., Fujita, M., Underwater Explosion Welding of Thin Magnesium Plate onto some Metal Plates, in: *Materials Science Forum*, vol. 566, pp. 303–308, Trans Tech Publications Ltd., 2008.
77. Hanliang, L., Ning, L., Xiaojie, L., Xin, S., Tao, S., Zhanguo, M., Joining of Zr60Ti17Cu12Ni11 bulk metallic glass and aluminum 1060 by underwater explosion welding method. *J. Manufac. Processes*, *45*, 115–122, 2019.
78. McKenney, C.R. and Banker, J.G., Explosion-bonded metals for marine structural applications. *Mar. Technol. SNAME News*, *8*, 03, 285–292, 1971.
79. Buijs, N.W., Explosion welding of metals in a vacuum environment. *Stainless Steel World*, *3*, 1–4, 2010.
80. Palmer, T.A., Elmer, J.W., Brasher, D., Butler, D., Riddle, R., Development of an explosion welding process for producing high-strength welds between niobium and 6061-T651 aluminum. *Weld. J. Newyork*, *85*, 11, 252, 2006.
81. Sherpa, B.B., Pal, D.K.B., Batra, U., Mishra, G.C., Singh, B.B., Study of the explosion welding process and applications, in: *Adv. Appl. Phys. Chem. Sci.-Sustain. Approach*, pp. 33–39, 2014.
82. Zlobin, B.S., Explosion welding of steel with aluminum. *Combust. Explos. Shock Waves*, *38*, 3, 374–377, 2002.
83. Matsui, Y., Otsuka, M., Hinata, T., Carton, E., Itoh, S., Explosion welding of light weight metal sheets, in: *8th International LS-DYNA Conference*, pp. 8–59, Detroit, 2004.

9

Hybrid Welding Processes

S. Ajithkumar*, B. Arulmurugan, D. Balaji, V. Manoj Mohan Prasath and S. Idhayaraja

Department of Mechanical Engineering, KPR Institute of Engineering and Technology, Coimbatore, Tamil Nadu, India

Abstract

Hybrid welding is a specially developed manufacturing process where two or more forms of welding are combined to derive maximum benefits from each of their characteristics. The most common form of hybrid welding uses the combination of laser with arc welding. This chapter addresses the current status, progress, and applications of hybrid laser arc welding (HLAW) technology. It is the simultaneous use of laser and arc welding techniques that gives faster welding speeds compared to when one uses either technology. The laser provides a deep penetration, and the arc welding fills up the junction properly to provide strength and excellent weld. The concentrated thermal energy provided by the laser drastically reduces the overall extent of the heat-affected zone, which consequently results in an enormous decrease in thermal distortion and residual stresses at the welded regions. This process produces welds with very high mechanical properties and lower possibilities of defects like porosity and cracking. HLAW finds massive applications across various industries. The automotive industry applies it to make lightweight highly strength-to-weight-proportioned components. Shipbuilding industries use this process in joining heavy plates without the distortion of members. The aerospace industries apply it in making highly critical structural components. It is also used in the construction and energy industries in the construction of strongly base metal structures and pipelines.

***Keywords*:** Hybrid laser arc welding, laser welding, arc welding, process parameters, applications

**Corresponding author*: ersajithkumarbe@gmail.com

Sandip Kunar and Gurudas Mandal (eds.) Advanced Welding Technologies, (177–196) © 2025 Scrivener Publishing LLC

9.1 Introduction

Hybrid laser arc welding (HLAW) unites the advantages of laser and arc welding in a weld pool. Thus, it realizes the cream of features from both methods. In view of such cooperation, the deepest and thinnest welds are achieved with improved production rates [1]. Laser welding has one key advantage related to its ability to produce very accurate welds using a directed beam. Such a beam produces high penetration but a very small width. The accuracy of laser welding enables faster welding speeds than most other processes because the laser quickly melts and fuses the material along the path of the intended weld [2]. Because of its concentrated shape, the laser beam transfers less heat to the pieces being welded producing less thermal distortion. This reduction in distortion is essential where the rigidity and dimension accuracy of the welded parts must be maintained, especially where high tolerances are specified in service applications [3]. However, laser welding equipment is very expensive and usually has poor electrical performance.

Laser welding does not allow for bridging of gaps and has problems where the processes involve very bright metals like copper, aluminum, and gold [4]. On the other hand, arc welding is effective in connecting spots, offers excellent electrical conductivity, and can weld metals that reflect light strongly. The ability of the welding process to bridge gaps is due to the proper utilization of filler materials and their ability to adapt the electric arc. This makes it accommodate joint fit-up variations and produce strong continuous welds [5, 6]. Arc welding methods easily transfer electrical energy into heat through the electric arc creating low-energy wastage and higher productivity. The high electrical efficiency permits fast heat generation, hence, handy for melting base and filler materials. This creates practical energy usage and restricted waste [7]. The possibility of precision in heat control offered by arc welding provides an ideal for welding highly reflective metals. It provides flexibility and quick contact with the metal surface producing perfect welding. It solves the problems induced by high reflectivity [8].

Compared with a laser welding instrument of the same ability, an arc welding machine is relatively cheap. However, the lower energy density of former welding processes results in lower welding speeds and higher heat input that possibly causes thermal distortion of the welded products [9]. HLAW solves these problems by integrating the advantages of arc and laser welding processes. Yu *et al.* [10] found that the hybridization of the arc and laser welding processes in the same weld pool reinforces the positive features of each process while controlling its drawbacks. The laser has smooth, deep penetration and quick welding, while the arc fills gaps and interacts with a variety of materials. Consequently, HLAW is very suitable for a large number of welding operations because it is productive in terms of quality, flexibility, and affordability.

9.2 Hybrid Laser–Arc Welding Process Development

As shown in Figure 9.1, HLAW combines a laser beam and an electric arc in two different ways. The first process, also referred to as laser-assisted arc welding, utilizes a low-power laser beam that heats the metallic parts before going ahead to conduct the actual arc welding process. In this case, the laser simply serves to preheat the parts before they are well ready enough for welding [11]. This makes the arc welding process easy and effects slight improvements on both the welding speed and penetration. This confirms that the materials have been perfectly prepared, thus ensuring the welding procedures are smooth and very effective. The second process uses a

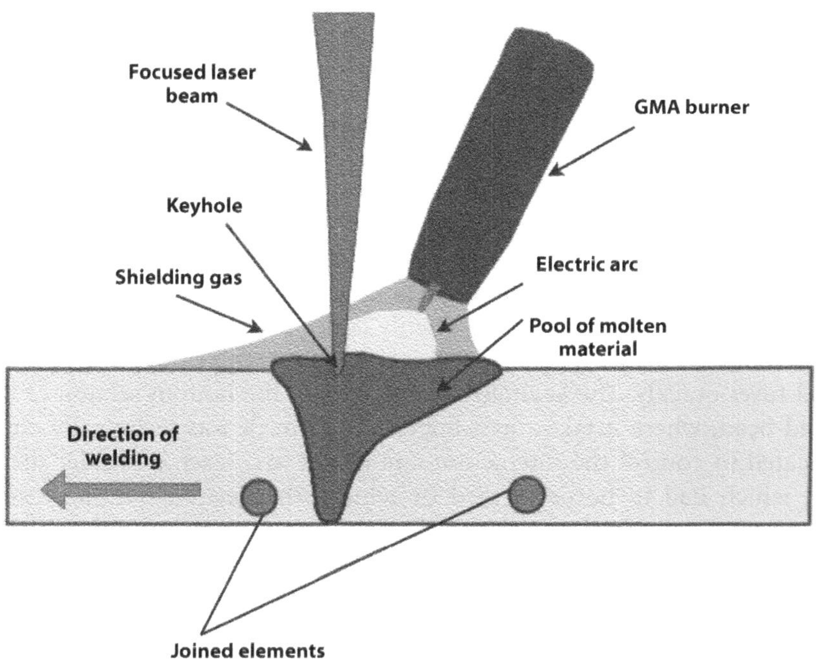

Figure 9.1 Hybrid laser arc welding.

high-intensity laser beam to effect deep penetration welding; provision for a keyhole can be provided at the joint. This process, also called arc-augmented laser welding or hybrid laser-arc welding [12], combines the high energy concentration of a laser beam with optimum energy use from an electric arc. This process provides precise and concentrated welding at high speeds using the deep penetration capability of the laser. The arc portion plays a key role in improving gap bridging and weld pool stabilization simultaneously so as to realize effective welding with highly reflective materials [13].

HLAW is characterized by dynamic interaction between three primary constituents: an electric arc, laser radiation, and a droplet of filler. Here, the influence of several factors defines the nature of weld bead form, dimensions, and durability and, hence, requires complex interactions [14]. None of the elements can be easily replaced since each plays a vital role: the electric arc provides heat and initiates the melting of base material. Simultaneously, laser irradiation delivers additional energy into the weld pool enhancing penetration and affecting the entire weld geometry [15]. The droplet of filler plays a key role in developing the weld and filling up the gaps between the two materials being joined. Ultimately, the complex interaction between these three components controls the formation and features of the weld bead in HLAW and gives it a very sophisticated and effective welding technique.

The weld bead produced in HLAW contains clearly defined fusion zones, and every area affected is distinguished by different reasons. The electric arc focuses on the top area of the weld bead and generates heat, thus initiating the melting of the base material. This zone typically takes a semi-hemisphere shape with a cup-like appearance, thereby revealing the contribution of the arc in the welding process. In contrast, the region affected by laser irradiation in the weld bead is significantly small [16]. The intensity of a laser beam is very highly concentrated, which significantly penetrates the material creating a "keyhole" effect where the material melts and fuses quickly. The keyhole effect occurs in the bottom section of the weld bead, where it takes on a finger-like form. It was technically complicated to control the interaction among the arc, laser, and filler droplet, which had to be performed to achieve the desired welding result. There are many advantages of the HLAW process compared to traditional welding processes coupled with some associated limitations that must be taken into consideration. A very important advantage is that it can realize greater depth of penetration, faster welding rates, and higher productivity compared to the traditional methods of laser welding and arc welding [17]. This increased efficiency results from the integration of laser and arc welding techniques, both of which realize faster and deeper welds adhering to quality. Liu *et al.* [18] observed that there are only a few welding passes in a hybrid process, which generates smaller weld seams and less heat-affected zones improving the quality of the weld and reducing distortion. Moreover, excellent gap-bridging ability has been proven to ensure good joining efficiency under difficult conditions with gaps between the materials to be welded.

More so, HLAW improves the uniformity of the weld seam and the sidewall fusion, together increasing the weld component's integrity. The other benefit includes, in some cases, reduced capital costs compared to using a high-power laser only. Better cost effectiveness of the setup is also achieved by the elimination of an increase in the laser power, thus eliminating the need for an expensive high-power laser through an additional heat source or an electric discharge [19]. Even though the aforementioned advantages are offered by this technique, HLAW has its own limits. The most glaring disadvantage of the hybrid welding process remains the high initial investment that is required for equipment and infrastructure. Adequate alignment and positioning of the components are necessary for a process to have the best possible results at each stage of the welding process. It makes the process even more cumbersome. In addition, the process parameters of HLAW should be controlled with the assistance of experienced operators or by employing the latest technologies. Many energy sources have safety problems and require more security measures and techniques to achieve safety for workers without any catastrophes.

9.3 Hybrid Laser Arc Welding Methods

Several HLAW procedures have been proposed to fully use the potentials of both the arc and laser welding processes. The most used laser sources in the devices are Nd and CO_2 lasers. CO_2 lasers are used because of their high continuous wave power, which varies from some watts up to more than 60 kW. The high-power output makes the CO_2 lasers suitable for use as a primary heat source for any welding process. Contrastingly, the maximum continuous wave power for Nd lasers is normally located in the few kilowatt region; these lasers are solid state. Nd lasers have much lower power compared to CO_2 lasers; however, they have narrower wavelengths of 1.064 mm and better beam uniformity [20]. It is in the nature of Nd lasers to be used in welding very reflective metals, such as stainless steel, aluminum, copper, and gold.

Other alternative laser options to HLAW, such as the disk and fiber lasers, have relatively loopholes compared to the traditional CO_2 and Nd lasers. With these lasers, some of the benefits that come along are advanced power output, small size, higher peak power, excellent quality of the optical output, and an enhanced plug efficiency of 18%–38% [21]. The output power and outstanding beam quality of the fiber laser make them increasingly common. It enables the welding of complex materials, like titanium and aluminum alloys, that are very challenging to be welded by CO_2 lasers. Even though diode lasers may be used, their application in HLAW is restricted because of the poor beam quality and low output power. According to research, high-power diode lasers used when welding needs to penetrate deep into the base materials, or workpieces provide more benefits than other types of welding methods. However, hybrid laser arc welding technologies use mainly gas–metal arc welding processes in addition to others such as plasma arc welding (PAW), gas–tungsten arc welding (GTAW), and last submerged arc welding (SAW).

9.3.1 Hybrid Laser GMAW Approach

The hybrid laser GMAW process has three primary heat sources: an Nd laser, a CO_2 laser, and a fiber laser. The second ancillary heat source is an electric arc created during GMAW, sometimes called MIG welding. The laser part of this hybrid process is responsible for creating a keyhole in the weld pool to achieve good penetration, especially at higher thicknesses for joining. This process produces strong and solid bonds with a lesser number of iterations than the traditional methods of welding [22]. The liquid filler material generated by the GMAW process fills up the principal space, thus allowing the joining of elements with higher gaps and fit-up problems. This is quite essential in industrial functions since the perfect alignments are hard to set up. Figure 9.2 shows the setup of weld bead generation.

When the GMAW arc is combined with a laser beam, uniformity in the arc increases, The laser also gives stability to the arc, without much

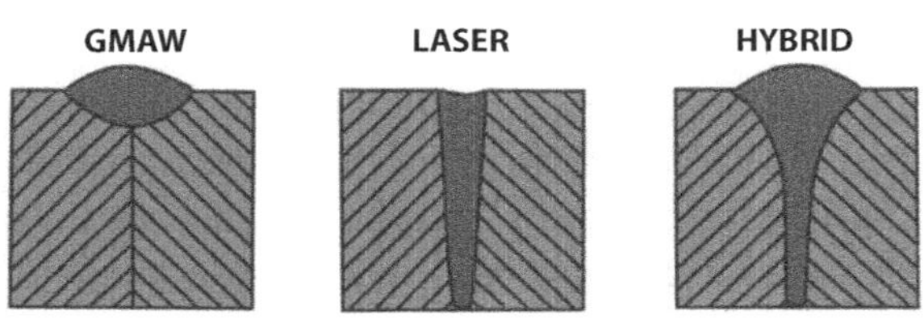

Figure 9.2 Patterns of weld bead formation.

fluctuation, to provide a uniform and regulated welding process [23]. The presence of filler materials introduces metallic components into the weld pool, which enhances the structural qualities and chemical constitution of the weldment. Thus, the welds have more tensile strength, hardness, and resistance to cracking. The larger magnitude of the molten pool created by using GMAW is more helpful for bridging weld gaps than GTAW and PAW techniques, along with other arc welding methods. Reliability should be guaranteed so that the weld integrity in components of various gap widths is maintained.

Shielding gases, such as argon, helium, nitrogen, or their mixtures, protect the molten pool of the weld from ambient contamination and thus prevent oxidation and related weld defects that may result in a weak weld. Quite often, a pulsed-based GMAW power source is employed in the hybrid laser GMAW process to enhance control over metal droplet development, arc length, transfer mode, arc intensity, and spatter generation. This refinement also aids in better control of the welding process, hence reducing post-weld cleaning [24]. The melting filler can further homogenize the structure of a weld by increasing dilution in the weld pool, which may penetrate a joint as tiny as 1.2 mm and makes it more uniform and dependable. Compared with laser beam welding and GMAW, hybrid laser GMAW utilizes more heat sources to increase the rate of welding, which decreases welding time and improves manufacturing efficiency.

9.3.2 Hybrid PAW Method

The hybrid laser plasma arc process uses a confined plasma arc as an auxiliary heat source considerably expanding the range of welding applications. The most important element is the nonexpendable tungsten electrode inside the welding torch. Inside the torch, the arc is restricted by a small-diameter, water-cooled constricting nozzle, thereby increasing the intensity far greater before and improving the stability and shape. This configuration provides very good control over the arc, which is further improved by passing inert gas, called orifice gas, through the constricting nozzle to form plasma [25]. In addition, shielding gas via the torch's outer ring is provided to enhance the shielding effect and protect the welding area from ambient pollution.

PAW offers other benefits through its versatility in its operation utilizing two distinct methods to generate the arcs: the non-transferred arc and the transferred arc. The non-transferred arc method involves the creation of an electric arc between the electrode and constricting nozzle [26]. This design ensures that the arc stays within the welding flame and does not touch the workpiece directly. This process is mainly used where limited heat input is required, and this process provides a considerable amount of control over the welding process. This process is best suited for welding thin sections and performing precise welding activities. The transfer arc process produces an electrical arc between the electrode and the work component [27]. This design enables the arc to directly transfer much of the heat to the workpiece; this process is, therefore, more suitable for welding thicker materials and for making deeper penetrations.

The transfer arc technique is normally utilized for production welding due to its high efficiency and capability of handling a large range of material thicknesses. The plasma arc in the PAW is surely much more stable than that in GTAW. Many attribute the stability to the restriction of the arc plasma, which generates a concentrated and steady heat source. Chen *et al.* [28] reported that this constricted arc ensures that the plasma jet remains stable even at the closest point to the base metal for enhanced precision and reliability in the welding process. During PAW, the arc plasma itself is constricted, which eventually leads to a very focused plasma jet. The jet action of the process into the weld pool allows arc penetration to go deep and maintain a constant temperature, hence a quality weld with minimal distortion, while the hybrid of plasma arc and laser technologies makes use of both these technologies in applying. The laser basically provides a concentrated heat source that augments both the welding speed and efficiency, in addition to stabilizing the plasma arc [29]. This hybrid approach can increase the speed of the welding process, greatly improving the stability and reliability of the arc plasma, hence leading to better weld quality and higher process efficiency.

9.3.3 Hybrid Laser GTAW Approach

In hybrid laser GTAW, the electric arc of GTAW provides additional heat. GTAW uses a non-consumable tungsten electrode, which differs from GMAW and requires that filler materials be used in welding. The nature of GTAW sets it apart from GMAW and demands special attention to the handling of filler material [30]. One major difference between hybrid laser GTAW and hybrid laser GMAW processes is the difference in the metal deposition rate. GTAW has a significantly lower rate of metal deposition compared to GMAW since only a portion of the arc's heat is used to melt the filler metal. The lower rate at which material is laid down has its advantages in certain applications, especially where there is a requirement for close control over the welding process. This is because filler wire operates independent of the hybrid laser GTAW setup, and thus, welders are able to have complete control over the rate that metal is laid down and the current used in welding. This will enable them to have better flexibility and accuracy in the welding process.

Although GTAW has several advantages, this process has some issues, such as electrode wear, contamination, and oxygen reactivity. The creation of the high-frequency arc causes erosion of the tungsten tip; as a result, arc stability will be reduced, and periodic maintenance is required. Metal vapor existing at a high temperature easily gets contaminated and causes impurities in the tungsten electrode leading to welding defects [31]. Besides, tungsten is oxidized at high temperatures, which results in a loss in electrode performance and contamination of the weld. To prevent the formation of oxides in the atmosphere and a contamination process of both the electrode and the weld pool, shielding gas is used. The shielding gas provides some protection around the weld pool, which aids in maintaining the strength and consistency of the weld. The base metal melting rate obtained in the hybrid laser GTAW process is found to be approximately twice that obtained independently in both the GTAW welding and laser welding processes. This increase in value could be attributed to the specific synergism between the GTAW arc and the laser beam.

When a GTAW arc is integrated with a laser beam at a common interface region, several major consequences ensue. First, the ability of the laser to absorb energy in the base material is enhanced due to the preheating caused by the leading GTAW arc. The preheating, therefore, raises the temperature of the base material such that it becomes readily prepared to melt on exposure to the laser beam. This makes the main material melt faster and leads to the overall increase in speed of the melting process that is observed with hybrid laser GTAW. Also, having the laser on the weld pool actually maintains the GTAW arc based on a number of synergistic effects, which improve the welding process. The high-power density of the laser beam quickly heats up the base metal causing localized melting and evaporation. It creates in its leading edge a liquified pool and a region of material that is vaporized above the area where the welding is going on. In effect, this would increase the electrical conductivity of the welding zone considerably. The course of the GTAW arc is thus improved by the molten and evaporated material having high conductivity [32]. The GTAW arc is automatically established in the contact zone of the laser, the maximum conductivity area, as it follows the path of least resistance. The process of being rooted has a stabilizing effect on the arc reducing the variations or abnormalities. The increased ionization of the vaporized material raises the conductivity in the arc column providing a more continuous electrical flow. This provides a stable, focused arc reducing the tendencies of arc wander and instability.

Arc stability is one of the most important criteria for a quality welding process. This can ensure a constant and uniform heat supply at the weld area. Consistency can hence be achieved to have continual welds with lesser imperfections. A consistent arc is a prerequisite to control the heat-affected zone, especially for materials that are hard to weld, such as stainless steels or aluminum–magnesium alloys. These materials bring out the challenges of welding owing to factors such as high heat conductivity and reactivity. Unless proper control of this is done, warping, cracking, and oxidation may occur [33]. The refined control afforded by a stabilized arc reduces these issues through consistent temperatures in HAZ. Besides, a consistent arc allows for smoother welding processes, thus reducing the need for a welder to make several adjustments and corrections. This, in turn, increases welding efficiency while reducing the chances of porosity, inclusions, or incomplete fusion defects. The role of the laser to create a consistent and reliable electric discharge allows for deeper penetration and a more consistent weld, which is critical for applications requiring high-strength welds with integrity.

9.4 Hybrid Laser Arc Welding Process Variables

In HLAW, the combination of the welding processes presents more variables, in addition to the already existing control variables, all of which must be controlled with precision. The process variable management is not only an addition of variables involved in the two independent processes; their values may greatly differ due to the interactions between the laser beam and the electric arc. A typical example in the HLAW is that the laser power normally reduces due to the presence of a second heat source when compared to autogenous laser welding.

9.4.1 Welding Speed

The welding speed in HLAW is one of the vital variables that significantly influence the efficiency and quality of the whole process. Higher welding rates in HLAW increase productivity by raising the welding speed. Fast welding reduces the heat input per unit length, which further reduces HAZ and thermal distortion [34]. However, it may be at the expense of a reduced width and reduced penetration depth with the increased welding speed. To maintain this constant volume of weld pool by counteracting such effects, additional measures have to be taken: increasing laser power and raising the filler wire feed rate.

Keyhole stability and overall process stability are ensured only if the ratio between the welding speed and the filler wire feed rate is maintained. Improper welding speed is capable of creating problems such as inadequate fusion or insufficient penetration, which may interfere with the weld's integrity. On the other hand, too low a speed is guaranteed to bring about an oversized weld pool characterized by too much penetration, which possibly leads to burn-through and partial degradation of the material. Mazar *et al.* [35] have observed that the higher rates of welding created rapid heating and cooling cycles resulting in steep temperature gradients close to the weld. These variations in gradients can lead to metallurgical flaws in the welded structure in terms of fractures or undesired microstructures. Enhancing the speed at which welding is done—as restricted—has a significant effect on reducing the concentration of thermal residual stress, hence improving the weld's overall mechanical properties.

9.4.2 Laser Power

The laser provides the primary heat in hybrid welding and makes a keyhole effect, thus enabling deep penetration welding. Strong welds of high quality are realized by such welding in thick sections. The depth of penetration of the weld is related directly to the laser power. With increasing laser power, the depth of penetration increases allowing single-pass welding of thicker sections [36, 37]. The depth of penetration is known to increase by approximately 2–4 mm/kW of laser power for materials, such as the AZ31 magnesium alloys, while for arc power, this is approximately 0.5 mm/kW.

The electric arc in hybrid welding enhances the absorption of laser energy within the metal [38]. The coupling of the laser beam to the workpiece is better. Inside the weld pool, it provides better circulation of molten metal, which results in uniform heat distribution and better quality of weld. With an increased power of laser, there will be a change in geometry of weld bead. For instance, a higher laser power in copper alloys increases the root width, face width, and height of the root reinforcement, while it decreases the face reinforcement height. It applies a high-power density to the weld area creating a keyhole that enhances the laser absorptivity and hence stabilizes the process of welding.

9.4.3 Laser Beam and Electrode Relative Positioning

In hybrid welding, the laser beam and electrode positioning in regard to one another is very crucial to weld quality, penetration, and stability. The main configurations include laser leading and arc leading. In the former, the laser beam precedes the welding gun, set up at a push angle that creates, from top view, a 'bullet-shaped' weld pool and, in cross-sectional area, a "cocktail cup-shaped" one. The setup provides a smaller laser area width, a larger arc area width, a deeper penetration depth, and reinforcement height, which improves the stability of the arc through stable keyhole formation [39]. This is particularly suitable in materials, like Mg alloys, stainless steel, and Al alloys, where deeper penetration and stable keyhole formation are desired.

The arc-leading configuration, on the other hand, consists of the drag angle welding torch leading and the laser beam following, thus creating a "gourd-shaped" weld pool in the top view and a "cone-shaped" cross-section. This is less stable than laser leading, but it has some particular advantages with chosen materials. For example, tensile strength and mechanical properties can be improved in high-speed steel with better weld penetration when a GMAW welding gun is placed behind a CO_2 laser beam. However, the combination generally creates another distribution in heat and material flow, which subsequently influences the form of the weld bead and its properties.

9.4.4 Wire Feed Rate

The optimum wire feed rate is required to obtain the correct amount of material to be deposited. Too small a feed rate can provide too little filler material producing weak welds, gaps, and lack of fusion; on the other hand, too large a feed rate may cause overfilling and spatter leading to poor weld quality again due to excessive material deposited. A higher wire feed rate allows the operator to more easily bridge greater gaps between parts. It provides for welding over varying joint gaps and inconsistent fit-up of parts [7, 32]. Higher WFR also accommodates higher welding speeds, which increase productivity and reduce cycle times. As the wire feed rate increases, the welding current must also increase to melt the additional filler material being fed into the weld pool. This, therefore, creates an extremely important relationship in ensuring that there will be a stable arc and consistent deposition of the material.

9.4.5 Selection of the Focus Area

The optimum level of penetration is usually attained when the focus point of the laser beam lies a little under the surface of the workpiece. Sub-surface focus allows the laser to concentrate energy further down the material, thus increasing the penetration and the toughness of the weld. During hybrid welding, the focal point of the laser beam inside the molten pool changes because of the varying properties and dynamics of the weld pool as it passes through. In contrast to laser welding alone, it is evident that the substantially larger volume of the HLAW weld pool will have effects on both laser beam focus and material interaction [40]. The HLAW cup-shaped weld distribution feature will enhance the curved shape of the top layer of the molten stream. A distorted focal point from the curvature reduces the effective penetration depth of the laser beam and its precision.

It was observed that the focal point of the laser beam is reduced by approximately 0.6 mm during hybrid CO_2 laser GMAW welding than during autogenous laser welding. This may be attributed to melting pool dynamics and surface curvature changes [41]. This ideal target of the laser beam varies from 1.5 to 3.5 mm when mild steel plates are welded with hybrid CO_2 laser GMAW welding at a welding rate of 0.6 m/min. Such adjustments need to be made to ensure adequate penetration and weld quality under various process conditions [42].

9.4.6 Laser-to-Electrode Distance

The laser-to-electrode separation impacts the quality and penetration of the weld, as well as the overall effectiveness of the operation. It normally changes from 0 mm in coaxial setups to up to 5 mm in paraxial setups; the distance at which the laser beam interacts with the arc determines thermal dynamics and weld characteristics [43]. This guarantees a synergistic effect between the laser and arc when the distance of separation is optimal since the laser will preheat the material to enable the easy melting and binding of the materials by the arc.

Proper spacing improves weld penetration and arc stability while reducing imperfections such as porosity and undercut. In coaxial configurations, the laser and arc are precisely aligned, which could outcome in a reduction in the penetration of laser energy by the arc plasma [44]. In contrast, a small gap, like 2 mm, usually leads to better results due to the avoidance of such interference, thus improving penetration. This distance even affects the morphology of the weld pool. As the spacing between welds is increased, the shape of the weld bead progresses from a "cocktail cup" to a "cone." Proper calibration, with real-time adjustments based on empirical testing, is necessary to maintain this separation distance for the best quality and efficient welds.

9.5 Industrial Applications of Hybrid Laser Arc Welding

HLAW has enormous advantages making it a very attractive option for many industries. It increases productivity and produces welds of higher quality compared to laser and ordinary arc welding. Due to the combination of laser and arc-welding technologies, improved efficiency is immediately achieved resulting in faster times in the processing of materials and better welding performance [45]. Although there is a fairly big upfront investment in HLAW systems, these are normally offset by some long-term benefits. On account of the employment of HLAW, there is an increase in welding speed along with greater depth of penetration, which in turn brings the cost per unit length of welding down. This makes the HLAW approach relatively cheaper in the long run. HLAW technology can bond a large number of materials, including many metallic materials such as steel, stainless steel, aluminum, nickel, magnesium, and titanium. It is potentially applicable to many industries, processing thin metallic sheets as well as metallic plates [46]. HLAW deep penetration and high welding speed minimize distortion and maximize the quality of the weld, hence lending themselves to applications requiring high strength and structural integrity.

Due to the efficiency in merging the advantages of laser and arc welding to produce excellent welds with efficient production methods, HLAW has enjoyed further popularity in shipbuilding, automotive, and tube manufacturing enterprises. This versatile process has also been used in the aerospace industry, rail stock, and container production to join different materials efficiently. A lot of research has been conducted to make HLAW more cost effective, automated, and compatible with the use of different materials. The possibility of HLAW to bridge gaps and fast process makes the technology very attractive to the automotive industry [47]. Even large industry players, such as Volkswagen, Audi, and Daimler, had already embraced and incorporated HLAW technology in their production lines. In Volkswagen, for instance, hybrid welds were used in the Phaeton body that involved complex junctions of elongated elements, moldings, and aluminum sheets. HLAW is also applied by Audi in the roof frame and other components of the A8 [48]. HLAW process in axle manufacturing by Daimler yields increased productivity due to increased welding speed, improved wire efficiency, and improved penetration and metallurgical qualities of welded joints.

In the building of aircraft fuselages and wings, HLAW ensures strong joints with minimal distortion; this is very critical in retaining the structural integrity and aero-performance of the aircraft. The process joins, with varying effectiveness, thin skin panels, and thick structural parts and hence accommodates the diversity of material thicknesses encountered in aerospace applications [49]. This process is also heavily used in the manufacture of engine parts, such as turbine blades and casings. The welds are exposed to very severe operating conditions; hence, high precision is required for producing defect-free, high-strength welds. This can be attained through HLAW since it is able to join dissimilar metals and has a very small heat-affected zone, which is quite useful in making complex assemblies. This method is further valuable in the production of leak-tight joints for critical systems, such as fuel tanks and hydraulic lines, thus ensuring aircraft safety and reliability [50]. Other advantages of HLAW include a high level of automation—the possibility for automation—improved production efficiency and consistency, and therefore reduced costs about human error. This makes it suitable for high-volume manufacturing and focused repairs that extend the life of high-value aerospace components [51].

HLAW is very useful in constructing ship hulls and bulkheads, as it guarantees the safety and strength of essential joints, all the while saving much welding time compared to conventional techniques. The process allows for a large range of material thicknesses and is therefore very appropriate for welding thin plates and thick sections used in ship structures [52]. This allows for minimization of thermal distortion and reduces the extent of extensive post-weld treatment required for large ship components so that their dimensional accuracy and performance are maintained [53]. Besides, HLAW offers deep, narrow welds with a rather small heat-affected zone, the feature that lets it improve the fatigue resistance of welded joints, thus increasing service life and durability for ships operating in these aggressive marine environments. This technique is also highly effective in joining dissimilar metals, which is helpful in making advanced ship designs featuring different materials to enable optimum performance and minimize weight.

9.6 Conclusion

HLAW is an advanced welding process that combines the capabilities of laser and arc welding to achieve outstanding performance over a wide industrial application spectrum. This hybrid technology combines the deep penetration and speed aspects of the laser welding process with the stability and variety of filler substances in arc welding. It thereby provides HLAW with higher welding speeds, excellent gap-bridging ability, increased penetration, higher production efficiency, and far improved process stability. The advantages associated with this process are very important in industries where superior and reliable welds are required, like aerospace, shipbuilding, automotive, and construction industries. The efficacy of the process relies on the proper setting of different variables such as welding speed, laser power, filler wire feed ratio, shielding gas composition, and laser-to-electrode distance. Proper control of these process variables is helpful to achieve the expected outcomes, like deep penetration and low defect density, through the primary heat source, laser power, in keeping a stable keyhole and strong weld.

HLAW offers considerable process improvements while welding a wide range of materials, from titanium and stainless steel to aluminum. This process helps in the efficient joining of thick metal plates in a single pass, thereby reducing the number of weld passes and increasing productivity and cost effectiveness. The limited heat-affected zone reduces thermal distortion and maintains the dimensional consistency required for structural integrity. Moreover, HLAW fabricates excellent welds with superb mechanical properties and high fatigue strength making the process very suitable for critical applications. It also lowers the formation of common welding defects like cracks and porosity; hence, it ensures higher reliability and a longer life for welded joints. Though HLAW has gained wide acceptance and has been widely applied, there is still immense scope for further research and development. Future research efforts may be targeted toward the welding of complicated materials, like copper, that remain difficult to weld using conventional welding processes. Research in varying process variables and their interaction, and further optimizing these variables, can increase the accuracy of the process. Detailed theoretical modeling and simulation must, therefore, be conducted to understand the dynamics of keyhole formation/collapse, laser–arc interaction, thermal cycles, cooling rates, dilution effects, etc. Such efforts will help in predicting and preventing defects while optimizing the welding parameters.

References

1. Subashini, L., Prabhakar, K.V.P., Ghosh, S., Joint design influence on hybrid laser arc welding of maraging steel. *Weld. World*, 68, 6, 2024.
2. Chen, G., Hirohata, M., Hyoma, K., Matsumoto, N., Inose, K., Deformation and residual stress of T-shaped joints fabricated by laser-arc hybrid welding for steel bridge members. *Weld. World*, 68, 2, 2024.

3. Aqeel, M., Shariff, S.M., Gautam, J.P., Padmanabham, G., Liquation cracking in Inconel 617 alloy by Laser and Laser-Arc Hybrid welding. *Mater. Manuf. Processes*, 36, 8, 2021.
4. Cai, J., Wei, Y., Ouyang, Z., Liu, X., Jin, H., Chen, J., Investigation on clockwise circular oscillating laser welding for the 5A06-H112 aluminum alloy: Energy distribution, seam appearance, microstructure, and mechanical properties. *Opt. Laser Technol.*, 176, 111026, 2024.
5. Pramod Kumar, G., Balasubramanian, K.R., Phani Prabhakar, K.V., Cheepu, M., Investigation of microstructure, mechanical, and corrosion properties of Inconel 617 joints welded by laser–MIG hybrid welding. *Proc. Inst. Mech. Eng., Part L: J. Mater.: Des. Appl.*, 237, 9, 2023.
6. Ning, J., Zhang, L.J., Na, S.J., *et al.*, Numerical study of the effect of laser-arc distance on laser energy coupling in pulsed Nd: YAG laser/TIG hybrid welding. *Int. J. Adv. Manuf. Technol.*, 91, 1–4, 2017.
7. Yu, J., Cai, C., Xie, J., Huang, J., Liu, Y., Chen, H., Weld formation, arc behavior, and droplet transfer in narrow-gap laser-arc hybrid welding of titanium alloy. *J. Manuf. Process.*, 91, 44–52, 2023.
8. Bunaziv, I., Wenner, S., Ren, X., Frostevarg, J., Kaplan, A.F.H., Akselsen, O.M., Filler metal distribution and processing stability in laser-arc hybrid welding of thick HSLA steel. *J. Manuf. Process.*, 54, 228–239, 2020.
9. Gao, X., Wang, L., Chen, Z., Zhang, Y., You, D., Process stability analysis and weld formation evaluation during disk laser–mag hybrid welding. *Opt. Lasers Eng.*, 124, 105835, 2020.
10. Yu, J., Cai, C., Xie, J., Chen, Z., Chen, H., Keyhole stability, arc behavior, and molten pool flow in narrow-gap oscillating laser-arc hybrid welding of titanium alloy. *Int. J. Heat Mass Transf.*, 220, 124922, 2024.
11. Ragavendran, M. and Vasudevan, M., Effect of Laser and Hybrid Laser Welding Processes on the Residual Stresses and Distortion in AISI Type 316L(N) Stainless Steel Weld Joints. *Metall. Mater. Trans. B*, 52, 4, 2021.
12. Aqeel, M., Gautam, J.P., Shariff, S.M., Influence of heat input on bead profile and microstructure characteristics in laser and laser-hybrid welding of Inconel 617 alloy. *Weld. Int.*, 36, 5, 2022.
13. Üstündağ, O., Gook, S., Gumenyuk, A., Rethmeier, M., Hybrid laser arc welding of thick high-strength pipeline steels of grade X120 with adapted heat input. *J. Mater. Process. Technol.*, 275, 116358, 2020.
14. Wang, D., Dong, Y., Liu, L., Zhu, M., Wang, H., Liu, C., Effect of pulsed laser and laser-arc hybrid on aluminum/steel riveting-welding hybrid bonding technology. *J. Mater. Res. Technol.*, 17, 1043–1053, 2022.
15. Lei, Z., Li, B., Wu, S., Chen, Y., Xiong, Y., Effects of MnN powder on the microstructure and properties of high nitrogen steel joint via laser-arc hybrid welding. *Opt. Laser Technol.*, 138, 106877, 2021.
16. Urbańczyk, M. and Adamiec, J., Hybrid welding (Laser–electric arc mag) of high yield point steel s960ql. *Materials*, 14, 18, 2021.

17. Gook, S., Gumenyuk, A., Rethmeier, M., Hybrid laser arc welding of X80 and X120 steel grade. *Sci. Technol. Weld. Joining*, 19, 1, 2014.
18. Liu, T., Yan, F., Liu, S., Li, R., Wang, C., Hu, X., Microstructure and mechanical properties of laser-arc hybrid welding joint of GH909 alloy. *Opt. Laser Technol.*, 80, 56–66, 2016.
19. Jia, Y., Wen, T., Huang, N., *et al.*, Research on aluminum alloy welding process based on high frequency and low power pulsed Laser-MIG hybrid welding. *Opt. Laser Technol.*, 150, 107899, 2022.
20. Zhu, Y., Cai, Y., Wang, M., Effects of He content in shielding gases on high-efficient hybrid laser arc welding with C-276 filler metal. *J. Mater. Process. Technol.*, 299, 117367, 2022.
21. Skowronska, B., Chmielewski, T., Golanski, D., Szulc, J., Weldability of S700MC steel welded with the hybrid plasma + MAG method. *Manuf. Rev. (Les Ulis).*, 7, 4, 2020.
22. Zhu, Y., Cai, Y., Dong, H., Wang, M., Tailoring droplet transfer and molten pool flow during hybrid laser arc welding of nickel base alloy. *Opt. Laser Technol.*, 147, 107620, 2022.
23. Yin, F., Li, X., Chen, C., Zhao, L., Peng, Y., Tian, Z., Microstructure and mechanical properties of weld metal in laser and gas metal arc hybrid welding of 440-MPa-grade high-strength steel. *J. Iron. Steel Res. Int.*, 28, 7, 2021.
24. Bang, H.S., Bang, H.S., Kim, Y.C., Oh, I.H., A study on mechanical and microstructure characteristics of the STS304L butt joints using hybrid CO2 laser-gas metal arc welding. *Mater. Des.*, 32, 4, 2328–2333, 2011.
25. Mu, Z., Chen, X., Zheng, Z., Huang, A., Pang, S., Laser cooling arc plasma effect in laser-arc hybrid welding of 316L stainless steel. *Int. J. Heat Mass Transf.*, 132, 861–870, 2019.
26. Zhou, H., Fu, F., Dai, Z., *et al.*, Effect of Laser Power on Hybrid Laser-Gas Metal Arc Welding (GMAW) of a 6061 Aluminum Alloy. *J. Korean Phys. Soc.*, 77, 11, 2020.
27. Liu, L., Xu, X., Zhu, Y., Study on synchronous induction of arc plasma by laser in laser-double TIG hybrid welding. *Opt. Lasers Eng.*, 133, 106130, 2020.
28. Chen, X., Mu, Z., Hu, R., Liang, L., Murphy, A.B., Pang, S., A unified model for coupling mesoscopic dynamics of keyhole, metal vapor, arc plasma, and weld pool in laser-arc hybrid welding. *J. Manuf. Process.*, 41, 119–134, 2019.
29. Gao, Z., Jiang, P., Shao, X., Cao, L., Mi, G., Wang, Y., Numerical analysis of hybrid plasma in fiber laser-arc welding. *J. Phys. D Appl. Phys.*, 52, 2, 2019.
30. Chen, M., Xin, L., Zhou, Q., He, L., Wu, F., Effect of laser pulse on alternative current arc discharge during laser-arc hybrid welding of magnesium alloy. *Opt. Lasers Eng.*, 100, 208–215, 2018.
31. Dai, X., Zhang, H., Liu, J., Feng, J., Microstructure and properties of Mg/Al joint welded by gas tungsten arc welding-assisted hybrid ultrasonic seam welding. *Mater. Des.*, 77, 65–71, 2015.

32. Hasanniah, A. and Movahedi, M., Gas tungsten arc lap welding of aluminum/steel hybrid structures. *Mar. struct.*, 64, 295–304, 2019.
33. Chen, C., Li, W., Fan, C., Du, W., Zhang, H., Understanding the changing mechanism of arc characteristics in ultrasound-magnetic field coaxial hybrid gas tungsten arc welding. *Plasma Sci. Technol.*, 24, 11, 2022.
34. Kuppusamy, M.V., Venkatraman, B., Albert, S.K., Athmalingam, S., Hybrid Welding Process in Lieu of Conventional Practices for Fabrication of Austenitic Stainless Steel AISI 304L Components in Nuclear Facility. *J. Mater. Eng. Perform.*, 32, 17, 2023.
35. Mazar Atabaki, M., Yazdian, N., Kovacevic, R., Hybrid laser/arc welding of thick high-strength steel in different configurations. *Adv. Manuf.*, 6, 2, 2018.
36. Sandnes, L., Grong, O., Welo, T., Berto, F., Fatigue properties of AA6060-T6 butt welds made by hybrid metal extrusion & bonding. *Fatigue Fract. Eng. Mater. Struct.*, 43, 10, 2020.
37. Ye, Z., Huang, J., Cheng, Z., *et al.*, Mechanisms of an innovative hybrid arc welding process in enhancing joint penetration and weld property control through resistive and induction heat. *J. Manuf. Process.*, 72, 500–514, 2021.
38. Hao, K., Gao, Y., Xu, L., *et al.*, Beam oscillating parameters on pore inhibition, recrystallization and grain boundary characteristics of laser-arc hybrid welded AZ31 magnesium alloy. *J. Magnesium Alloys*, 12, 2489–2502, 2022.
39. Fulong, Z., Beibei, M., Shuangyu, L., *et al.*, Effect of a nickel-based alloy cladding layer on the strength and toughness of the high-strength steel laser-MAG hybrid welding joint. *Mater. Res. Express*, 7, 7, 2020.
40. Han, X., Yang, Z., Ma, Y., Shi, C., Xin, Z., Comparative study of laser-arc hybrid welding for aa6082-t6 aluminum alloy with two different arc modes. *Met. (Basel).*, 10, 3, 2020.
41. Vorontsov, A., Zykova, A., Chumaevskii, A., *et al.*, Advanced high-strength AA5083 welds by high-speed hybrid laser-arc welding. *Mater. Lett.*, 291, 129594, 2021.
42. Zhang, X., Mi, G., Wang, C., Microstructure and performance of hybrid laser-arc welded high-strength low alloy steel and austenitic stainless steel dissimilar joint. *Opt. Laser Technol.*, 122, 105878, 2020.
43. Bunaziv, I., Dørum, C., Nielsen, S.E., *et al.*, Laser-arc hybrid welding of 12- and 15-mm thick structural steel. *Int. J. Adv. Manuf. Technol.*, 107, 5–6, 2020.
44. Hammad, A., Churiaque, C., Sánchez-Amaya, J.M., Abdel-Nasser, Y., Experimental and numerical investigation of hybrid laser arc welding process and the influence of welding sequence on the manufacture of stiffened flat panels. *J. Manuf. Process.*, 61, 527–538, 2021.
45. Bunaziv, I., Akselsen, O.M., Ren, X., Nyhus, B., Eriksson, M., Laser beam and laser-arc hybrid welding of aluminium alloys. *Met. (Basel).*, 11, 8, 2021.
46. Jeong, Y.C., Cho, Y.T., Jung, Y.G., Design of stitch welded shape with laser-ARC hybrid welding for ultra-high strength steel. *Int. J. Precis. Eng. Manuf. Green Technol.*, 3, 2, 2016.

47. Wang, H., Feng, B., Song, G., Liu, L., Laser–arc hybrid welding of high-strength steel and aluminum alloy joints with brass filler. *Mater. Manuf. Processes*, 33, 7, 2018.
48. Staufer, H., Laser hybrid welding and laser brazing at AUDI and VW. *Weld. World*, 50, 7–8, 2006.
49. Zhang, C., Gao, M., Zeng, X., Influences of synergy effect between laser and arc on laser-arc hybrid welding of aluminum alloys. *Opt. Laser Technol.*, 120, 105766, 2019.
50. Casalino, G., Mortello, M., Leo, P., Benyounis, K.Y., Olabi, A.G., Study on arc and laser powers in the hybrid welding of AA5754 Al-alloy. *Mater. Des.*, 61, 2014.
51. Unt, A., Poutiainen, I., Salminen, A., Influence of Filler Wire Feed Rate in Laser-Arc Hybrid Welding of T-butt Joint in Shipbuilding Steel with Different Optical Setups. In. *Physics Procedia*, 78, 45–52, 2015.
52. Gook, S., Midik, A., Biegler, M., Gumenyuk, A., Rethmeier, M., Joining 30 mm Thick Shipbuilding Steel Plates EH36 Using a Process Combination of Hybrid Laser Arc Welding and Submerged Arc Welding. *J. Manuf. Mater. Process.*, 6, 4, 2022.
53. Churiaque, C., Sánchez-Amaya, J.M., Üstündağ, Porrua-Lara, M., Gumenyuk, A., Rethmeier, M., Improvements of hybrid laser arc welding for shipbuilding T-joints with 2F position of 8 mm thick steel. *Opt. Laser Technol.*, 143, 107284, 2021.

10

Ultrasonic Welding: Foundations, Influential Factors, and Material Applications

Marxim Rahula Bharathi B.[1]*, Akhilesh Kumar Singh[1], D.V.S.S.S.V. Prasad[1], A. Ramesh[2], P. V. Elumalai[3,4] and N.S. Balaji[5]

[1]*Department of Mechanical Engineering, Aditya College of Engineering and Technology, Surampalem, Andhra Pradesh, India*
[2]*Department of Electrical Engineering, Aditya College of Engineering and Technology, Surampalem, Andhra Pradesh, India*
[3]*Department of Mechanical Engineering, Aditya University, Surampalem, Andhra Pradesh, India*
[4]*Faculty of Engineering, Shinawatra University, Bang Toei, Thailand*
[5]*Department of Mechanical Engineering, SRM Institute of Science and Technology, Tiruchirappalli Campus, Tiruchirappalli, Tamil Nadu, India*

Abstract

This book chapter provides an in-depth insight into ultrasonic welding, a state-of-the-art technology widely used in a variety of industries for its ability to produce high-quality joints without the need for additional materials or adhesives. The chapter explores the core concepts of ultrasonic welding, including the functions of crucial components such as the ultrasonic generator, transducer, horn, and booster. The study also investigates the impact of ultrasonic welding on several materials, including aluminum alloys, steels, titanium, magnesium, and fiber-reinforced composites. It specifically emphasizes the benefits and difficulties connected with each material. In addition, the chapter examines the key elements that affect the quality of welds, such as the calculation of bond area, and provides valuable information on how to optimize welding settings to improve performance. This chapter is a significant resource for scholars and professionals who want to comprehend and utilize ultrasonic welding in modern manufacturing

**Corresponding author*: prof.bmrbharathi@gmail.com

Sandip Kunar and Gurudas Mandal (eds.) Advanced Welding Technologies, (197–218) © 2025 Scrivener Publishing LLC

processes. It offers extensive analysis and practical examples to aid in understanding and application.

Keywords: Ultrasonic welding, components, dissimilar materials, factors influencing welding, applications of ultrasonic welding

10.1 Introduction

This is a world in which miniature parts are connected in a very complex way, but without the need for excessive amounts of glue and huge screws. This is the promise of ultrasonic welding, a revolutionary technique that uses sound waves to fuse materials in a near-perfect and perfect manner. In comparison with the classic methods of welding, which use direct application of heat and open, ultrasonic welding gives heat through high-frequency vibrations, which are faster, will last longer, and are more reliable. In case of ultrasonic welding, one advantage is that it moves and welds thin metal and thermoplastic without a filler material. This technique works through converting electrical signals of high frequencies (with a frequency band of 20–40 kHz) into ultrasonic. With such waves, mechanical vibrations are produced, and these are then magnified to provide the heat that is used in joining the parts. The product is a highly secure and clear joint that is done in a third of the typical time that construction joint is completed with regular construction methods.

Thus, the importance of ultrasonic welding is not only technological but (also) humanitarian. It has significant implications for several sectors, offering solutions that conventional welding techniques cannot solve. For instance, in medical areas, assorted surgical instruments and medical equipment are welded, through ultrasonic welding to avoid contamination from substances such as adhesives. It is also used in the medical textile market products, like gowns, covering drapes, and masks, where it provides strong and hygienic stitching without the use of very durable materials needed [1, 2]. For instance, in the electronics industry, ultrasonic welding enables one to join components, which are tender, and to do this without putting the other sensitive circuits at risk. Indeed, the automotive industry gains from the method as it forms stiff and lightweight joints that increase the safety and efficiency of the automobiles. Due to less reliance on mechanical fasteners, ultrasonic welding enables the development of slender models, lighter vehicles with improved fuel economy and acceleration.

The history of ultrasonic welding dates to the 1960s when it was first applied to weld hard plastics. Robert Soloff, a lab manager at Branson

Instrument, played a pivotal role in its development [4]. He discovered that ultrasonic probes could weld thermoplastic parts without direct contact leading to the creation of the first ultrasonic press. This innovation revolutionized the toy industry by enabling the mass production of plastic toys with strong, aesthetically pleasing joints. Over the decades, the technology has evolved, and its applications have expanded to include a wide range of materials and industries. Figure 10.1 illustrates various types of bench-top ultrasonic welding machines.

In this chapter, we will delve into the fascinating world of ultrasonic welding. You will learn about the instrumentation involved, including the ultrasonic generator, transducer, horn, and booster, which are crucial for the welding process. The ultrasonic generator converts electrical energy into high-frequency signals. These signals are then transformed into mechanical vibrations by the transducer. The horn and booster amplify these vibrations directing them precisely where the heat is needed to create the weld. This intricate interplay of components ensures that ultrasonic welding can produce joints with high precision and strength. We will explore the welding of various materials—such as aluminum alloys, steel, titanium, magnesium, fiber-reinforced composites, and thermoplastics—and examine how ultrasonic welding impacts their mechanical properties. Each material presents unique challenges and opportunities for ultrasonic welding. For instance, aluminum alloys benefit from the process' ability to create strong joints without melting the base material, while fiber-reinforced composites require careful control of vibration parameters to avoid damaging the fibers. By understanding these nuances, you can appreciate the versatility and adaptability of ultrasonic welding.

Furthermore, the paper will address the ultrasonic welding of dissimilar materials including one such application of welding 3003 aluminum with 304 stainless steels. The possibility of connecting two and more different

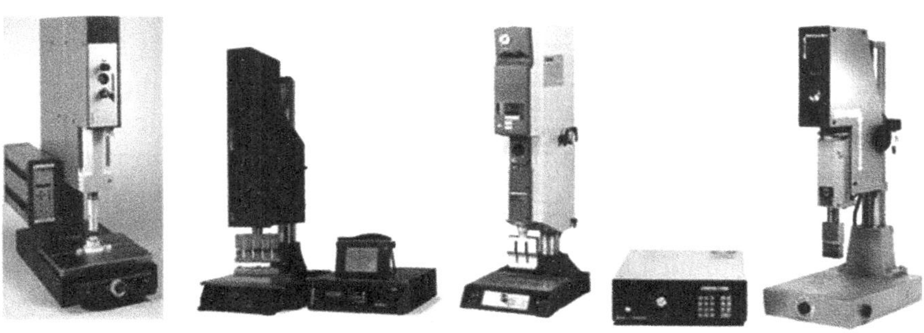

Figure 10.1 Various bench-top ultrasonic welding machines [3].

materials creates new opportunities in imaginative and practical design of new things making it possible to produce components whose mechanical characteristics are better than the characteristics of the connected materials. For instance, joining that hardness of stainless steel with the light aspect of aluminum will give high-performance structures that are strong and light. At the end of this chapter, you will need to have an understanding of the ultrasonic welding and its work principle, its usage, and importance in welding. You will find out why they have adopted it as a go-to method in industries such as medical device manufacturing, automotive industries, and many more, and why it is still leading in the way material joining is perceived and practiced. This chapter is for the engineer interested in implementing this technology as well as for the student studying novel manufacturing technologies since the reader will be equipped with the required information and understanding to explain and use ultrasonic welding.

Starting from this chapter, it is possible to move further to the explanation of the principles of the ultrasonic welding providing the reader with information about the basic instruments including an ultrasonic generator, a transducer, a horn, and a booster. It also looks at how ultrasonic welding can affect different types of material—aluminum alloys, steels, titanium, magnesium, and fiber-reinforced composites—as well as the problem and potential of welding dissimilar materials.

At the end of this chapter, the reader will have broad knowledge of how ultrasonic welding works and on what materials, and benefits over conventional welding. It will equip engineers, researchers, and students who are in employment or learning institutions aspiring to participate in the development of multi-disciplinary advanced manufacturing processes such as ultrasonic welding-relevant knowledge.

10.2 Principle of Ultrasonic Welding

Ultrasonic welding is a type of welding in solid phase where the electrical energy is converted to mechanical ones. These are ultrasonic vibrations usually ranging from 15 to 70 kHz, which are transmitted to the workpieces by the use of a sonotrode or horn. The vibrations result in generation of heat at the contact interface of the workpieces due to friction. This heat is enough to melt the materials that come in contact with each other making them stick when pressure is applied. As soon as the ultrasound vibrations are turned off, the materials cool down and make a strong weld

Figure 10.2 Ultrasonic welding process.

with each other. A schematic representation of the ultrasonic welding process is made in the form of a flowchart in Figure 10.2.

The ultrasonic welding procedure can be categorized into multiple separate phases as follows:

1. **Initial Phase:** These include the ultrasonic generator, which uses power in the form of standard electrical energy to produce high-frequency electrical signals. These signals are then converted to mechanical vibrations by the transducer that is integrated in the system.
2. **Energy Transfer:** Mechanical vibrations produced by the transducer are conducted by the booster to the sonotrode. The high-frequency vibrations are then transmitted to the workpieces at the contact interface through the sonotrode.
3. **Frictional Heating:** When the sonotrode is placed in contact with the surfaces of the workpieces, the latter are stimulated to undergo relative motion. The motion in this case produces heat as a result of friction at the point of contact, that is, localized heat. The heat generated is adequate to make the materials at the interfaces become soft or even melt.
4. **Melting and Bonding:** Such localized heating results in softening or melting of materials in a contact interface to a plastic state. On the pressure application, the material initially undergoes a phase of softening and then fuses with the other material present. It puts pressure that would cause the substances to get in close contact, and this results to increased intermolecular forces being experienced.
5. **Solidification:** Upon attaining a designated welding period or creating the requisite bond, the ultrasonic vibrations are ceased. Pressure is then applied for a rather short time to enable the materials to cool and harden. Upon solidification of the materials, one gets a metallurgical bond of high quality.

This welding has the following benefits when compared with the other welding procedures: least energy input, low temperatures in welding, and short welding time.

10.2.1 Equipment for Ultrasonic Welding

A typical ultrasonic system comprises four main components: siren's components include the generator, transducer, booster, and horn. When assembled, these elements generate and emit ultrasonic waves at a precise location on the substrate during the course of a production process. These components include the dielectric materials, coupling agents, and other elements in the design. Selection of these components ensures that power transfer is enhanced, and the components' desirable characteristics for the ultrasonic application in question are achieved. Therefore, if one was to get out the designs of every facet and get to the ultimate result, it is of vital importance that one has adequate understanding in those various elements. The schematic diagram of ultrasonic welding is illustrated in Figure 10.3.

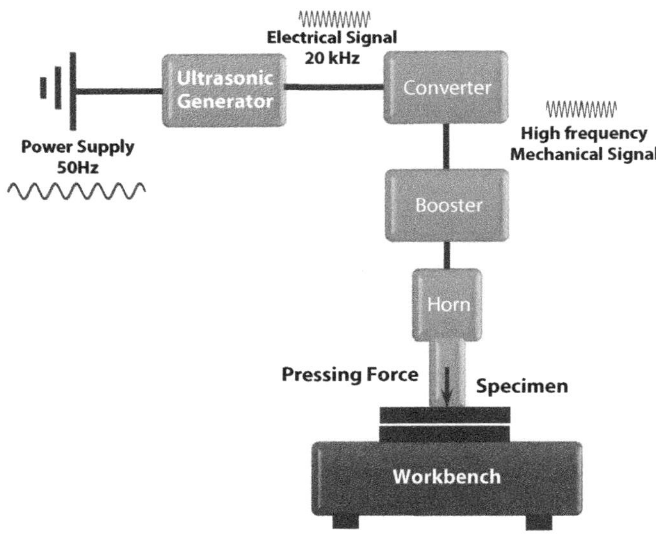

Figure 10.3 Schematic diagram of ultrasonic welding.

10.2.1.1 Ultrasonic Generator

The ultrasonic generator generates an electric signal having a high voltage typically in the range of 800—1,000 V and works at a frequency of 20 kHz. This small electronic device converts the normal electrical supply of 220 V AC 50 Hz to the required high-frequency output. The system has an electronic programmer, which supervises the functions, protective circuits for the parts used, and a microprocessor that controls all the processes. All these components are packaged in a power supply cabinet. The power rating of the generator is not constant depending on the particular purpose it is used, for example, plastic welding or metal welding. Figure 10.4 illustrates the commercially available ultrasonic generator.

Ultrasonic generators can be classified into two distinct categories: manual and fully automated. The difference depends on the scope and degree of the computer rendering of engineering solutions. The ordinary generators require manual adjustments to achieve the resonance, while the automatic will adjust the frequency of the output to match the resonating frequency of the horn/tool. The modern ultrasonic generators whose frequency is 20 kHz are designed to work not at a fixed frequency but within the range of 19–21 kHz. Ultrasonic frequencies are displayed on computer monitors and can therefore be monitored and adjusted using Digital Ultrasonic Generators. The "power variation" system of automatic generators allows consumers to change both the power and the amplitude as per their preference. Automatic generators also have superior technologies that correct small errors resulting from wear and other peculiarities that may develop and cut the energy loss. Some of the protection measures of this product are the auto switch-off function when the horn or tool is on and has a failure/overload.

Figure 10.4 Ultrasonic generator.

10.2.1.2 Transducer and Booster

The ultrasonic transducer or further known as the converter is that part of the probe in charge of converting the electrical signal generated by the generator into mechanical vibrations. However, the structures upon which they work successfully resemble the principles used in radar or sonar systems, and transducers can produce vibrations in alternate longitudinal and compressive ways. There are two primary types of transducers. It is divided into the following two main categories: magneto strictive and piezoelectric. Magneto strictive transducers are versatile and can be readily adjusted to minimize wear and tear. They can create movement in a range of frequencies normally between 17 and 23 kHz when used together with a 20-kHz generator. But they still need other systems to cool the losses in the form of heat to complete their circuits. Piezoelectric transducers with efficiencies of between 90% and 97% consist of piezoelectric disks, which are sandwiched between metal plates. Although they may also require cooling systems, they are characterized by high thermal shock resistance and easy to fashion. Figure 10.5 shows a diagrammatic representation of a transducer and booster.

The booster/or more commonly known as an amplifier, is an element placed at the end of the horn to boost the amplitude if necessary. Its main role is rather to change the amplitude of vibrations taking place between the transducer and the horn. The booster is specifically designed in a manner whereby the amplitude fed into the horn will either be amplified or

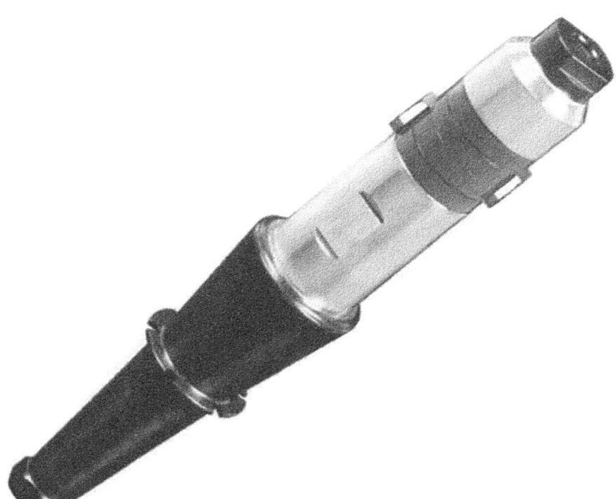

Figure 10.5 Ultrasonic welding transducer with booster.

attenuated depending on the need. When inserted at their nodal point, booster horns therefore offer a secure and firm stand for the converter horn assembly.

10.2.1.3 Ultrasonic Horn

The ultrasonic horn is also called an acoustic coupler, velocity/mechanical transformer, tool holder, concentrator, stub, or sonotrode, and its function is to amplify the ultrasonic vibrations created by the transducers. The choice of the horn material depends with the horn application that is needed or intended for. Aluminum is appropriate for standard uses, steel or titanium for high pressure, and intricate uses. For this application, the selected material must solder and braid well; possess good sound quality and durability; and fatigue, corrosion, heat, and wear resistant. Some of the horn materials from which horns are made are titanium alloy, stainless steel, PCD, aluminum, and Monel aluminum bronze. Horns can be fixed to the machining tool through different techniques, such as welding, including soldering and brazing, and mechanical fastening such as screws and tapers fitting. They can be machined to conform to the shape and size of the respective tool depending on the degree of detail that is desired. One thing with threaded attachments is that they allow for fast and convenient ways in which the horn can be connected to the tool; however, problems such as loosening, fatigue failure, and poor levels of acoustic power can manifest. Figure 10.6 describes some kinds of horn's profile, which is used in ultrasonic welding.

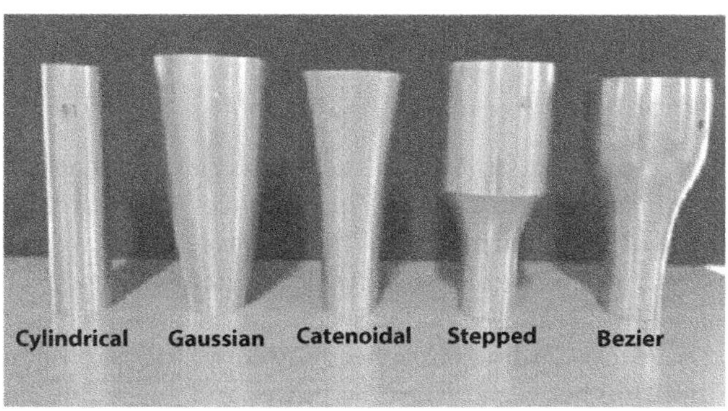

Figure 10.6 Ultrasonic horns with different profiles [5].

10.3 Factors Influencing Ultrasonic Welding

The ultrasonic welding is a fairly elaborate and delicate process in that it is very sensitive to a number of parameters that dictate the quality and integrity of the welded seam. Knowledge of these factors is important to be able to fine-tune the welding parameters to the best effect. Some of the parameters that affect ultrasonic welding include ultrasonic welding power, amplitude, welding time, welding pressure, and the use of energy directors.

10.3.1 Welding Power

The most important parameter of ultrasonic welding is the welding power because it dictates the energy applied to the weld zone. The power required to do this also depends with the number of layers being welded and the hardness of the materials being welded. A higher welding power can quickly melt the polymer and, in the initial stages of welding, will increase the rate of welding. Conversely, at the end of the welding, a high amount of heat can cause the material to melt and undergo thermal degradation within the weld zone, hence, having poor mechanical properties and containing develop cavities.

This can be computed with the help of the following formula for welding power [6]:

$$P = k \cdot H \cdot t$$

where P is the welding power (W), k is the welding coefficient, H is the hardness of the material, and t is the thickness of the composite plates (mm). The control of the welding power is critical in mitigating the degree of melting to the required level while at the same time not degrading the material.

10.3.2 Amplitude

The extent of movement applied to these materials is what is referred to as amplitude in ultrasonic welding. It influences the heat generation by friction as well as the deformation of the interface of the welded material. Substances vary in electrical conductivity; therefore, they need different amounts of amplitude to get a good weld. For example, materials such as metals that have low hardness may need low amplitudes to not damage it.

The relationship between ultrasonic power and amplitude can be given as follows [7]:

$$A = \frac{P}{4SF\mu f}$$

where A is the amplitude (μm), P is the ultrasonic power (W), F is the static pressure (N), S is the solder joint area (mm^2), μ is the friction coefficient, and f *is* the vibration frequency (kHz). There is a need to regulate the amplitude in an appropriate way to get the right heat being produced and flow of material at the weld zone.

10.3.3 Welding Time

Welding time is another important variable used in the context of the present study, which explains the period for which ultrasonic vibrations are exercised while welding. The duration of the welding process has to be carefully calibrated. Long welding durations may not generate enough heat to form a strong bond, and short welding durations may be indicative of a generation of a lot of heat damaging the material [8]. There is a need to find the most appropriate welding time to make sure there is enough melting and joining while at the same avoiding overflow or spoiling the material. The quality of the weld, as well as the temperature, which is achievable when working with CF-PEEK composites, depends on welding time.

10.3.4 Welding Pressure

Pressure applied during welding is of great importance to ensure that there is closure of the joints in the welded materials. Too low pressure results in poor jointness and too high pressure in splashing, extrusion of molten material, and even damages of the welded parts. There is also a relationship of the required welding pressure with respect to the properties of the material and welding conditions [8]. Appropriate pressure plays a great role in ensuring that at the weld area, the polymer is well dispersed; hence, a stronger bond is created. Scholars have also discovered that different comprehensive parameters, like the welding pressure, welding time, and amplitude, yield the greatest focal welding power of composite materials.

10.3.5 Energy Directors

Energy directors (ED) are small, raised structures on the surface of the material being welded, which help in focusing the ultrasonic energy at some individual points so that they can weld quickly. The structure, as well as the size of the energy directors, affects the welding process in one way or the other. There are several common structures with regard to the energy director, and these are triangular, semi-cylindrical, rectangular/flat, and trapezoidal. These structures assist in dissipating the heat uniformly over the welding zone, and so, there is good adhesion. However, it is not easy to machine energy directors on large or complex parts, and they are also less effective when the material thickness is large because of loss of energy while it is being transmitted.

10.4 Ultrasonic Welding on Various Materials

10.4.1 The Impact of Ultrasonic Welding on Aluminum Alloys

Reducing expenses and weight in the aviation industry is very important to improve on the fuel consumption and efficiency of the aircrafts. Ultrasonic metal welding is preferable to other methods, such as spot welding, because it requires lesser energy than self-piercing riveting and is therefore cheaper. This technique produces comparatively low heat; thus, there is no deformation and destruction of the aluminum alloy characteristic and improved handling of the components. Aluminum ultrasonic welds are characterized by a similar behavior in quasi-static and dynamic tests. There are various configurations that are used in ultrasonic spot-welding systems, and they include wedge, diagonal, and torsion. Transducers are machines that change an electrical signal of high frequency into a mechanical signal by making sound waves. Li *et al.* did a study to establish the correlation between ultrasonic welding parameters and the mechanical properties, microstructure, and diffusion behavior at the weld zone [9]. This paper has shown that the welding of Ti6Al4V and A6061 aluminum alloy by ultrasonic welding increases the hardness of the matrices. Optimal welding pressure and soldering time must be used. Applying 4-MPa pressure and 170-ms duration of pressure enables maximum joint strength. The aluminum alloys and copper joints are preferred in a battery pack and electrical manufacturing industries because of the high conductivity of the material. However, the formation of intermetallic compounds (IMCs) during fusion welding of these metals is a challenge that occurs at a very fast rate. Ultrasonic welding can do this by lowering bonding time and heat input. USW is capable of delivering high-strength welds in less than a second, regardless of the joining of dissimilar materials, such as aluminum to steel. The clamping forces are controlled, and the welding time also helps in improving the strength of the joint.

10.4.2 Steel and Its Alloys

Ultrasonic welding is a very efficient solid-state process for joining steel and aluminum. The study by Watanabe *et al.* demonstrated successful welding of magnesium-containing A5052 aluminum alloy to SS400 mild steel [10, 11]. The strength of the joint is greatly influenced by the optimum clamping force and welding time. For example, applying a clamping force of 588 N and a welding time of 2.5 s results in stronger joints, whereas longer reaction times can result in the formation of brittle Fe2Al5 intermetallic compounds, to loosen the joints.

Commonly used in aluminum and steel in the automotive and aerospace industries, ultrasonic spot welding provides an efficient method for producing high-strength joints Research shows that parameters, such as tensile force, vibration moment, and interface structure, play an important role in determining joint strength. Furthermore, the elevation in temperature during ultrasonic welding augments bonding strength, since phase transitions in titanium alloys diminish interface gaps and facilitate robust bonding.

10.4.3 Magnesium

Magnesium, the most common alkaline metal, melts at approximately 650°C. Ultrasonic welding (USW) systems for magnesium alloys have several important features: the electrical supply, the vibration wedge is large, a pneumatic cylinder that creates clamping pressure, and a piezoelectric transducer that converts electrical energy into mechanical vibration to enable the welding process.

A study on ultrasonic welding of AZ31BH24 magnesium alloy described the effect of welding energy on microstructure, crystal formation, hardness, and lap shear strength (LSS) [12, 13]. Their findings detected recrystallization in the nugget zone (NZ), which enabled slip systems, low stacking defect energy, and enhancement of grain boundary dispersion as the reasons for a similar study by Leon *et al.* It focused on the effects of vibration amplitude and welding time on joint quality, strength, and fracture morphology. The results showed that higher vibration amplitudes and surface friction forces contribute to improved weld strength and increased failure loads.

10.4.4 Titanium

Ultrasonic welding is widely used in the automotive industry to join dissimilar metals such as aluminum and titanium. This process requires careful control of interface mechanisms to control cracking and precipitation, which is important for optimal mechanical properties of the weld [14, 15]. It was found that silicon cracking in Al–Ti welding provides mechanical properties, such as bend angles, which are increased compared to welds without such cracks. Temperature during Al–Ti ultrasonic welding can increase at approximately 450°C facilitating rapid formation of magnesium oxide on the surface.

When welding dissimilar materials, such as Ti6Al4V and AA6111, significant accumulation of silicon, magnesium, and oxygen is observed at the interface. The high chemical reactivity of titanium poses challenges during manufacturing, as it can cause mechanical failure. To mitigate this, ultrasonic-assisted machining (USAM) is used to evaluate the performance of titanium alloys produced with different tool materials. Ultrasonic welding is proving effective for commercially pure titanium and titanium alloys, such as Ti15 and Ti31, enabling strong joint formation without excessive damage to the base material.

10.4.5 Fiber-Reinforced Composites

Efficient methods for bonding lightweight steels and fiber-reinforced polymers (FRP) in many applications are needed. Ultrasonic metal welding technology has shown success in forming high-strength joints between aluminum alloys and carbon fiber-reinforced polymers (CFRPs). Characterization of bonding techniques using photooptical microscopy and scanning electron microscopy revealed a strong relationship between metal surfaces and between carbon fibers in CFRP, and a tensile shear strength of up to 58 MPa [16, 17].

In addition, ultrasonic metal soldering provides a unique method for joining soft materials, such as glass and ceramic and metal parts, especially in the automotive and aerospace industries. This method provides vacuum-sealed welds, achieving tensile shear strengths of up to 50 MPa for steel/glass joints and 120 MPa for steel/ceramic joints.

10.4.6 Thermoplastics

From a kinetic point of view, ultrasonic welding of thermoplastics is very different from metal welding. Especially when the ultrasonic oscillation has the energy of an ultrasonic pulse, sonotrode uses high welding energy on materials up to approximately a ton. Unlike metal welding, where heat direct handling is important, the ultrasonic waves of thermoplastic welding initiate the welding effect causing the material to soften and bond under high-frequency pressure [18].

In ultrasonic welding processes for thermoplastics, high-amplitude vibrations are applied to the parts raising the temperature to a sufficient level for bonding Rapid heating near the interface is required to form a strong bond. This method is known for its speed, cost effectiveness, and ease of automation making it the preferred method for applications such as heat pump coupling.

Near-field ultrasonic welding is a technique in which the distance between the weld kernel and the joint interface is less than 6 mm. Pulsation energy increases increasing energy dissipation and subsequently improving solder strength. Although long soldering times benefit from the combined strength of quasi-crystalline polymers, amorphous polymers require optimal, often short, soldering times. Monitoring energy dissipation and static displacement is important for optimizing welding parameters to achieve the best of the best bond properties.

10.5 Applications

Ultrasonic welding is a versatile assembly technique used in a variety of industries. Its ability to reliably produce strong connections without the need for additional materials or extensive postprocessing is especially valuable in areas where accuracy and cleanliness are paramount This section explores the use of ultrasonic welding species emphasizing its importance in large-scale projects. Figure 10.7 gives examples of ultrasonically welded components.

10.5.1 Medical Devices and Equipment

In the medical field, ultrasonic welding is increasingly used to join soft and hard components without compromising the indestructible reproduction. This process is important for manufacturing devices such as catheters, blood filters, IV tubing, and surgical instruments. The minimum heat generated during the procedure preserves the materials' characteristics, hence ensuring their bio-compatibility and structural integrity. The lack of adhesives or solvents reduces the risk of contamination, which is essential for medical applications.

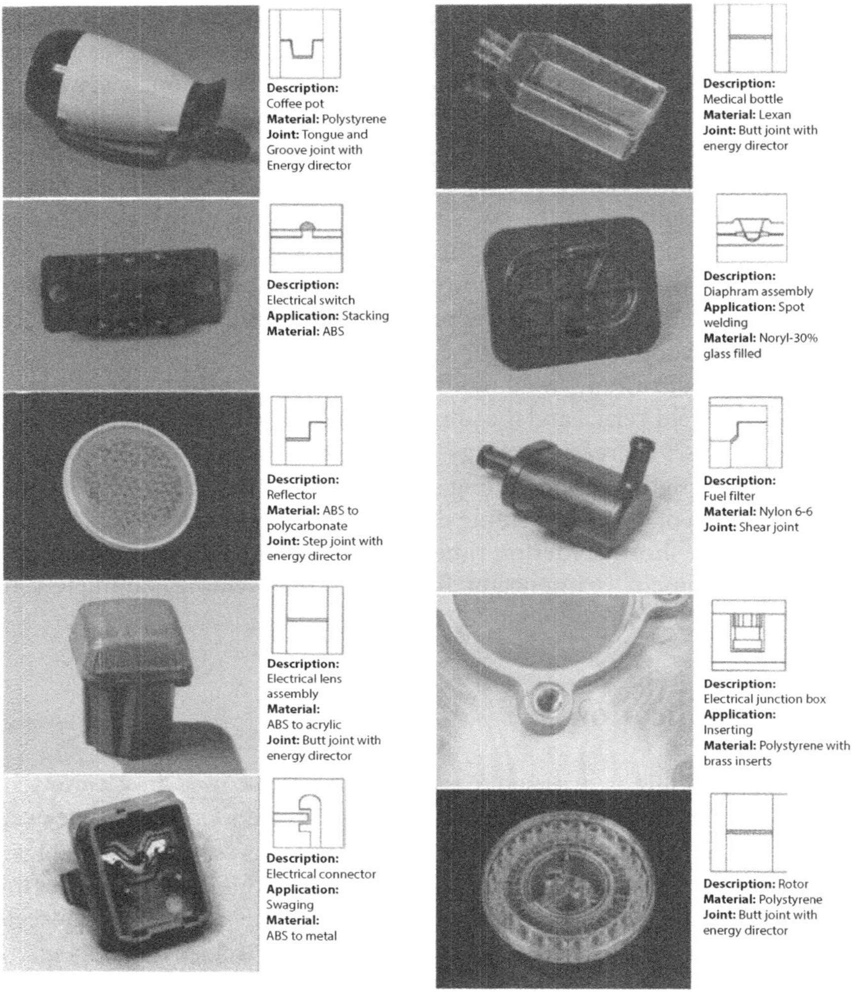

Figure 10.7 Examples of ultrasonic-welded items [19].

10.5.2 Automotive Industry

The automotive industry benefits greatly from ultrasonic welding due to its ability to join lightweight components helping to reduce the overall weight of a vehicle and improving fuel efficiency. Welded components are commonly made with this technique as wire harnesses, interior panels, instrument clusters, and airbag housings. Ultrasonic welding allows for fast cycles and provides durable joints that withstand the mechanical stresses experienced in automotive conditions and also enables reliable connections between battery packs and electric motor assemblies between species supporting the trend toward electricity.

10.5.3 Electronics and Electrical Components

Ultrasonic welding is widely employed in the electronics industry to join delicate and sensitive components without damaging them. It is used to assemble connectors, circuit boards, sensors, and microelectronics. The precision and control offered by ultrasonic welding ensure that connections are made without affecting the functionality of the electronic components. Moreover, the process is highly repeatable ensuring consistent quality in mass production environments.

10.5.4 Packaging Industry

In packaging industry, ultrasonic welding is used to coat tubes, bladders, containers, and packaging, especially for food and pharmaceuticals. The process provides an airtight and waterproof seal that protects ingredients from contamination and extends shelf life. Ultrasonic welding is preferred over traditional sealing methods because it produces a solid seal more quickly and efficiently, reduces waste and energy consumption, and allows water to remove powders and other contaminants as it can occur on the seal.

10.5.5 Textile and Apparel Industry

The textile and apparel industry uses ultrasonic welding to join synthetic fabrics and woven materials without the need for threads or adhesives, and this process is used to manufacture products, such as sportswear, outdoor equipment, medical textiles, and filters median. The benefits include the formation of smooth, flexible, and waterproof boundaries. Also, ultrasonic welding increases production speed and efficiency because multiple layers of fabric can be welded simultaneously.

10.5.6 Consumer Electronics and Appliances

The consumer electronics uses ultrasonic welding to join plastic parts and casings. Devices, such as smartphones, laptops, home appliances, and toys, often include ultrasonically welded parts. This process produces a clean, strong joint that gives the final product a durable and aesthetically pleasing finish. In addition, the high accuracy of ultrasonic welding complements the miniaturization trend of consumer electronics by enabling the assembly of small, complex parts.

10.5.7 Renewable Energy

In renewable energy, ultrasonic welding is used to manufacture photovoltaic (solar) panels and wind turbine components. It is used to combine thin metal and electrical connections for solar panels providing efficient and long-term reliable power transfer. Ultrasonic welding is used to join composites and components in wind turbines helping to improve the overall capacity and efficiency of the turbines.

10.5.8 Aerospace Industry

The aerospace industry uses ultrasonic welding to create lightweight, high-strength joints in critical applications and to assemble parts, such as fuel tanks, ducts, and structural elements. The accuracy and reliability of this method is essential to meet the stringent safety and performance standards required in the aerospace industry. Also, ultrasonic welding helps to join advanced materials, such as high-performance composites and alloys, which are widely used in modern aircraft.

In summary, the applications of ultrasonic welding are large and diverse spanning a wide range of industries that require accuracy, reliability, and efficiency. Its ability to combine a wide range of materials without compromising their quality makes it an indispensable option in modern architecture.

10.6 Advantages and Limitations

10.6.1 Advantages

1. **Energy Efficiency**
    - Ultrasonic welding consumes significantly less energy compared to other welding techniques, such as resistance or laser welding, resulting in cost savings and a lower environmental impact.

2. **Low Heat Generation**
    - The process generates minimal heat, prevents thermal distortion, material quality degradation, and eliminates the need for long cooling times. This is particularly beneficial for heat-sensitive materials.

3. **Rapid Processing**
    - The welding cycle is very short, usually completed in less than a second. These rapid processes increase overall productivity and manufacturing efficiency.

4. **Clean and Eco-Friendly**
    - Because ultrasonic welding requires no additional materials, such as adhesives or solvents, it produces clean joints free of contaminants making it an environmentally friendly alternative.

5. **Strong and Consistent Joints**
    - The method provides robust and reliable joints with repeatable capacity, which is important for applications such as aerospace, automotive, and electronics where joint integrity is critical.

6. **No Consumables Needed**
    - Unlike other welding techniques that require filler material, flow, or shielding gas, ultrasonic welding requires no consumables, reduces material costs, and simplifies the process.

7. **Versatility**
    - Ultrasonic welding is a versatile technique capable of joining a wide range of materials including metals, plastics, and composites. Especially effective for welding dissimilar materials, such as aluminum to metal or titanium, providing strong and portable joints reliable in a variety of areas.

8. **Automation Compatibility**
    - The process can be easily integrated into automated manufacturing processes to increase efficiency and ensure continuous quality control.

10.6.2 Limitations

1. **Material Restrictions**
   - Not all materials are suitable for ultrasonic welding. Very hard materials, large differences in hardness, or very soft materials may not be chewed with this method.

2. **Thickness Constraints**
   - Ultrasonic welding is usually limited to thin materials. Creating complex parts may require additional or multiple steps complicating the process and reducing productivity.

3. **High Initial Equipment Cost**
   - Initial investment can be higher for ultrasonic welding equipment. While long-term savings in energy and materials can offset these costs, upfront costs can be a barrier for some manufacturers.

4. **Precision Requirements**
   - Obtaining high-quality welds by ultrasonic welding requires precise alignment and careful observation of welding parameters. Any change in these parameters can result in poor joints or incomplete welds compromising the integrity of the overall bond.

5. **Tool Wear and Maintenance**
   - Sonotrodes (horns) and other equipment parts are subject to damage and may require regular maintenance or replacement. This ongoing maintenance can increase operating costs and downtime.

6. **Joint Design Constraints**
   - Ultrasonic welding may require specific joints to ensure force transmission and effective bonding. This can limit the failure rate and requires parts to be redesigned to match the welding process.

7. **Noise Generation**
   - The high levels of vibration used in ultrasonic welding can cause noise, which may necessitate workplace soundproofing measures to protect operators and reduce noise damage.

8. **Limited to Small-Scale Applications**:
   - Ultrasonic welding is more suitable for small jobs or precision components than larger structural welding, and its applications are tailored to specific jobs and applications.

Understanding these advantages and limitations will allow manufacturers to assess whether ultrasonic welding suits their specific needs and how best to incorporate this technology into their manufacturing processes.

References

1. Kumar, S., Wu, C.S., Padhy, G.K., Ding, W., Application of ultrasonic vibrations in welding and metal processing: A status review. *J. Manuf. Process*, 26, 295–322, Apr. 2017. 10.1016/J.JMAPRO.2017.02.027.
2. Tilahun, S., Vijayakumar, M.D., Ramesh Kannan, C., Manivannan, S., Vairamuthu, J., Manoj Kumar, K.P., A Review on Ultrasonic Welding of Various Materials and Their Mechanical Properties. *IOP Conf. Ser. Mater. Sci. Eng.*, 988, PP 012113, Dec. 2020. 10.1088/1757-899X/988/1/012113.
3. Yeh, H.J., Ultrasonic welding of medical plastics. *Joining Assembly Med. Mater. Devices*, 296–323e, Jan. 2013. 10.1533/9780857096425.3.296.
4. Márton, S., Anna, K.T., Márton, S., Anna, K.T., Producing a Dissimilar Joint of Copper to Austenitic Stainless Steel by Ultrasonic Welding. *AcMT*, 4, 2, 109–112, Oct. 2021. 10.33924/AMT-2021-02-10.
5. Roopa Rani, M. and Rudramoorthy, R., Computational modeling and experimental studies of the dynamic performance of ultrasonic horn profiles used in plastic welding. *Ultrasonics*, 53, 3, 763–772, Mar. 2013. 10.1016/J.ULTRAS.2012.11.003.
6. Senders, F., van Beurden, M., Palardy, G., Villegas, I.F., Zero-flow: a novel approach to continuous ultrasonic welding of CF/PPS thermoplastic composite plates. *Adv. Manuf. Polym. Compos. Sci.*, 2, 3–4, 83–92, Oct. 2016. 10.1080/20550340.2016.1253968.
7. Ganesh M. and Praba Rajathi R., Experimental study on ultrasonic welding of aluminum sheet to copper sheet. *Int. J. Res. Eng. Technol.* 2, 12, 161–166, 2013.
8. Liu, S.J., Chang, I.T., Hung, S.W., Factors affecting the joint strength of ultrasonically welded polypropylene composites. *Polym. Compos.*, 22, 1, 132–141, Feb. 2001. 10.1002/PC.10525.
9. Li, M., Zhu, Z., Xiao, Q., Zhang, Y., Mechanical properties and microstructure evolution of dissimilar Mg and Al alloys welded using ultrasonic spot welding. *Mater. Res. Express*, 6, 8, 086588, May 2019. 10.1088/2053-1591/AB2010.

10. Watanabe, T., Sakuyama, H., Yanagisawa, A., Ultrasonic welding between mild steel sheet and Al–Mg alloy sheet. *J. Mater. Process. Technol.*, 209, 15–16, 5475–5480, Aug. 2009. 10.1016/J.JMATPROTEC.2009.05.006.
11. Watanabe, A., Watanabe, T., Sasaki, T., University, N., Ultrasonic Welding Mild Steel Sheet to Al-Mg Alloy Sheet. *Adv. Mat. Res.*, 89–91, 627–632, 2010. 10.4028/WWW.SCIENTIFIC.NET/AMR.89-91.627.
12. Rubino, F., Parmar, H., Esperto, V., Carlone, P., Ultrasonic welding of magnesium alloys: a review. *Mater. Manuf. Processes*, 35, 10, 1051–1068, Jul. 2020. 10.1080/10426914.2020.1758330.
13. Huang, H., *et al.*, Heat generation and deformation in ultrasonic welding of magnesium alloy AZ31. *J. Mater. Process. Technol.*, 272, 125–136, Oct. 2019. 10.1016/J.JMATPROTEC.2019.05.016.
14. Zhang, C.Q., Robson, J.D., Ciuca, O., Prangnell, P.B., Microstructural characterization and mechanical properties of high power ultrasonic spot welded aluminum alloy AA6111–TiAl6V4 dissimilar joints. *Mater. Charact.*, 97, 83–91, Nov. 2014. 10.1016/J.MATCHAR.2014.09.001.
15. Zhang, C., Robson, J.D., Haigh, S.J., Prangnell, P.B., Interfacial Segregation of Alloying Elements During Dissimilar Ultrasonic Welding of AA6111 Aluminum and Ti6Al4V Titanium. *Metall. Mater. Trans. A Phys. Metall. Mater. Sci.*, 50, 11, 5143–5152, Nov. 2019. 10.1007/S11661-019-05395-7/TABLES/3.
16. Zhi, Q., Tan, L.X., Zhongxia, Cancel edits My Library Add REFERENCE TYPE Journal Article. *Welding J.*, 96, 429S–438S, 2017. Accessed: Aug. 11, 2024, [Online], Available: https://s3.amazonaws.com/WJ-www.aws.org/supplement/WJ_2017_11_s429.pdf.
17. Gao, Y.H., Zhi, Q., Lu, L., Liu, Z.X., Wang, P.C., Ultrasonic welding of carbon fiber reinforced nylon 66 composite without energy director. *J. Manuf. Sci. Engineering Trans. ASME*, 140, 5, 051009, May 2018. 10.1115/1.4039113/366552.
18. Raza, S.F., Khan, S.A., Mughal, M.P., Optimizing the weld factors affecting ultrasonic welding of thermoplastics. *Int. J. Adv. Manuf. Technol.*, 103, 5–8, 2053–2067, Aug. 2019. 10.1007/S00170-019-03681-7/METRICS.
19. Ultrasonic Welding, in: *Handbook of Plastics Joining*, pp. 15–35, Jan. 2009, 10.1016/B978-0-8155-1581-4.50004-4.

Part II

APPLICATIONS OF COMPUTATIONAL TECHNIQUES AND SUSTAINABILITY IN WELDING

11

Review and Analysis of Intelligent Welding Using Automation and Machine Learning Tools

Protyasha Kundu[1], Swarnadeep Saha[1], Sumanta Banerjee[2*] and Anindita Kundu[3]

[1]*Department of Computer Science and Engineering, Heritage Institute of Technology, Kolkata, West Bengal, India*

[2]*Department of Mechanical Engineering, Heritage Institute of Technology, Kolkata, West Bengal, India*

[3]*School of Computer Science and Engineering (SCOPE), VIT, Vellore, Tamil Nadu, India*

Abstract

This book chapter, as a research survey, investigates the recent advancements in automation and machine learning applications within the domain of welding technology. The focus centers on the presentation and analysis of methodologies, whereby these novel computational tools can be combined with prevalent welding techniques, such as tungsten inert gas (TIG) welding, friction stir welding (FSW), laser welding, resistance welding, metal inert gas welding, and plasma arc welding. The reviewed state-of-the-art studies, within the realm of these metal-joining processes, highlight the significance of automation and intelligent learning in enhancing welding consistency and efficiency. This book chapter presents a welding case study of machine learning (ML) model application using the kernel ridge regression, which achieves a root mean square error (RMSE) of 2.10 in predicting current flow during automated TIG welding. In a second study, the ML model applied for void formation prediction in FSW achieved an accuracy of 90.91%, which exceeds the accuracy levels of existing models in surveyed studies. The dataset, for both applications, has been collected from various open literature sources in this field. The results of the analysis show that machine learning has the

**Corresponding author*: sumanta.banerjee@heritageit.edu

Sandip Kunar and Gurudas Mandal (eds.) Advanced Welding Technologies, (221–260) © 2025 Scrivener Publishing LLC

potential to optimize the various welding parameters, besides ensuring that the welds are consistently of high quality.

Keywords: TIG welding, FSW welding, machine learning, kernel ridge regression, Gaussian process classifier

11.1 Introduction

Welding practice, the dynamic process and technology of fusing metals, has served as a critical component in shaping the infrastructure of modern human societies. From towering skyscrapers to delicate medical instruments, the expertise and accomplishments of skilled welders have left indelible, positive impacts on countless projects of regional, national, or global importance. Over the decades, this traditional workshop practice has witnessed unprecedented improvements and expansion to cope with global industrial demands. Presently, this technical art is undergoing significant transformations, owing to the cross-influence and infusion of automation and artificial intelligence (AI) tools at a growing pace. Apart from altering the fundamental practices of welding, this technological infusion has brought about improvements in quality, efficiency, and predictive maintenance. The AI and ML algorithms can evaluate real-time data, from welding operations, to identify flaws like porosity, cracks, and incomplete fusion. Machine vision systems use cameras and sensors to capture images and videos of welding processes, which are then analyzed by AI models to detect and classify defects. Early detection and correction minimize the chances of rework and wastage. Also, the AI-powered systems fine-tune welding parameters like current, voltage, travel speed, and torch angle. Leveraging extensive datasets from past welding operations, the ML models recommend ideal settings for various materials and conditions. This ensures better weld quality, consistency, and increased productivity. AI and ML can also foresee equipment malfunctions by examining feedback data from sensors that track the condition of welding machines. By detecting patterns and irregularities, these systems can predict potential failures, aiding preventative maintenance. This approach minimizes downtime and prolongs the equipment's lifespan.

Among others, one salient change in the automated welding scenario is the emergence of automated robotic welding machines. These machines, programmed for precision and accuracy, can perform repetitive welds with high consistency and speed, making them invaluable for high-volume production works that require identical welds on several parts. Robotic welding virtually eliminates the potential for human error in repetitive tasks. This leads to more consistent and reliable weld quality and reduces the need for rework. However, complex welding tasks demand adaptability and decision-making skills that most robotic welding machines currently lack. This creates an opportunity to implement AI in welding. Utilizing AI and ML algorithms, robots can execute intricate welding tasks with exceptional accuracy. These robots learn from (human) welders and can adapt to changing conditions in real time. These machines are AI-empowered to make prompt decisions, adjust parameters instantly, and manage tasks requiring high levels of dexterity and precision.

The AI-powered welding systems introduce higher adaptability and dynamic decision-making in real time. By analyzing data from sensors such as torch position, metal temperature, and weld pool dynamics, AI can tune the parameters dynamically to ensure optimal quality under varying working conditions. This adaptability is particularly valuable for more complex projects, where precision control is crucial. AI and ML also assist in efficiently managing the inventory for welding consumables and equipment. Predictive algorithms estimate the need for welding materials, which ensure timely restocking and prevent shortages. This optimizes the supply chain and lowers operational expenses. Moreover, AI and ML can improve and customize the basic welding simulation software by forecasting the results of welding processes under various conditions. These models, then, assist in planning and optimizing specific welding operations, thereby minimizing the necessity for numerous physical trials.

The AI models are evolving into powerful tools, designed to augment human capabilities. This collaboration between a human and his machine creates a new, unprecedented level of efficiency and precision, expanding the possibilities and horizons of welding processes. AI-powered virtual welding simulators offer training for novice welders through real-time feedback and assessments. Trainees can practice and enhance skills in a controlled setting, with the AI analyzing their performance and, thereafter, giving personalized suggestions for improvement.

There are several compelling reasons for the essential, and not just optional, incorporation of automation and AI in the present-day welding and manufacturing industry. Automation frees up skilled welders from repetitive jobs, which allows them to focus on intricate and complex tasks. Safety is another major concern. Welding can create potential safety risks for workers, which include exposure to harmful fumes, intense heat, and electromagnetic radiation. AI-enabled robotic welding can remove humans from these dangerous environments, which is crucial for human safety. Automation and AI can also greatly increase production speeds and consistency, leading to faster project completion times and lower production costs. Robots and AI can produce welds with higher degrees of accuracy and consistency than what is achievable by hand. This leads to stronger and more reliable welds, especially important for infrastructural projects.

The future of welding is one of collaboration, where skilled welders need to team up with intelligent machines, creating a new generation of human-machine collaboration that excels in technical skills and problem-solving. Automation and AI are, therefore, transforming welding, making it faster, more efficient, and more reliable.

11.2 Review of Welding Techniques and Literature Survey

11.2.1 Tungsten Inert Gas (TIG) Welding: An Overview

TIG welding is a technique for joining metals using a special electrode made from tungsten. Tungsten is an ideal choice since it has the highest melting point of any element (at about 3422°C). Unlike other welding methods, the tungsten electrode itself does not melt during use. In the TIG process, the weld area is usually shielded by argon (an inert gas), which prevents other gases from entering the weld zone and causing defects. The TIG welding produces clean, high-quality welds and provides precise control over the amount of heat generation rate, making the operations easier to manage and shape. The TIG process can be used for welding thin materials and complex geometries, where precision and weld quality are of paramount importance (such as aerospace and automotive sectors).

One of the major drawbacks of TIG welding in practice is its overreliance on the skills and competence of technicians, unlike some other welding methods. This aspect makes it one of the costliest welding techniques. Therefore, scopes and avenues for automation and/or applying AI and ML tools in TIG welding are sought after. Introducing these novel techniques can simultaneously render high precision and reduce process costs.

To train an automated (welding) model using ML, data from industry and workshops are required. Sophisticated TIG welding algorithms, informed by human expertise, have been created to optimize input parameters, which improves the precision of welding tasks.

Let us cite a specific instance in this regard. In TIG welding, it is well known that the weld bead shape significantly influences the mechanical properties of the assembly [1, 2]. To gather data, a robotic TIG welding system is used to conduct controlled experiments on aluminum welds. Data on the settings used (weld parameters) followed by the quality of the weld (weld-bead quality measurements) are automatically collected from the system to build a comprehensive dataset. This is followed by applying different ML models for analysis [3]. To model the relationships between system parameters and the resulting output, the Gaussian process regression (GPR) algorithm has been applied in [3]. This procedure enables accurate predictions of the necessary adjustments in welding current and/or operator skills in order to achieve the desired weld bead thickness.

Other instances of intelligent welding practice are now cited. Variations in the root gap cause a significant impact on the weld quality of materials. Root gap (see Figure 11.1) refers to the deliberate space left between the joining edges of two metals and needs to be carefully controlled. If too small, it can be difficult to achieve full penetration; if the gap is too large, it can waste filler metal and potentially weaken the weld due to excessive melting of base metal. An appropriate neural network architecture is designed in [4] in order to forecast the dimensions of the weld's rear side and the height of its upper side, by examining welding parameters and the upper weld form. In mechanized TIG welding, weld penetration is an important control parameter [5, 6]. In both [5] and [6], initially, a dynamic direct model is established and its results are analyzed. Thereafter, this model is enhanced by incorporating a nonlinear operating point using an adaptive neuro-fuzzy inference system (ANFIS), which effectively mimics the intelligence of experienced craftsmen with an accuracy delay of 1.5 units.

The incorporation of AI into manufacturing is expected to resolve a number of automation challenges, encountered by production departments globally that utilize advanced methods. A system that watches welds in real-time analyzes the data with intelligent models and catches defects with human-level accuracy is the future of welding. To cite an example, high welding speeds can drive deeper penetration but can also introduce flaws. Advanced camera sensors can identify these issues on the fly. The key to a perfect weld (of desired strength and thermomechanical properties)

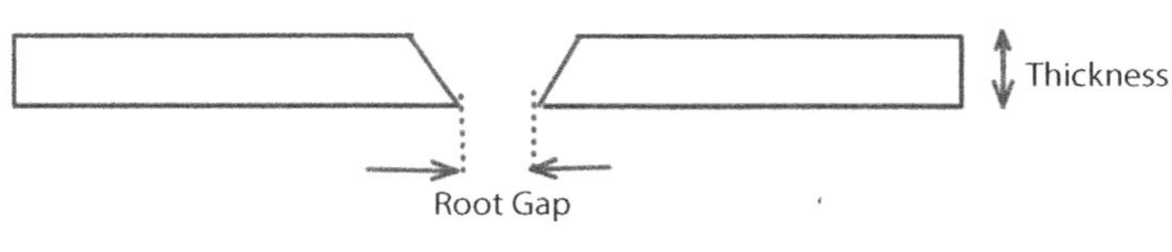

Figure 11.1 Root gap between metals.

lies in fine-tuning controllable settings like arc length, speed, and current, as these factors significantly influence the weld's precision.

11.2.2 Friction Stir Welding (FSW): An Overview

The FSW is a solid-state joining process for metals. Unlike traditional methods that melt the workpiece, the FSW process relies on friction and stirring in order to create a strong intermetallic bond. The technique makes use of a specialized rotating tool that is pressed against the joint (interfacial) area between the two pieces of material. The tool generates frictional heat, which softens the material without reaching its melting point. The tool used for FSW typically constitutes a shoulder and a pin (refer to Figure 11.2). The shoulder applies force on the material surface, while the pin generates the necessary heat that stirs (softens) the material to form a permanent joint. The FSW offers several advantages, which include slower distortion, reduced residual stresses, and improved mechanical properties. Additionally, the FSW technique enables the welding of materials that are traditionally difficult to join with conventional methods, such as aluminum alloys [7]. This makes the FSW process particularly well-suited for applications involving copper and aluminum (or magnesium) alloys, as well as the fabrication of heat exchangers and other complex geometries (spanning across a range of length scales).

In the FSW process, the formation of voids (or air gaps) inside the weld is a major concern. Voids can significantly deteriorate the quality of the weld and reduce its strength [8]. The abrupt change in material geometry (caused by voids) disrupts the flow of stress in the weld. This concentration of stress can initiate and propagate cracks at the edge of the void, often

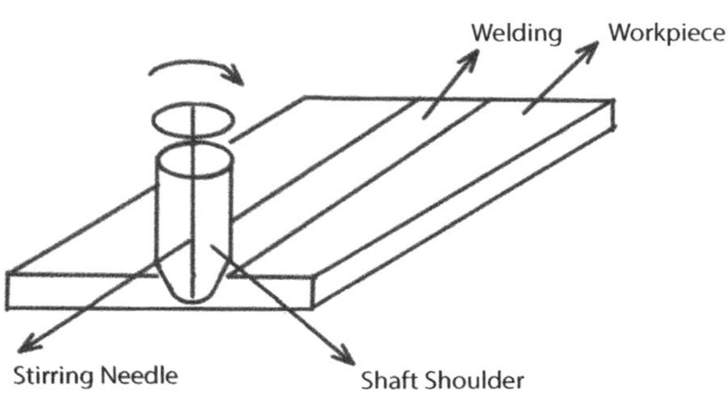

Figure 11.2 Working of friction stirs welding.

referred to as the "weld toe." Thus, it is necessary to identify and eliminate the "conducive" conditions in which voids are formed.

The quality assessment can be performed using data from previous friction stir-aided welds, conducted under varying conditions with different component speeds [9, 10]. An advanced monitoring framework can provide a quantitative assessment of the weld joint, which can potentially surpass traditional visual inspection. The strength of a weld is also linked to the condition of the surfaces being joined, as rough or damaged surfaces can create weaker joints.

In [11], the FSW process uses an ANFIS (adaptive neuro-fuzzy inference system), a technology that blends Fuzzy logic and Artificial Neural Networks (ANN), in order to estimate the strength and quality of a joint. This extended model, which uses inputs such as rotational speed and load, simplifies the modeling process. Fuzzy logic, integrated within the ANFIS framework, is versatile and straightforward and allows simultaneous modeling and online management.

The FSW process relies on a complex interplay between heat distribution, material flow velocities, deformation rates, and various physical and chemical characteristics of the material. This is in order to determine the final weld bead geometry and minimize void formation [12]. Void formation is a crucial issue that determines weld strength, and ML algorithms offer promising forecasting techniques to address this problem [13]. However, intelligent forecasting alone cannot completely resolve void formation issues. Machine learning, coupled with AI tools, plays a crucial role in optimizing process parameters to mitigate void formation, suggesting substantial scope for further exploration in this field.

11.2.3 Laser Welding: An Overview

Laser welding has emerged as a prominent craft for joining metals and thermoplastics. The process utilizes concentrated laser beams to generate narrow, deep welds at high speeds [14]. This method is well-suited for automated applications across industries, owing to its versatility in handling materials like carbon steel, aluminum, and even hard metals like titanium [15]. The quality of laser welds can rival those produced by electron beam welding, with weld speeds depending on the power used and the type of materials being joined [16]. Gas lasers are particularly favored for high-volume applications, making them prevalent in the automotive industry [17].

Ensuring flawless welds in the laser welding process requires close monitoring. In [18], researchers have used convolutional neural networks

(CNN) to develop a deep learning system that identifies defects in high-power laser welds. This intelligent system acts like a visual inspector and uses cameras, spectroscopes, and other sensors to detect process flaws in real time. To eliminate interference and improve training efficiency, the system incorporates filters and a special data-reduction algorithm. Another innovative approach [19] involves a self-learning ML system that analyzes complex data from laser welding cameras and adjusts welding power accordingly. Additionally, a technology called optical coherence tomography allows for real-time measurement of the weld zone, providing valuable insights into weld quality [20].

Artificial intelligence has been employed in [21] to optimize weld parameters, particularly in laser welding of aluminum-magnesium sheets. In [22], researchers have compared machine learning methods with traditional approaches to determine the most accurate settings for factors like oxygen pressure and pulse characteristics. This analysis helps assess the effectiveness of machine learning for optimizing laser welding processes [23].

11.2.4 Resistance Welding: An Overview

Resistance welding offers a rapid and efficient method for joining metallic components [24]. This technique utilizes the inherent resistance of metals to electrical current, which generates localized heating at the desired joint location (refer to Figure 11.3). A number of factors influence the resulting weld quality and strength. These include the dimensional characteristics of the workpieces, the presence (or absence) of any surface treatments on the metals, the geometry of the electrodes and their material properties, the applied clamping force, the parameters of the electrical current, and the (precisely controlled) welding time [25].

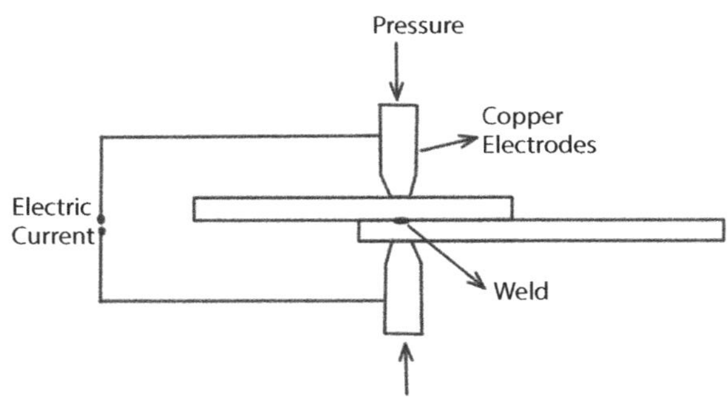

Figure 11.3 Resistance welding.

While resistance welding offers numerous advantages, it is important to address the complexity of the equipment involved. These systems often require high levels of technical expertise for both operation and maintenance, and their large industrial capacity may not be suitable for all applications. One of the major challenges is the "mushroom effect", characterized by the electrode wearing down, thereby increasing its diameter. The electrical current can be adjusted in order to counter this (adverse) effect. In addition, the implementation of non-destructive testing methods for quality assurance can be challenging with resistance welds [23, 25].

To address some of these limitations, intelligent welding systems have been developed and implemented. AI techniques are employed to monitor the quality of the weld in resistance spot welding processes [26]. To continuously monitor weld quality without damaging the weld itself, an ANN is equipped to analyze several key features, as reported in [27, 28]. These features include the temporal lag between two specific points on a signal curve (intonation and beta points), the corresponding difference in signal strength between these points, the rate of change in the curve as it relates to the applied weld heat, and the voltage signal [29]. The network employs additional means to filter out unwanted noise. By analyzing these combined features, the ANN can effectively assess the quality of the weld [28, 29].

In [30], an anomaly detection model has been developed for unbalanced data classification, which proves more effective than alternative methods such as one-class SVM and local outlier detection. The model categorizes data into normal and abnormal classes that address the significant difference between the 'minority' and 'majority' classes [31–34]. This approach is crucial for monitoring the quality of the welding process, particularly for enameled wire manufacturing. This method offers a practical approach to managing resistance and power waveforms [35]. Additionally, it enables intelligent Non-Destructive Testing (I-NDT) for defect identification.

11.2.5 Metal Inert Gas Welding (MIG): An Overview

MIG welding, also called Gas Metal Arc Welding (GMAW), is an established arc welding process, valued for its versatility, speed, and ability to produce clean, strong welds [32]. Unlike some other arc welding methods, where the electrode is non-consumable, MIG welding uses a continuously-fed solid wire electrode. This wire acts both as filler metal and a heat source, which eliminates the need for separate filler rods. As the electrode wire feeds through the welding torch, an arc forms between it and the workpiece. This generates intense heat that melts the wire as well as the base metal (Figure 11.4).

The metal inert gas welding employs argon or helium as a shielding gas, which segregates the weld pool from the surrounding atmosphere. This inert gas creates a protective barrier, preventing (atmospheric) nitrogen and oxygen from compromising the weld quality [33]. A continuous wire feed and inert gas shielding contribute significantly to the speed and efficiency of MIG welding. The process is considered relatively easy to train and learn, as compared to other arc welding techniques. It stands out as a preeminent arc welding process due to its remarkable adaptability [34]. This adaptability translates into a prominent role across a vast array of industries. For instance, one can cite the automotive sector [36], where MIG welding seamlessly joins car body panels and frames. Similarly, in shipyards, the MIG welding plays an important role in constructing decks and hulls.

In addition to these prominent applications, MIG welding finds itself equally adept in sheet metal fabrication, where it excels at joining thin metal sheets [37]. Construction projects heavily rely on MIG welding for tasks such as building bridges and pipelines [38]. However, there are inherent limitations to MIG welding. The dependence on inert gas shielding renders the process susceptible to disruptions caused by wind and convection currents, particularly in outdoor applications where proper safeguards are not implemented. Additionally, highly reactive metals like magnesium or titanium pose a challenge due to their inherent affinity for oxygen [39]. This affinity can deteriorate the weld quality.

In MIG welding, the following three parameters play key roles in achieving optimal weld strength and hardness: gas pressure, welding current, and welding speed. Process optimization can be attained using techniques like

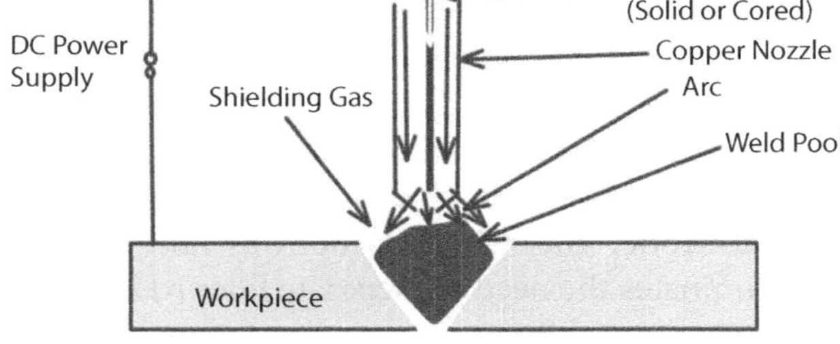

Figure 11.4 The MIG welding.

the Analysis of Variance (ANOVA), which identifies the salient (influential) parameters in a dataset. Genetic Algorithms can also offer further refinement for precise optimization [40]. Finally, by using experimental data to train Artificial Neural Networks (ANNs), researchers have developed a system capable of predicting the ideal combination of welding current, gas pressure, and welding speed in order to achieve the requisite hardness and tensile strength [41].

Gas pressure significantly influences the hardness of the weld, and the welding current enables it to attain optimum tensile strength in the welds. The Response Surface Methodology (RSM) is another mathematical tool utilized to model and analyze processes where multiple parameters affect the desired response, thereby enhancing prediction accuracy [42]. Through a series of planned experiments under RSM, the optimal parameter values are identified to construct precise ANN models, which can determine tensile strength [43].

Addressing MIG welding challenges like inadequate penetration [44], fusion deficiency [45], and adept control of weld profile requires advanced machine learning algorithms. These technologies offer solutions to enhance welding outcomes by refining parameter settings and process control strategies.

11.2.6 Plasma Arc Welding (PAW): An Overview

The plasma arc welding (PAW) process, also known as Plasma Metal Arc Welding (PMAW), is a sophisticated arc welding process and stands out for its precision. It has the ability to handle a wide variety of metals. Unlike traditional (arc welding) methods that rely on a consumable electrode, the PAW technique utilizes a non-consumable tungsten electrode and a constricted plasma arc for heat generation [46]. This plasma arc is formed by ionizing an inert gas, typically argon or helium, as it travels through a narrow nozzle within the welding torch. The high-velocity, high-temperature plasma stream offers several advantages over conventional arc welding [47].

The most notable benefit of PAW lies in its exceptional heat concentration. The constricted plasma arc allows for deeper penetration into the workpiece, as compared to other arc welding methods [48]. This capability makes PAW ideal for welding thick sections of metal in a single pass, improving both efficiency and weld quality. Additionally, the focused nature of the plasma arc minimizes heat distortion in the surrounding metal, leading to cleaner and more precise welds. The PAW's versatility extends to its ability to weld a broader range of metals, compared to many other processes. The high-temperature, concentrated plasma arc effectively melts even challenging materials like stainless steel, aluminum, and other reactive metals [49]. This adaptability makes PAW a viable tool across diverse industries, from automotive manufacturing and aerospace applications to construction and shipbuilding.

However, it is important to acknowledge that PAW also presents some limitations. Compared to other arc welding methods, the complex equipment and precise control requirements can make it a cost-intensive process. Additionally, the high operating temperatures associated with PAW necessitate the use of specialized shielding gases and proper operator training. This is to ensure safety and optimal weld quality [50].

The intricacies of PAW have attracted a number of studies devoted to ML applications. Machine Learning has been used to analyze acoustic features of the keyhole, which enables regression analysis for weld quality prediction [51, 52]. Another method involves the Taguchi experiment-based fuzzy neural networks, which can optimize weld quality by enabling better decision-making capabilities [53]. The Taguchi method is an optimization process that involves eight established steps. By proposing, managing, and analyzing matrix experiments, it identifies optimal settings for keyhole control factors like the current, speed, gas flow rate, and voltage [51]. These optimized parameters then serve as valuable inputs for the neural network design. Existing research demonstrates the effectiveness of this approach, with successful control and optimization of welding parameters achieved through neural networks and utilizing the Taguchi method [54]. The future of ML in PAW is promising, with the potential for further improvements in weld quality and employability for lesser-skilled operators [55].

11.3 Machine Learning Models

11.3.1 Decision Trees

In the realm of machine learning, where algorithms grapple with complex patterns and predictions, decision trees offer a unique and novel approach to solutions. It is like a flowchart, a roadmap of choices, where each twist and turn leads closer to a solution. A decision tree, in essence, is a supervised learning method that unravels insights from data through a series of branching questions [56, 57]. At the heart of a decision tree lies a hierarchical structure, mimicking the branches of a tree [56]. The journey begins at the root node, which holds the most critical question, the one that best divides the data into meaningful subsets. Each internal node that follows poses another question, further refining the data, based on the answer to the previous one. This continues until one reaches the leaf nodes (or the terminal points) where the final predictions or classifications are made (refer to Figure 11.5).

The construction of a decision tree is akin to a meticulous interrogation of the data [58]. The algorithm meticulously sifts through the available features, identifying the most informative ones at each step. This informative feature is the one that best separates the data points into distinct categories, bringing one closer to the desired outcome. The process is often referred to as a "greedy search", where the algorithm relentlessly pursues the split that yields the most significant improvement at each stage [59].

Decision trees excel in their interpretability. Unlike some complex algorithms that hide their inner working, decision trees offer a clear view of the logic behind their predictions. One can comprehend the sequence of questions that lead to the final outcome, by revisiting the path from the root node to the leaf node [60]. This transparency makes them a valuable tool for understanding not only what the model predicts, but also why it makes those predictions.

The versatility of decision trees is another strength. Decision trees are effectively employed for both classification tasks, where the category of data must be predicted, and regression tasks, where the objective is to estimate a continuous value [61]. Decision trees can evaluate past welding data to pinpoint the best parameters for different welding conditions and can assist in finding the optimal combinations of voltage, current, travel speed, and other factors, ensuring consistent high-quality welds. For defect detection and classification, decision trees can identify welding flaws such as

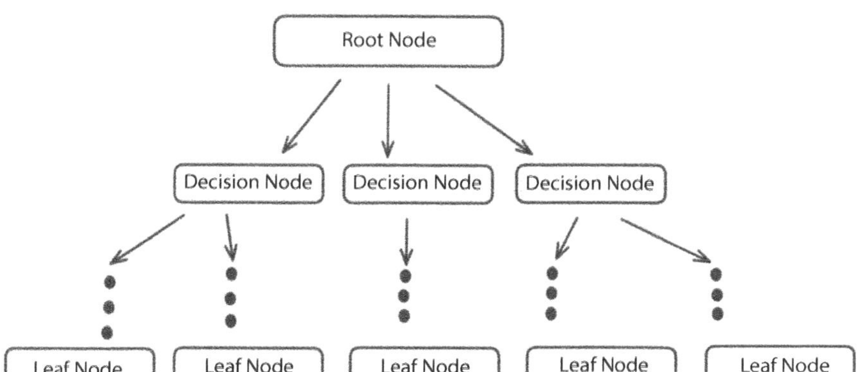

Figure 11.5 Schematic of decision tree architecture.

cracks, porosity, and incomplete fusion. In predictive maintenance, decision trees can forecast equipment failures by examining data from sensors monitoring welding equipment health. They detect signs of wear and tear, predicting when maintenance is required to avoid unexpected downtime. Quality control is enhanced by decision trees through the analysis of correlations between welding parameters and weld quality. In automated welding systems, decision trees make real-time decisions about process adjustments, making modifications in speed, angle, or heat input based on current data, ensuring optimal performance throughout the welding operation. For root cause analysis, decision trees systematically break down problems and evaluate the impact of different variables. By optimizing welding processes and reducing defects, decision trees contribute to cost savings. They promote efficient material use, extend equipment longevity, and minimize rework and scrap.

However, decision trees are not without their limitations. One potential pitfall is overfitting, where the tree becomes highly attuned to specific training data and fails to generalize well to unseen examples. To overcome this limitation, techniques like pruning can be used to remove redundant branches and prevent the tree from becoming overly complex.

Another consideration is the inherent bias introduced by the splitting process at each node. The order of feature evaluation can influence the final structure of the tree. To address this issue, ensemble methods like random forests, which combine predictions from multiple, slightly different trees, can be employed to create a more robust model.

11.3.2 Random Forest Classifier

Random forests offer a powerful approach that leverages the wisdom of crowds. Unlike a single decision tree, which can be susceptible to biases and overfitting, random forests harness the collective strength of an ensemble of decision trees, which yields more robust and accurate results [62].

A random forest classifier integrates a multitude of decision trees, with each tree trained on a subset of the original data [63]. These subsets are generated through a technique called bootstrapping, where data points are sampled with replacement, meaning some points may appear multiple times in a single tree's training data, while others are omitted entirely.

The diversity introduced by bootstrapping is a cornerstone of random forests' strength [64, 65]. Each tree in the ensemble learns a slightly different set of rules, based on its unique training data. This heterogeneity prevents the forest from becoming overly reliant on any specific feature or pattern within the data, which is a common pitfall of single-decision trees. When a new, unseen data point enters the forest, it traverses each individual decision tree, encountering a series of questions at each node. These questions delve into the data point's features, ultimately leading to a classification prediction at the leaf node using its given parameters [64]. The final output generated by a random forest classifier is determined by a democratic process, i.e. the class that receives the most "votes" from the individual trees is declared the winner. This "voting mechanism" offers a significant advantage; by combining the predictions of multiple trees, random forests are less susceptible to biases or errors that can plague individual trees [66]. If one tree makes a mistake due to a quirk in its training data, the other trees can potentially compensate for it, leading to a more accurate overall prediction. Another strength of random forests lies in their ability to handle missing data [67]. Unlike some algorithms that struggle with incomplete data points, random forests can gracefully navigate such scenarios. When a data point is missing a particular feature during its journey through a tree, the algorithm can consider alternative splitting rules based on the available features, ensuring a robust prediction even with incomplete information.

As the literature survey shows, the use of random forest classifiers significantly benefits welding process optimization. They enable the analysis of welding data to determine the optimal parameters for various scenarios and offer precise recommendations for voltage, current, travel speed, and other system variables to ensure consistent, high-quality welds. These classifiers identify defects like cracks, porosity, and incomplete fusion by training on labeled datasets, allowing them to recognize patterns and predict potential defects in real time for immediate corrective action. Integrating random forest classifiers into predictive maintenance systems forecasts equipment failures by analyzing sensor data for signs of wear, enabling timely maintenance and preventing downtime. They also analyze the relationship between welding parameters and weld quality, helping establish stricter control processes to ensure each weld meets quality standards. In case of defects, random forest classifiers perform root cause analysis by evaluating various variables, identifying underlying causes, and leading to effective solutions. Optimizing welding processes and reducing defects with these classifiers results in cost savings through efficient material use, improved equipment longevity, and minimized rework and scrap. They enhance data-driven decision-making by providing insights from vast datasets, enabling manufacturers to make informed decisions that improve efficiency, quality, and productivity.

However, random forests are not without limitations [68]. Tuning the hyperparameters, which control the construction of the individual trees, can be crucial for optimal performance. Finding the right balance between the number of trees and their complexity requires careful experimentation with the data [69]. Additionally, while random forests offer interpretability for individual trees, understanding the logic behind the entire ensemble's decision can be more challenging [70].

11.3.3 Support Vector Machine

The Support Vector Machines (SVMs) are unique in their approach to handling data. In a high-dimensional space where data points reside, the SVMs aim to construct a special dividing boundary, called a hyperplane, within this space (refer to Figure 11.6). This hyperplane is a decision boundary; it is meticulously crafted to achieve a specific goal. The goal (objective) is to maximize the separation between different classes of data points [71].

The core concept of SVMs revolves around the notion of a margin. The margin signifies the separation between the hyperplane and the closest data points on either side, known as support vectors [72]. The SVMs strive to create a hyperplane with the widest possible margin, essentially aiming for the most clear-cut division between classes. Intuitively, a wider margin translates to better generalization, meaning the model is more likely to perform well on unseen data [73].

The process of constructing an SVM is similar to carving a path through a dense forest. The algorithm meticulously analyzes the data, searching for

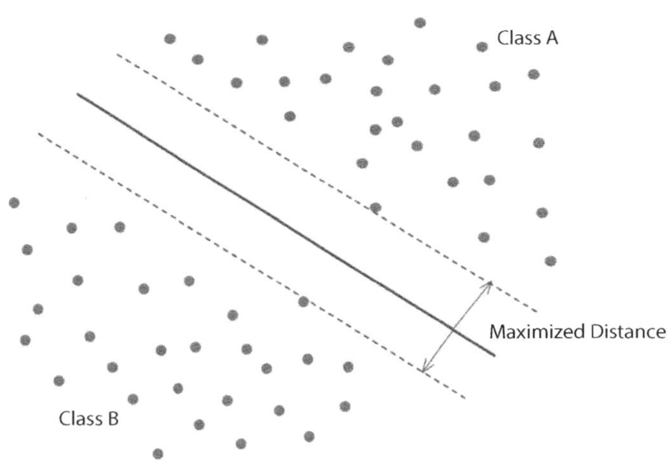

Figure 11.6 The support vector machines method: The optimal hyperplane (solid line) separating class A and class B parallel to the bounding hyperplanes (dotted lines).

the optimal hyperplane that separates the classes with the greatest margin. This search considers various factors, which include the number of dimensions of the data as well as the inherent complexity of the relationships between features [74]. One of the significant advantages of SVMs lies in their effectiveness in high-dimensional spaces. Unlike some algorithms that struggle with the "curse of dimensionality", SVMs can handle data with many features without a significant performance drop [75]. This makes them particularly well-suited for tasks where data is inherently complex and multifaceted [76]. Furthermore, SVMs are equipped with a certain degree of elegance in their ability to handle nonlinear data [77]. While the core concept revolves around hyperplanes, which are inherently linear, SVMs are generally adapted to work in nonlinear scenarios through the use of kernel functions [78]. These functions essentially project the data into a higher-dimensional space, where a linear separation becomes possible. This allows SVMs to tackle problems where a clear-cut linear separation in the original space may not exist [79].

However, it is worthwhile to note that SVMs are not without their limitations. One potential challenge lies in the computational cost associated with training the model. With increasing dataset size, the time and resources required to train an SVM effectively proportionately increase [80]. Additionally, choosing the optimal kernel function for nonlinear problems can be an art form, requiring careful consideration and experimentation, which is (more often) an outcome of experience with handling data [81]. Despite these limitations, SVMs remain a powerful tool in the machine learning toolbox. Their ability to handle high-dimensional data, maximize margins for robust classification, and extend to nonlinear problems through kernels makes them a versatile choice for various tasks. From text classification [82] and image recognition [83] to bioinformatics [84] and anomaly detection, SVMs continue to be a mainstay in the machine learning landscape.

11.3.4 Artificial Neural Networks (ANNs)

Unlike traditional algorithms with rigid rules, ANNs are designed to learn and adapt through a process that loosely mimics the way biological neurons process information [85]. In a web of interconnected nodes, each represents an artificial neuron (Figure 11.7). These artificial neurons are said to be loosely modeled after their biological counterparts, receiving inputs from neighboring nodes, performing simple calculations, and transmitting a signal to others. The connections between these nodes hold weights [86], which signify the importance of each input in influencing the final output.

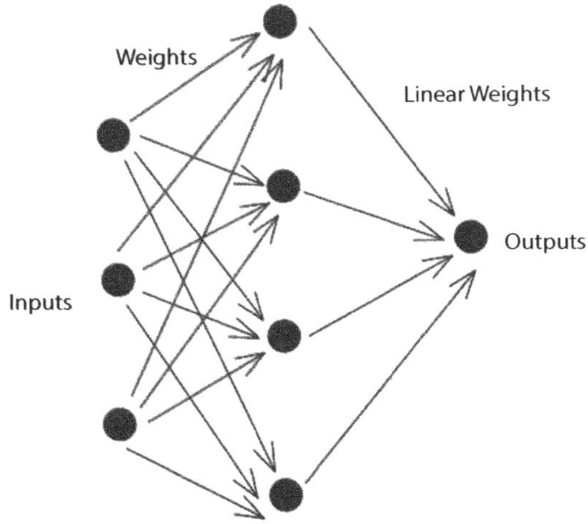

Figure 11.7 Schematic of artificial neural network (ANN) architecture.

The learning process in ANNs is akin to training the brain. The network is exposed to data, and based on the desired outcome, it adjusts the weights of the connections [87]. This adjustment can be likened to strengthening or weakening the connections between neurons in the biological brain. Through a series of iterations, the network iteratively refines the weights, striving to minimize the loss function, which is a variant of predicted values subtracted from the real outputs [88]. Training involves adjusting the weights and biases to minimize the difference between the predicted and actual outputs. This is typically done using a process called backpropagation [89], which involves the following:

- Forward Pass: Computing the output of the network with the current weights.
- Loss Calculation: Measuring the error between the predicted and actual output using a loss function.
- Backward Pass (Backpropagation): Calculating the gradient of the loss function with respect to each weight and updating the weights to reduce the error.

There are several key components that define the architecture of an ANN. The input layer is where the raw (field) data reside [90]. The hidden layers, often stacked upon each other, perform the core computations. These layers are where the transformation happens—the intricate web of connections and weighted sums allows the network to extract complex patterns and relationships within the data [91]. Finally, the output layer translates the internal processing into a concrete prediction, such as a classification or a numerical value [92].

The power of ANNs lies in their ability to learn intricate, nonlinear relationships within data [93]. Unlike simpler algorithms that rely on predefined rules, ANNs can uncover hidden patterns through the complex interplay between neurons and weighted connections [94]. This makes them particularly well-suited for tasks like image recognition, where the underlying patterns can be highly nuanced and difficult to express with explicit rules [92].

Furthermore, ANNs boast a remarkable degree of flexibility [95]. By adjusting the number of hidden layers and the number of neurons within each layer, the network's capacity for learning and adaptation can be tailored to the specific problem at hand. This allows ANNs to tackle a wide range of tasks, from natural language processing and speech recognition to complex time series forecasting and financial modeling.

The literature survey shows a growing R&D focus on applications of several variants of ANNs in a wide range of applications in welding technology, significantly enhancing various aspects of the process. ANNs are used to optimize welding parameters (e.g., current, voltage, travel speed, and torch angle) by learning from extensive datasets of previous welding operations, which play a crucial role in defect detection (e.g., cracks, porosity, and incomplete fusion) and ensures consistent, high-quality welds. In predictive maintenance, ANNs analyze sensor data to forecast equipment failures. Additionally, ANNs improve the efficacy of welding simulation software by accurately predicting the outcomes of different welding scenarios, reducing the need for extensive physical trials. With the aid of ANNs, virtual welding simulators (that provide real-time feedback and personalized improvement recommendations) can be implemented in industrial workshops for training new welders and/or helping trainees refine their skills in a controlled environment.

However, ANNs are not without their limitations. One key challenge lies in the inherent complexity of these networks [96]. Training large ANNs can be computationally expensive, requiring significant processing power and vast amounts of data. Additionally, interpreting the inner workings of a complex ANN can be difficult [97], making it challenging to understand how the network arrives at its conclusions.

11.4 Predicting Current in Tungsten Inert Gas (TIG) Welding

11.4.1 Methodology

A human welder has to assess the properties of a workpiece material before commencing the process of welding. A skilled welder can determine the quality of the weld bead by visual inspection of the thickness of the workpiece material. After initiating the welding process, the welder must dynamically change the weld-bead quality at time 't', based on the observed quality of the weld-bead put in place at time (t-Δt), where 'Δt' is a pre-assigned time step. The dynamic updation is done by changing the regulating of current flowing into the welding torch [3]. However, the primary difficulty lies in determining the amount of current over each time step 'Δt'.

In order to automate the process of TIG welding, one needs to determine the current (I) at a particular time instant. This warrants a functional I-t relation. The dependence between the input parameters and the final output current is a nonlinear one, and the multiparametric dependence implies that the mathematical relation is not simple. From the foregoing discussion, one realizes the merit of a machine learning model to accurately predict the values of desired parameters. In the present context, the input parameters are the width of the weld material at the previous time instant, the width of the weld at the current (time) instant, and the current (at the previous time instant) passed onto the welding torch to achieve the desired width. The objective is to determine the current that should be passed onto the welding site to produce a good-quality weld bead.

11.4.2 Gaussian Process Regression

The outline of the GPR is as follows. When a large number of random points are plotted on a graph, a pattern is generally observed: most of the data tends to cluster around a central point, with fewer points farther away. This familiar bell-shaped curve is known as a normal distribution. A normal distribution is defined by its mean, which represents the center of the data, and its variance, which measures how spread out the data is. Instead of plotting individual points, if the values are sampled from an infinite vector space, it leads to the definition of the Gaussian process.

When the analytical form of a function, based on limited data points, needs to be predicted, one can assume that this function is a sample from a Gaussian process. By treating many possible functions as random variables within this process, one can identify the function most likely to fit the observed data (based on calculated probability). In the Gaussian process, the best-fitting function is selected from several other possible functions.

The Gaussian process is used in conjunction with a kernel function, which mathematically quantifies the similarity between data points, and can also map the data into a higher-dimensional space without explicitly doing the conversion. Because of this, the GPR does not depend on the number of model parameters, and hence, it is nonparametric in nature. In other words, one can estimate a model of infinite parameters. The GPR is particularly a powerful tool in estimating functions in small- to moderate-sized datasets that exhibit smooth and continuous functions.

11.4.3 Kernel Ridge Regression

The standard linear regression works on the principle of reducing the "sum of squares of the error" term. The "error" term is calculated by taking the difference between the actual output value $\{y_i\}$ and the predicted output value $\{w^T x_i\}$. The primary drawback of this technique, especially on a smaller dataset, is "overfitting." "Overfitting" is when the model performs well on training data but fails to perform well on the testing data [98, 99]. The model learns the training data very well, which might not be as useful when one applies the model to other data of the same distribution. If there is multicollinearity, i.e., multiple features are correlated with each other, then ridge regression can be applied. Ridge regression adds a penalty term to the sum of squares of the error, which discourages large values of coefficients. The cost function hence becomes

$$C = \frac{1}{2}\Sigma_i \left(y_i - w^T x_i\right)^2 + \frac{1}{2}\lambda |w|^2 \tag{11.1}$$

In order to minimize the cost function, its derivative must be equated to zero

$$\Sigma_i \left(y_i - w^T x_i\right) x_i = \lambda w \Rightarrow w = \left(\lambda I + \Sigma_i x_i x_i^T\right)^{-1} \left(\Sigma_i y_j x_j\right) \tag{11.2}$$

This prevents the model from excessively relying on any one feature and makes the coefficients more spread out. Along with reducing the effects of overfitting, the ridge regression technique also helps improve the stability of the model by making it less sensitive to small changes in data. An unstable model is highly sensitive to even small changes in (input) data, whereby large changes in output values may be obtained.

For nonlinear data, it becomes necessary to project the data to a higher-dimensional space to obtain linear correlations and then apply the ridge regression technique to them [98]. The "kernel trick" in kernel ridge regression takes two data points as inputs and calculates their similarity in a higher-dimension data space implicitly, without performing the mapping explicitly. Using the kernel trick, the objective is to replace the datasets with their feature vector $x_i \rightarrow \phi_i = \phi(x_i)$. Applying the method, the solution can be obtained as in [100, 101]:

$$w = \left(\lambda I_d + \phi\phi^T\right)^{-1} \phi y = \phi\left(\phi^T\phi + \lambda I_n\right)^{-1} y \tag{11.3}$$

By projecting the value of the calculated dataset to the solution (w), we get

$$y = w^T\phi(x) = y\left(\phi^T\phi + \lambda I_n\right)^{-1} \phi^T\phi(x) = y(K + \lambda I_n)^{-1}\kappa(x) \tag{11.4}$$

In Eq. 11.4, $K\left(bx_i, bx_j\right) = \phi(x_i)^T\phi\left(x_j\right)$ and $\kappa(x) = K(x_i, x)$.

By restricting the analysis to linear models, the kernel trick enables us to avoid the "curse of dimensionality" problem. Instead of working with high-dimensional data points [100], one can simply compute the kernel function (computationally more efficient).

11.4.4 Results

The dataset has been collected from the published study by Huff *et al.* [3]. Table 11.1 depicts an instance of the utilized dataset. The dataset consists of input parameters that had been collected from an industrial automated robot, and then trained using machine learning algorithms. The setup incorporates several key components. An ABB IRB140 industrial robot equipped with an IRC5 controller provides the requisite precise and controlled movement required for the (automated) welding process. The actual welding is performed using a Coplay Norstar T200 AC/DC TIG welding machine. To ensure a safe and clean working environment, an air ventilation system is also included. Finally, a 100% argon gas environment is utilized to shield the weld pool from contamination, essential for achieving high-quality welds. In [3], the GPR has been used. For an overview of the GPR, the reader is referred to Section 11.4.2.

Table 11.1 Instance of the dataset [3].

Exp. no.	Current 1 (A)	Mean width 1,2,3 (in)	σ^2(in × 10^{-6})	Current 2 (A)	Mean width 4,5,6 (in)	σ^2(in × 10^{-6})
...	...	...	...	...	...	...
70	68	0.1697	4.333	75	0.1933	37.333
71	79	0.2073	6.333	83	0.122	37
72	59	0.1367	8.333	69	0.1683	2.333
73	81	0.2127	6.333	52	0.1267	8.333
74	53	0.1153	2.333	83	0.21	25
75	52	0.1203	20.333	76	0.1893	26.333
...	...	...	...	...	...	...

The referred dataset consists of only three input parameters: weld width, previous current (i.e., current at the previous time instant), and previous weld width. The expected result is to detect the current (in A) with sufficient accuracy. The estimation of process parameters such as voltage and/or current is important for (welding) process monitoring and automation, whereby weld quality, welding defects, and arc stability can be assessed. In the present analysis, the models have been trained using an open-source machine learning toolbox, using Python programming language. Several algorithms were run to predict better accuracy of the data. Kernel ridge regression (KRR) performed significantly better than other models, with an accuracy of more than 95%. The RMSE between the predicted and the real value is 2.104 A; this margin is lesser than 2.26 A achieved by Huff *et al.* [3] using the GPR tool.

The models employed in the present study have significantly outperformed other models applied to the dataset. The authors have randomly selected 90% data as the training set, and the remaining 10% as the testing dataset. The model predicts the testing dataset with sufficient accuracy, as shown in Figure 11.8.

Five (5) data points were also taken as validation testing for the chosen model, as shown in Table 11.2. Overall, the chosen model performed better than the applied model in [3], and it exhibited relatively higher accuracy, as shown in Figure 11.9.

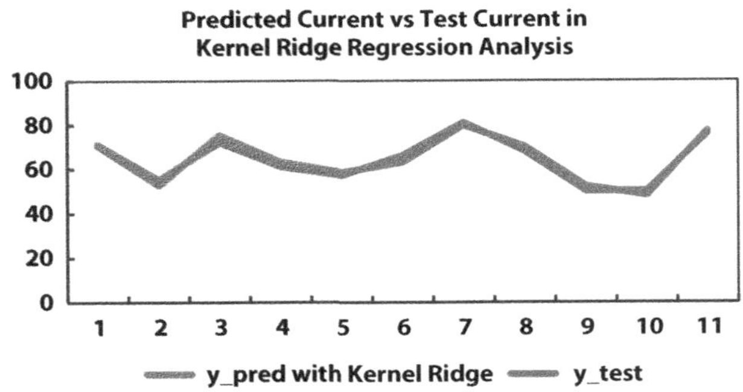

Figure 11.8 Predicted current plot versus test current plot in kernel ridge regression analysis.

Table 11.2 Validation dataset (##: Validation trials).

Exp.##	Previous width (inches)	Previous current (A)	Width (inches)	Test current (A)	Predicted current using Gaussian process regression (A)	Predicted current using kernel ridge (A)
v1	56	0.14	400	73	72.1065	74.51737
v2	81	0.2407	12.333	64	68.1439	60.59227
v3	78	0.232	4	63	64.141	61.49977
v4	83	0.2443	5.333	58	56.185	56.98631
v5	68	0.18	16	52	54.8713	51.61526

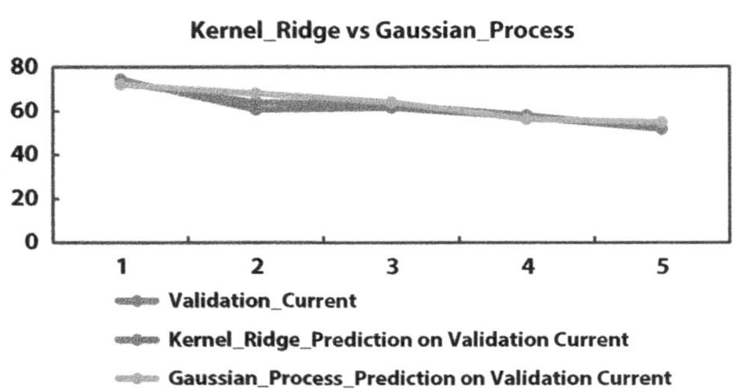

Figure 11.9 Plots of validation current, kernel ridge prediction on validation current, and Gaussian process prediction on validation current.

In this analysis, the KRR performed (relatively) better than the other models owing to the following reasons:

(i) The utilized dataset had a hundred (100) data points, which is significantly less than what is usually employed in machine learning models. An ML model captures the general trend of the distribution of a dataset. Consequently, when a dataset consists of limited data points the probability of seeking a good representation of the general trend decreases. This means that the model may be trained to learn unique quirks specific to the training set, rather than recognizing the general patterns that apply to new, unseen data. In addition, splitting the (primary) dataset into training and testing data exacerbates the problem and (often) leads to overfitting. While the model may perform well on the training data, when tested on a different dataset, the model displays relatively poor performance. The method of ridge regression addresses this problem by using regularized parameter control to stop overfitting of the dataset. It penalizes large weights and prevents the model from being overly complex, minimizing the chances of capturing noise in the data. This is particularly useful in high-dimensional spaces, where models have a high capacity to accommodate even minor fluctuations in the data. When the weights are constrained, the model is forced to learn more robust and general patterns that are less prone to random noise. If the overfitting is kept in check, one can obtain higher accuracies in training samples.

(ii) The kernel trick in KRR makes the dataset fit a nonlinear function model while being computationally more effective than other models. It has been forestated that the dataset is nonlinear, and hence, it cannot be represented by a simple mathematical equation. This aspect places the requirement of data in higher dimensions, in order to capture the nonlinearity aspect in the output value. In other models, one needs to explicitly take the data into a higher dimension and fit a linear function. Whereas, in KRR, the data can be transformed into higher-dimensional feature space using a kernel function, such as the Gaussian radial basis function (RBF), polynomial, or linear kernel. This transformation enables the model to capture nonlinear relationships between the features that would be virtually impossible to detect in the original feature space. In the present analysis, a polynomial kernel (with a learning rate of 0.5) was found to perform the best and with the least error. The use of kernel functions (along with the kernel trick) in KRR means it can handle high-dimensional data efficiently, without explicitly computing the high-dimensional feature space. This aspect simplifies computations and reduces the computational burden. This combination of nonlinear mapping and regularization makes KRR particularly effective for complex datasets, where inter-variable relationships are intricate and linear models fall short.

(iii) Other models like the XGBoost regressor and the random forest regressor were also tried out, but the results were not as effective (refer to Figure 11.10). In complex models like the XGBoost regressor, the model is capable enough to fit every single (data) point, which includes the noise. This high capacity, relative to the amount of data, increases the risk of overfitting. The random forest regressor, on the other hand, relies on creating diverse decision trees by sampling data and features. With fewer data points and attributes, there is limited scope for creating diverse trees, which leads to diminished benefits as compared to the ensemble approach. The trees might end up

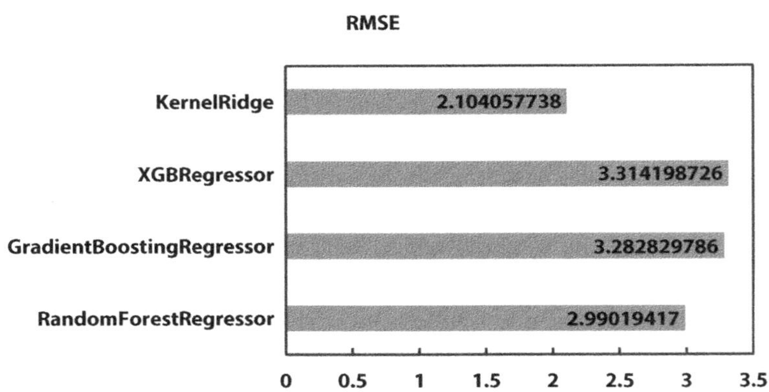

Figure 11.10 Comparison of kernel ridge with other models.

being too similar, reducing the overall model's robustness. Also, fewer attributes mean that each decision tree in the forest has fewer options for making splits. This limitation can result in less informative splits and less powerful trees, reducing the model's overall performance.

11.5 Predicting Conditions for Void Formation in Friction Stir Welding

11.5.1 Methodology

In the FSW process, metals are joined without melting them. It is done by rotating metals at a very high speed and bringing the metals in contact with each other. At the point of contact, the metals lose their hardness and become soft. The spinning pin then stirs this softened metal, much like mixing dough. The friction heats up the metal intensely, but not to the point of melting. One of the major drawbacks of FSW is void formation. Voids are air gaps that are trapped within the metals and compromise the strength of the weld. Moreover, voids can act as stress concentrators. When pressure is applied, these voids can become focal points where cracks are more likely to initiate and propagate. This can lead to unexpected weld failure under load. Therefore, the formation of voids needs to be minimized during FSW. It can only be minimized by avoiding the welding conditions which lead to the formation of void. In our study, the Gaussian process classifier (GPC) could determine the probability of formation of void with an accuracy of 90.91% percentage, higher than most traditional machine learning algorithms. The dataset had been collected from a literature survey of friction stir welding of three aluminum alloys: AA2024, AA2219, and AA6061 [102].

11.5.2 Gaussian Process Classifiers

The GPCs are a powerful ML technique for categorizing data points. Unlike many classification algorithms that learn a single, definitive decision boundary, the GPC takes a probabilistic approach [103]. It assumes that the relationship between the data points and their corresponding class labels can be modeled as a continuous function, with some inherent uncertainty. For any given input, this function is essentially a probability distribution over possible outputs.

The core concept behind GPC lies in Gaussian processes, which are a special kind of statistical model. They represent collections of functions where any finite set of points follows a joint Gaussian distribution. In simpler terms, it allows the user to make predictions about unseen data points based on how similar they are to the data points that have been already observed.

To achieve classification, GPC employs a kernel function [104]. This function defines how similar two data points are based on their input features. Similar data points are expected to have a higher probability of belonging to the same class. The kernel function plays a vital role in shaping the decision boundary and the model's overall performance. The GPC offers a variety of kernel options, each with its own strengths and weaknesses, allowing for customization to the specific problem at hand [105].

One of the significant advantages of GPC is its ability to provide not only class labels but also class probabilities. This indicates the confidence that the model offers in its predictions, especially where quantification of certainty of an outcome is crucial [106]. Additionally, GPC can handle situations with limited data, unlike some classification algorithms that perform adversely when training data is scarce.

11.5.3 Results

The dataset chosen for analysis has 108 entries. Out of these, 90% of the data (which are randomly selected) have been used as training datasets. The remaining 10% have been used as a testing dataset. From analysis, it is observed that the chosen model has predicted the testing data fairly accurately, an instance of which has been shown in Figure 11.11. Out of several different classifiers that were tried out, the GPC was found to be the most accurate (refer to Figure 11.12).

The present dataset contains 11 attributes for each data point, which complicates the prediction process and reduces the efficiency of simpler models. The complexity arises as the high dimensionality of the data creates intricate relationships among the attributes that simpler models struggle to capture. This is where Gaussian process classification proves to be advantageous.

The confusion matrix serves as a performance evaluation tool to visualize the accuracy of the GPC model (Figure 11.13). It is a table that compares the predicted classifications to the actual classifications; the rows represent actual classes and the columns represent predicted classes. Each cell in the matrix shows the count of instances classified into each category.

For a GPC, which predicts probabilities of class membership, the confusion matrix identifies true positives, true negatives, false positives, and false negatives, offering insights into the classifier's precision, recall, and overall accuracy. This breakdown provides a comprehensive understanding of the model's strengths and weaknesses in differentiating between various classes.

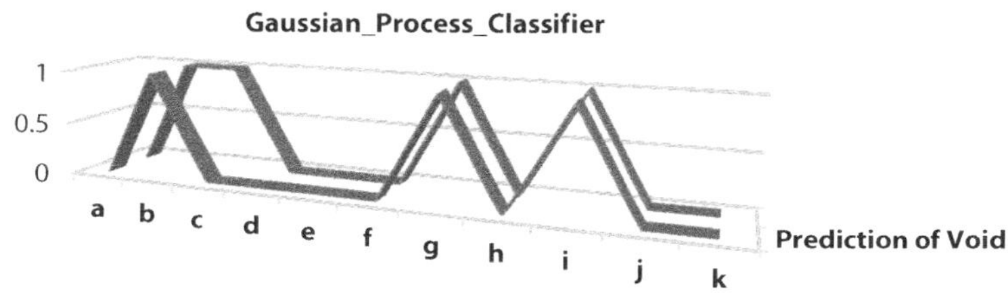

	a	b	c	d	e	f	g	h	i	j	k
Prediction of Void	0	1	0	0	0	0	1	0	1	0	0
Test of Void	0	1	1	0	0	0	1	0	1	0	0

Prediction of Void Test of Void

Figure 11.11 Testing versus predicted data.

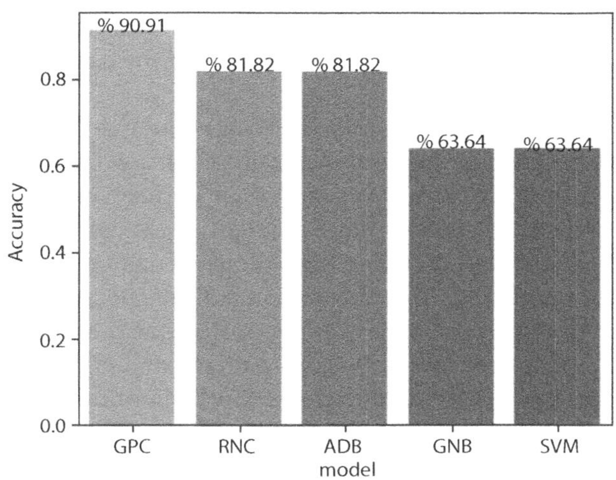

Figure 11.12 Comparison with other classifiers.

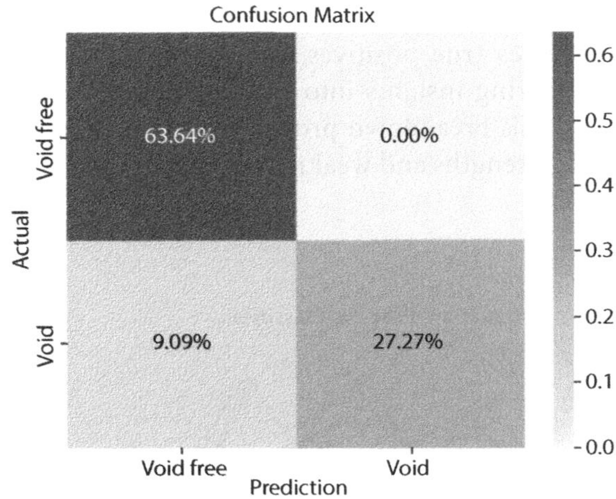

Figure 11.13 Confusion matrix of GPC.

The GPC employs a kernel function, which plays a crucial role in making it easier to separate complex data characterized by multiple attributes. Essentially, a kernel function defines the similarity between data points based on their attributes. When a dataset has numerous attributes, the relationships between data points can become quite intricate. A well-chosen kernel can capture these complex relationships effectively. By combining the similarities defined by the kernel, GPCs can model nonlinear patterns and uncover hidden dependencies among multiple features. This capability allows GPC to handle the high dimensionality of the data more effectively than simpler models.

In simpler models, dealing with (welding) data that have high dimensionality is challenging. These models fail to manage the complexity and interactions of multiple attributes, and their performance is suboptimal. On the other hand, the present book chapter has discussed that more complex models (capable of handling high-dimensional data) suffer from the "curse of dimensionality." This phenomenon occurs because, on increasing the number of dimensions, the volume of space also increases exponentially, which causes the model to learn patterns and classify data inaccurately. This feature is attributed to the lack of a significantly large number of data points [107, 108]. The GPC model addresses this issue by employing a technique known as the "kernel trick." Instead of transforming the data into a higher-dimension space, GPC uses a kernel function to simulate the effects of this transformation. The kernel trick allows GPC to effectively model the complex relationships between attributes without actually increasing the dimensionality of the data. This approach reduces the dependency on a large number of data points, mitigating the curse of dimensionality. The kernel trick is a mathematical technique that enables to perform operations in the original space of the data while implicitly considering a higher-dimensional feature space. By doing so, it captures the same complexity and nonlinear patterns that would be obtained by explicitly transforming the data to a higher dimension [107]. This makes GPC particularly powerful for handling datasets with numerous attributes, as it can model the underlying patterns and relationships more effectively than simpler models.

11.6 Conclusion and Future Scope

The research survey and analysis of the welding dataset, presented in this book chapter, have explored the dynamic landscape of automation and ML applications in various welding techniques. An updated literature review was conducted, with a focus on the integration of these advanced intelligent tools with established practices like TIG welding, laser welding, friction stir welding, metal inert gas welding, resistance welding, and PAW. Each technique presents unique opportunities for automation and ML implementation, thereby paving the way for a future of intelligent welding processes.

The reviewed studies showcased the remarkable potential of automation in enhancing welding efficiency and consistency. Robotic welding systems, for instance, offer significant advantages by mitigating human error and enabling precise, repeatable weld paths. This is particularly beneficial for working complex geometries or high-volume production lines. Furthermore, advancements in sensor technology allow for real-time process monitoring, enabling adjustments to welding parameters based on the dynamic conditions of the weld pool.

The integration of ML algorithms further elevates the capabilities of modern welding systems. For instance, the ML-based analysis presented here exemplifies this potential. By employing the KRR, the research group could achieve significant accuracy, obtaining an RMSE of 2.10, in predicting current flow during automated TIG welding, surpassing the performance of other models (in published studies) applied to the same dataset. This accomplishment highlights the dynamic proliferation of sophisticated ML models, which can optimize welding parameters, leading to improved weld quality and reduced production costs. Similarly, exploration into void formation prediction in FSW (using ML algorithms) also yielded promising results. The developed model achieved a remarkable accuracy of 90.91%, exceeding the performance of previously applied models (cited in the literature). This achievement underlines the potential of ML to identify critical process parameters that influence weld integrity, ultimately leading to the prevention of defects and ensuring consistent weld quality. Looking ahead, the synergy between automation and ML holds promise for the future of welding. The development of self-learning welding systems, capable of autonomously adapting to changing weld pool dynamics and material properties, is a potential possibility. Additionally, advancements in cloud computing and big data analytics could pave the way for the creation of intelligent welding platforms that leverage real-time data from various sensors to optimize welding processes across geographically dispersed locations.

R&D efforts directed toward developing algorithms that can analyze real-time sensor data and modify system parameters based on the dynamic characteristics of the weld pool, are crucial. This could involve incorporating reinforcement learning techniques, where the system receives feedback on its performance and continuously refines its approach. The integration of big data analytics and cloud computing also holds immense potential for creating a network of intelligent welding systems. This outlines briefly the scope for future work in the fascinating domain of welding.

References

1. Rodríguez-Gonzálvez, P. and Rodríguez-Martín, M., Weld bead detection based on 3D geometric features and machine learning approaches. *IEEE Access*, 7, 14714–14727, 2019.
2. Mahadevan, R., Jagan, A., Pavithran, L., *et al.*, Intelligent welding by using machine learning techniques. *Mater. Today Proc.*, 46, 7402–7410, 2021.
3. Huff, S.A., TIG welding skill extraction using a machine learning algorithm, Masters Thesis, Texas State University, 2017.
4. Kesse, M.A., Buah, E., Handroos, H., *et al.*, Development of an artificial intelligence powered TIG welding algorithm for the prediction of bead geometry for TIG welding processes using hybrid deep learning. *Metals*, 10, 4, 451, 2020.
5. Lin, T., Chen, H.B., Li, W.H., *et al.*, Intelligent methodology for sensing, modeling, and control of weld penetration in robotic welding system. *Ind. Robot*, 36, 6, 585–593, 2009.
6. Li, C., Weld penetration identification based on convolutional neural network, Doctoral Thesis, University of Kentucky, 2018.

7. Cai, W., Daehn, G., Vivek, A., *et al.*, A state-of-the-art review on solid-state metal joining. *J. Manuf. Sci. Eng.*, 141, 3, 031012, 2019.
8. Zhao, Y., Han, J., Domblesky, *et al.*, Investigation of void formation in friction stir welding of 7N01 aluminum alloy. *J. Manuf. Processes*, 37, 139–149, 2019.
9. Sudhagar, S., Sakthivel, M., Ganeshkumar, P., Monitoring of friction stir welding based on vision system coupled with Machine learning algorithm. *Measurement*, 144, 135–143, 2019.
10. Nasir, T., Asmaela, M., Zeeshana, Q., *et al.*, Applications of machine learning to friction stir welding process optimization. *J. Kejuruter.*, 32, 1, 171–186, 2020.
11. Gill, S.S. and Singh, J., Artificial intelligent modeling to predict tensile strength of inertia friction-welded pipe joints. *Int. J. Adv. Manuf. Technol.*, 69, 2001–2009, 2013.
12. Kumar, H.A. and Ramana, V.V., An overview of friction stir welding (FSW): a new perspective. *Int. J. Eng. Sci.*, 4, 6, 1–4, 2014.
13. Zeng, X.H., Xue, P., *et al.*, Material flow and void defect formation in friction stir welding of aluminium alloys. *Sci. Technol. Weld. Joining*, 23, 8, 677–686, 2018.
14. Tsukamoto, S., Laser welding. *Weld. Int.*, 17, 10, 767–774, 2003.
15. Klemens, P.G., Heat balance and flow conditions for electron beam and laser welding. *J. Appl. Phys.*, 47, 5, 2165–2174, 1976.
16. Knaak, C., Thombansen, U., Abels, P., *et al.*, Machine learning as a comparative tool to determine the relevance of signal features in laser welding. *Procedia CIRP*, 74, 623–627, 2018.
17. Assunção, E., Quintino, L., Miranda, R., Comparative study of laser welding in tailor blanks for the automotive industry. *Int. J. Adv. Manuf. Technol.*, 49, 123–131, 2013.
18. Zhang, Y., You, D., Gao, X., *et al.*, Welding defects detection based on deep learning with multiple optical sensors during disk laser welding of thick plates. *J. Manuf. Syst.*, 51, 87–94, 2019.
19. Günther, J., Pilarski, P.M., Helfrich, G., *et al.*, Intelligent laser welding through representation, prediction, and control learning: An architecture with deep neural networks and reinforcement learning. *Mechatronics*, 34, 1–11, 2016.
20. Stadter, C., Schmoeller, M., von Rhein, L., *et al.*, Real-time prediction of quality characteristics in laser beam welding using optical coherence tomography and machine learning. *J. Laser Appl.*, 32, 022046(1–9), 2020.
21. Kumar, M.M. and Tripathi, M.S., Artificial Intelligence based optimization of weld bead geometry in laser welding of Al-Mg alloy sheet. *IJMER*, 4, 3, 63–68, 2014.
22. Günther, J., Pilarski, P.M., Helfrich, G., *et al.*, First steps towards an intelligent laser welding architecture using deep neural networks and reinforcement learning. *Procedia Technol.*, 15, 474–483, 2014.

23. Thornton, M., Han, L., Shergold, M., Progress in NDT of resistance spot welding of aluminium using ultrasonic C-scan. *NDT & E Int.*, 48, 30–38, 2012.
24. Nayak, N. and Ray, A., An integrated system for intelligent seam tracking in robotic welding. Conceptual and analytical development, in: *Proceedings, IEEE International Conference on Robotics and Automation*, IEEE, pp. 1892–1897, 1990.
25. Weman, K., *Welding processes handbook*, Woodhead Publishing, United Kingdom, 2011.
26. El-Banna, M., Filev, D., Chinnam, R.B., Intelligent constant current control for resistance spot welding, in: *2006 IEEE International Conference on Fuzzy Systems*, IEEE, pp. 1570–1577, 2006.
27. El Ouafi, A., Bélanger, R., Méthot, J.F., An on-line ANN-based approach for quality estimation in resistance spot welding. *Adv. Mater. Res.*, 112, 141–148, 2010.
28. Arunchai, T., Sonthipermpoon, K., Apichayakul, P., *et al.*, Resistance spot welding optimization based on artificial neural network. *Int. J. Manuf. Eng.*, 2014, 1, 154784, 2014.
29. Zhao, D., Wang, Y., Lin, Z., *et al.*, Quality monitoring research of small scale resistance spot welding based on voltage signal. *ISIJ Int.*, 53, 2, 240–244, 2013.
30. Marić, D., Duspara, M., Šolić, T., *et al.*, Application of SVM models for classification of welded joints. *Tehnički Vjesnik*, 26, 2, 533–538, 2019.
31. Zhou, L., Zheng, W., Li, T., Zhang, *et al.*, A material stack-up combination identification method for resistance spot welding based on dynamic resistance. *J. Manuf. Processes*, 56, 796–805, 2020.
32. Sen, R., Choudhury, S.P., Kumar, R., *et al.*, A comprehensive review on the feasibility study of metal inert gas welding. *Mater. Today Proc.*, 5, 9, 17792–17801, 2018.
33. Madavi, K.R., Jogi, B.F., Lohar, G.S., Metal inert gas (MIG) welding process: A study of effect of welding parameters. *Mater. Today Proc.*, 51, 690–698, 2022.
34. Wan, X., Wang, Y., Zhao, D., Huang, Y., *et al.*, Weld quality monitoring research in small scale resistance spot welding by dynamic resistance and neural network. *Measurement*, 99, 120–127, 2017.
35. Yu, J., New methods of resistance spot welding using reference waveforms of welding power. *Int. J. Precis. Eng. Manuf.*, 17, 1313–1321, 2016.
36. Ogbonna, O.S., Akinlabi, S.A., Madushele, N., *et al.*, Application of MIG and TIG welding in automobile industry. *J. Phys. Conf. Ser.*, 1378, 4, 042065, 2019, IOP Publishing.
37. Muzakki, H., Prasetyo, T., Umam, M.S.U., *et al.*, Effect of metal inert gas welding process parameters to tensile strength on ST 37 steel sheet joint. *J. Phys. Conf. Ser.*, 1569, 3, 032055, 2020, IOP Publishing.

38. Misra, R.K. and Porwal, R.K., Optimization of Metal Inert Gas Welding Process during Joining of Structural Steels-An Overview. *E3S Web Conf.*, 184, 01028, 2020, EDP Sciences.
39. Uhrlandt, D., Diagnostics of metal inert gas and metal active gas welding processes. *J. Phys. D*, 49, 31, 313001, 2016.
40. Pal, S., Pal, S.K., Samantaray, A.K., Determination of optimal pulse metal inert gas welding parameters with a neuro-GA technique. *Mater. Manuf. Processes*, 25, 7, 606–615, 2010.
41. Pal, S., Pal, S.K., Samantaray, A.K., Artificial neural network modeling of weld joint strength prediction of a pulsed metal inert gas welding process using arc signals. *J. Mater. Process. Technol.*, 202, 1-3, 464–474, 2008.
42. Korra, N.N., Vasudevan, M., Balasubramanian, K.R., Multi-objective optimization of activated tungsten inert gas welding of duplex stainless steel using response surface methodology. *Int. J. Adv. Manuf. Technol.*, 77, 67–81, 2015.
43. Pal, S., Pal, S.K., Samantaray, A.K., Prediction of the quality of pulsed metal inert gas welding using statistical parameters of arc signals in artificial neural network. *Int. J. Comput. Integr. Manuf.*, 23, 5, 453–465, 2010.
44. Lee, J.H., Park, S.H., Kwon, H.S., *et al.*, Laser, tungsten inert gas, and metal active gas welding of DP780 steel: comparison of hardness, tensile properties and fatigue resistance. *Mater. Des.*, 64, 559–565, 2014.
45. Arunkumar, N., Duraisamy, P., Veeramanikandan, S., Evaluation of mechanical properties of dissimilar metal tube welded joints using inert gas welding. *Int. J. Eng. Res. Appl.*, 2, 5, 1709–1717, 2012.
46. Kah, P., Latifi, H., Suoranta, R., *et al.*, Usability of arc types in industrial welding. *Int. J. Mech. Mater. Eng.*, 9, 1–12, 2014.
47. Tanaka, M., An introduction to physical phenomena in arc welding processes. *Weld. Int.*, 18, 11, 845–851, 2004.
48. Sahoo, A. and Tripathy, S., Development in plasma arc welding process: a review. *Mater. Today Proc.*, 41, 363–368, 2021.
49. Prasad, K.S., Rao, C.S., Rao, D.N., Application of design of experiments to plasma arc welding process: a review. *J. Braz. Soc. Mech. Sci.*, 34, 75–81, 2012.
50. Metcalfe, J.C. and Quigley, M.B.C., Heat transfer in plasma-arc welding. *Weld. J.*, 54, 3, 99–103, 1975.
51. Karna, S.K. and Sahai, R., An overview on Taguchi method. *Int. J. Eng. Sci. Math.*, 1, 1, 1–7, 2012.
52. Wu, D., Chen, H., *et al.*, A prediction model for keyhole geometry and acoustic signatures during variable polarity plasma arc welding based on extreme learning machine. *Sens. Rev.*, 36, 3, 257–266, 2016.
53. Chi, S.C. and Hsu, L.C., A fuzzy radial basis function neural network for predicting multiple quality characteristics of plasma arc welding, in: *Proceedings Joint 9th IFSA World Congress and 20th NAFIPS International Conference (Cat. No. 01TH8569)*, IEEE, pp. 2807–2812, 2001.

54. Tarng, Y.S., Yang, W.H., Juang, S.C., The use of fuzzy logic in the Taguchi method for the optimisation of the submerged arc welding process. *Int. J. Adv. Manuf. Technol.*, 16, 688–694, 2000.
55. Wu, D., Huang, Y., Zhang, P., *et al.*, Visual-acoustic penetration recognition in variable polarity plasma arc welding process using hybrid deep learning approach. *IEEE Access*, 8, 120417–120428, 2020.
56. Navada, A., Ansari, A.N., Patil, S., *et al.*, Overview of use of decision tree algorithms in machine learning, in: *2011 IEEE control and system graduate research colloquium*, IEEE, pp. 37–42, 2011.
57. Kingsford, C. and Salzberg, S.L., What are decision trees? *Nat. Biotechnol.*, 26, 9, 1011–1013, 2008.
58. Kotsiantis, S.B., Decision trees: a recent overview. *Artif. Intell. Rev.*, 39, 261–283, 2013.
59. Alkhalid, A., Chikalov, I., Moshkov, M., Comparison of greedy algorithms for decision tree optimization, in: *Rough Sets and Intelligent Systems-Professor Zdzisław Pawlak in Memoriam*, vol. 2, pp. 21–40, 2013.
60. Song, Y.Y. and Ying, L.U., Decision tree methods: applications for classification and prediction. *Shanghai Arch. Psychiatry*, 27, 2, 130, 2015.
61. Hernández, V.A.S., Monroy, R., Medina-Pérez, M.A., *et al.*, A practical tutorial for decision tree induction: Evaluation measures for candidate splits and opportunities. *ACM Comput. Surv. (CSUR)*, 54, 1, 1–38, 2021.
62. Biau, G. and Scornet, E., A random forest guided tour. *Test*, 25, 197–227, 2016.
63. Boulesteix, A.L., Janitza, S., Kruppa, J., König, I.R., Overview of random forest methodology and practical guidance with emphasis on computational biology and bioinformatics. *Wiley Interdiscip. Rev. Data Min. Knowl. Discovery*, 2, 6, 493–507, 2012.
64. Probst, P., Wright, M.N., Boulesteix, A.L., Hyperparameters and tuning strategies for random forest. *Wiley Interdiscip. Rev. Data Min. Knowl. Discov.*, 9, 3, e1301, 2019.
65. Edeh, M.O., Dalal, S., Obagbuwa, I.C., *et al.*, Bootstrapping random forest and CHAID for prediction of white spot disease among shrimp farmers. *Sci. Rep.*, 12, 1, 20876, 2022.
66. Oshiro, T.M., Perez, P.S., Baranauskas, J.A., How many trees in a random forest?, in: *Machine Learning and Data Mining in Pattern Recognition: 8th International Conference, MLDM 2012, Proceedings 8*, Berlin, Germany, July 13-20, 2012, Springer Berlin Heidelberg, pp. 154–168, 2012.
67. Tang, F. and Ishwaran, H., Random forest missing data algorithms. *Stat. Anal. Data Min. Data Sci. J.*, 10, 6, 363–377, 2017.
68. Rigatti, S.J., Random forest. *J. Insur. Med.*, 47, 1, 31–39, 2017.
69. Probst, P. and Boulesteix, A.L., To tune or not to tune the number of trees in random forest. *J. Mach. Learn. Res.*, 18, 181, 1–18, 2018.
70. Louppe, G., Understanding random forests: From theory to practice, PhD Dissertation, University of Liège, 2014, arXiv preprint arXiv:1407.7502.

71. Salcedo-Sanz, S., Rojo-Álvarez, J.L., Martínez-Ramón, M., *et al.*, Support vector machines in engineering: an overview. *Wiley Interdiscip. Rev. Data Min. Knowl. Discovery*, 4, 3, 234–267, 2014.
72. Deris, A.M., Zain, A.M., Sallehuddin, R., Overview of support vector machine in modeling machining performances. *Procedia Eng.*, 24, 308–312, 2011.
73. Pisner, D.A. and Schnyer, D.M., Support vector machine, in: *Machine learning*, pp. 101–121, Academic Press, United Kingdom, 2020.
74. Moguerza, J.M. and Muñoz, A., Support vector machines with applications. *Stat. Sci.*, 21, 3, 322–336, 2006.
75. Salimi, A., Ziaii, M., Amiri, A., *et al.*, Using a Feature Subset Selection method and Support Vector Machine to address curse of dimensionality and redundancy in Hyperion hyperspectral data classification. *Egypt. J. Remote Sens. Space. Sci.*, 21, 1, 27–36, 2018.
76. Stitson, M.O., Weston, J.A.E., Gammerman, A., *et al.*, Theory of support vector machines. *University of London*, 117, 827, 188–191, 1996.
77. Suykens, J.A., Nonlinear modelling and support vector machines, in: IMTC 2001. Proceedings of the 18th IEEE instrumentation and measurement technology conference. Rediscovering measurement in the age of informatics (Cat. No. 01CH 37188), vol. 1, IEEE, pp. 287–294, 2001.
78. Suykens, J.A., Support vector machines: a nonlinear modelling and control perspective. *Eur. J. Control*, 7, 2–3, 311–327, 2001.
79. Kecman, V., Support vector machines–an introduction, in: *Support vector machines: theory and applications*, pp. 1–47, Springer Berlin Heidelberg, Berlin, Heidelberg, 2005.
80. Basudhar, A., Dribusch, C., Lacaze, S., *et al.*, Constrained efficient global optimization with support vector machines. *Struct. Multidiscip. Optim.*, 46, 201–221, 2012.
81. Kavzoglu, T. and Colkesen, I., A kernel functions analysis for support vector machines for land cover classification. *Int. J. Appl. Earth Obs. Geoinf.*, 11, 5, 352–359, 2009.
82. Burges, C.J., A tutorial on support vector machines for pattern recognition. *Data Min. Knowl. Discov.*, 2, 2, 121–167, 1998.
83. Guo, G., Li, S.Z., Chan, K., Face recognition by support vector machines, in: *Proceedings fourth IEEE international conference on automatic face and gesture recognition (cat. no. PR00580)*, IEEE, pp. 196–201, 2000.
84. Yang, Z.R., Biological applications of support vector machines. *Briefings Bioinf.*, 5, 4, 328–338, 2004.
85. Zou, J., Han, Y., So, S.S., Overview of artificial neural networks, in: *Artificial neural networks: methods and applications*, pp. 14–22, 2009.
86. Rosa, J.P., Guerra, D.J., Horta, N.C., *et al.*, Overview of artificial neural networks, in: *Using artificial neural networks for analog integrated circuit design automation*, pp. 21–44, 2020.

87. Buscema, M., A brief overview and introduction to artificial neural networks. *Subst. Use Misuse*, 37, 8–10, 1093–1148, 2002.
88. Jansson, P.A., Neural networks: An overview. *Anal. Chem.*, 63, 6, 357A–362A, 1991.
89. Hecht-Nielsen, R., Theory of the backpropagation neural network, in: *Neural networks for perception*, pp. 65–93, Academic Press, New York, 1992.
90. Abraham, A., Artificial neural networks, in: *Handbook of measuring system design*, 2005.
91. Dias, F.M., Antunes, A., Mota, A.M., Artificial neural networks: a review of commercial hardware. *Eng. Appl. Artif. Intell.*, 17, 8, 945–952, 2004.
92. Micheli-Tzanakou, E., Artificial neural networks: an overview. *Network Comput. Neural Syst.*, 22, 1–4, 208–230, 2011.
93. Suykens, J.A., Vandewalle, J.P., De Moor, B.L., *Artificial neural networks for modelling and control of non-linear systems*, Springer US, Switzerland, 2012.
94. Almeida, J.S., Predictive non-linear modeling of complex data by artificial neural networks. *Curr. Opin. Biotechnol.*, 13, 1, 72–76, 2002.
95. Curry, B., Morgan, P., Beynon, M., Neural networks and flexible approximations. *IMA J. Manage. Math.*, 11, 1, 19–35, 2000.
96. Livingstone, D.J., Manallack, D.T., Tetko, I.V., Data modelling with neural networks: Advantages and limitations. *J. Comput.-Aided Mol. Des.*, 11, 135–142, 1997.
97. Mishra, M. and Srivastava, M., A view of artificial neural network, in: *2014 international conference on advances in engineering & technology research (ICAETR-2014)*, IEEE, pp. 1–3, 2014.
98. Haworth, J., Shawe-Taylor, J., Cheng, T., *et al.*, Local online kernel ridge regression for forecasting of urban travel times. *Transp. Res. Part C Emerg. Technol.*, 46, 151–178, 2014.
99. Tsigler, A. and Bartlett, P.L., Benign overfitting in ridge regression. *J. Mach. Learn. Res.*, 24, 123, 1–76, 2023.
100. Exterkate, P., Groenen, P.J., Heij, C., *et al.*, Nonlinear forecasting with many predictors using kernel ridge regression. *Int. J. Forecast.*, 32, 3, 736–753, 2016.
101. Welling, M., Kernel ridge regression, in: *Max Welling's classnotes in machine learning*, pp. 1–3, 2013.
102. Dhanraj, J.A., Lingampalli, B., Prabhakar, M., *et al.*, A credal decision tree classifier approach for surface condition monitoring of friction stir weldment through vibration patterns. *Mater. Today Proc.*, 46, 1127–1133, 2021.
103. Pérez-Cruz, F., Van Vaerenbergh, S., Murillo-Fuentes, J.J., *et al.*, Gaussian processes for nonlinear signal processing: An overview of recent advances. *IEEE Signal Process Mag.*, 30, 4, 40–50, 2013.
104. Seeger, M., Bayesian model selection for support vector machines, Gaussian processes and other kernel classifiers. *Adv. Neural Inf. Process. Syst.*, 12, 603–609, 1999.

105. Kuss, M., Rasmussen, C.E., Herbrich, R., Assessing Approximate Inference for Binary Gaussian Process Classification. *J. Mach. Learn. Res.*, 6, 10, 1679–1704, 2005.
106. Kapoor, A., Grauman, K., Urtasun, R., *et al.*, Gaussian processes for object categorization. *Int. J. Comput. Vis.*, 88, 169–188, 2010.
107. Liu, H., Ong, Y.S., Yu, Z., *et al.*, Scalable Gaussian process classification with additive noise for non-Gaussian likelihoods. *IEEE Trans. Cybern.*, 52, 7, 5842–5854, 2021.
108. Zhang, N., Xiong, J., Zhong, J., *et al.*, Gaussian process regression method for classification for high-dimensional data with limited samples, in: *2018 Eighth International Conference on Information science and technology (ICIST)*, IEEE, pp. 358–363, 2018.

12

Machine Learning in Welding Technologies

D.V.S.S.S.V. Prasad[1]*, Akhilesh Kumar Singh[1], Marxim Rahula Bharathi B.[1], Yerrapragada K.S.S. Rao[1] and V.V. Kamesh[2]

[1]Department of Mechanical Engineering, Aditya College of Engineering and Technology (A), Surampalem, Andhra Pradesh, India
[2]Department of Mechanical Engineering, Aditya University, Surampalem, Andhra Pradesh, India

Abstract
Pivotal in the augmentation of accuracy, effectiveness, and weld improvement of welding technologies is the use of machine learning (ML). Using sensors and welding machines feeds the right parameters to the ML algorithms where the algorithms then determine the right welding parameters that will give good weld quality and consistency. Real-time and adaptive control systems vary its settings based on the context to reduce on defects. ML is also applied in maintenance prediction, where one is able to estimate the likelihood of an equipment failing and thereby minimize on the time it pulls out. Indeed, refined statistical patterns and anomalies shall help identify some problems in advance. Traditionally, welding process also makes use of robots whereby a programmer is always essential, while in robotic welding, machine learning algorithms help the robot learn from data and thus are useful in handling new tasks and thus increase productivity while at the same reducing costs of labor. Controlling and monitoring of real-life welding situations and practices through ML-derived Integrated Master Models and making use of the information acquired enable better welding instructions and practices by means of situations and procedures used in real-life welding processes. In addition, high accuracy, less scrap, and ultimately lower cost is added to high productivity realized by the industries that have embraced the use of ML in welding processes. Nevertheless, the following difficulties can be mentioned: data quality, the ability to explain the algorithms that will be employed, and issues to do with human capital. However, ML has doubled to bring breakthrough changes to the welding technologies and improve the manufacturing procedures.

**Corresponding author*: myselfdvsprasad@gmail.com

Sandip Kunar and Gurudas Mandal (eds.) Advanced Welding Technologies, (261–276) © 2025 Scrivener Publishing LLC

***Keywords*:** Welding, machine learning, algorithms, productivity, simulation, industrial precision

12.1 Introduction

Welding is one of the key processes in the fabrication trades and construction in which materials, like metals and thermoplastics, are joined by the application of heat, force, or both. Traditional processes include the physical method, while advanced technologies include the use of software. SMAW, GMAW, and GTAW are the methods known as the traditional ones each of which has its field of usage and advantages. The resistance welding methods include the spot and the seam welding and are highly used where there is large-scale production. There are special services offered in the provision of Laser Welding, Electron Beam Welding (EBW), and Friction Stir Welding (FSW). Robots and automation have improved efficiency in the production and assembly process, and future advancements, like machine learning automation and artificial intelligence, are designed to optimize and define defects. Nevertheless, some problems are always left unsolved, for example, thermal problems, problems related to welds, and integration of enhanced technologies to the existing system. Discussing the shifts in the welding technologies, it must be pointed out that the extraction has never ceased looking for ways to increase the productivity and to improve the quality of the products to be welded. With the further growth of the industry, it is possible that new staking and new technologies will appear, and therefore, welding will not stand still.

Joining processes, such as fusion, take place in welding where heat and/or pressure is used to join materials such as metals and thermoplastics. There are broad categories of SMAW GMAW GTAW Laser, EBW, and FSW. Robotics and automation make work accurate and precise, while novel technologies, such as machine learning, enhance the processes. However, the issues remain to some extent to make continuous enhancements in the welding technologies.

12.1.1 The Evolution of Welding Technologies

Welding can be defined as a process of joining of materials, such as metals and non-metals, and especially thermoplastic using heat or force or both. The various processes include SMAW, GMAW, and GTAW among the conventional welding procedures and then Laser, EBW, and FSW among the finish—innovative technology. It emphasizes two important favorable

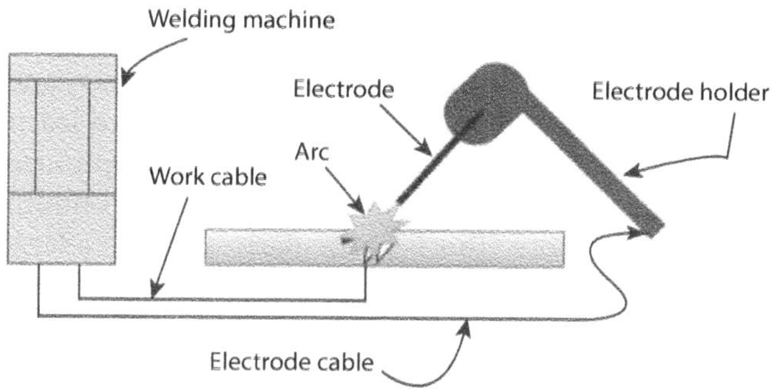

Figure 12.1 Arc welding process.

changes: accuracy and usefulness of robotization, and the procedure's improvement by new technologies such as machine learning. Still, the difficulties are not eliminated, which leads to further advancement of the welding technologies. The block diagram of basic arc welding process is shown in Figure 12.1.

12.1.2 Importance of Welding in Industry

Welding technology plays a critical role in industries such as construction, vehicles, aircrafts, naval architectures, and manufacturing. It connects components in the construction of structures and other parts with a view of providing strength and stability. In construction, it is used in the production of frames and structures, such as by fabricating steel in welding, and in the automotive industry, it is used to assemble car bodies and other body parts. It is widely used in aerospace to manufacture light yet strong structures, as well as shipbuilding to create and repair ships and submarines. Welding is also used in the manufacturing of machinery equipment and other consumer products, for instance, in the automotive field. Improvement in welding technology has enhanced efficiency in welding, and high technology in welding has improved accuracy. For the consequence of heat management and defect avoidance issues, welding cannot be replaced, as it is an irreplaceable tool, which provides accurate, reliable, and cheap outcomes. Continuous improvement in the techniques and technologies of this process makes it an essential factor that facilitates the growth of modern industries and various technologies. In so many years to come, welding will continue to be a critical technology in the growth and development of industries.

12.1.3 Machine Learning: An Overview

Artificial intelligence deals with complex control systems, and machine learning (ML) is an application that allows computers to learn from the data it processes. While ordinary programming just optimizes code to operate faster, machine learning optimizes performance by finding patterns in data and predicting those. This is particularly useful in a large environment, such as in welding, where numerous cables may be required. There are three main types of machine learning algorithms: supervised learning (performs training on some training data, which is labeled), unsupervised learning (estimates the relationship between the inputs and the outputs without any prior knowledge of training labels), and reinforcement learning (learns from experience; the essence of a good experience is determined by the reward achieved). Machine learning is used in welding for recommendations about the processes, the projected failures of the equipment, and the standard of welding. Relating these algorithms, the welding industries are further able to increase the performance ratio, productivity, as well as use innovative solutions.

12.1.4 Relevance of Machine Learning to Industrial Applications

A revolution of industrial applicability and effectiveness is being observed with the development of methods of artificial intelligence, such as ML. Using innovative algorithms and large volume of data, ML results in predictive maintenance thereby cutting on costs and time. The use of IT aids in improving the quality of deliverables, and improves the supply chain and control over energy consumption. In talent acquisition, automation and even in exceptional cyber security, and identifying threats and even anomalies, ML plays a remarkable part. The application of ML in research and development work expedites the discovery of insights as well as patterns in the datasheets and prediction of the consequence accordingly. Furthermore, ML serves customer demographic analysis, finance and operation enhancement, and better and more predictive information. Some of the challenges that make ML difficult include data quality and skills needed to implement it; however, constant innovation indicates continuous increase in the ability of the technology. Implementation of ML can help industries experience an increase in operational performance, novelty, and competition. The reach of ML is seen in cross-sectional areas such as manufacturing, supply chain, energy management, human resource, research, and development among others. When the novel industries become mature, ML will be the engine for growth, innovation, and sustainability. There is no doubt that it has revolutionary features, and that its further implementation will define the development of industrial uses.

12.1.5 Machine Learning in Welding Technologies

There are several advantages of including machine learning with the technologies of welding such as enhancing the process control, inspecting the defects in real time, and scheduled maintenance. Many welding parameters can be learned from the large data sets. The use of AI and ML will help predict the weld quality and adjust the process parameters in real time. For instance, Zhang *et al.* [1] explained that there is a possibility of using machine learning in welding for geometry weld bead predictions that enhance the consistency of the welded beads. Further, it facilitates quality control by performing image analysis that would address crack and porosity issues beyond the achieved human capacity. Liu *et al.* [2] showed that a convolutional neural network (CNN) is efficient for weld defect real-time detection. In addition, due to machine learning, there is modelling of equipment failure through predictive maintenance, cutting on downtime and costs of maintenance. Sharma *et al.* [3] showed how to use machine learning to forecast the useful years of welding equipment to perform prescriptive maintenance.

12.2 Fundamentals of Welding Technologies

Welding is an essential procedure in fabricating and constructing industries because it involves the use of heat to bring two or more materials together to join them. The key fundamentals of welding include the following:

- Heat Generation: The adequate generation of heat required to fuse the work piece by using energy tools such as electrical arcs, gas flames or lasers.
- Material Fusion: To be able to end up with a fused form needed melting the material, control of temperature, and energy besides skill.
- Cooling and Solidification: Cooling and solidifying of the materials to make sound joint, which can be influenced by rates of cooling and other forms of solidification.

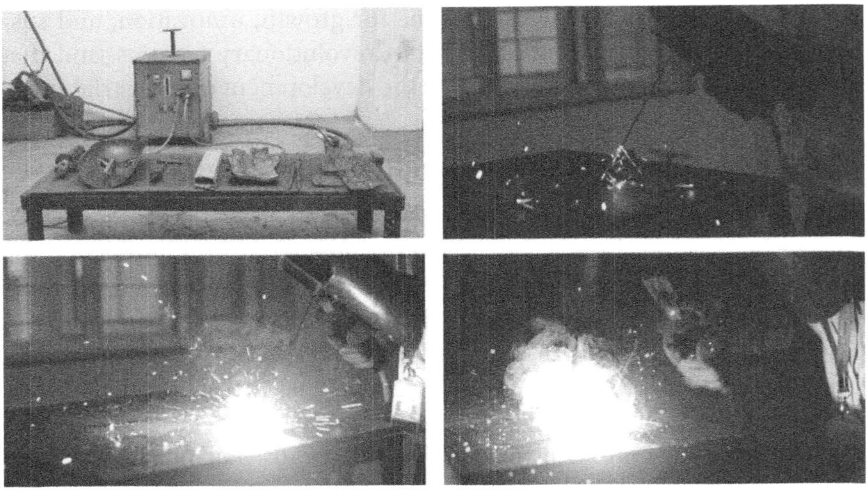

Figure 12.2 Welder performing arc welding in the laboratory.

- Filler Materials: Inserting materials of the improved joint characteristics, which depend on the base materials and the kind of welding.

A few images of the welding process captured in a research laboratory are presented in Figure 12.2.

12.2.1 Types of Welding Processes

The conventional techniques of welding involve arc welding, including MIG, TIG welding, and MMA, in which there is use of an electric arc or heat source with a filler material. Some of the sophisticated methods, including laser welding, electron beam welding, and friction stir welding, are more accurate, quicker, and produce stronger and less distorted welds, which are more appropriate for strategic sectors. Simpler techniques are portable and economical than the complex sophisticated methods, and the latter permits complex and superior performance. The classification of welding processes is shown in Figure 12.3.

Different welding techniques are mentioned in Figure 12.3. In SMAW, or what is referred to as stick welding, a flux layer and an electric arc protect the weld. MIG welding uses a constant-feed wire and an inert gas for its fast and flexible welding procedure. Gas Shielded Arc Welding (GSAW),

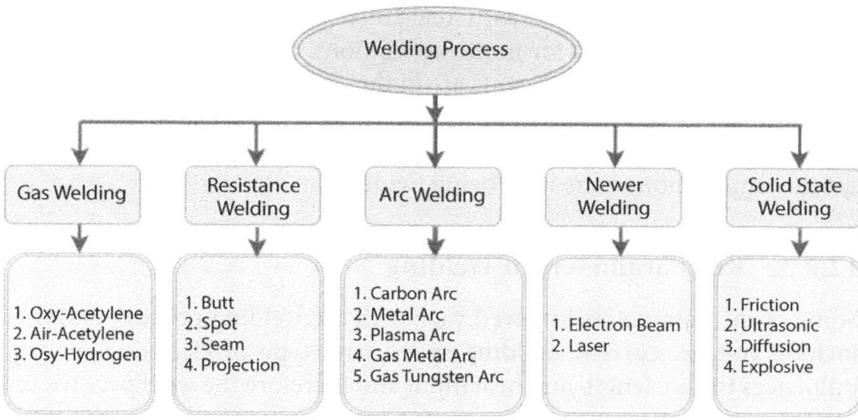

Figure 12.3 Classification of welding processes.

Manual Metal Arc Welding (MMAW), and Tungsten Inert Gas (TIG) Welding use a tungsten electrode and separate filler for accurate high standard welds. Flux Cored Arc Welding (FCAW) involves the use of a tubular wire containing flux, which is ideal for use outdoors. Submerged Arc Welding (SAW) has continuous electrode wire and flux to minimize spatter and hence the quality of the joint. Plasma Arc Welding (PAW) is also referred to as the plasma torch welding, which is a process that is precise. Laser Welding offers the deep weld profile and low distortion. Electron Beam Welding (EBW) in vacuum uses a high-velocity electron beam, and it provides deep penetration. Using a rotating tool for solid-state welding is called Friction Stir Welding (FSW). Ultrasonic welding uses pressure-based high-frequency sound waves to combine comparable plastics and thin metal foils. Every process has its advantages regarding particular tasks and problems.

12.2.2 Recent Advances

Recent developments of welding have specialized in aspects of accuracy, speed, and compatibility of their materials. Robots make certain welds that are done correctly to ensure the best output also in cases that the assemblies are complicated. Laser and electron beam welding is characterized by precise high-speed work with comparatively low thermal effect. These are techniques where laser is used together with MIG or any other form of welding to suit the material and the job in question. Modern technology delivery systems integrated with real-time monitoring improves welding quality and provides for multi-use options, are lightweight, and have high strength. New developments are in the areas of combination welding, advanced robot and automation welding, and 3D printing welding. Each process is suitable for its specific material and type of weld desired, and the specific application of the weld being made.

12.2.3 Key Parameters in Welding

Some of the parameters involved vary with the welding process, and they include voltage, current, welding speed, and type of electrode. Voltage influences the arc length and heat input and therefore the weld penetration as well as the appearance of the bead. Current determines the amount of heat produced and the size and stability of the arc, while welding speed controls the heat input and how the materials to be welded melt and overall. Welding speed controls the heat input and how the materials to be welded melt, and weld deposition, and the appearance of the weld. Depending on the selected electrode or filler metal, weld strength, corrosion, and the ability to unite with base materials are determined. Precise regulation of these variables is an indispensable requirement to produce sound, a well-developed bead, and secure the appropriate characteristics and behavior of the weld.

12.2.4 Challenges in Welding

Welding directly brings several problems such as control of heat input, control of consistent quality, and protection from safety hazards. It is important to note that raw materials that are used to make welds fit-up when used in joining of the materials, and the environment can influence the quality of the welds. Due to the issue of spatter, control of defects, and proper fusion, accurate technique and equipment calibration are important. Each material and thickness is very different from the other and has specific issues that need to be addressed through various processes and methods. The reason is that new techniques are introduced, and sometimes, the change in industry, set quality standards, and new materials come up. Concerns that are central to effective function are the ability to maintain the quality of what is produced, to be able to identify defects early and to optimize the function. Areas, such as scrutiny, non-destructive testing, and the monitoring of weld parameters, are critical in the achievement of the above goals and objectives.

12.3 Basic Concepts of Machine Learning

Some of the complexities of welding include heat control and management, standardization of the weld quality, and safety measures. Because of both the metal type and quality of fit-up, differing environments in the assembly area influence the quality of the weld. Accuracy in the technique, and also calibration of the precise equipment, is important. Thicker and thinner materials will need other processes, hence the need for specialization. Skill enhancement and technology update processes are to be ongoing all the time. Machine learning algorithms can be classified as follows:

1) Supervised Learning: Capable of causing a change in a real-world event or outcome after being trained with labeled data for classification and regression.
2) Unsupervised Learning: Learned from supervised data for clustering and dimensionality reduction.
3) Reinforcement Learning: Adapts from the experiences within an environment for the purpose of getting the highest returns.

Knowledge of these ML techniques allows us in choosing the appropriate methods and algorithms of machine learning in practically applied tasks. The classification of machine learning and its algorithms are given in Figure 12.4.

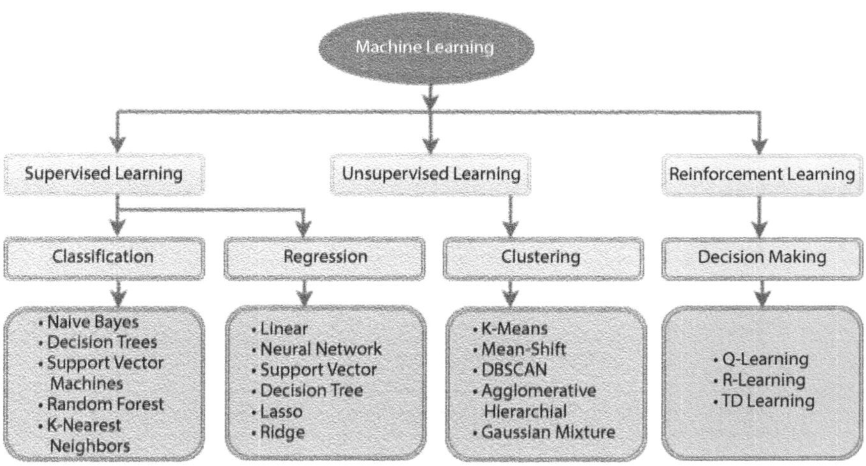

Figure 12.4 Classification of machine learning and its algorithms.

12.3.3 Machine Learning Workflow

The machine learning workflow can be described as the systematized process of going from the idea to the delivery of a model, including important milestones that define the building of the correct model. All these phases play an important role in developing good models that will perform well and give good results in real-life problems. The ML workflow is displayed in Figure 12.5.

The machine learning workflow is a systematic process that spans five stages which include the following: Data Acquisition, Data Cleaning, Building the Models, Model Assessment and Validation, and lastly, the models' implementation.

Data Collection involves the collection of simple and raw data from a Source, which is relevant and contains adequate Volume with Quality. Data Preprocessing entails data cleaning that involves removing missing data and other irrelevant information, data transformation that involves converting data into an appropriate form for analysis, feature engineering that involves creating new features from existing features, and data splitting that is the separation of the data set into a training and testing data set. Model Training takes in the training data and chooses the corresponding algorithm to execute, the parameters to use, and how the algorithm should be tuned. Evaluation also approves the generalization of the model with the help of validation and test data, along with the selection of the appropriate metrics and considering fairness. Deployment takes the trained model into the production environment where the performance of the model

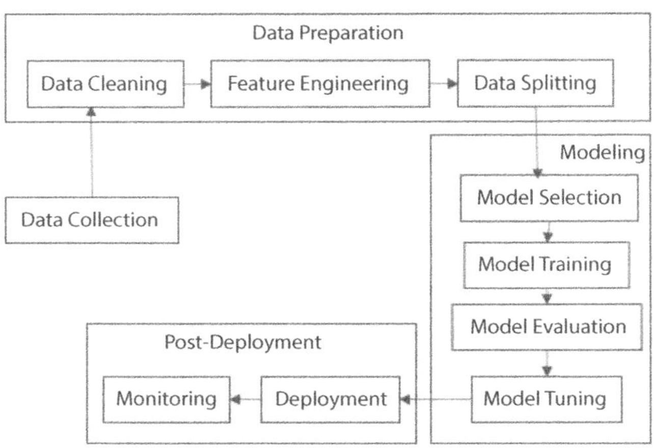

Figure 12.5 Machine learning workflow.

is checked, updates must be made, and the entire model must be scaled, secured, and must accommodate the feedback taken from the users.

Each of them is important for achieving the goals of creating the models, which can be efficient in practical applications. These include factors such as data quality; performance of the model; the number of computations, which the model can handle; and bias. This way, the developers can design and implement sound machine learning to select the right solution set to achieve specific criteria and be accurate.

12.4 Applications of Machine Learning in Welding

Concerning the welding industry, ML is being applied with benefits in process optimization, quality assurance, as well as in training [4–6]. The following is a detailed overview of ML applications in welding:

Process Optimization

- Nonparametric predictions for the best parameters
- Real-time change of the settings of the welder

Acceptance Sampling Plans and Control of Defects

- Picture analysis for weld examination
- Defect detection in the quality of welded structures
- Maintenance of welding equipment that would prevent equipment failure.

Data-Driven Insights

- Interpretation of the data of the welding process
- Different patterns and trends
- Enhancing education procedures for the welders

ML Enhances Welding Operations by the following:

- Increasing the performance and quality
- Incoming virtual patient families, and virtual patient paths may come from other frameworks with less focus on defects and rework
- Optimizing training programs

- Saving the chance to make the right decision supported by statistical data

Applications Include the Following:

- Predictive maintenance
- Automated inspection systems
- Adaptive welding systems
- Training programs
- Continuous improvement initiatives

Through hiring the services of ML, the standard of the welding within the organization is enhanced, hence reducing the cost, hence enhancing productivity of welders within an organization and competitiveness of the organization [7, 8].

12.5 Challenges and Future Directions

12.5.1 Challenges in Implementing Machine Learning in Welding

Incorporation of ML in the welding technologies is promising, but it comes with several aspects that must be addressed with a view of enhancing the efficiency of ML in the welding technologies [9, 10]. This section provides a brief examination of the primary hurdles and future directions in the adoption of ML for welding.

Key Challenges

- Data quality and quantity
- From the aspect of model interpretability and explainability
- Compatibility with current processes of welding
- Managing variation in weld activities
- Forced sharing of transportation: This aspect itself will catalyze forced sharing, which is imperative for safety and reliability.

Future Directions

- Recent developments in the collection and analysis of data

- Increased complexity of a range of new and existing machine learning algorithms
- A shift of concern toward cooperative activity with machines
- The utilization of ML in welding has been increased gradually.

Knowing these challenges and future trends, organizations enhance response toward the application of ML in welding technologies and its potential to contribute to further technological advancement and development of the technique, performances, and quality of welded products.

12.5.2 Future Trends in Machine Learning for Welding

Machine learning (ML) is a dynamic field, and therefore, it is evident that there are major breakthroughs that could be implemented in the field of welding technologies in the future.

Key Trends

- Algorithmic Advancements: New advances in deep learning, such as CNNs and GANs, will improve weld defect detection as well as process control. Reinforcement learning will allow the building of a control system for welding parameters to change with progress, and transfer learning will enhance passing of knowledge from one task to another and even from one domain to another.

Future Directions

- Optimization of the algorithm that has to be used to utilize the limited resources in the best manner possible.
- Models that incorporate several forms of the ML and AI to increase efficiency
- Fairness to ensure that decisions are made clearly—Explainable AI

These will open the possibilities for the fully automated welding systems with the use of ML within Industry 4.0 technologies. Thus, the development of ML can be expected to bring many improvements in welding technologies in terms of accuracy and production efficiency. If these developments are adopted, organizations can open various other opportunities in welding as well as return with improved positions in the competitive environment.

12.5.3 Ethical and Practical Considerations in Machine Learning for Welding

With the incorporation of machine learning (ML) into welding technologies, it is vital to pay attention to ethical and practical elements of its inclusion. It has two primary subjects: data protection and human–robot interaction.

Some of the privacy issues that relate to data are securing one's identity, or credit card details, or medical history, regulatory requirements, and securing the data. To address these concerns, future directions include the following:

- The advanced state-of-the-art encryption mechanisms should be adopted for practicing severe security measures. For this reason, access controls must be strict and authentication mechanisms put in place
- Exploring privacy-enhancing technologies
- Human–machine collaboration is also critical, with a focus on the following:
  - Curing workplace hazards via fail–safes and emergent stop systems
  - Designing training setups for operators, which would enable them to manage machine learning systems
  - Establishing the users' confidence and trust in the ML models through a human–centered approach with clear decision making

Future Directions

- How to build better interfaces for machine learning systems
- Training of the operators, carrying out of overall substantial training programs
- An innovative approach to examining the development of cooperative robotics

It is highlighted some of the ethical and practical considerations that organizations need to overcome to apply ML in welding technologies safely, efficiently, and ethically.

12.6 Conclusion

The future of ML in welding is showing a positive sign, and therefore, more technological updates are looked forward to achieving delicate and efficient systems in welding. It is promising for the industry stakeholders for they should accept new trends, fund experimental development, and implement advanced ML perspectives. Cooperation and knowledge exchange will be key success factors in the given case for addressing the challenges and developing the utilizations of AI/ML. We believe that over time, ML will further enhance the performance of welding procedures, quality assurance, along with automation anywhere, decreasing time, expense, and enhancing weld quality. For the full leveraging of ML to enhance the business sectors' performance, the industry must do additional research, innovations, and adoption. In this way, the stakeholders can free the next generation of welding technologies and develop sustainable competitive advantage in the emerging market environment. Overall, ML is revolutionizing the welding industry, and its evolution will be a key to future improvement.

References

1. Zhang, Y., Li, J., Chen, X., Wu, H., Prediction of weld bead geometry in arc welding using machine learning algorithms. *J. Manuf. Processes*, 49, 18–28, 2020.
2. Liu, Q., Gao, H., Wang, Y., Zhang, L., Real-time weld defect detection based on convolutional neural networks. *IEEE Trans. Ind. Inf.*, 17, 5, 3497–3505, 2021.
3. Sharma, R., Gupta, A., Singh, H., Predictive maintenance of welding equipment using machine learning techniques. *Int. J. Adv. Manuf. Technol.*, 105, 1-4, 1111–1121, 2019.
4. Zhao, H., Zhang, L., Wang, Q., Weld quality monitoring using machine learning: A comparative study. *Mater. Des.*, 162, 47–56, 2018.
5. Wang, B., Zhou, J., Tang, Y., Application of deep learning in welding defect detection. *IEEE Access*, 7, 17885–17893, 2019.
6. Khan, M.A., Islam, S., Akhtar, R., Hybrid welding processes and their applications: A review. *J. Manuf. Processes*, 35, 336–352, 2018.
7. Smith, D. and Brown, K., Machine learning algorithms for predictive maintenance in welding technology. *J. Intell. Manuf.*, 30, 3, 1225–1236, 2019.
8. Chen, Y., Li, Q., Wang, J., Advanced machine learning techniques for weld defect detection and classification. *NDT & E Int.*, 107, 102123, 2019.

9. Patel, P., Kumar, R., Bhattacharya, S., Integration of AI and machine learning in automated welding systems. *Rob. Comput. Integr. Manuf.*, 59, 238–251, 2020.
10. Dong, L., Liu, H., Zhang, G., Data-driven approaches for weld quality prediction using machine learning. *J. Manuf. Syst.*, 54, 45–56, 2020.

13

Artificial Intelligence and Machine Learning in Welding Technologies

Thavasilingam K.[1*], Giridharan K.[1], Yuvaraj S.[2], Gopi Kannan K.[3] and Prashanth M.[3]

[1]*Department of Mechanical Engineering, Easwari Engineering College, Chennai, India*
[2]*National Institute of Ocean Technology, Chennai, India*
[3]*Department of Mechanical Engineering, Indian Institute of Technology (ISM), Dhanbad, Jharhand, India*

Abstract

Artificial Intelligence (AI) and Machine Learning (ML) are futuristic platforms for all engineering applications, especially in welding technologies. Precision, efficiency, and quality control are more important factors to consider and improve by developing the concepts through AI and ML. This section provides a detailed exploration of the application of AI and ML in welding applications over to the conventional welding techniques. AI algorithms reduce human errors and ensure quality by enhancing real-time monitoring and adaptive control of welding processes, whereas ML models trained on numerous data sets estimate feasible welding parameters, identify the defects and errors, and recommend corrective actions, by the way downtime and wastage of materials can be reduced. This section also describes the development of advanced welding robots that possess machine vision and sensors to execute complex tasks with high accuracy and speed. These enhancements equip the welding automation in challenging environments, such as welding underwater, in space, and where human-aided tools are not suitable. By going through the case studies and industrial applications, this section describes the ability of AI and ML on welding technologies to establish the right way into future smart manufacturing and to define the new standards in welding technologies.

**Corresponding author*: thavasi.k.89@gmail.com

Sandip Kunar and Gurudas Mandal (eds.) Advanced Welding Technologies, (277–286) © 2025 Scrivener Publishing LLC

Keywords: Artificial intelligence, machine learning, welding technologies, real-time monitoring, intelligent welding robots

13.1 Introduction to AI and ML in Welding

Welding technology has evolved important development over the past few decades, from simple manual processes to highly advanced techniques. These developments have been providing need for the highly precise, effective, and top-quality welding operations. Welding is an essential process in various industries, not limited to automobile, construction, manufacturing, and aerospace, where the role of the welding process ensures safety and accuracy.

Traditional welding techniques merely needed a skilled welder. Even many skilled welders can encounter many barriers to maintaining consistency and quality, especially in critical and repetitive welding tasks. To approach these challenges, recent welding techniques have been initiated to enhance advanced computational techniques that incorporate the control and monitoring of welding processes.

The tool ML is the technology under Artificial Intelligence (AI) that provides the platforms for analyzing the data, identifying the patterns, and assuming the process predictions. In welding, ML is used to correct the welding factors, predict welding defects, and suggest corrective actions. This section explores the integration of ML in welding techniques and describes the importance of the industry. By examining various applications and case studies, we will demonstrate how ML is transforming welding from a manual art to a precise science.

ML methods can aid in waste reduction and encourage a more prosperous and sustainable manufacturing sector. The excellent modeling capabilities and absence of explicit programming required by ML technologies are drawing the attention of modern industrial sectors. Improvements in pattern recognition and computational learning methods for artificial intelligence have led to the development of Machine Learning technology. Existing innovative techniques, with developed computer applications, set modern ML algorithms apart from its usual other technologies. The manufacturing sector has been using ML and data science methods extensively for a long time to enhance the fabrication and structural properties of manufactured or machined components [1]. Welding systems are evolving because of the emergence of modern information disciplines including Artificial Intelligence (AI), the Internet of Things (IoT), Cloud Computing, Big Data, with smart manufacturing. Intelligent Welding Systems (IWS) are displaying interest from the industry and academics, thanks to these technologies. Smart welding is the process of reproducing, aiding, and/or adding human workers in a variety of tasks, including sensing, acquisition, tracking, decision making, and control. These processes are completed by fusing the benefits of the resources of humans and material systems into smart cyber systems. A complete investigation of the elements, applications, and achievable next-level advancements of intelligent welding, as illustrated in Figure 13.1, will aid in the creation of a standard word for IWS [2].

The processes of welding produce accuracy and efficiency and have been demonstrated to increase significantly with the application of ML algorithms. Industrial robots with AI will deal with a variety of hard problems in the manufacturing sector. Many welding techniques rely on human judgment to determine the best settings, which reduces their effectiveness and increases the likelihood of human error. Robots and machine-driven methods are disciplined in implementing Neural Networks, which can make invariable quality welding and increase productivity, to reduce this trust. ML also helps in visualizing welding, as visual inspection is important in assessing the caliber of the weld. Another tool that can be used in coincidence with these approaches to measure the root causes of different medical issues is regression investigation [4]. A range of observations and objectives, such as procedure factors optimizing performance, weld

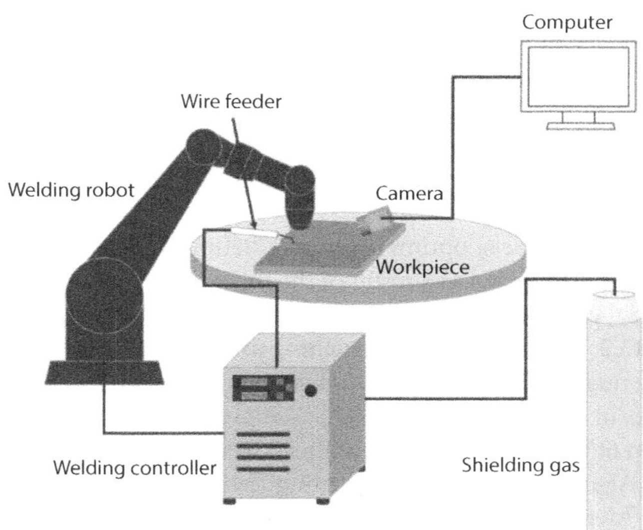

Figure 13.1 State of the system during welding [3].

seam observation, categorizing faults in welding, and workflow control by feedback, are described, and the latest practical applications based on AI are employed to achieve them. Keep an eye on potential investigation questions and disputes related to real-time smart surveillance, including adaptive control technology, data depth fusion, intelligent multi-sensor signaling systems, and others [5].

Regression models were created by conducting experiments using data collected through statistical design techniques. Numerical models that demonstrate the state between weld bead size and welding process parameters are crucial for functional purposes. The numerical model allows for the discovery of the weld bead sizing while calculating particular welding process factors. The results of an experimental investigation prevailed to evolve numerical models that show the state between welding method factors and weld bead dimensions in TIG, MIG, and SMAW welding procedures [6]. Techniques based on Machine Learning (ML) will help reduce waste and promote a more lucrative and sustainable manufacturing industry. Developments in pattern recognition and computational methods for AI have led to the growth of ML algorithms. A large number of industries, including financial institutes, learning or educational institutes, metro cities, and other technologies, have enriched their use of ML techniques in the last several years. An addition of innovative, cutting-edge computer approaches set modern Machine Learning apart from its predecessors. The manufacturing sector has been utilizing data science and Machine Learning methods for a long time to enhance the mechanical and microstructure properties of manufactured mechanical assemblies [7].

13.2 Quality Control and Defect Detection in Welding

Developing the models, optimization, predicting, and regulating complex systems have all changed as a result of the introduction of AI approaches in an industrial environment, especially Solid-State Welding (SSW), as shown in Figure 13.2. Due to less material melting using the SSW process, the property of the nugget region is cured making it an excellent joining technique. it examines in detail the effects of AI-based predictions on SSW through an application of techniques such as Artificial Neural Networks (ANN), Meta-Heuristic Algorithms, Machine Learning (ML), Fuzzy Logic (FL), as well as Hybrid Methods (HM) to Diffusion Bonding (DB), Ultrasonic Welding (UW), and Friction Stir Welding (FSW). By combining AI methods with

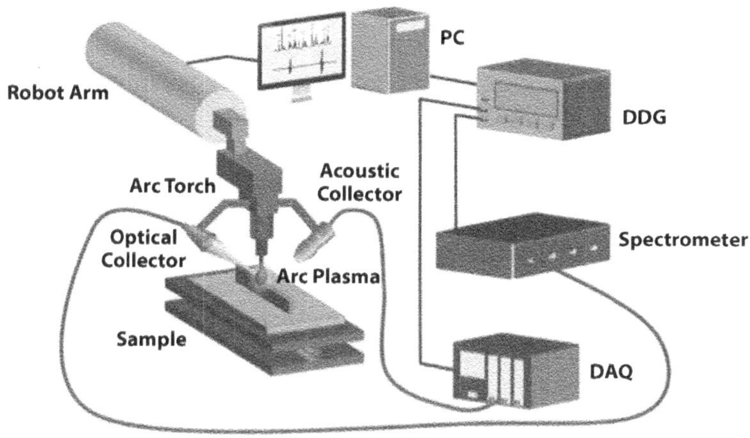

Figure 13.2 Laser opto-ultrasonic setup for online monitoring [16].

optimization algorithms, such as GA and PSO, one may optimize UW processes and enhance parameter prediction with a significant increase in accuracy. When predicting welding parameters, ANN outperforms other methods, like FL and ML, with a high accuracy of over 95% [8].

With the integration of new information technologies, like ML and intelligent manufacturing, welding systems have become smarter in the modern era. To develop an Intelligent Welding System (IWS), Machine Learning (ML) is applied across a broad range of weld-joining applications. Researchers and business people are giving the IWS, which uses Machine Learning, a lot of thought to produce high-quality products at reasonable prices. Intelligent welding reduces or eliminates human intervention using contemporary computers for sensing, learning, decision making, tracking, and control. The main uses of Machine Learning-integrated welding are in multiple variable modeling, identification, optimization, and control [9]. Welding engineers are now more fascinated by employing modern welding applications than traditional processes of welding because of the various benefits of futuristic welding processes, including their cost effectiveness. Because it improves the weld's Heat-Affected Zone (HAZ) micro-structure and bonding strength, the application of an innovative welding approach was found to improve the welding characteristics for these observed with welding using the traditional welding formalities. Fusion welding is one of the most often used processes, and it is also widely employed to combine different kinds of metals. Fusion welding techniques come in a variety of forms; each operation has a distinct weld thermal cycle based on the weld deposition rate, shielding environment, and welding parameters [10].

A spectroscopic study was used to determine the range of wavelengths that is less prone to arc light while taking weld pool photos, and a 1,000-nm band-pass filter was chosen. Weld pool photos were taken with numerous exposure time frames since each feature point has a different brightness distribution that is suited for detection. Artificial Intelligence (AI) technology, namely, the pose prediction model Dark Pose, which may increase the precision of detection by adding training data, was utilized to accurately understand the feature points of different weld pool photos. Upon evaluating the detection models, it was discovered that there was excellent accuracy in detecting the electrode, wire, and groove wall. The newly developed feature point recognition technique was used to create the torch orientation control system [11].

13.3 Automation in Challenging Environments in Welding

This focuses on using an artificial Neural Network and Online Programming (OP) together with visual control techniques to classify the type of weld quality in robotic Tungsten Inert Gas (TIG) welding on Stainless Steel (SS) 304. Automated welding planning is made possible by feature matching algorithms in conjunction with continuous sensory-guided methodologies and IoT high-level operator interface. Passive vision real-time pictures of the weld species have been captured for defect classification using area scan-based Complementary Metal Oxide Semiconductor (CMOS) cameras. Discrete reference workpieces and picture feature extraction procedures appear to be challenging in an unorganized welding environment in real time [12].

One of the most important aspects of production engineering is welding engineering, which entails connecting two distinct materials using a variety of techniques. Welders continue to face challenges due to the complexity of welding engineering, and the best way to measure weld quality will be through the application of Machine Learning, as shown in Figure 13.3. The inspections of weld joints are difficult because welders might not be able to identify the most fundamental problems within the welded junction using sensors and ideas from image processing to identify problems in the weld while supporting real-time monitoring with Machine Learning models applied in several research projects [13].

The manufacturing industries use TIG welding, which needs highly skilled workers. The field of AI is evaluated quickly since researchers are

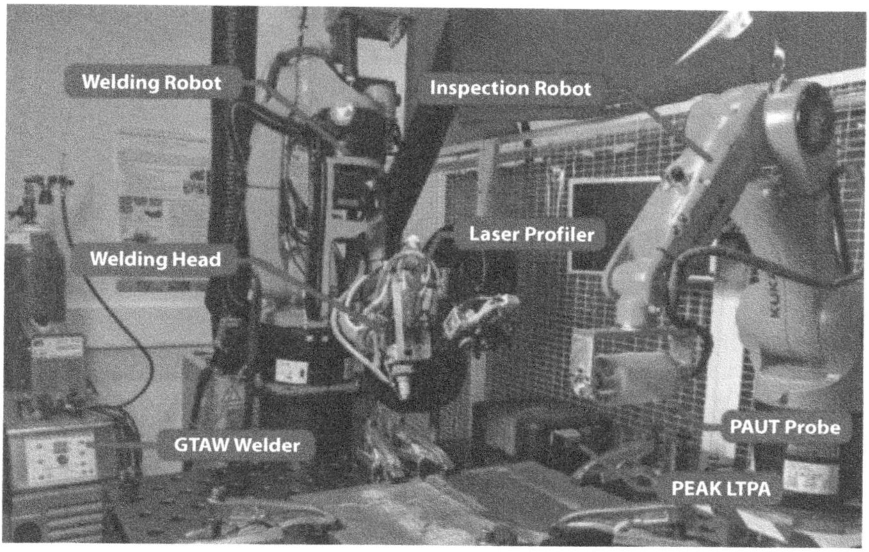

Figure 13.3 Sensor-enabled multi-robot welding and in-process NDE system [16].

always coming up with new ideas and applications for smart robots that benefit their respective industries. The benefits of an AI-based welding system include higher production rates, increased safety, and less need for human input. An important step in the TIG welding process is weld monitoring and an effective AI system involved in the process that allows failure prediction and preventive remedial measures [14].

13.4 Data-Driven Approach in Welding

An A/B testing technique is used in a data-driven manner to estimate and compare the quality performance of pipe-welding operators for industrial construction projects. To show the viability and application of the suggested technique, existing engineering design and quality-management data from a pipe fabrication company are processed and examined. To account for uncertainty, operator welding capabilities are quantified as Bayesian distributions of posteriors using a customized version of the Metropolis–Hastings algorithm. The A/B testing algorithm is used to compute probabilistic differences across operator-quality performance distributions to assess their quality performance quantitatively, as shown in Figure 13.4 [15].

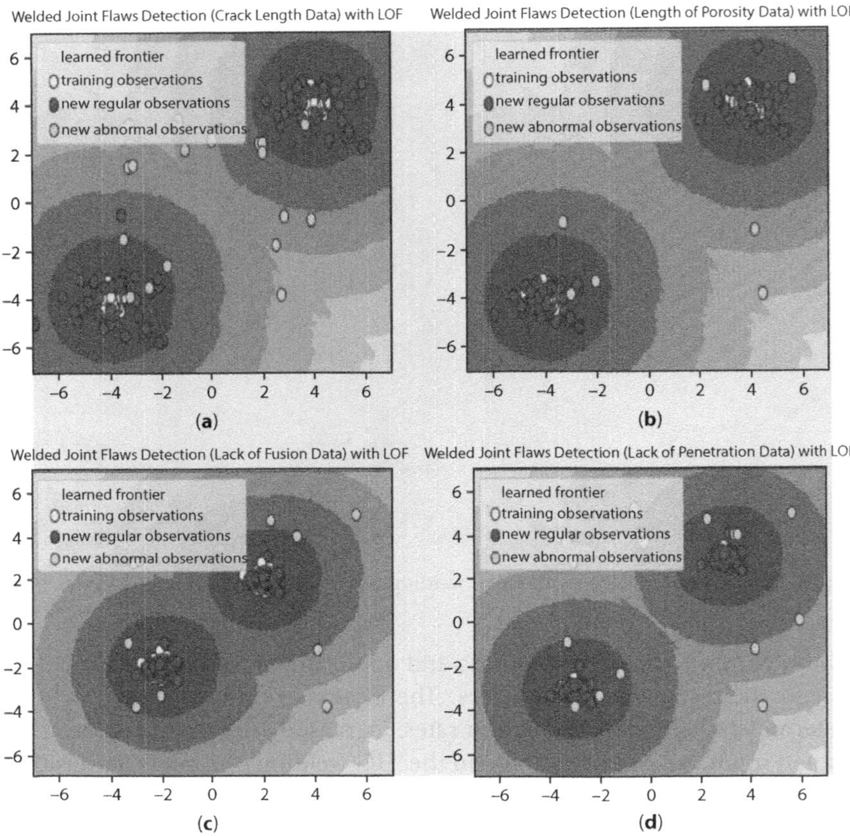

Figure 13.4 Detection of flaws in welding [17].

13.5 Conclusion

An important aspect of the improvements in welding technologies has been attained with the help of Machine Learning (ML). As this section has described, Machine Learning (ML) has a lot of new entities in terms of defect prediction, welding parameter optimization, and frequent quality assurance. These improvements exist in real-world applications that increase the rate of production and decrease material waste.

Welding processes provide more accuracy and precision because the AI database derive significant insights. This kind of welding process is very crucial in many industries, like construction, automotive, and aerospace, where welding ensures safety and precautions. Day-by-day monitoring and adaptive control are made possible by ML-driven techniques ensuring effective and excellent welding processes. Furthermore, in the field of automation under challenging conditions, computer vision and sensor technology are used to detect and evaluate welding processes, along with the development of intelligent welding robots. This section elaborately discusses human ability and how it reduces the dangers connected with manual welding under challenging environments.

The process of involving ML in welding is still in progress. There are many issues to be resolved like the requirement for detailed datasets, the right algorithms, and handling safety and ethical issues. The expectation in development in the welding domain will probably concentrate on overcoming these challenges and executing the potential of ML in the welding process.

In conclusion, the welding industries are setting new rules and norms with the inclusion of ML into welding technology. By using reinforced computational methods and datasets, welding processes are evolving from skilled or semi-skilled manual processes to highly accurate and efficient ones. The welding processes have a bright future in the next decade, with ML having the biggest influence on how it develops and is used in many other industries.

References

1. Goka, S., Narayana, G.S., Jyothi, G.D., Shaik, H.S., Moinuddin, S.Q., AI and ML in Welding Technologies, in: *Automation in Welding Industry: Incorporating Artificial Intelligence, Machine Learning and Other Technologies*, pp. 73–90, Scrivener Publishing LLC, 2024
2. Wang, B., Hu, S.J., Sun, L., Freiheit, T., Intelligent welding system technologies: State-of-the-art review and perspectives. *J. Manuf. Syst.*, *56*, 373–391, 2020, https://doi.org/10.1016/j.jmsy.2020.06.020.
3. Lu, J., Yang, A., Chen, X., Xu, X., Lv, R., Zhao, Z., A Seam Tracking Method Based on an Image Segmentation Deep Convolutional Neural Network. *Metals*, *12*, 8, 1365, 2022, https://doi.org/10.3390/met12081365.
4. Mahadevan, R.R., Jagan, A., Pavithran, L., Shrivastava, A., Selvaraj, S.K., Intelligent welding by using machine learning techniques. *Mater. Today Proc.*, 46, 7402–7410, 2021, https://doi.org/10.1016/j.matpr.2020.12.1149.
5. Cai, W., Wang, J., Jiang, P., Cao, L., Mi, G., Zhou, Q., Application of sensing techniques and artificial intelligence-based methods to laser welding real-time monitoring: A critical review of recent literature. *J. Manuf. Syst.*, *57*, 1–18, 2020, https://doi.org/10.1016/j.jmsy.2020.07.021.

6. Mishra, A., Artificial Intelligence Algorithms for the Analysis of Mechanical Property of Friction Stir Welded Joints by using Python Programming. *Weld. Technol. Rev.*, *92*, 6, 7–16, 2020, https://doi.org/10.26628/wtr.v92i6.1120.
7. Tran, N.H., Bui, V.H., Hoang, V.T., Development of an Artificial Intelligence-Based System for Predicting Weld Bead Geometry. *Appl. Sci.*, *13*, 7, 4232, 2023, https://doi.org/10.3390/app13074232.
8. Yaknesh, S., Rajamurugu, N., Babu, P.K., Subramaniyan, S., Khan, S.A., Saleel, C.A., Nur-E-Alam, M., Soudagar, M.E.M., A technical perspective on integrating artificial intelligence to solid-state welding. *Int. J. Adv. Manuf. Technol.*, 2024, https://doi.org/10.1007/s00170-024-13524-9.
9. Kumar, S., Gaur, V., Wu, C., Machine learning for intelligent welding and manufacturing systems: research progress and perspective review. *Int. J. Adv. Manuf. Technol.*, *123*, 11–12, 3737–3765, 2022, https://doi.org/10.1007/s00170-022-10403-z.
10. Meena, P., Farhan, A.M., Anant, R., Saheb, S.H., Advancements in Welding Technologies, in: *Automation in Welding Industry: Incorporating Artificial Intelligence, Machine Learning and Other Technologies*, Scrivener Publishing LLC, pp. 13–35, 2024, https://doi.org/10.1002/9781394172948.ch2.
11. Amano, S., Tsujimura, Y., Ogawa, T., Shibata, T., Development of in-process welding torch position control system using AI technology. *Weld. World*, *67*, 5, 1223–1234, 2023, https://doi.org/10.1007/s40194-023-01486-7.
12. Thangavel, S., Maheswari, C., Priyanka, E.B., Passive machine vision-based defect classification in tungsten inert gas welding on SS304 using AI-based gradient descent algorithm. *Proc. Inst. Mech. Eng. Part E J. Process Mech. Eng.*, *237*, 5, 2102–2114, 2023, https://doi.org/10.1177/09544089231193564.
13. Raj, A., Chadha, U., Chadha, A., Mahadevan, R.R., Sai, B.R., Chaudhary, D., Selvaraj, S.K., Lokeshkumar, R., Das, S., Karthikeyan, B., Nagalakshmi, R., Chandramohan, V., Hadidi, H., Weld quality monitoring via machine learning-enabled approaches. *Int. J. Interact. Des. Manuf. (IJIDeM)*, 17. 6, 2023, https://doi.org/10.1007/s12008-022-01165-9.
14. Kuppuswamy, R., Calo, K., Ramakumar, J., Use of ResNet modelling for TIG weld feature digitization and correlation : a technique for AI based welding system. *Manuf. Technol. Today (MTT)*, *22*, 1, 25–32, 2023, https://doi.org/10.58368/mtt.22.1.2023.25-32.
15. Ji, W., Li, Y., AbouRizk, S.M., Integrated data-driven approach for analyzing pipe welding operator-quality performance. *Autom. Constr.*, *106*, 102814, 2019, https://doi.org/10.1016/j.autcon.2019.04.009.
16. Shaloo, M., Schnall, M., Klein, T., Huber, N., Reitinger, B., A Review of Non-Destructive Testing (NDT) Techniques for Defect Detection: Application to Fusion Welding and Future Wire Arc Additive Manufacturing Processes, *Materials*, *15*, 10, 3697, 2022a. https://doi.org/10.3390/ma15103697.
17. Oleka, C.J., Aikhuele, D.O., Omorogiuwa, E., Data-Driven Intelligent Model for the Classification, Identification, and Determination of Data Clusters and Defect Location in a Welded Joint. *Processes*, *10*, 10, 1923, 2022, https://doi.org/10.3390/pr10101923.

14

Internet of Things in Welding Industries

Alok Kumar[1]* and Ravi Shankar Rai[2]

[1]Department of Mechanical Engineering, Indian Institute of Technology, Jodhpur, Rajasthan, India
[2]Department of Automation and Robotics, Sandip Institute of Technology and Research Center, Nashik, Maharashtra, India

Abstract

Welding is an essential industrial process that plays an important role in the different manufacturing sectors. The welding process is considered unsatisfactory and requires the entire procedure to be repeated. Deploying sensors at various locations is essential for tracking and detecting immediate difficulties during the welding process. These sensors are positioned in the vicinity of the welding area. Sensors and controllers can be utilized to collect and transmit data from the arc and column interface. By evaluating the signals captured by the sensors, the data can be utilized to ascertain the impact of the arc on the weld's quality and scrutinize the characteristics of any imperfections. A sophisticated welding system can be created by harnessing the power of the Internet of Things (IoT) and employing many approaches and strategies to monitor the welding process. Intelligent welding is the outcome of IoT integration in welding and offers an adaptive control approach. Changes in current and inaccurate measurements from the equipment could jeopardize the quality of the joint. The microstructure's properties can change in response to temperature changes or a rise to a molten state. IoT-based temperature sensor technologies are used to monitor the temperature of molten metal during welding processes. Temperature sensors (LAN/wireless connections, servers, and server) are used to increase automated temperature monitoring of the molten metal during the welding process. During the welding process, data are gathered and analyzed by IoT welding systems. Close monitoring of welding parameters (temperature, voltage, and current) is made feasible by applying a data-driven methodology. IoT enables the implementation of predictive maintenance, which can detect equipment issues in advance and reduce repair expenses

**Corresponding author*: dohareyalok@gmail.com

Sandip Kunar and Gurudas Mandal (eds.) Advanced Welding Technologies, (287–322) © 2025 Scrivener Publishing LLC

by minimizing equipment downtime. Integrating the IoT-based welding sector brings about a revolutionary period of smart manufacturing and substantial improvements in productivity, quality control, and worker safety.

Keywords: IoT-based welding, sensors, current–voltage monitoring

14.1 Introduction

The welding industry is at the forefront of the worldwide industrial transformation by adopting the IoT, a network of networked devices that share data and communicate with each other. This enables continuous monitoring, collecting of data, and analysis in real time. The integration of connection is propelling notable advancements in welding methodologies, including high efficacy, enhanced precision, and the assurance of safety and proactive maintenance. Precision, performance, and efficiency are essential for upholding the highest level of safety and quality in the welding business. Conventional welding techniques frequently encounter difficulties such as variable weld quality, equipment malfunctions, and safety hazards. IoT technology tackles these difficulties by offering instantaneous information about welding operations, equipment position, and ambient factors. The IoT allows for the ongoing surveillance and regulation of welding settings guaranteeing optimal functionality and uniform excellence through the utilization of sensors, data analysis, and networking. Welding systems utilizing the IoT have several benefits such as the capability to perform predictive maintenance. This feature allows for the prompt detection of possible problems, hence preventing equipment failure and resulting in shorter periods of downtime and decreased maintenance expenses. Through the continuous real-time monitoring of welding parameters, it becomes feasible to rapidly implement remedial actions, ensuring the integrity of the weld and attaining enhanced quality control. The IoT enables the remote oversight and manipulation of activities enabling supervisors and engineers to monitor and make adjustments from any geographical location. This enhances the system's ability to adjust and respond rapidly. The incorporation of IoT technology in welding enhances safety by continuously monitoring the working circumstances and health metrics of welders, identifying potential risks, and promptly adopting safety regulations. It facilitates the effective distribution of resources by closely monitoring the use of supplies and equipment guaranteeing timely restocking and optimal utilization. The welding sector is utilizing IoT technology to bring forth a new era of smart manufacturing marked by increased efficiency, greater quality, and improved safety protocols. This introduction emphasizes the vast capacity of the IoT to bring about significant changes in the welding industry. This facilitates a thorough examination of the various uses, advantages, and future advancements of IoT in this field. The sensor device utilized the IoT to accurately monitor body temperature [1]. Processes were automated by utilizing the various IoT architectures and protocols [2]. Soil conditions were monitored using temperature sensors. The function of temperature sensor is used in temperature regulation. It examined how smart sensors fit with microelectromechanical systems. Analysis was done on the duty cycle temperature sensor and its parametric influence [3]. A smart system enhances the process performance. A microcontroller was used to examine the system's signal output [4].

The utilization of sensor technology is playing a crucial role in effectively addressing the inherent constraints associated with the manufacture of a new generation of highly accurate products. Process monitoring is necessary to improve cost, quality, and productivity qualities. Several elements influence the quality of a weld in any welding process, pinch effect, arc-(initiation, energy density, column stability), shielding effect, electron emission rate, thermal ionization rate, and dimensional correctness. Sensors provide a series of methods for the acquisition of knowledge and the development of a reliable diagnostic system for industrial operations [5]. The recorded sensor signals are utilized for the development of monitoring and control techniques in all welding processes. The monitoring method can be classed based on the technology used to assimilate response data, such as mechanical, thermal, electrical, magnetic, and radiant energy. Various sensors possess varying levels of accuracy and precision. They can be employed to create feedback control techniques aimed at restoring optimal welding conditions [6]. Sensors (acoustic sensors, temperature, power, current–voltage, force) are employed to monitor and comprehend the physics behavior of welding in the welding process. The user determines the selection and positioning of the sensor in the welding, deformation, and machining region based on the features of the process [5]. Sensors have a crucial function in helping to make items more affordable for the mass consumer market. Ensuring a consistent arc is necessary to regulate the form and dimensions of the liquified pool and the characteristics of the solidified weld. Temperature profiles of the arc column in the welding process play a main role in determining the weld bead quality. These profiles are influenced by several process parameters and the stability of the arc. The impact in weld bead geometry of the above parameters is crucial for achieving optimal outcomes. This includes carefully choosing and managing their influence on the form of the deposited bead [5]. If the current increases, constriction leads to greater arc force, which in turn leads to increased penetration. The arc current magnitude is directly proportional to the depth of penetration. The separation of the molten droplet from the electrode facilitated arc force and determined the mechanism of metal transfer. The arc condition varies along the length of the welding process, and the gravitational influence of the spatter has a significant impact on the transfer of metal.

14.1.1 IoT-Based System

The temperature sensor system based on the IoT proved more cost effective in comparison to alternative systems. Sensors (wireless communication and LAN) were employed to streamline the process of temperature measurement of the molten metal in the WM and IoT-based temperature-sensing system as shown in Figure 14.1 [7]. For example, stainless steel was welded using an arc welding (AW) machine with a power output of 5 kVA [7]. The welding temperature zone was calculated, and the various parameters of the AW process (current, voltage, and welding time) were measured. The mild steel plate melting occurs when the temperature exceeds the solidus resulting in the formation of a fusion zone. In the fusion zone of mild steel plate, the microstructure exhibits a coarse morphology. The welding zone temperature depends on the input elements of arc welding. Metal fusion has been enhanced because of the precise manipulation of temperature and control factors during the welding process. The welding zone temperature has been elevated through the augmentation of input parameters (time between 50 and 70 s, the current between 70 and 90 A, and the voltage between 30 and 40 V) [7].

14.1.2 Internet of Things (IoT)

The IoT is a new technology with a lot of new features. The IoT is a physical network equipped with software, sensors, and technologies sharing data with other Internet devices and systems. These high-tech industrial tools include simple home devices [8]. The IoT is made up of sensors and

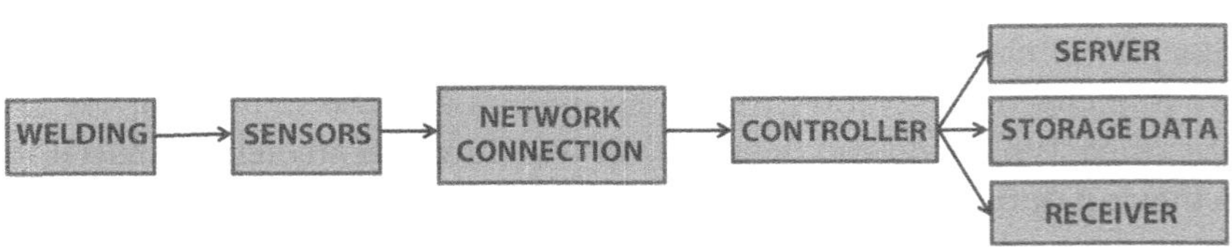

Figure 14.1 IoT-based temperature welding sensing system.

tools that are connected to wired and wireless networks, and the Internet connects using the same Internet Protocol (IP) [4]. The different technologies that are often misunderstood for the IoT are shown in Table 14.1 [4]. In 2000, L.G. Electronics made the first refrigerator that could connect to the Internet. People could use the site to make video calls and buy food.

Table 14.1 IoT-based technologies.

S. no.	Technologies	Different IoT
1	Machine-to-machine communication (M2M)	The telecommunications industry has been utilizing this mode of communication for more than 10 years [9]. M2M communication involves a direct connection between two units forming a one-to-one relationship. The increasing number of cell phone connections suggests that we are now able to transport data connected to a far wider range of devices over the same system of IP networks
2	Web of Things	Its express focus is on software architecture resulting in a narrower comparison to related technologies [2]
3	Industrial-based IoT	It strongly promotes the concept of Industrial IoT. However, it extends beyond M2M communication as it also includes human interactions, rather than only communications between machines
4	Industry 4.0	Industry 4.0 is strongly promoted by the German government, which is limited to industrial ecosystems or industrial Internet. When compared, the concepts of Industry 4.0 possesses the most extensive range of applications and implications. It provides the fundamental concepts for the upcoming industrial revolution, which will integrate many network concepts in the same workplace. It involves tangible modifications to the physical changes, and implementation of novel augmented reality (AR) or 3D printing technology [4]

In 2005, a robot was made and looked like a bunny. The latest newest news, weather forecasts, and changes in the stock market, all these were thanks to that robot Benny. In 2008, Switzerland hosted an IoT-based first international conference [4].

14.1.3 Welding-Based Revolution Industry 4.0

The systematic adoption of Industry 4.0 impacts the concept of the future factory, which will evolve into smart factory. For welding applications, a smart factory to operate with minimal human interaction and certain requirements must be fulfilled [9]. The expertise in welding must be digitized and formatted to enable a computer to make accurate selections comparable to those of a skilled welding specialist. The advanced welding equipment should be equipped with more information and communication technology and specialized sensors to enable the digitization of all production-related information and ensure the system operates in real time. Effective management of extensive data necessitates robust network infrastructures and ample storage capacity. It is necessary to develop and use principles for ensuring the security of data and communication. During the initial phases of welding technology, the primary focus was on the electrical energy converted into welding voltages and currents. During the 1990s, there was a focus on the digital welding process and implementing digital control [9]. Modern welding systems possess rapid internal and extensive data storage capabilities, robust IT security, external data communication channels, precise real-time control, and intelligent machine–human–machine collaboration [9]. The fully automated welding systems will work without human interaction and the creation of new parts and tap into the extensive knowledge stored in a central storage data system.

14.2 Functioning of Arc Welding

AW, referred to as stick welding, is a prevalent form of welding that involves the use of an electric power source and fixed electrode to fuse two metal parts [10]. AW generates a heat-based electric arc between the workpiece and the electrode. Between two electrodes, the electric arc is the emission based on ionized gas. Typically, a power supply in AW involves alternating current (AC) or direct current (DC), along with a welding electrode, electric wires, and a workpiece. In the process of AW, the circuit is closed and generates an electric arc between the workpiece and the electrode. During this process, the molten metal of the workpiece might react with the surrounding atmosphere leading to contamination from nitride and oxide present in the atmosphere. The quality of the workpiece and weld can degrade. A protective gas (shielding flux or argon gas) is employed to safeguard the weld pool against ambient impurities. AW is a welding procedure of joining two metal workpieces and energy generated from the welding arc creating a molten weld pool on the workpiece [11]. An appropriate composition required for the shielding flux is applicable electrodes. The metallic core of the electrode fuse, and the weld pool will be protected by gas and shield slag. The flux provides certain alloy components in the weld, purifying the metal surface, stabilizing the arc, and shielding the molten metal from oxidation. The slag produced in the solidification process will be eliminated. To get a seamless and impeccable welding result, it is crucial to meticulously adhere to these five key aspects. Figure 14.2 depicts the schematic representation of the operational mechanism of AW [12].

14.2.1 Voltage Effect in Welding Machine (WM)

Voltage is a crucial quantity that is frequently monitored to ensure a seamless welding operation, as it directly influences the length of the arc. While welding voltage does have an impact on arc duration, its influence on weld penetration is minor compared to other factors (travel speed and current) [13]. Increasing the voltage increases the arc length assuming the wire speed remains constant. Conversely, decreasing the voltage leads to

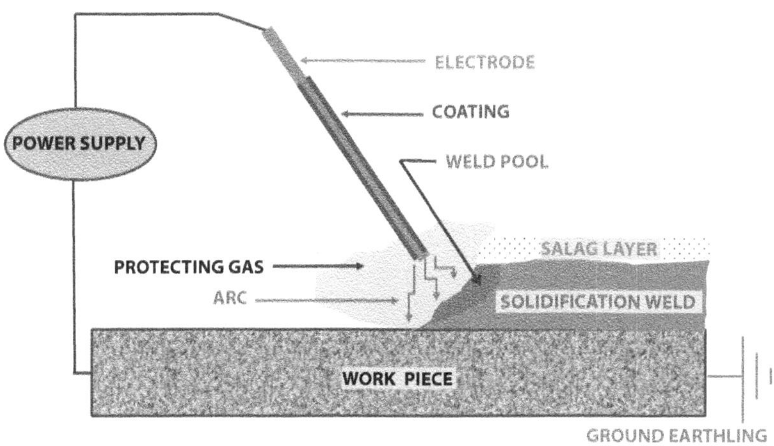

Figure 14.2 The schematic operational mechanism of AW.

a shorter arc length. Figure 14.3 illustrates the variation in arc length and cone width resulting from different voltage levels. The arc length determines both the magnitude and breadth of the arc cone. As the arc length diminishes, the arc cone will shrink and concentrate leading to a narrower and more elongated weld bead with a modest reduction in weld penetration. The arc length extends, the arc cone expands, and the arc will become more extensive. The outcome will be a broader and more level weld bead, which could potentially lead to a minor increase in welding penetration [12]. Ultimately, the voltage applied during welding impacts the formation of the weld bead, but it does not significantly alter the weld penetration depth. Figures 14.3(a), (b) illustrate the impact of employing higher and lower voltage ratings on welding while maintaining travel speed, electrode diameter, and consistent amperage [12].

14.2.2 Functioning of Voltage Sensors (VS)

A VS is a crucial component of an electrical system. It enables the monitoring, measurement, and determination of the voltage supply in both DC and AC systems [14]. In addition to basic voltage measurement, a VS is utilized for other purposes (power demand control, temperature management, power failure detection, load sensing, defect detection, and safety switching). The input of the VS is received from the system; the output can manifest in several signal formats.

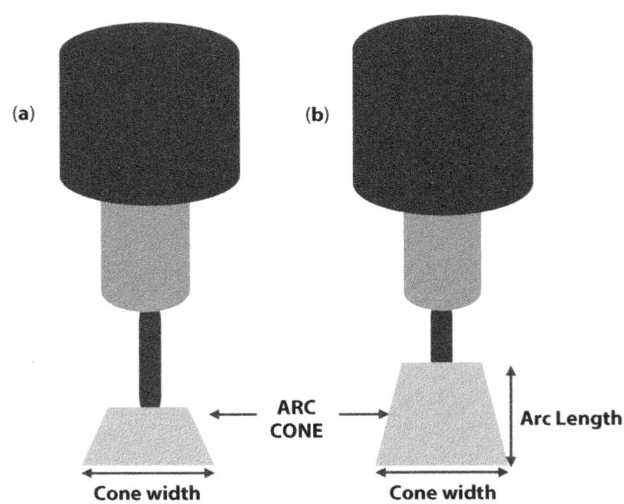

Figure 14.3 Arc length and cone. (a) Lower arc voltage. (b) Higher arc voltage.

The signal forms encompass analog current, analog voltage, switches, auditory signals, frequency, and frequency-modulated output. In addition, certain VS can carry out Amplitude or Frequency Modulation, Pulse Width Modulation, and generate pulse sine trains as their output [15]. The voltage divider operational mechanism of a VS typically relies on. There are two prevalent types of VS—capacitive and resistive VS. Figures 14.4(a), (b) depict the diagram of a VS that utilizes both capacitive and resistive components [12].

14.2.3 Different Types of Probes

Two types of probes were employed in this system to measure the voltage of the WM as follows:

a) Multimeter sensor
b) Voltage sensor

Both probes operate based on distinct principles. The functioning principle of a multimeter is based on the utilization of a shunt resistance to measure the voltage generated across it. The VS of the microcontroller operates in a manner akin to a voltage divider. Analysis of the acquired data reveals that the utilization of various types of probes has resulted in disparate findings. The difference in probe values can be influenced by several external influences. These external influences may occur when the probe is created in a subjected environment. The exogenous variables that can induce perturbations to the measurements are (a) Variables used in the application:

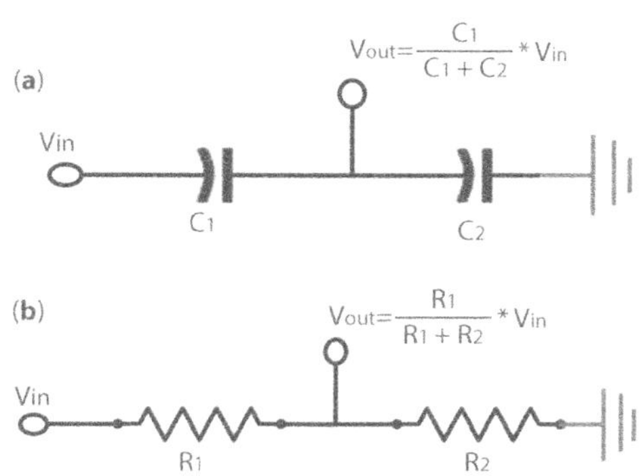

Figure 14.4 VS (a) capacitor type and (b) resistance type.

The initial external factor comprises application variables (specific gravity, barometric pressure, shock, vibration, overpressure, obstacles, and temperature). The temperature was the most prominent independent variable in this experiment. The transmission of data over temperature-sensitive wires may potentially disrupt the sensor data [16]. (b) Improper installation: The second element encompasses inadequate installation, such as poor wiring, incompatible wire, and inaccurate specifications. The system's wiring may not be optimal due to the excessive use of jumper wires. The occurrence of a voltage drop in the circuit can lead to readings that are not trustworthy [12]. Natural drift refers to the movement or displacement of objects or substances in a natural environment. Natural drift arises from the progressive degradation of wires caused by both mechanical abrasion and corrosion resulting from exposure to severe environmental conditions [17].

14.2.4 Voltage Data Collection From Internet Cloud

The data are obtained in a comma-separated value (CSV) file format, which is a sort of file compatible with Excel. The Excel file retrieved from the website was used for monitoring. The CSV file contains four columns (voltage reading, entry number, welding usage, and time stamp). The time stamps are expressed in universal time coordinates (UTC) [12]. The logging data are presented in the format of a CSV file. Subsequently, the data logging was analyzed and presented as a graph illustrating the utilization of the WM. The efficiency of the utilization of the WM has also improved. Welding usage does not follow a specific pattern since it is primarily determined by ongoing tasks assigned to pupils by their instructors. The data logging, which was to gather voltage readings from the Internet cloud, was achieved as the data were successfully acquired and interpreted from the internet cloud.

14.2.5 Welding Manufacture Based on IoT

The advent of IoT has significantly transformed the production methods employed in the welding manufacturing industry. After conducting an extensive study on the impact of IoT on the manufacturing business, a concise framework for implementing smart factories using IoT technology was developed [18]. The smart factory is comprised of four distinct objects within the framework as follows:

(i) Physical resource object,
(ii) Industrial network object,

(iii) Cloud object,
(iv) Terminal control supervision object[19].

In Figure 14.5, The physical resource object relies on intelligent devices that can communicate with one another over an industrial network. Manufacturing execution systems (MES), Cloud-based information systems, and enterprise resource planning (ERP) may collect large amounts of data from physical resources and communicate with users through terminals. This creates a cyber-physical system (CPS), which is for physical objects. The information entities are closely interconnected [20]. The intelligent framework of the factory relies on the IoT. This framework includes programmable logic controllers (PLC) with finding field devices (actuators and sensors) [19].

14.2.6 Sensing and Analyzing of Welding Information

With the advancement of the IoT, acquiring and integrating welding process data have become significantly more convenient. There is a significant amount of research on the technique of sensing information in welding—a

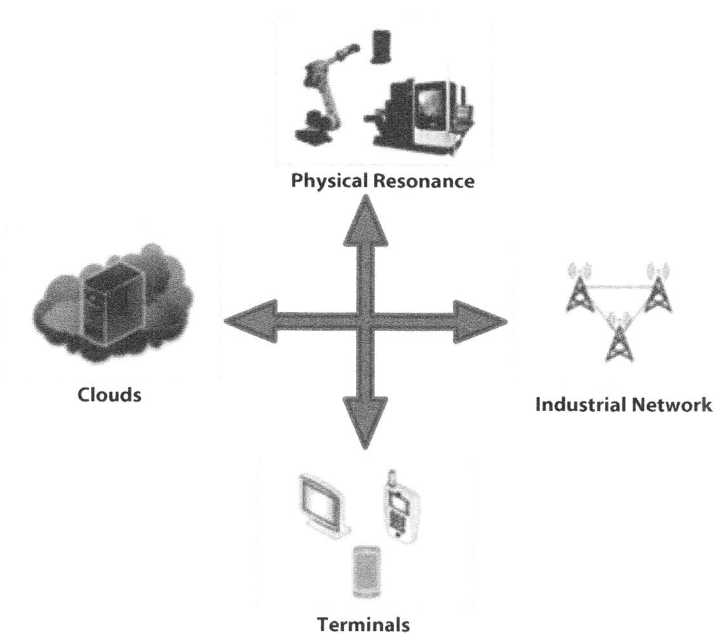

Figure 14.5 Interconnection of IoT smart factory structure.

comprehensive analysis of the multi-information sensing technology used in the dynamic process of AW [21]. A technique for combining multiple sensory data is demonstrated, along with its practical use in monitoring the welding process. The next section discusses a range of techniques for sensing welding information.

14.2.6.1 Electrical Information

The parameters most frequently observed are welding current and arc voltage. Both possess a strong capability to identify aberrant welding conditions, such as excessive or insufficient welding gas, mismatched welding feed rate, and other similar issues [6]. However, it remains challenging to identify the precise parameters that only affect the standard welding condition through the detection and analysis of welding current and arc voltage.

14.2.6.2 Optical Information With Spectrum Emission

The utilization of optical spectrum-based emission information has numerous benefits. The extensive information on welding processes encompasses spectral lines of different particles in arc plasma. The protective gases and black-body radiation follow spectrum lines of electrodes and molten metal. The optical spectrum-based emission information has a remarkable level of sensitivity and precision. The acquisition procedure of optical spectrum emission information does not have any direct impact on the welding system. Ensuring the correctness of measuring information is essential. There are two primary techniques for extracting features as follows:

(a) The physical characteristics of plasma (electron density, arc temperature, intensity of specific spectrum lines, and wavelength) [22].
(b) Involves extracting statistical parameters, geometric, and morphological, such as spectral linewidth, peak value, and mean intensity by evaluating specific or multiple spectral lines or sections of spectral lines [23].

14.2.6.3 Vision Information (VI)

VI in welding mostly comes from visible light, which immediately reflects the dynamic changes in the weld seam and welding pool. The information provided is extensive and includes details (status of the arc form, welding pool, connection type, and weld seam position). The welding VI can be categorized into two groups based on the variation in the objective light source. The modes mentioned are active and passive mode [23]. The visual-based sensor system can be categorized into two modes—(a) 2D mode and (b) 3D mode, based on the picture features acquired by the sensor [24]. The active VI requires an additional light source, such as a laser and weld seam, with the reflected laser stripes, weld seam, and pool geometrical information [19]. They developed a welding pool visual detection system that combines a dot matrix laser with an electronic-based high-resolution shutter camera. The detecting technology allows for the acquisition of a clear image of the welding pool in a dot matrix reflection. The purpose of this system is to determine an accurate dimensional location of the weld seam. It can be utilized to obtain precise information on the dimensional position of weld seams in a welding robot system.

14.2.6.4 Welding Sound Information

The sound information signal has two categories as follows:

(a) Acoustic Emission (AE)
(b) Audible Sound (AS)

The AE signal is a signal consisting of elastic stress waves that arise during plastic deformation within a material [25]. The frequency can reach several million utilized acoustic emission sensor technology to monitor the welding status in both AW and laser welding processes. The sound sensor may convert the AS signal into a voltage signal using a vibrating membrane. The processing approach may be categorized into two main groups as follows:

(i) Involves extracting features of the frequency domain, time domain, and frequency–time domain.
(ii) Involves extracting features by constructing a mathematical model of an arc sound channel using the linear predictive coding (LPC) method.

There is a significant amount of research on the acoustic emissions produced during the welding process to extract the acoustic characteristics of welding sound. Acoustic sound is produced during welding due to the influence of welding settings.

14.2.6.5 Other Welding Information

In general welding, there exist numerous other categories of welding information, including temperature data, ultrasonic data, and more. They provide a comprehensive analysis of the dynamic welding process from many viewpoints, including both direct and indirect insights into the quality of the welding. In the welding process, real-time monitoring is observed by an infrared sensing approach. The ultrasonic testing system developed utilizes resistance spot welding to detect and accurately identify welding broken spots with a 100% success rate [19]. Due to the intricate nature of the welding process, the welding sensing technique possesses constraints of the sensing medium with its limitations.

14.3 Structure of Welding

14.3.1 Structure

In welding IoT-based systems, there are six crucial components involved in the advancement prototype (hardware and software). Each of these components had a crucial function in the successful implementation of the system in this project. The six components consist of LEDs, a microcontroller, VS, IOT, a Wi-Fi module, and web pages as shown in Figure 14.6 [12]. AVS has been employed to quantify the voltage passing through the WM wire while it is being used. The microcontroller's functions (controller) are to receive and transmit input data from the IoT sensor. An Internet connection is necessary to transmit the data to IoT. The Wi-Fi module is utilized to establish a connection between a microcontroller and the wireless access point nearby. After the data (input data) are uploaded to a web page, the Internet is utilized to present the user with the current sensor reading in real time. The subsequent elucidation delineates the operational concept of the entire system as follows:

(a) The welding operational monitoring system is activated, and the green signal in the LED will illuminate.
(b) The VS will be connected to the welding station for monitoring.
(c) The Internet is used to observe the real-time voltage and utilization of the welding equipment.
(d) The red signal in the LED will illuminate if the voltage surpasses the threshold value of the sensor.
(e) All data are in an Excel file format.

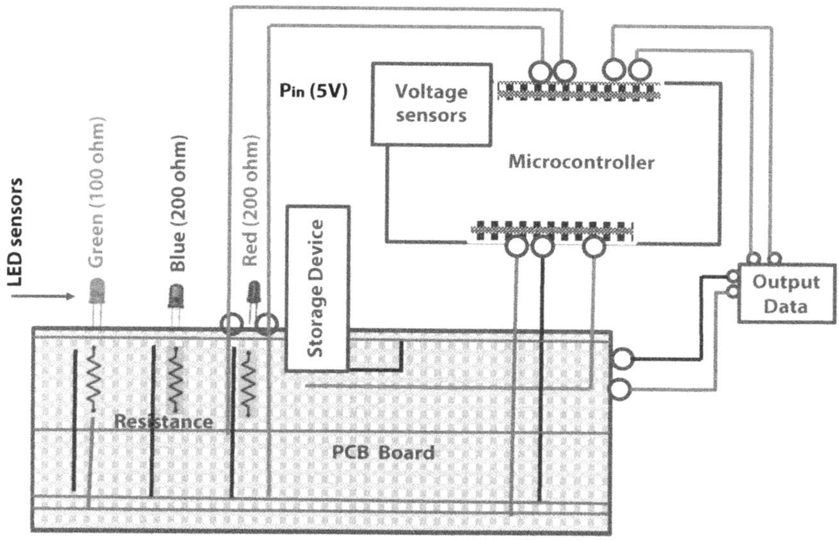

Figure 14.6 Circuit diagram of a welding monitoring system.

14.3.2 Circuit Diagram

A circuit diagram was constructed and developed for the seamless progression of the system development. A circuit diagram serves as a navigational tool for the user during the development stage. To mitigate the risk of incorrect wiring and subsequent short circuits or damage to electrical components, a circuit design can be employed to address issues stemming from a lack of concentration or human mistakes. The Fritzing software, an open-source application, was utilized for creating a circuit diagram. According to the circuit diagram depicted in Figure 14.6, three LEDs are utilized: green, red, and blue [5]. The green LED, utilized for indicating the system power status, is powered by a 3.3-V supply from the microcontroller [12]. To prevent excessive current that could damage the LED, it is safeguarded with a 100-Ohm resistor. The blue and red light in the form of light-emitting diodes (LEDs) were linked to digital pins 8 and 9 on the microcontroller and safeguarded with a resistor (220 Ohms). The Wi-Fi module is linked to digital pin 3, and digital pin 2 is linked to a 5-V power source and the ground on the microcontroller [12]. The Wi-Fi module utilizes digital pins to establish a connection with the microcontroller using serial communication.

14.3.3 Expert Welding System

An expert system (computer-based software system) utilizes knowledge to address practical problems. It is a part of an Artificial Intelligence (AI) application [26]. It is distinguished by the capacity to incorporate expert knowledge of AI-based many domains. Operators can use the relative inference approach to start inputting data from the expert welding system application. The expert system can provide the operator with a conclusive assessment at an expert level, and Figure 14.7 displays the conventional basic structure of an expert system [19].

Due to the complexity of the welding process and the necessity for several decision-making factors, other welding parameters are to be determined using quantitative functional equations alone. A qualitative assessment of the welding state is necessary to arrive at a quantitative result. Currently, the welding area is witnessing numerous instances where the expert system is being utilized.

14.3.4 Data Collected From IoT-Based Welding Expert System

The welding monitor process, thanks to its advanced welding sensor technology, generates around 0.5 GB of data for every weld seam on average. In the heavy industry and mass production sector, a substantial amount of data, up to dozens of terabytes, will be generated daily [19]. With the

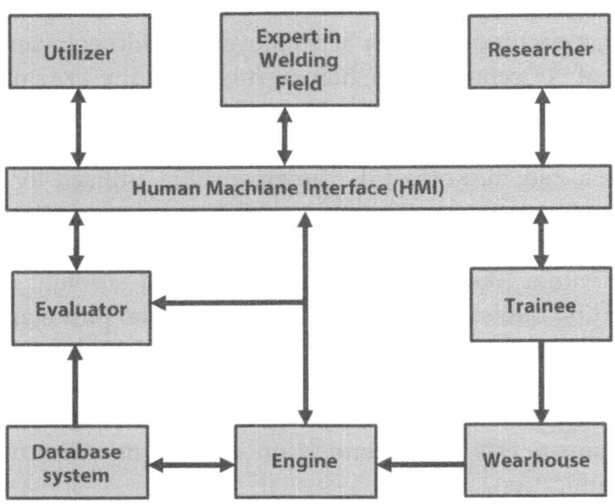

Figure 14.7 The basic structure of an expert system.

advancement of IoT, the obtaining of data will be more analyzed to streamline these methodologies, which are used for data analysis. In the welding process, the amount of statistical features of welding data will become more prominent. The reservoir of the conventional welding expert system, depicted in Figure 14.8, is incapable of acquiring information from the actual manufacturing data and sensor data. The learning module and human–machine interface module for welding expert systems facilitate the conversion of welding expert information into a format that is comprehensible to the computer. The welding expert system can collect data from IoT-based WM and analyze and summarize welding data.

14.3.5 Data Design Structure of a Welding Expert System

There are two types of expert systems on display as follows:

(i) A welding process design system, an expert system (conventional, welding monitoring, and data-driven welding), can predict the system module by assimilating information from welding manufacturing data. The Regression Model replaces the conventional inference engine, which is limited to comprehending human expert knowledge. The fundamental component of the knowledge of data-driven

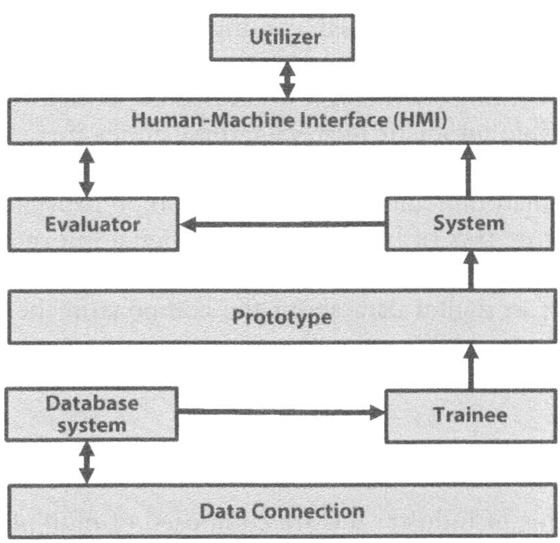

Figure 14.8 Data collected from human–machine interface with IoT-based expert system.

processing data approach in welding systems can analyze raw data and deduce expert rules. The utilization of the machine learning (ML) technique is crucial in the development of a data-design welding system. ML is a branch of computer technology that acquires this knowledge to improve their performance without the need for programming [19].

(ii) ML methods can be categorized into three groups based on their learning signal and feedback. The three types of machine learning are as follows:

 (a) Reinforcement ML,
 (b) Unsupervised ML,
 (c) and Supervised ML.

Supervised learning is a task in ML where a function is inferred from training data that have been labeled [27].

14.4 Welding Technologies and Resources

14.4.1 Digitalized Welding Science

The proper welding parameters and consumables need to be stocked in a networked, modular welding cell, contingent upon the task that the production controls. The onus of selecting the proper welding parameters now rests with the welding technology. It is essential to digitize current information and enable automatic retrieval if these duties are to be supported in the future by intelligent software systems. All information in welding technology is extremely difficult to quantify and comprehensible to digitization to the challenging task of Industry 4.0 [9]. Digital information of WM is a requirement automatically for self-cells to identify the best set of welding parameters in digital data about the component, the protective gas, the filler metal, etc. At this point, the smart factory-integrated networking system needs to perform well and provide timely access to all information.

14.4.2 Modern Welding Power Sources

Welding systems of Industry 4.0 are composed of multiple microprocessors that are interconnected by bus systems creating an internal network. This architectural design enables the versatile construction of many system configurations using a limited number of distinct base devices. These power sources are groups of components that carry out specific functions in applications. The digital arc welding parameter characteristics are processed to achieve target signal curves for the voltage and current. They capture and store recent data recordings related to welding. The welding power is automatically limited based on the allowable load capacity of the attached hardware components. The provision of networking capabilities includes support for cable networks such as wireless networks, Ethernet, Bluetooth, and near-field communication [9].The system is capable of autonomously identifying and diagnosing installed components (cooling unit, welding torch, display, and wire feeder). To fulfill these criteria, welding equipment is comprised of transformers with power electronics components, ferrite cores, copper cabling for the distribution and regulation of the welding current, and numerous electronic control boards that are interconnected by digital interfaces. In WM, a welding power source is a sophisticated power electronic equipment that must not cause any adverse effects on itself or other systems through electromagnetic interactions in any working situation. It is imperative to provide a dependable safeguarding of the delicate electronic components from the adverse conditions prevalent in the industrial setting (harsh, unclean, and dusty nature).

14.4.3 Data Storage Communication

One of the fundamental concepts of Industry 4.0 is undoubtedly communication. The power sources of required machine communication can be used using a simplified calculation. The short arc techniques function with up to 150 Hz of droplet stripping frequency resulting in the removal of one droplet approximately every 7 ms [9]. The voltage and current signals can be modulated 50-100 times during the drop separation cycle, providing optimal control over drop removal in the welding circuit [9]. The outcome is a remaining communication time of approximately 30 μs between the arc and the control computer. During this period, several signal-processing procedures need to be executed. The physical data must be converted into digital form and transmitted to several microprocessors. All essential calculations required to decide the appropriate control action must be performed. All generated control signals are returned to an actuator of the system. The data transport and signal-processing requirements in real time provide a significant technical barrier. To address this issue of Intranet, it is beneficial to differentiate between internal data transmission, which forms the foundation of the entire control system, and external data communication. The Internet is utilized for documentation purposes.

14.4.4 Data Storage Welding Parameter

The primary objective of Industry 4.0 is to achieve a manufacturing cell that operates independently and flexibly. This cell is composed of several interconnected subsystems. To operate autonomously, the manufacturing cell needs to be capable of recognizing both the raw materials required for the product and the specific order for manufacturing [9]. The separate systems are thereafter given instructions. In the welding system, the welding storage parameters are transmitted. These are needed, inspected, and rectified throughout the ongoing welding process. It is essential to achieve a robot welding power source to equip real-time data communication with robust microprocessors, with control algorithms. The selection of a welding process and welding parameter characteristics based on factors (sheet thickness, base material, potential gaps, and layer structure) is separate from the actual production welding. This process can only be successful if the essential welding expertise has been appropriately digitized. There is a strong inclination among users to record and preserve data at the highest possible resolution because it has not been accurately determined at what degree of detail the data should be available for creating suitable algorithms. Due to frequent Internet connection speeds at client sites and data security issues unresolved on cloud systems, it is also possible to store and analyze production data locally.

14.4.5 Data Security

The fundamental concept of Industry 4.0 is to facilitate the sharing of data between different products or systems and to enhance the effectiveness of solutions for a diverse range of tasks within a broader framework. This analysis can be done with the help of software systems or by domain experts. To execute this technique, it is necessary to have a high level of computerization production facilities with the integration of local networks—globally accepted standards for data communication. Open network structures, data transfer, and interfaces are also required. Cyberattacks could potentially occur on local corporate networks through production facilities [9]. Therefore, both the machine producer and the user must thoroughly address the issue of cybersecurity for their production facilities and use specialized protective software and equipment. The most effective security measures can be achieved through a combination of electronic hardware and security software.

14.4.6 Identification of Welding Torch Position

The alignment of the torch about the workpiece along the welding trajectory is a significant determinant of welding quality. For automatic robot welding systems, the accurate assessment of the part surface topography and the precise positioning of the welding flame may already be accomplished with confidence. Laser scanners are typically employed to gather data, which are then utilized for the implementation of adaptive welding procedures. These operations adjust the position of the torch and the characteristics of the power source in response to any variations in the part's topology. Furthermore, variations in the robot's real movement, such as welding speed, can also be considered. The applications currently cannot determine the position of the flame during manual welding jobs. The absence of a suitable measurement solution for this operation is mostly due to the elevated temperatures, extensive exposure to arcs, metallic part surfaces, welding fumes, and restricted accessibility.

14.4.7 Interpretation of WM and Humans

The utilization of Industry 4.0 ideas brings us closer to achieving the objective of a completely autonomous welding cell that can independently complete welding activities. Implementing the requirements mentioned above in commercial applications may involve a significant investment of time and effort due to their complexity. Therefore, both the welding technologist and the welder will continue to have the responsibility of ensuring a high standard of welding in intricate welding products for the foreseeable future. The communication methods between humans and WMs must remain crucial. The current trends in interface development are heavily affected by the gadgets that welders use in their personal lives. Smartphones and tablets are widely available and can serve as a standard for simplifying operations. Capabilities, like being able to speak many languages, operate through touch, and control through voice, are already being introduced into the challenging workshop setting. Advancements in telecommunications, shown by devices (Google Glass or the Microsoft HoloLens), are leading to the welding helmet being recognized as a communication platform between humans and machines [9]. The concept of displaying welding parameters on the protective glass of the helmet is gaining increasing attention. The helmet might be outfitted with voice-activated

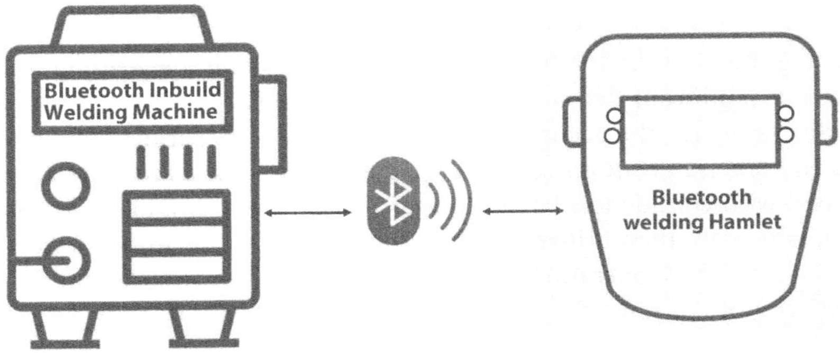

Figure 14.9 Bluetooth-connected WM and helmet.

control, which is linked to the power supply. To facilitate this, the helmet must have the capability to establish a wireless connection with the power source. Recent advancements have already made it feasible to establish a connection between a helmet and a power source WM as shown in Figure 14.9. The power source transmits information regarding the arc status to the helmet via this connection. The helmet uses these data to regulate the automatic dimming feature of the protective glass accordingly. This feature prevents the standard auto-darkening from failing to respond in situations when there is either too much light or the arcs are obscured.

14.4.8 Virtual Welding

To produce an actual component by an Industry 4.0 welding cell, it is imperative to develop the welding program in an offline mode. This involves determining and verifying the movement of the welding torch to prevent any unnecessary rejections. This procedure is necessary, particularly when it pertains to the production of individual pieces. In the future, specialized software systems will be required to simulate the welding process. The foundation for their advancement will rely on the existing software programs utilized by the robot programmers and the training of welders to virtually simulate the accurate programming of torch positioning and movement on the robot.

14.5 Monitoring System of IoT-Based Welding

14.5.1 Online Process Monitoring

The main obstacle in the AW method is the absence of a visible arc and the difficulty in determining its exact placement on the workpiece. The behavior of an arc is influenced by the level of ionization complexity and the current voltage between the workpiece and electrode. Inspection of welded components occurs after the completion of product welding resulting in significant delays, additional inspection expenses, and material wastage in the event of a bad weld. Traditional methods for monitoring processes include destructive testing (gas flow rate, arc voltage, current, and speed) and non-destructive testing (ultrasonic testing and radiographic examination). Welding defects occur due to differences in the welding process, including changes in the dimensions and geometry of the arc, the intensity of the arc, the way the metal is transferred, and the movement of the molten pool [5]. Online process monitoring is beneficial because it allows for the early and swift discovery of abnormalities in the arc [5]. The arc exhibits fluctuations in temperature, sound level, and length, which can be measured and converted into desired arc outputs (energy intensity, acoustic pressure, temperature, and column length). These measurements can be obtained using sensors and utilized for analysis purposes. The impact of arc and speed characteristics on weld quality and attributes is also significant [28]. The different input parameters are optimized to obtain maximum or minimum replies. To ensure the proper melting zone and prevent defects during welding, the process parameters have an influence on the thickness of graphite sheet and insulation brick. The contact between the insulation brick surfaces are better in the microwave hybrid heating process. Factorial approaches and analysis of variance are valuable in laser welding for assessing the impact of process parameters on output variables and arc welding's suitability for titanium and aluminum materials. A regression model was utilized to predict excessive penetration and burn through, specifically to determine the penetration depth [29]. Conventional and unconventional techniques can be used to monitor the welding process, as seen in Figure 14.10. An online monitoring process with adaptive control is advantageous because it enables prompt and accurate detection of abnormalities in the arc. The implementation of IoT in AM offers a remote platform accessible through mobile devices and computers. This platform allows for the monitoring of the welding process and the identification of any abnormalities. Process monitoring can be conducted by analyzing current, voltage,

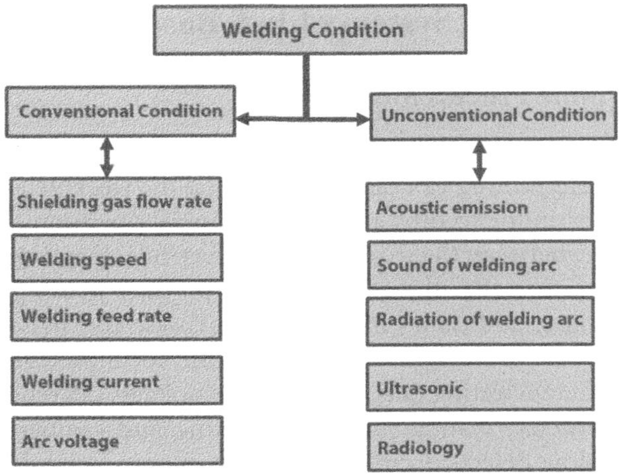

Figure 14.10 Welding monitoring different conditions.

temperature, and auditory signals [5]. Variations in arc behavior led to varied temperature gradient profiles, which in turn affected the welding pool, and ultimately impacted the quality of the weld. Utilizing current, voltage, temperature, and acoustic measurements for process monitoring enables timely decision making to regulate the quality of welds. Automated systems utilize a range of sensory and control techniques, including weld pool oscillation, radiography, heat, and vision [30]. Monitoring of the robotic welding process is conducted by employing several sensors from diverse sources and can be applied in many civilian scenarios [31].

14.5.2 Temperature Measurement Monitoring

The integration of sensor systems with welding expert systems is essential for the advancement of smart welding. This integration allows for a reduction in human involvement during the welding process resulting in a lower rate of rejection and optimal cost efficiency [5]. It is crucial to categorize the sensors utilized in different welding procedures due to the varying temperature generation and its impact on weld quality. The placement of sensors in the vicinity of the welding arc area is crucial since the precision and functionality of the sensors are influenced by factors such as the temperature of the weld pool, the creation of magnetic fields owing to arc current voltage, arc spatter, and the heating and cooling cycles. There are three types of temperature sensor monitoring as follows:

(i) For monitoring welding temperature, contact probe sensors with geometry-oriented designs are used such as limit switch types and potentiometers.
(ii) Non-contact design-based geometry sensors—ultrasonic, electromagnetic, laser stripe, and eddy current.
(iii) Process-oriented sensors (arc electrical, current-VS) and phenomena-oriented sensors (pyranometers, thermocouples, IR sensors, and CCD cameras) [5].

The incorporation of online data on arc column behavior and the implementation of an adaptive control approach for weld quality control need the utilization of IoT technology. In addition, infrared detection is a viable method for measuring temperature. Temperature monitoring throughout the welding process facilitates the analysis of several aspects such as weld bead size, microstructure, distortion, weld quality, and heat-impacted zone.

14.5.3 Current–Voltage Monitoring

The impact of weld beam on travel speed and arc voltage current on the porosity was done by utilizing a high-speed camera and mobile camera to observe the welding process and analyze the relevant welding parameters. The relationship between welding speed and current is crucial for understanding weld pool properties and creating a dynamic prediction algorithm [15]. The procedure of compensating for fluctuations in the welding process involves selecting a fixed current and manually managing the speed, arc length, and torch dimension as the primary means. Control algorithms are used to assist in this process [5]. A system is created in constant-voltage AW to detect defects using current and voltage signals. Weld monitoring systems are very responsive to instances of melt through caused by thin sections, loss of shielding gas, and other factors that result in the presence of surface and subsurface porosity. The impact of welding parameters on weld quality in AW is crucial. This can be done by monitoring the current–voltage waveform in real time using the Huawei automatic walking control device. It has been discovered that voltage stability can be employed to monitor the stability of arc length [32]. A thermal imaging system, based on a focal plane array, is utilized for the detection and analysis of infrared radiation. This method accurately defines the heat distribution on the surface of plates. The stability of the welding process can be affected by the welding arc, and this can be assessed by digitally analyzing the acoustical welding parameters and detecting disturbance-related events [33].

14.5.4 Acoustic Monitoring

A sound wave is formed when there is a change in pressure within a liquid, solid, or gaseous media. A skilled operator is responsible for detecting any irregularities in the manufacturing setting [5]. During the process of welding, acoustic sound waves are produced because of changes in phase, arc behavior, and the environment of the arc column. These sound waves include valuable information about the physics of the arc and the quality of the weld [32]. By utilizing a microphone and strategically positioning it in the laser welding area, it has been shown that it is possible to gather sound data to effectively monitor the depth of penetration during laser welding. AE sensors are used in the laser welding area to collect sound data. The sound impulse and turbulent noise are affected and polluted by overlapping background noise. Analyzing the effect of background noises on acoustic intensity pressure levels, both with and without welding, is necessary [34]. Identifying and calculating the shape of the pool is crucial for the advancement of an intelligent WM and gaining a deeper grasp of the operation through the utilization of image processing [24]. Major issues in welding (joint edge recognition, joint seam tracking, penetration control, and measurement profile of a junction) can be efficiently addressed by utilizing sensory feedback signals from the weldment [35].

14.5.5 IoT Monitoring

This is the analysis of process monitoring using Artificial Intelligence and IoT technology. They utilize Artificial Intelligence (AI) and IoT to enhance quality control while minimizing costs. The perception of sensor facts, which include optical, electrical, temperature, and auditory characteristics, allows for the monitoring of welding [5]. By using IoT-based monitoring, it is possible to detect most instabilities for faults in the welding process. This not only allows for immediate identification of issues but also enhances the overall efficiency of the process. There is no requirement for a distinct post-welding inspection. By predicting weld quality, the overall quality of the resulting weld can be improved. The proliferation of intelligent products is leading to the emergence of intelligent living, which necessitates a shift toward very cost-conscious markets operating on shorter cycles compared to previous periods [20]. An IoT system is useful for gathering and combining welding data to support data-driven learning and extract knowledge from sensor data when there is no interaction with welding experts. The monitoring of appliances is accomplished by remote data capture using wireless sensor technologies and IoT apps. The application field of IoT is vast and not limited. It can completely transform several aspects of the world (cities, environments, smart devices, manufacturing, health technologies, and automation). An IoT creates a better perception to develop a consisting of many layers including resources, services applications [36]. Smart manufacturing is an effective approach for Industry 4.0 that incorporates AI, IoT, and other technologies. It involves decision support and real-time monitoring ultimately resulting in the development of intelligent welding systems [37].

14.5.6 Sensor Process Monitoring

The integration of sensor systems with welding expert systems is crucial for the advancement of smart welding. This integration aims to minimize human involvement during the welding process resulting in a lower rejection rate and optimal cost efficiency [32]. Identifying the types of sensors utilized for different welding processes is crucial due to the distinct physics involved in each operation. The positioning of sensors in the vicinity of the welding arc area is crucial because the precision and functionality of the sensors are influenced by factors such as the temperature of the weld pool, the creation of magnetic fields from the arc's current voltage in the presence of arc spatter in the heating and cooling cycles [5]. The temperature sensor-based profile data are essential for conducting a controlled system analysis. The sensors suitable for temperature monitoring in the welding process are as follows: thermography high-resolution camera, infrared thermometer, fetoscope, temple stick (thermocouple, electronic surface thermometer), ultrasonic sensor, optical sensor laser strip sensor, an electromagnetic sensor [5]. Current monitoring can be conducted using hall sensors, split-core transformers, and current series transformers. Voltage monitoring can be accomplished by utilizing a voltage transducer. The unidirectional microphone is an affordable option for acoustic monitoring, while a piezoelectric sensor with data-collecting modules allows for precise capturing of acoustic signals. The typical specifications of a sensor are as follows: image resolution, repeatability, response time, resolution, relative humidity, stability, sampling rate, weight dimension (including battery), power requirements, temperature measurement range, battery life, measuring modes, power consumption, calibration technique, time between measurements, display type, pixel size, accuracy, measuring range, sensitivity, lift-off distance, spectral response, and display hold.

14.5.7 Website System Monitoring

The World Wide Web (WWW) is a compilation of webpages located on a network of interconnected computers accessed via the internet [12]. Due to the rapid and continual improvement of network technology, the WWW is growing at an astonishing pace and remains one of the most widely used applications. The WWW was conceived by Tim Berners-Lee, an English scientist, in 1989 [12]. The WWW serves as the fundamental infrastructure of the information age and functions as the primary medium through which individuals worldwide communicate and engage with one another on the Internet. Web pages on the WWW can be viewed using a web browser, which can display and interpret the content. Google and Yahoo are web search engines, which are software systems designed specifically for searching for content on the WWW [12]. The search results are often displayed in a list format and may include a combination of websites, photos, and other types of content. The WWW is seen in the format of web pages. Web page documents have formatted and annotated text using Hypertext Markup Language (HTML). HTML is a programming language used to create websites that can be accessed by anybody with Internet connectivity. In addition to textual information, web pages can also incorporate images, videos, audio, and other forms of multimedia to enhance the visual appeal of the website. A website can be either publisher maintained or open source allowing viewers to change its content. Websites are frequently utilized for the objectives of acquiring and sharing information, providing entertainment, conducting commercial activities, and serving the needs of governmental or non-governmental organizations. A web-hosting service provider storing the data on the provider's server enables website accessibility through the WWW.

14.6 Applications of IoT-Based Welding

14.6.1 Ongoing Applications of the IoT

The IoT has numerous applications. IoT can be utilized by industries in a multitude of applications. The IoT has led to remarkable advancements in the market. We adopted a comprehensive methodology categorizing 10 prominent figures and innovators in the field of IoT. The primary emphasis is on enterprises that are adopting and utilizing IoT applications and technologies to advance, rather than on suppliers that are focused on selling [4].

14.6.2 Smart Materials

The utilization of various materials has been in practice for many years leading to enhanced living conditions. Civilizations have been categorized according to their identification and utilization of resources, such as the earliest time of the Stone Age. The bronze age is renowned for its remarkable inventiveness, primarily because of the superior durability and intricacy of bronze in comparison to stone. This advancement in metallurgy marked the onset of a new era paving the way for the discovery of additional materials with similar properties. Advancements in science and technology have greatly improved the ability to create novel materials in the past 20 years. The materials are classified into ceramics, metals, polymers, and intelligent materials. Smart materials are objects that exhibit a property that external stimuli can alter. These modifications may involve electrical, chemical, thermal, mechanical, and magnetic alterations. The welding sector benefits from the revolutionary uses of the IoT, which improve safety, quality, and efficiency. The following are a few primary applications:

(a) Planned Maintenance Monitoring Equipment Health—IoT sensors can monitor the state of welding equipment by identifying problems like overheating, excessive vibration, or component wear. These data facilitate proactive maintenance minimizing operational interruptions and prolonging the operational lifespan of equipment. (Automated Notifications)—If sensors discover irregularities, automated warnings can be dispatched to maintenance crews to proactively address the issue and prevent a potential breakdown.

(b) Quality Control—The real-time monitoring of IoT devices enables the continuous monitoring of welding parameters, including temperature, voltage, current, and gas flow. This guarantees that certain parameters consistently remain within the set ranges. This aids in maintaining a consistent level of welding quality.

(c) Defect Detection—By employing advanced sensors and machine learning algorithms, it is possible to examine welding patterns with precision to identify mistakes or abnormalities. This allows for immediate corrective actions to be implemented.

(d) Process Optimization—By employing data analytics to gather and analyze data from welding operations, companies can identify inefficiencies and optimize parameters to improve productivity and save costs.

(e) Adaptive Welding—IoT-enabled systems have the capability to adaptively alter welding parameters based on real-time data enabling adjustments to be made in accordance with changes in material, thickness, or environmental circumstances.

(f) Remote Surveillance and Management (Remote Access)—Supervisors and engineers can remotely monitor the welding processes of IoT-enabled equipment. This enables them to offer supervision and support from any geographical location. (Control Systems)—The Internet of Things (IoT) can facilitate remote alterations to Wireless Modules (WMs) enabling quick responses to changing situations without the need for physical presence.

(g) Safety Improvements (Occupational Safety)—Wearable technology Internet of Things (IoT) devices have the capability to monitor and evaluate the physical health and safety of welders. They accomplish this by analyzing many factors including their exposure to dangerous gases, levels of fatigue, and the surrounding environmental conditions. (Hazard Detection)—Internet of Things (IoT) sensors have the capability to identify and detect possible risks, such as gas leaks or fire hazards. Upon detection of these threats, the sensors promptly activate safety protocols to protect personnel.

(h) Inventory Management (Consumables Tracking)—The IoT has the capability to monitor and document the utilization of consumable products such as welding rods, gases, and protective gear. This allows for effective replenishment and reduces the amount of time lost due to insufficient supply. Asset Management—By deploying IoT sensors on welding equipment, companies can track the location and usage of tools and machines, thereby improving the efficiency of asset utilization.

(i) Training and Skill Development [Virtual Reality (VR) and Augmented Reality (AR)]—Integrating IoT technologies can enhance the immersive training experiences provided by VR and AR systems. This allows novice welders to hone their skills in a controlled and supervised setting, while also obtaining instant feedback.

(j) Skill Assessment—IoT devices have the capability to assess the skill level of welders identifying precise areas that require enhancement and tailoring training programs accordingly.
(k) Energy Management (Energy Consumption)—IoT sensors are used to track and analyze the energy usage of welding equipment. These sensors can identify any areas where energy is being wasted and provide insights on how to save energy. Sustainable Practices—The Internet of Things (IoT) can aid companies in attaining their sustainability goals and reducing their carbon emissions by enhancing welding processes and lowering energy use.
(l) Interoperability With Other Systems—Integrating IoT-enabled welding equipment with Enterprise Resource Planning (ERP) and Manufacturing Execution equipment (MES) enables seamless data transfer and enhances overall operational efficiency.
(m) Supply Chain Management—Utilizing real-time data from IoT devices can improve the transparency of the supply chain ensuring the timely delivery of materials and components needed for welding operations.
(n) Tailored Welding Solutions (Customized Welding Programs)—The data from IoT can be used to develop customized welding programs that are specifically intended to meet the specific requirements of a project, including the materials used and the ambient conditions. This ensures optimal results for every job. The Internet of Things (IoT) enables flexible manufacturing processes that can quickly adjust to new product designs or changes in production schedules.

By implementing IoT technology, the welding industry may achieve higher levels of automation, precision, and efficiency, leading to improved productivity, quality, and safety

14.7 Conclusions

The use of IoT technology in the welding sector represents a substantial advancement toward improved efficiency, quality, safety, and environmental responsibility. Welding systems that are equipped with IoT technology can monitor in real time, do predictive maintenance, and utilize advanced data analytics. These attributes work together to guarantee constant weld quality, minimize machine downtime, and enhance operational efficiency. By utilizing IoT technology, the welding sector may enhance precision and dependability through continuous monitoring of welding parameters enabling prompt identification of any anomalies. As a result, the quality of the product increases, and the need for rework lowers leading to cost savings and higher production. Integrating IoT technology into predictive maintenance reduces unexpected downtime and prolongs the lifespan of welding equipment, hence enhancing operational efficiency. The IoT has enabled the ability to remotely monitor and control welding processes providing flexibility and improving oversight. This enables prompt responses to dynamic circumstances and enhances well-informed decision making. IoT-based wearables and sensors significantly improve safety by monitoring the working environment and the health indicators of welders. This guarantees prompt intervention in case of any hazards. Efficient resource and inventory management through the monitoring of supply and machinery usage ensure optimal utilization and prompt replenishment. This reduces inefficiencies and operational disruptions, hence maximizing production efficiency. The use of IoT in the welding industry facilitates a smart, integrated, and data-driven approach to manufacturing. It enhances both operational efficiency and quality while also promoting safety and sustainability. The continuous advancement of this technology is facilitated by IoT-based welding.

The temperature-detecting mechanism was enhanced by employing sensors and a microcontroller. The temperature of the welding zone was measured by monitoring the variations in input parameters such as current, voltage, and time. Intelligent smart welding can be achieved by utilizing various sensors to monitor the arc column, collecting information about the arc column, and analyzing the sensor data to determine the quality of the weld, whether it is sound or defective. Relevant data can be utilized to facilitate the advancement of a control rule-based expert system. An IoT solution that utilizes rule-based training and real-time monitoring allows for optimal output weld conditions by facilitating communication with the welding environment. The performance of the sensor may be influenced by the magnetic field produced by the arc and the spatter formed by the arc pool. This can be avoided by placing the sensor in a ceramic or other insulating enclosure. The welding station monitoring system successfully identified welding activity, measured welding voltage values, and transmitted the data to the IoT for monitoring purposes. The voltage reading of the WM was measured using a VS module. The data were then processed by a microcontroller and uploaded to the IoT using a Wi-Fi module. The development of the prototype was hindered by other issues, including a malfunctioning sensor and incorrect circuit connections. Supervisors utilized the system to oversee welding operations resulting in enhanced production within the industry.

The advancement of welding techniques has led to numerous innovations in the manufacturing industry. The robust sensor technology enhances the ease of obtaining welding process data (optical information, welding current, welding sound, and arc voltage). The advent of the IoT expedites the procedure of gathering and merging data. As the amount of data rises, the statistical characteristics of welding data are unable to match the requirements of WM. AI is increasingly crucial in the welding manufacturing business due to its ability to analyze welding data and offer better information and functionality. The system, a potentially dangerous form of AI, can efficiently aid in the process. The primary objective of the welding system is to regulate the formation and characteristics of welding. The data-driven system will be more prominent in the ability from raw welding cases and consolidate its expert knowledge to enhance its knowledge repository. The fundamental issue of a data-driven system lies in the selection of an inference engine. It is imperative to successfully integrate the system with data-driven approaches of ML.

References

1. Ali, M.M., Haxha, S., Alam, M.M., Nwibor, C., Sakel, M., Design of Internet of Things (IoT) and Android Based Low Cost Health Monitoring Embedded System Wearable Sensor for Measuring SpO2, Heart Rate and Body Temperature Simultaneously. *Wirel. Pers. Commun.*, 111, 4, 2449–2463, 2020, doi: 10.1007/s11277-019-06995-7.
2. Sobin, C.C., *A Survey on Architecture, Protocols and Challenges in IoT*, Wireless Personal Communications, vol. 112, 1383–1429, Springer US, 2020, doi: 10.1007/s11277-020-07108-5.
3. Wang, G. and Zhang, L., Synthesis, self-assembly and pH sensitivity of PDEAEMA-PEG-PDEAEMA triblock copolymer micelles for drug delivery. *React. Funct. Polym.*, 107, 1–10, 2016, doi: 10.1016/j.reactfunctpolym.2016.08.001.
4. Ammar, M., *et al.*, Significant applications of smart materials and Internet of Things (IoT) in the automotive industry. *Mater. Today Proc.*, 68, 1542–1549, 2022, doi: 10.1016/j.matpr.2022.07.180.

5. Barot, R.S. and Patel, V.J., Process monitoring and internet of things feasibility for submerged arc welding: State of art. *Mater. Today Proc.*, 45, 4441–4446, 2021, doi: 10.1016/j.matpr.2020.12.782.
6. Quinn, T.P., Smith, C., McCowan, C.N., Blachowiak, E., Madigan, R.B., Arc Sensing for Defects in Constant-Voltage Gas Metal Arc Welding. *Weld. J. (Miami Fla)*, 78, 9, 322–328, 1999.
7. Sanjith, S., Balaji, N., Anand, L.D.V., Anne, W.R., Shanmugapriya, P., Rani, S.S., IoT enabled temperature sensing system of molten metal in welding. *Mater. Today Proc.*, 45, 2514–2517, 2021, doi: 10.1016/j.matpr.2020.11.124.
8. Fatima, S., Haleem, A., Bahl, S., Javaid, M., Mahla, S.K., Singh, S., Exploring the significant applications of Internet of Things (IoT) with 3D printing using advanced materials in medical field. *Mater. Today Proc.*, 45, 4844–4851, 2021, doi: 10.1016/j.matpr.2021.01.305.
9. Ennsbrunner, H., Bruckner, J., Posch, G., INDUSTRY 4.0 IN WELDING White Paper, Fronius International, 2018.
10. Paul, A.K., Practical realization of scalar optimization function of shielded metal arc welding process, *IFAC Proceedings,* Volumes, 47, 1, 684–691, 2014 doi: 10.3182/ 20140313-3-IN-3024.00216.
11. Paul, A.K., Robust Product Design Using SOSM for Control of Shielded Metal Arc-Welding (SMAW) Process. *IEEE Trans. Ind. Electron.*, 63, 6, 3717–3724, 2016, doi: 10.1109/TIE.2016.2522945.
12. Yusof, E.M.M., Yusof, M., II, Ali, R., Harjimi, I.H., Bahrin, Q.K., Welding station monitoring system using internet of thing (IoT). *Indones. J. Electr. Eng. Comput. Sci.*, 18, 3, 1319–1330, 2020, doi: 10.11591/ijeecs.v18.i3.pp1319-1330.
13. Ghazvinloo, H.R., Honarbakhsh-Raouf, A., Shadfar, N., Effect of arc voltage, welding current and welding speed on fatigue life, impact energy and bead penetration of AA6061 joints produced by robotic MIG welding. *Indian J. Sci. Technol.*, 3, 2, 156–162, 2010, doi: 10.17485/ijst/2010/v3i2/29669.
14. Zhou, Q., He, W., Li, S., Hou, X., Research and experiments on a unipolar capacitive voltage sensor. *Sens. (Switzerland)*, 15, 8, 20678–20697, 2015, doi: 10.3390/s150820678.
15. Chen, P., Wang, L., Wu, X., Lu, S., High-voltage Isolation Current Sensor by using Signal Modulation Method. *Int. J. Control Autom.*, 7, 3, 333–348, 2014, doi: 10.14257/ijca.2014.7.3.32.
16. Abu-Mahfouz, I., Drilling wear detection and classification using vibration signals and artificial neural network. *Int. J. Mach. Tools Manuf.*, 43, 7, 707–720, 2003, doi: 10.1016/S0890-6955(03)00023-3.
17. Rehorn, A.G., Jiang, J., Orban, P.E., State-of-the-art methods and results in tool condition monitoring: A review. *Int. J. Adv. Manuf. Technol.*, 26, 7–8, 693–710, 2005, doi: 10.1007/s00170-004-2038-2.
18. Wang, S., Wan, J., Li, D., Zhang, C., Implementing Smart Factory of Industrie 4.0: An Outlook. *Int. J. Distrib. Sens. Netw.*, 12, 1, 2016, doi: 10.1155/2016/3159805.

19. Chen, C., Lv, N., Chen, S., Data-Driven Welding Expert System Structure Based on Internet of Things. *Trans. Intell. Weld. Manuf.*, volume I No. 3 2017 (pp. 45–60) Springer Singapore, 2018, doi: 10.1007/978-981-10-8330-3_3.
20. Zuehlke, D., SmartFactory-Towards a factory-of-things. *Annu. Rev. Control*, 34, 1, 129–138, 2010, doi: 10.1016/j.arcontrol.2010.02.008.
21. Chen, S.B. and Lv, N., Research evolution on intelligentized technologies for arc welding process. *J. Manuf. Process.*, 16, 1, 109–122, 2014, doi: 10.1016/j.jmapro.2013.07.002.
22. Mirapeix, J., Cobo, A., García-Allende, P.B., Conde, O.M., Lopez-Higuera, J.M., Welding diagnostics based on feature selection and optimization algorithms. *Opt. Sens. Detect.*, 7726, 77260F, 2010, doi: 10.1117/12.854697.
23. Shea, J.E. and Gardner, C.S., Spectroscopic measurement of hydrogen contamination in weld arc plasmas. *J. Appl. Phys.*, 54, 9, 4928–4938, 1983, doi: 10.1063/1.332759.
24. Song, H.S. and Zhang, Y.M., Three-dimensional reconstruction of specular surface for a gas tungsten arc weld pool. *Meas. Sci. Technol.*, 18, 12, 3751–3767, 2007, doi: 10.1088/0957-0233/18/12/010.
25. Pal, K., Bhattacharya, S., Pal, S.K., Prediction of metal deposition from arc sound and weld temperature signatures in pulsed MIG welding. *Int. J. Adv. Manuf. Technol.*, 45, 11–12, 1113–1130, 2009, doi: 10.1007/s00170-009-2052-5.
26. Liao, S.H., Expert system methodologies and applications-a decade review from 1995 to 2004. *Expert Syst. Appl.*, 28, 1, 93–103, 2005, doi: 10.1016/j.eswa.2004.08.003.
27. Goar, V. and Yadav, N.S., *Foundations of machine learning*, Intelligent Optimization Techniques for Business Analytics, 25-48, IGI Global, 2024, doi: 10.4018/979-8-3693-1598-9.ch002.
28. Sen, R., Choudhury, S.P., Kumar, R., Panda, A., A comprehensive review on the feasibility study of metal inert gas welding. *Mater. Today Proc.*, 5, 9, 17792–17801, 2018, doi: 10.1016/j.matpr.2018.06.104.
29. Nomura, K., Fukushima, K., Matsumura, T., Asai, S., Burn-through prediction and weld depth estimation by deep learning model monitoring the molten pool in gas metal arc welding with gap fluctuation. *J. Manuf. Process.*, 61, October 2020, 590–600, 2021, doi: 10.1016/j.jmapro.2020.10.019.
30. Liu, Y.K. and Zhang, Y.M., Control of 3D weld pool surface. *Control Eng. Pract.*, 21, 11, 1469–1480, 2013, doi: 10.1016/j.conengprac.2013.06.019.
31. Zhang, Z., Wen, G., Chen, S., Multisensory data fusion technique and its application to welding process monitoring. *Proc. IEEE Work. Adv. Robot. Soc. Impacts ARSO*, 2016-Novem, 294–298, 2016, doi: 10.1109/ARSO.2016.7736298.
32. Wang, J.F., Chen, B., Chen, H.B., Chen, S.B., Analysis of arc sound characteristics for gas tungsten argon welding. *Sens. Rev.*, 29, 3, 240–249, 2009, doi: 10.1108/02602280910967657.

33. Cayo, E.H. and Alfaro, S.C.A., A non-intrusive GMA welding process quality monitoring system using acoustic sensing. *Sensors*, 9, 9, 7150–7166, 2009, doi: 10.3390/s90907150.
34. Horvat, J., Prezelj, J., Polajnar, I., Čudina, M., Monitoring gas metal arc welding process by using audible sound signal. *Stroj. Vestnik/Journal Mech. Eng.*, 57, 3, 267–278, 2011, doi: 10.5545/sv-jme.2010.181.
35. Kah, P., Shrestha, M., Hiltunen, E., Martikainen, J., Robotic arc welding sensors and programming in industrial applications. *Int. J. Mech. Mater. Eng.*, 10, 1–16, 2015, doi: 10.1186/s40712-015-0042-y.
36. Tao, F., Zuo, Y., Da Xu, L., Zhang, L., IoT-Based intelligent perception and access of manufacturing resource toward cloud manufacturing. *IEEE Trans. Ind. Inf.*, 10, 2, 1547–1557, 2014, doi: 10.1109/TII.2014.2306397.
37. Febriani, R.A., Park, H.S., Lee, C.M., An approach for designing a platform of smart welding station system. *Int. J. Adv. Manuf. Technol.*, 106, 7–8, 3437–3450, 2020, doi: 10.1007/s00170-019-04808-6.

15

Evolution of Technology in Testing and Inspection of Welding: From Traditional to Modern Methods

Akhilesh Kumar Singh[1]*, Sandip Kunar[2], M. Zubairuddin[2], Marxim Rahula Bharathi B.[1], D.V.S.S.S.V. Prasad[1] and Yarrapragada K.S.S. Rao[1]

[1]*Department of Mechanical Engineering, Aditya College of Engineering and Technology, Surampalem, Andhra Pradesh, India*
[2]*Department of Mechanical Engineering, Aditya University, Surampalem, Andhra Pradesh, India*

Abstract
Technological development in welding inspection describes the progression from the early method used to the modern ones. Traditionally, welding inspecting was done through L. P. inspection testing, E. C. testing, M. P. testing, X-ray and γ-ray radiographic testing, and Ultrasonic inspection testing. While helpful, these methods were not very efficient in identifying mild flaws and gave quick as well as precise results. The incorporation of new techniques, including the use of PAUT and TOFD, was a revolution since it led to the development of better defect detection and analysis. The foregoing methods were strengthened by the advance in the digital age through digital radiography and Real-Time Radiographic Testing (RTRT), as well as 3D laser scanning, which gave detailed surface and volumetric measurements. Advanced technology has continued to advance the field with Automated Ultrasonic Testing (AUT), drones besides the traditional Smart Maintenance. This chapter summarizes these advancements and their operative effects on industry practices, and potential projections for the future, thus giving an insight to the current state of affairs of welding quality assurance.

Keywords: Welding inspection, NDT, PAUT, TOFD, RTRT, AI in welding inspection

**Corresponding author*: akhileshkr.singh@hotmail.com

Sandip Kunar and Gurudas Mandal (eds.) Advanced Welding Technologies, (323–338) © 2025 Scrivener Publishing LLC

15.1 Introduction

Many significant structures, like the systems of bridges, pipelines, and buildings, consist of welded joints that are vital to possess service-related capacities. It is with these kinds of loadings that simple or complicated stress is placed upon the joints and the welds [1]. Hence, the quality of a weld is very crucial in the safety and integrity of such structures. Nonetheless, the appearance of a weld does not often depict the quality of the weld, and that is why there is a need to undergo through standard examinations and tests to determine the fitness of the weld [2].

Inspection and testing of welds are two separate but nearly related processes; their aim is to guarantee the high quality of the weld joints. Inspection, on the other hand, is concerned with assessment of the manufacturing process and the weldments to conform to the required characteristics or attributes. It can be as simple as a "visual inspection to confirm the dimensions and the surface finish of the component as well as any other visible defects to something more substantive" [3]. In contrast, testing involves making physical tests so as to determine certain quantitative measures of certain characteristics such as mechanical strength, results which give a physical assessment of the strength of the weld. While there is an attempt to reveal the characteristics of the weld in the course of testing, there is an attempt to regulate and sustain quality in the course of inspection using set standards and sending back low-quality products [9]. These two processes are important to enable assessment of the performance of the welded joints in as far as they meet the set standards for their applications [4].

Welding is very important in fabrication and structures; thus, quality control is important to avoid compromise on strength and safety. In the past, the test and inspection techniques were simple and used mostly visible examination and some mechanical tests. These conventional techniques, while being standard, had constraints in effectively identifying and evaluating internal weld flaws, or causing substandard safety and efficacy of the welded equipment. But with evolution in technologies, newer techniques have emerged, and the assessment of welding quality has evolved a new era. These developments have not only helped to introduce improvements in the weld inspections and tests results but also in the weld inspections and test efficiency and economic benefits [5].

Technology applied to welding has relatively evolved overtime and has even revolutionized manufacturing and construction industries. Weld quality control is important for structures in construction and parts of equipment for manufacturing in many businesses. As a result, the testing and inspection techniques have, therefore, similarly adopted to these technological developments. The current chapter examines the history of transformation from conventional to advanced technologies in welding testing and inspection [6]. In it, the author begins from the introduction of history and then outlines ordinary methods including visual inspection and radiographic testing, then outlines more updated methods of digital radiography, 3D laser scanning, and automated inspection systems. Through analyzing these developments, the chapter presents a clear vision of the ways the modern applied technologies are improving weld quality assurance, and how further development is going to form the future of weld inspection. Welding is an exceptional procedure in numerous industries and entails automotive, aerospace, construction, and oil and gas industries. The stability, efficiency, and durability of structures and their components and also their ability to carry out their intended service depend largely on the quality and strength of welds [7]. New innovations in welding test and inspection have greatly enhanced the test and inspection to provide the highest accuracy, effectiveness, and affordability. These are some of the advanced approaches to flaw detection that have rendered the traditional approaches effective in the following ways: Better identification of defects, documentation, and even improved prediction of defects. Various advances in the testing and inspection of welding; NDT, AVI, RT, and AEI; and artificial intelligence and machine learning have been leveled in this chapter. Therefore, it is the intent of this chapter to give a comprehensive discussion of the history of the specific testing and inspection technologies to derive the significance of such developments to achieve the performance, reliability, and safety of welded structures [8]. After reviewing both conventional and current practices and bringing out the importance of testing and inspection in the welding industry, and explaining how the technology has revolutionized the area, the readers of this paper will be in a position to be enlightened on the importance of testing and inspection in the field of welding.

15.2 Stages of Weld Inspection and Testing

Inspection of welded structures occurs in three stages as follows [9–11]:

1. **Before Welding**—Welding checks ensure that the specifications, material, and equipment required meet certain

necessary requirements. There is a crucial role of this stage for controlling the problems that may arise at the welding stage and to check all the components for compliance with the design specifications.

2. **During Welding**—During welding, for instance, the inspection emphasizes on procedures and cleanness, which are essential in the production of the welded joints. They are employed to provide consistent monitoring so that the weld parameters are implemented as required and that no contamination occurs to the weld.
3. **Post Welding**—Post-welding examination evaluates the characteristics of the welded joint for their Fitness for Purpose and Performance Characteristics. This step entails examination in a bid to be certain that the welds are good and possess the right strength and resilience.

Inspection of welded structures occurs in three stages as shown in Figure 15.1.

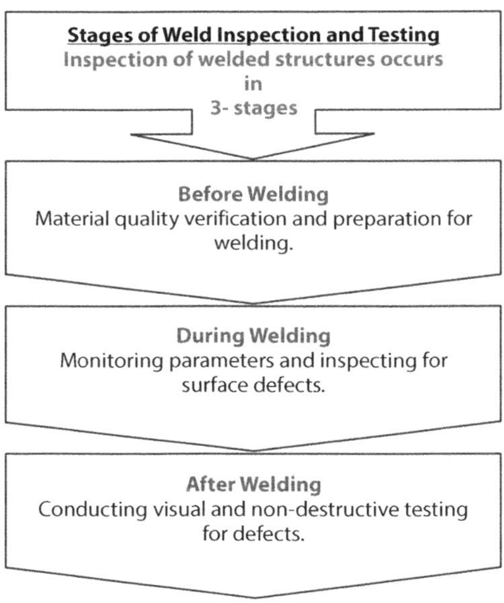

Figure 15.1 Weld inspection and testing involve several critical stages to ensure quality and integrity.

15.3 Weld Testing and Inspection After Fabrication Can Be Classified Into Two Main Categories

(i) **Destructive Testing**—This refers to tests that actually damage a portion of the weld to establish its characteristics and strength [12].
— The testing in this category is very destructive, and after the experiments are over, the component or specimen is rendered useless. If put to test, a component collapses or can no longer be used to its capacity. Vivid examples include tensile testing that is destructive testing (DT), which has an objective of pulling off the specimen required to establish physical and mechanical signs such as strength, tolerance, durability, toughness, and ductility. About it is more efficient for the times of mass-produced items since the cost of demolishing a few pieces is irrelevant. There are tensile, compressive, shear, hardness, impact, fatigue, creep, bending, and torsion tests as well as chemical test as shown in Figure 15.2. Some of them include confirmation

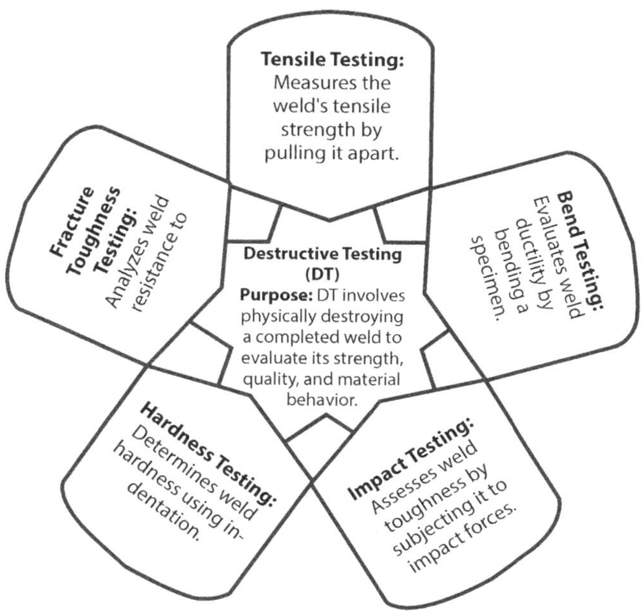

Figure 15.2 Different classes of destructive testing (DT) of welding.

of material properties, assessment of the weld quality, decline in the number of failures and cost, and cognizance with the government regulations prevailing.

(ii) **Nondestructive Testing**—This type of testing technique examines the weld without causing any harm to it; the methods include ultrasonic test and radiography test [13]. Details: Non-destructive testing (NDT) is of paramount importance for assessing materials, components, or systems with the purpose of not to destroy them for further use. It makes certain that parts, equipment, and vessels do not suffer component failure after inspection; hence, it is useful for manufacture, fabrication, or in-service container inspections. NDT is used to identify subsurface or defects on the surface of the material and to measure some of its physical or chemical characteristics as shown in Figure 15.3. Some of the uses include quality assurance of products, limiting the procedures involved in production, and keeping cost low among other objectives of allaying quality consistency. The principal advantages are the capability to examine lots or samples of production as well as utilize various tests either successively or concurrently and on-site examination of the changes not being permanent on a specimen.

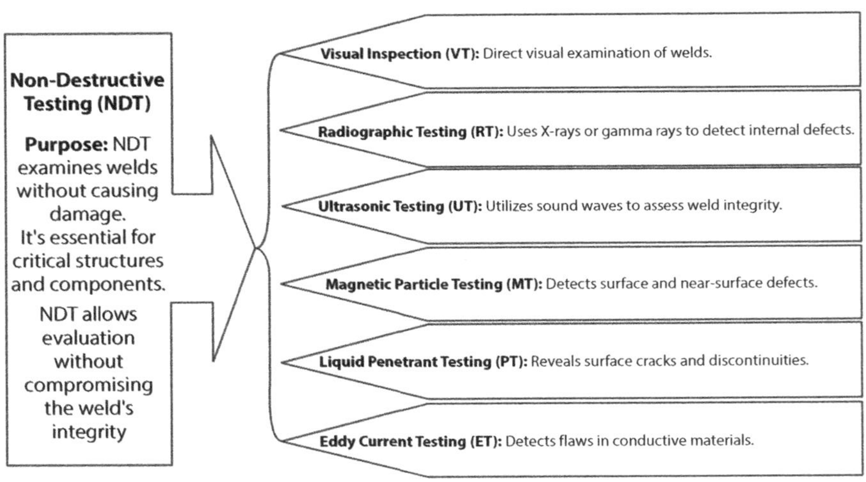

Figure 15.3 Different classes of non-destructive testing (NDT) of welding.

— It has been observed that both DT and NDT play crucial roles in ensuring weld quality and safety [12–14].

15.4 Purposes of Material Testing [15]

- **Evaluate Material Properties (Quantitatively and Qualitatively):** This involves assessing mechanical, thermal, electrical, and other material properties.
- **Detect Surface or Subsurface Defects:** Testing helps identify cracks, voids, or other imperfections.
- **Check Chemical Composition:** Analyzing the material's composition ensures it meets specifications.
- **Assess Material Suitability for Application:** Determine if the material is suitable for its intended use.
- **Predict Strength and Serviceability:** Testing provides insights into material behavior under load and over time.

15.5 Overview of Welding as a Critical Process in Manufacturing and Construction

— Welding is a basic forte in many industries and construction, and plays an important role in providing structural strength and durability on various applications. It is the joining of metals or thermoplastics by means of heat and pressure or their combination with an aim of producing strong and seamless joints indispensable to accurate construction of structures and parts. This is mainly in industries such as automobile, aircraft, marine, civil construction, oil and gas, as well as other civil construction projects [16].

— In automotive industry fabrication, welding is used extensively in constructing car body and frames: as a result, safety and performance are closely tied to the welding processes employed. Special for the aerospace industry, the accuracy of the weld performed is imperative for the structural integrity and stability of an aircraft part. In construction, welding is used in making strong frameworks, pipes, and some other basic structures that frame our development. As welding determines the stability and durability of such structures, it becomes clearer why welding is relevant to so many processes [17].

— New developments in industries make newer welding technologies to be in the market. A number of improvements have been made over the years regarding testing and inspecting of the welds, and this has led to better quality of welds for safety and high performance standards. Thus, showing the transition from the basic methods to developing and making use of advanced methods, aspects that seem to underline the stringent relevance welding holds when it comes to construction and maintenance of the world's structures and machines [18].

15.6 Importance of Testing and Inspection in Ensuring Weld Quality and Integrity

As we have seen the importance of welding, it becomes very important to determine the quality and strength of the welded joints. Weld defects may cause catastrophic failures with major economic consequences, risks to people, and are life threatening. Hence, sound testing and inspection procedures must be adopted to ensure compliance with laid down specifications and standards before endorsing welds [19].

Test and inspection are important in the following ways: They are used in identification of defects like cracks, porosity, inclusions, and poor fusion that are likely to affect the strength and serviceability of a weld. Furthermore, the above processes are a guarantee that the welded structures adhere to such regulations and standards and therefore capable of performing under stresses and other environmental factors [20].

Such line of traditional inspection includes the following: visual inspection, dye penetrant testing, magnetic particle testing, radiographic testing, and ultrasonic testing that have been crucial for the quality of welds. However, these methods have certain demerits as concerns the ability to accurately detect a target, the extent of penetration possible, and speed [19].

The development of new technologies has played a major role in changing the perspective of the welding inspection as shown in Figure 15.4. Newer and sophisticated NDT procedures, digital radiography, advanced 3D laser scanning, automated inspection systems, and use of artificial intelligence and machine learning have improved the capacity of identifying

Figure 15.4 Advanced non-destructive testing (NDT) techniques of welding.

weld defects. These innovations have enhanced the accuracy, speed, and credibility in the welding inspection, which in one way or another has enhanced the safety and quality remits in the manufacturing and construction industry [21].

15.6.1 Evolution of Non-Destructive Testing (NDT) Technology in Testing and Inspection of Welding

Phased Array Ultrasonic Testing (PAUT)

The application of PAUT means phased array ultrasonic testing is a dramatic advancement in ultrasonic exam equipment. It employs an array of ultrasonic transducers and electrical time-driven delay line to accurately control and focus the ultrasonic waves and is used to capture images of weld geometry to identify weld defects especially in compound shapes. This technique is very important in the assessment of important structures like pipelines and pressure vessels in detail to establish their safety and structural integrity. New developments in PAUT are based on an enhancement of the image clarity as well as the detection accuracy [22].

The Time-of-Flight Diffraction (TOFD)

Time-of-Flight Diffraction (TOFD) is a more advanced technique in UT that gauges the time that will take for the ultrasonic waves to travel round the weld flaws. This method is appreciated for its efficiency in determination of dimensions of weld imperfections and provision of information regarding their position and direction. The developments in the TOFD technique have been directed toward enhancing the efficiency of the technique with regard to the detection of the flaws as well as the reliability of the results in recent years [23].

Digital Radiography (DR)

Digital Radiography (DR) is a new generation radiographic technique in which digital X-ray sensors are employed for radiographic examination of weldment microstructures. DR, as a novel technology to the traditional film-based techniques, has the advantages of faster image acquisition, better image quality, as well as improved defect detection. This method is popular in welding inspections in manufacturing industries as it provides the best visuals of inner problems like porosity and cracks. Ongoing research endeavors have focused on the improvement of DR technology to yield better weld assessment [24].

Computed Tomography (CT)

Computed Tomography (CT) is a more complex form of NDT where X-ray technology is incorporated with computer programs to produce 3D images of the weld structures. This approach provides an outline of the welds, hence enabling a proper inspection and identification of the defects. CT is advantageous in complicated weld patterns as well as important parts in which conventional radiography could be inadequate. Over the years, new technological improvements have seen CT's usability in weld inspection advanced to enable recognition of the tiniest of flaws in welds with the highest levels of image quality [25].

Infrared Thermography (IT)

Infrared Thermography (IT) is one of the high tech and non-destructive testing techniques that are used to detect as well as identify both the surface and the subsurface flaws on the basis of thermal wave patterns. This is a non-destructive technique that involves applying thermal images over a surface of a material for one to be able to infer some problems such as cracks, voids, or delamination. They are most useful for inspecting weld areas and other important parts where general techniques can often overlook some possible flaws. Soh *et al.* concluded that due to the development in Infrared Thermography technology, its effectiveness in detecting defects has improved enormously [26].

Acoustic Emission Testing (AET)

Welding Acoustic Emission Testing or AET is a technique that involves measurement of high frequencies of sound produced by the defect that forms when welding. In this case, AE sensors are bond to the material, and this method records transient waves that may be an alert on the possibility of the presence of a crack, inclusion, or void in the material. AET is most beneficial in real time; thus, it is possible to detect the defect formation process in its progress. Recent developments in AE technology have therefore centered on improving the sensitivity of the sensors used as well as the accuracy of their detection capability [27].

Magnetic Flux Leakage (MFL)

Magnetic Flux Leakage (MFL) is a method that belongs to the group of NDT methods and used to define material discontinuities on the basis of changes in the magnetic field. This method entails giving the material a magnetic field and determining the fluctuations in the magnetic field through the presence of flaws such as cracks or voids. MFL is very effective for inspecting ferromagnetic material and widely used to verify the condition of pipelines, tanks, and structural members. Recent development focused on MFL technology mainly on increasing the sensitivity and accuracy [28].

Eddy Current Array (ECA)

Multiple-turn coil Eddy Current Array (ECA) is a complex NDT technique for flaw detection on a conductive material and is capable of detecting surface as well as subsurface flaws. This method creates existing eddy currents that counter current altering the conductivity of the material to see flaws like invisible cracks, corrosion, or inclusion. ECA also has many benefits such as defect identification and large-area deposit examination with high resolution. New development has enhanced the efficiency of ECA in the following way: the technique offers more refined examinations than when it was first introduced [29].

Laser Shearography (LS)

Laser Shearography (LS) is one of the most innovative Non-Destructive Testing (NDT) method, which primarily deals with identification of surface contour with respect to applied load and deduction of the flaw. In using this method, stresses are applied to the material, and the consequent surface changes are measured with the help of laser interferometry. In as much as comparing with other conventional methods of inspections, Laser Shearography can identify the outlined deformations as cracks, delaminations, or voids when none is visible. Some of the innovations in this regard are as follows: Recent technological advancement acted as a plus to its sensitiveness and accuracy [30].

Guided Wave Testing (GWT)

Guided Wave Testing is a sophisticated NDT technique where low frequency ultrasonic waves are employed for the objective of detecting defects at long ranges. This technique uses the transmission of ultrasonic waves along a structure, for instance, pipelines or welds, and then bounce off the material irregularities back to the sensors. Still, GWT is efficient for examining vast territories or constructions with restricted accessibility, where extensive sectors will require one and only placing of the test equipment. The improvements in GWT technology in the recent past have been geared toward enhancing the wave propagation and the level of sensitivity of wave detection [31].

15.7 Impact of Technological Advancements

The transition from traditional to modern methods in welding inspection has brought about significant improvements in several areas as follows:

- **Accuracy and Reliability:** High-tech produces have higher accuracy in detecting possible defects compared to traditional methods; thus, there are fewer undetected defects [32].
- **Efficiency:** They improve the rate of issuing inspections in manners that reduce turnaround time and improve its productivity [33].
- **Safety:** Sophisticated approaches reduce unsafe condition thereby promoting safer work environments [34].
- **Cost Effectiveness:** There are many advantages associated with the use of modern equipment some of which include whether it is a short term or a long-term investment, modern

equipment is cheap in the long run since the little costs are compensated for by the higher efficiency, and better quality of the products to be produced [35].

15.8 Conclusions

A comprehensive review of welding inspection methods, especially in terms of traditional to modern approaches. Substantial progress has been achieved. This transition has significantly enhanced both the quality and safety of welded structures. The welding industry can benefit from implementing new technologies that can improve the degree of precision, productivity, and reliability of the welded structures, thus promoting the progress of the manufacturing and construction processes. Future advancements in this area are expected from future explorations and innovations in this field making welded inspection and testing even more powerful.

References

1. Oberg, E., Jones, F.D., Horton, H.L., Ryffel, H.H., Mccauley, C.J., Heald, R., Hussain, M.I., *Machinery's Handbook* (Vol. 6). Industrial Press, New York, 1914.
2. Davies, A.C., *The Science and Practice of Welding* (Vol. 2). Cambridge University Press, 1992.
3. Raghavan, V., *Physical metallurgy: principles and practice.* PHI Learning Pvt. Ltd., New Delhi, 2015.
4. Lincoln Electric, *The Procedure Handbook of Arc Welding.* Lincoln Electric Company, Cleveland, Ohio, 2011.
5. Olson, D.L., Siewert, T.A., Liu, S., Edwards, G.R., *ASM Handbook: Welding, Brazing, and Soldering,* Vol 6. ASM International, 1993.
6. Althouse, A.D., Turnquist, C.H., Bowditch, W.A., Bowditch, K.E., Bowditch, M.A., *Modern Welding.* The Goodheart-Willcox Company, Inc., Tinley Park, IL, 2003.
7. Gourd, L.M. and Gourd, L., *Principles of Welding Technology.* Butterworth-Heinemann, Oxford, United Kingdom, 1998.
8. Hughes, S.E. (Ed.), *A quick guide to welding and weld inspection.* American Society of Mechanical Engineers, 2009.
9. Smith, A., Jones, B., Clark, C., Pre-Welding Inspection Practices. *Manuf. Weld. Technol.,* 29, 1, 45–58, 2017.
10. Johnson, R., Precision Welding and Inspection Methods. *Weld. Eng. Rev.,* 30, 4, 178–190, 2018.

11. Miller, A. and Brown, K., Post-Weld Inspection and Quality Control. *Struct. Integr. J.*, 42, 6, 122–135, 2019.
12. Anderson, T. and Morris, R., Advances in Weld Testing and Inspection Technologies. *J. Weld. Fabrication*, 45, 2, 234–248, 2022.
13. Li, H., Wang, Y., Zhang, Q., Advances in Welding Technology for Manufacturing. *Manuf. Eng. J.*, 30, 2, 100–115, 2020.
14. Gordon, M., The Evolution of Welding Technologies. *Adv. Manuf. Rev.*, 27, 1, 56–72, 2019.
15. Kumar, A. and Raj, S., Welding Processes: Theory and Applications. *J. Mater. Sci.*, 47, 8, 1520–1535, 2019.
16. Smith, A. and Brown, K., Automotive Welding: Techniques and Innovations. *Automot. Eng. J.*, 29, 7, 678–690, 2016.
17. Li, H., Wang, Y., Zhang, Q., Non-Destructive Testing Techniques for Welds. *J. Mater. Sci. Eng.*, 34, 3, 567–579, 2020.
18. Zhang, L., Zhao, Y., Liu, J., Modern Approaches to Non-Destructive Weld Testing. *J. Ind. Eng.*, 33, 4, 467–482, 2021.
19. Baddoo, N.R., *Inspection and Testing of Welding*. John Wiley & Sons, Hoboken, New Jersey, U.S., 2010.
20. Peruń, G., Advances in Non-Destructive Testing Methods. *Materials*, 17, 3, 554, 2024.
21. Wang, B., Hu, S.J., Sun, L., Freiheit, T., Intelligent welding system technologies: State-of-the-art review and perspectives. *J. Manuf. Syst.*, 56, 373–391, 2020.
22. Liu, L., Zhang, Y., Chen, J., Enhanced Defect Detection in Welding: Advances in Phased Array Ultrasonic Testing. *J. Non-Destr. Test. Eval.*, 52, 1, 45–59, 2023.
23. Kumar, R., Patel, V., Singh, A., Advanced Data Acquisition and Signal Processing in TOFD for Enhanced Weld Inspection. *J. Nondestr. Eval.*, 40, 2, 123–134, 2021.
24. Smith, R., Zhang, L., Brown, T., Advancements in Digital Radiography for Weld Inspection: Enhancing Image Processing Techniques. *J. Nondestr. Test.*, 45, 2, 88–101, 2023.
25. Smith, L., Anderson, J., Lee, M., High-Resolution Computed Tomography for Automotive Weld Inspection. *J. Adv. Non-Destr. Test.*, 38, 1, 45–59, 2023.
26. Ciampa, F., Mahmoodi, P., Pinto, F., Meo, M., Recent advances in active infrared thermography for non-destructive testing of aerospace components. *Sensors*, 18, 2, 609, 2018.
27. Grosse, C.U., Ohtsu, M., Aggelis, D.G., Shiotani, T. (Eds.), *Acoustic emission testing: Basics for research–applications in engineering*. Springer Nature, 2021.
28. Feng, B., Wu, J., Tu, H., Tang, J., Kang, Y., A review of magnetic flux leakage nondestructive testing. *Materials*, 15, 20, 7362, 2022.

29. García-Martín, J., Gómez-Gil, J., Vázquez-Sánchez, E., Non-destructive techniques based on eddy current testing. *Sensors*, 11, 3, 2525–2565, 2011.
30. Yang, L., Chen, F., Steinchen, W., Hung, M.Y., Digital shearography for nondestructive testing: potentials, limitations, and applications. *J. Hologr. Speckle*, 1, 2, 69–79, 2004.
31. Zang, X., Xu, Z. D., Lu, H., Zhu, C., Zhang, Z., Ultrasonic guided wave techniques and applications in pipeline defect detection: A review. *Int. J. Press. Vessels Pip.*, 105033, 2023.
32. Singh, A.K., Dey, V., Rai, R.N., Techniques to improveweld penetration in TIG welding (A review). *Mater. Today: Proc.*, 4, 2, 1252–1259, 2017.
33. Kumar, P., Sinha, A.N., Hirwani, C.K., Singh, A.K., Pathak, P.K., Murugan, M. Saravanan, A., Investigation of TIG cladding of NiTi wire on substrate 304L to study the effect of applied current on microstructure and mechanical properties. *Trans. Indian Inst. Met.*, 74, 1333–1348, 2021.
34. Singh, A.K., Debnath, T., Dey, V. Rai, R.N., A study on effect of preheating and post weld heat treatment (PWHT) of grade P91 steel. *J. Mater. Sci. Mech. Eng.*, 2, 8, 57–62, 2015.
35. Kunar, S., Jagadeesha, T., Mandal, G., Singh, A.K. Rama Sree, S., Fundamentals and Applications of Additive Manufacturing. *Adv. Addit. Manuf.*, 1–14, 2024.

16

Sustainability in Welding Industries

P. Sivasankaran

Department of Mechanical Engineering, Christ College of Engineering and Technology, Pondicherry, India

Abstract

Welding is one of the hardest jobs carried out in manufacturing industries for many decades. For every production part, welding serves as the base operation for all parts. Today, sustainable manufacturing is one of the important areas to focus on for better performance. In this book chapter, sustainability of welding operations is considered to be the main focus of study. The term sustainability means environmentally friendly welding and processing methods are those that use the least amount of energy, have the highest process efficiency, produce better results with fewer resources, and generate the least amount of waste material during operation. There are numerous industries that depend on welding, including the automobile, construction, aviation, and more. Many structures, fences, gates, tiny kitchen appliances, cars, and even space travel would not be possible without this type of metal construction. Sustainability in the welding process is influenced by a wide range of variables. Not all of these are listed here; some of them include filler metals, welding speed, system costs, raw resources, processing time, and seam quality. Gaining a general understanding requires examining the entire process chain. Energy is needed, all things considered, for the extraction of parent material raw materials, filler metal welding, and the production of auxiliary materials. For this reason, getting resources is not as important as manufacturing each portion of a welding system.

***Keywords*:** Sustainability, ecofriendly, automotive, efficiency, space craft, filler metals

Email: sivasankaranpaneerselvam83@gmail.com

Sandip Kunar and Gurudas Mandal (eds.) Advanced Welding Technologies, (339–352) © 2025 Scrivener Publishing LLC

16.1 Introduction

16.1.1 In Practice: Sustainable Welding

Achieving sustainable production also requires the application of cutting-edge welding technologies. Production operations are more productive when the welding system is dependable and adaptable. Operators can accelerate welding while continuously maintaining excellent quality and optimal circumstances by utilizing the appropriate technologies. Therefore, it may be feasible to eliminate entire production cells. Additionally, it helps to lower the amount of energy used and the amount of industrial resources, such as filler and auxiliary materials. What are the strategies a welder can employ to reduce their environmental footprint while still creating joints that are both high quality and cheaply priced? First, we need to make a distinction between the different welding techniques used in this instance. When employing manual arc welding or manual metal arc welding, for instance, the fact that roughly 25 to 35 mm of each welded electrode is left behind must be taken into account since it has a negative impact on sustainability. It takes time for the welder to clean the bead and change the rod electrode. For the duration of this operation, the welding power source is in idle mode and cannot be used.

16.1.2 Improving Effort Using Welding Procedure

One of the most environmentally friendly welding techniques has been shown to be MAG (metal active gas) welding. In practical use, it is extremely effective, tidy, quick, accurate, and adaptable. Another ecologically friendly welding technique is gas–metal arc laser welding. The two advantages of this approach are its fast speed and relatively low filler material use. However, using one of these two welding methods does not mean that you are doing something particularly eco-friendly. It is essential to decide ahead of time which approach is most suitable for each application. The equilibrium between technical system effort and resource consumption, quality, and cost effectiveness is necessary for welding to be sustainable.

16.1.3 Expertise of Endurance in Welding

Welder training can be more sustainably conducted by starting with the appropriate resources. Fronius offers a virtual welding simulation technology that allows novice welders to complete virtual instruction in basic methods and torch control. Once the user has mastered the use of virtual reality (VR), they can experiment with various welding techniques. Virtual training is less expensive and uses fewer resources because it does not require any consumables. It implies that training of this nature promotes sustainability. Sustainable production requires many factors to be considered: higher welding speeds, the right welding method for the job, and a low rate of rejection and rework are all necessary for sustainable action in welding technology. However, the initial step is to choose the proper welding system.

16.2 Strategies for Implementing Environmentally Friendly Welding

The elements that can assist us in making the transition to sustainable welding include welding technology, proper welding procedures, faster welding speeds, and reduced rework. One of the key technologies and procedures that is widely employed in the industrial sector is welding. By consuming necessary resources like electricity and other materials, the welding processes produce fumes and harmful pollutants. As environmental concerns over the effects of these industries rise, the following are five strategies for adopting sustainable welding to transition to more environmentally friendly and greener welding procedures.

a) **Virtual Training:** Through the use of virtual reality, simulators, and other virtual tools, training can be conducted for students, employees, or trainees without the need to use and waste raw materials, electricity, or release toxins into the air. The new welders are assisted in making sustainable decisions throughout their careers by being trained in environmentally friendly methods of their craft.

b) **Using Sustainable Tools:** High-quality, long-lasting tools are what we mean when we talk about sustainable tools. Typically, inexpensive products are made of inferior materials that degrade easily, cannot be fixed, and are thus effectively thrown away. High-quality instruments, like those made by Fronius, which preserve heat balance, extend product life, use fewer raw materials, etc., should be preferred over these.

c) **Various Welding Methods:** There are a lot of innovative and environmentally friendly welding methods available that can lower manufacturing or operational expenses as well as environmental costs. These welding methods include vacuum soldering, diffusion welding, magnetic pulse welding (MPW), friction stir welding (FSW), and so forth.

d) **Reusing Scrap:** Recycling is the most well-liked and straightforward approach to switching to environmentally friendly products and cutting expenses associated with the environment, regardless of industry. Waste reduction and scrap material recycling are the first steps toward switching to sustainable welding.

e) For recycling purposes, metal shavings, for instance, can be sent back to the raw material manufacturer. Before being brought back, the chips are separated using a centrifuge from the lubricant needed for the production process, and the lubricant is then brought back into the production hall. As such, the lubricant can be reused multiple times.

f) **Appropriate Disposal of Garbage:** Recycling is one way to properly dispose of waste, as was mentioned in the previous point. Waste needs to be disposed of properly if it cannot be recycled. It is necessary to ascertain whether the material can be applied to any other industrial division. It is important to utilize fume management technologies to appropriately decrease and dispose of the fumes. The global economy is transitioning to a "green economy," which focuses on lowering ecological risks and scarcity while pursuing sustainable development without causing environmental degradation. Consequently, achieving this bigger objective requires sustainability. Welding, as an industry, is vital to infrastructure development, economic growth, and other manufacturing sectors; therefore, adopting sustainable welding practices is critical to achieving a green economy.

16.3 Eco-Friendly Welding Procedures

The possible risks welding brings to one's health and safety are known to professional welders. In the absence of appropriate personal protective equipment (PPE), individuals run the risk of burns, electrical shock, lung damage from fumes, and threats from compressed and liquid gases. Also, they run the risk of irreversible eye damage. Through the use of appropriate safety procedures, advised attire, and any available safety gear and equipment, welders can reduce or eliminate these risks. There is little disagreement about the significance of health and safety concerns for welders because the trade is regarded as one of the riskiest. To implement eco-friendly welding methods, follow these eight steps:

a) **Modify the Settings of the Welding Fumes:** There is a clear correlation between the welder settings and the fumes emitted during welding. Elevated levels of welding fumes may result from welding operators using high voltage and amperage. Making the lowest amperage decision instead of starting with high power and heat might be the easiest way to reduce emissions. The penetration needed for a strong weld junction can still be guaranteed by operators.
b) **Select the Shielding Gas Wisely:** The heat of the arc increases with a 100% CO_2 shielding gas increasing vaporization and fumes. Use an argon gas combination instead.
c) **Prior to Welding, Clean the Metal:** As previously indicated, impurities or coatings on the metal surface can also contribute to welding fumes. These contaminants emit fresh welding fumes when they are heated or come into contact with UV light. Contaminants include paint and solvent residue, rust inhibitors, oil, and zinc on galvanized steel. To reduce welding fumes, use a stripping agent to remove them and any other residue before welding.
d) **Apply the Welding Technique Generally Making Fewer Fumes:** The two most used techniques are MIG welding (GMAW) and TIG welding (GTAW). Though the weld quality is not as good, MIG is still a popular option because it is simpler to learn and produces more detailed welds than TIG. Environmentally speaking, TIG welding is more environmentally benign than MIG welding since it generates less emissions, such as nitrogen oxides and ozone, than MIG welding.
e) **Select Another Mode of Welding:** Friction welding and diffusive welding are two other environmentally friendly welding techniques to think about if TIG welding is not practical. Friction force is used in friction welding. Enough

heat is produced by the friction of a revolving tool on the work piece to fuse the metals together. With friction welding, you can create a motor with just the necessary energy and no need for flux, thanks to this environmentally friendly technology. Cleaner air is the result of this technique's reduction of the carbon footprint.

f) **Do Vacuum Soldering Project:** By preparing the metal to accept the solder and preventing oxidation at the connection, flux is necessary for traditional soldering. Sadly, the flow that is having these consequences is endangering the ecosystem in particular. After the solder point is completed, corrosive agents must be cleaned out. However, the solvents used to do so may be hazardous to health and may alter the environment releasing harmful gases into the atmosphere.

g) **Use Proper Recycling for Welding Waste:** One of the first rules of environmental conservation is recycling. You save the useless scraps so they might be utilized in future projects, saving your business money, as opposed to throwing them away. To guarantee proper handling, it is also a good idea to bring them to a qualified recycler.

h) **Select Metals That Are Friendly to the Environment for Your Next Project:** Nowadays, environmental responsibility is demanded of every firm. Thus, if you are planning a welding project, think about utilizing environmentally friendly metals. Aluminum, among the most recycled metals, usually yields a high-quality product after recycling. Its popularity stems from the fact that it is strong, lightweight, adaptable, and suitable for a wide range of welding applications.

16.4 Research Advancement in Sustainability in Welding Industries

The manufacturing industry is thought to be the backbone of any industrialized, developed nation. The long-term development of the country is greatly influenced by welding and joining as a component of industry. To stay afloat in the highly competitive global marketplaces of today, welding companies are always looking for new and creative ways to accomplish the same objective, whether by drawing inspiration from established goods or natural events. These include lowering expenses and energy use, minimizing waste, optimizing settings, and being environmentally friendly, but they also include introducing more sophisticated computer technologies, like Industry 4.0, cyber-physical systems, cloud computing, big data, digital manufacturing, smart factories, or factories of the future. This work describes in-process weld quality inspection, defect identification and control in fusion and solid-state welding processes, and weld quality assessment [1]. The development of novel technologies and management ideas has been greatly influenced by sustainability, which has garnered a great deal of attention in manufacturing research in recent decades. With an emphasis on manufacturing technology and management, the current chapter seeks to present a structured overview of the broad field of study in sustainable manufacturing. In addition to outlining the complementary strategies required for a shift to sustainable manufacturing, it aims to explain the role that manufacturing plays in sustainability and highlight the importance of doing interdisciplinary research [2]. The research proposes a methodology for welding process map creation within a sustainability context. Our two-step method involves acquiring many near-optimal solutions using meta-heuristic optimization, based on realistic quality features. Then, we apply a fuzzy classifier to assess multiple options according to sustainability. A real thick weld is used to illustrate the approach, which takes into account electrical power and material use as sustainability factors in categorization and weld requirements as optimization goals. It is possible to create qualitatively viable sustainable process maps for complex engineering systems, the results show, before thinking about changes to equipment, materials, or processes [3]. An in-depth evaluation and analysis were conducted on the environmental impact of the friction stir welding method in relation to welding settings. For this aim, various rotating and welding rates were used to create butt joints in sheets of aluminum alloy AA5754. The environmental impact index was calculated by gathering and analyzing all input and output data about materials, energy, and emissions using life cycle assessment software. The sound weld functioned as the functional unit upon which all material and energy flows were dependent. The results of the life cycle assessment research show that welding and rotating speeds have a substantial impact on the environmental impact of friction stir welding [4]. It is believed to be essential that the welding processes used to unite metal parts or structural elements be carried out correctly for these commodities to remain dependable throughout their useful lives. In a workstation with sufficient design, an operator can safely do these duties. To ensure the quality of the welding seams that are used to unite metal parts and structural elements, a number of standards have been devised. These standards lay out specifications to guarantee the execution of welding procedures and the inspection of the final metal welds. Since the 5S model has proven to be an effective tool for improving workplaces in both the industrial and service sectors, companies adopting lean production approaches frequently employ it as a cornerstone. This study provides a foundation for using the 5S approach in metal welding work settings. It creates an index to evaluate how much the 5S technique and the essential performance analysis methodology (IPA) are being used [5]. Shielded metal arc welding (SMAW) is a commonly utilized material joining technology in a wide range of industrial applications, including construction, petrochemicals, automotive, aerospace, and marine. The manufacturing industry, which is under increasing pressure, wants to use an environmentally friendly welding method. The SMAW method makes use of many input and output stream types. Sustainability concerns surrounding the SMAW process are linked to the various input and output streams, including the need for electrical energy, the consumption of input materials, the formation of slag, the emission of fumes, and the hazardous working conditions that jeopardize worker health and safety. Improving the process's environmental performance requires characterizing the sustainability of the SMAW welding process within the broad definition of sustainability. The paper reviews the triple bottom line (economic, environmental, and social) sustainability approach in relation to the SMAW process, since most existing literature focuses on the technical and financial aspects of welding with little attention paid to the social and environmental aspects of the process. The study concluded with recommendations for developing a sustainable and economical SMAW welding technique [6].

We evaluate single-wire gas metal arc welding (SGMAW) and high-power tandem GMAW (TGMAW) in terms of energy efficiency. To represent GMAW energy efficiency across a range of material transfer techniques, the electrical deposition efficiency KPI is employed. The equipment's wall-plug efficiency is also tested to determine the total amount of energy used [7].

Many variables have a significant impact on the outcome when welding methods are applied. In the majority of manufacturing processes, the lead designer designates the protocols and techniques to be used based on his expertise or the quantitative impact he can determine from the variables. To optimize the development of the application, many models have been presented throughout the manufacturing process to implement a selection model for welding processes. These models can include, in a weighted manner, the effects or benefits of quality, prices, or skills. This work compiled the works in chronological order and systematically grouped them using the most pertinent methods that were presented. It also generated a brief description for each work and grouped the authors. Additionally, it analyzed some bibliographic factors that enable the estimation of the level of impact that these scientific productions have generated based on indicators like the quantity of citations or the application publisher. Last, this paper presents a suggestion for further research that, while still in the planning stages, anticipates novel avenues and implications within the selected field of study [8]. While sustainable operations are an essential component of modern manufacturing, optimizing a process based on sustainability has huge benefits for the manufacturing process. This article offers a brand-new, codified paradigm for enhancing industrial processes' sustainability. As opposed to earlier methods, the suggested strategy integrates a multi-objective optimization process for enhancing the economy, environment, and society with a methodology for choosing sustainability indicators. The pareto front, or set of non-dominated solutions, is returned by applying nature-inspired heuristics to optimize the process. Depending on the particulars of the situation, the decision maker selects the most convenient option from this set. Optimizing a submerged arc welding process serves as an example of how to use the suggested framework. The computed optimal solution satisfies the social effect while surpassing the sustainability of the economy and environment when compared to the welding settings currently in use [9]. "Sustainability" is a phrase that has acquired popularity recently in an attempt to go beyond the idea of being "green." This article presents general methods for sustainability evaluations. After that, by establishing a thorough framework and measuring and assessing the sustainability of welding processes, the best welding technique for a particular application was identified. Using this approach, information on the welding processes would be collected and categorized into four groups as follows: physical performance, social impact, economic effect, and environmental impact. Subsequently, the outcomes of every category would be integrated to generate a single sustainability score. The case studies were designed to demonstrate the capabilities of the technique for three different welding processes. Friction stir welding scored the highest overall sustainability when compared to gas tungsten arc welding and gas metal arc welding, respectively [10].

Sustainable manufacturing is a philosophy and environmental movement that aims to protect and enhance natural systems and resources over time and to promote social and economic practices that uphold human well-being within natural systems. It takes into account the implications of manufacturing practices for the environment, economy, and energy.

Changing how we utilize resources and reducing our reliance on consumption are frequently crucial steps toward achieving sustainability, especially when it comes to lowering our environmental impact [11].

Limited resources and experience present a problem for small- and medium-sized firms (SMEs) looking to integrate industrial robotics into their manufacturing processes. But because industries are moving toward automated processes and newer technologies, there is still a lot of economic opportunity for industrial robotics. There is a dearth of studies on sustainable decision making for SMEs when switching from traditional to robotic processes. To evaluate the viability of integrating robotic techniques into important operations that might help SMEs, this study offers a comprehensive framework. Throughout the transition period, the framework highlights a number of important technological, managerial, and economic decision-making considerations [12].

Two or more materials, usually metals, can be fused together with or without the use of pressure in one of the main methods for combining metals: welding. The necessity of sustainability in the welding process stems from the fact that welding applications have a significant influence on a number of environmental, social, economic, and physical performance factors. A few of the elements that are included in the study are the use of different control methods for different welding processes and sustainable welding techniques [13].

When compared to human labor, the automobile sector can benefit greatly from the use of automated technology to perform jobs. When using automation systems, quality, dependability, and safety stand out as significant benefits. Smaller businesses that specialize in a certain vehicle genre produce a variety of vehicles, including buses, ambulances, and garbage trucks. It is not possible to use fully automated manufacturing lines due to the small quantities of the production series and the high rate of customization per client [14].

There are a number of variables that significantly impact the outcome while applying welding procedures. Based on his expertise or the quantifiable impact he can determine from the variables, the lead designer determines the methods and procedures to be followed in the majority of manufacturing processes. A variety of models have been put out to implement a welding process selection model throughout the history of manufacturing. These models can weigh the effects of quality, cost, or skill development to improve application development. This study assembled the works in chronological order and arranged them in a systematic manner according to the most pertinent approaches offered. It also produced a synopsis of each work and categorized the authors. Additionally, it examined certain bibliographic elements that enable the assessment of the degree of influence these scientific works have had based on metrics like the quantity of citations or the application publisher. Ultimately, this work presents a research proposal that, since it is in the process of being developed, anticipates new directions and implications for the selected field of study [15].

16.4.1 Automation in Sustainable Welding

Automation of this process is vital for manufacturing to remain competitive given the challenging working conditions and the lack of skilled welders. The International Federation of Robotics estimates that approximately 50% of all robot applications include welding. In this case study, friction stir welding is presented by addressing challenges. The mass of an automobile is a significant consideration when buying one in the automotive business, where fuel economy is king. New technologies are always being developed to lower energy and fuel consumption because it requires more fuel and energy to move a larger vehicle. Modern electrical vehicles demand lighter bodies to run on electricity, despite some using little to no fuel due to the rise in car production. Because each component of a car adds weight, it is critical to take into account every manufacturing method used to create it and focus on specific processes that can be changed to produce a lighter, more fuel-efficient form of transportation.

Consider the welding process, where the joining of metals is essential to many aspects of the automobile production process. Metals are fused together using high heat and a molten filler metal in classical arc, MIG, and TIG welding. All of the ingredients become bonded together as the metals cool and solidify. Although the weld produced by this method is robust, the original material gains weight. In addition, the heat produced during arc, MIG, and TIG welding is extremely energy intensive and is used to melt metal. There are instances where the heat used to fuse two surfaces warps the product leading to its rejection and ultimately wasting money, effort, and materials. Welding techniques must adapt as the globe shifts to greener forms of transportation. Two facing surfaces can be joined using a solid-state technique called friction stir welding (FSW). When the weld head spins into the metal and applies more pressure, friction between the two surfaces creates heat that softens them sufficiently to mix. Clay-like joints are made from the softened metal. FSW completes the welding process with less heat than other welding techniques by moving the material into a plastic condition rather than melting it. As the welding tip never wears out, there are no consumables needed for this procedure. It is easier to control, regulate, and anticipate. The strongest, highest-quality welds are produced by friction stir welding, which may also unite materials that are not compatible. A number of sectors are already using FSW (friction stir welding) to generate strong, seamless welds on structures—aluminum, for instance, needs a high level of weld strength without a heat treatment—after years of research and practice.

16.5 Future Scope of Sustainability in Welding

Sustainability can be promoted, and the environmental effects of the sector can be reduced by putting eco-friendly practices into place, such as using renewable energy sources, managing materials efficiently to reduce waste, and using ecologically friendly welding technology. Using a laser to immediately melt metal and fuse two pieces together is one of the newest developments in welding. Even when working with incredibly complicated items, the laser beam's exceptional accuracy allows for exquisite welding.

Industries of all kinds are seeing dramatic changes in their operational models because of the unparalleled speed at which technology is developing. Exciting new developments that are revolutionizing the joining of metals and other materials are also appearing in the welding sector. The field of welding is expected to become extremely efficient and cutting edge in the future due to developments in automation, robots, artificial intelligence, and virtual reality. Because of these modifications, technicians and welders will have access to a wide range of effective instruments and methods that will allow them to operate more efficiently, quickly, and precisely than in the past. Metal-joining procedures underwent major alterations as a result of technical advancements. Because of this, the newest metals and alloys, automated welding, enhanced arc welding, and manual welding are gradually replacing traditional metals, techniques, and welding methods in the coming years. Furthermore, to encourage a secure welding environment and practical experience, welding safety and training have also been advanced. The welding sector, which is sometimes seen as the foundation of building and manufacturing, faces numerous difficulties that call for careful consideration and creative solutions. Welding industry experts need to be aware of these changes and take proactive measures to adapt when environmental concerns, technology, and worker dynamics change. In a time of unprecedented change and unpredictability, the welding sector must embrace innovation and sustainability while adjusting to new difficulties. Stakeholders can create a robust and prosperous future by tackling challenges including workforce development, environmental responsibility, supply chain resilience, safety, and technological progression. Fostering an environment that is favorable to growth and innovation in the welding sector requires cooperation between industry participants, educational institutions, and governmental organizations.

16.6 Conclusion

The sustainability ideas in the welding industry are covered in full in this chapter. It covers a range of sophisticated welding techniques, such as metal arc gas welding, submerged arc welding, tungsten gas welding, and others, that have been characterized in the literature as sustainable welding techniques.

These techniques are considered to be safe, low in pollution, have no negative effects on the environment, and are environmentally friendly. Additionally, one could consider a welding simulator to be a crucial assessment instrument for welder certification. Analytical Hierarchy Process (AHP)-based assessment and statistical methods have been employed in prior research on the sustainability of industrial businesses. A review of the welding industry's sustainability like this offers fascinating new perspectives on the fourth industrial revolution. In an effort to mitigate the effects and partially offset losses, cost cutting, downsizing, and mass layoffs have become commonplace tactics employed globally in reaction to the present economic difficulties. However, there has not been an assurance of a constant degree of performance and quality. Those who were previously hesitant had even more reason to exclude any mention of minimizing environmental damage from their development plans in light of the crisis. The global effects could be disastrous for the environment. So, a new concept is required to offer a more thorough and efficient way to assess products, companies, processes, etc. Sustainability makes certain that companies, procedures, and goods operate properly within environmental restrictions to maximize revenues and guarantee society's satisfaction.

References

1. Gadakh, V.S. and Badheka, V.J., Sustainability of fusion and solid-state welding process in the era of Industry 4.0, Handbook of smart materials, in: *Technologies & Devices*, pp. 1637–1654, Springer, 2022.
2. Stark, R., Seliger, G., Bonvoisin, J., *Sustainable manufacturing challenges, solutions and implementation perspectives*, pp. 1–286, Springer, 2017.

3. Sharma, A., A fundamental study on qualitatively viable sustainable welding process maps. *J. Manuf. Syst.*, 46, 221–230, 2018.
4. Bevilacqua, M., Ciarapicaa, F.E., D'Orazioa, A., Forcellesea, A., Simoncini, M., Sustainability analysis of friction stir welding of AA5754 sheets. *Procedia CIRP*, 62, 529–534, 2017.
5. Manzanares-Cañizares, C., Sánchez-Lite, A., *et al.*, A 5S lean strategy for a sustainable welding process. *Sustainability*, 14, 6499, 2022.
6. Alkahla, I. and Pervaiz, S., Sustainability assessment of shielded metal arc welding (SMAW) process. *IOP Conf. Series Mater. Sci. Eng.*, 244, 012001, 2017.
7. Sproesser, G., Chang, Y.J., *et al.*, Energy efficiency and environmental impacts of high-power gas metal arc welding. *Int. J. Adv. Manuf. Technol.*, 91, 3503–3513, 2017.
8. Bohórquez, O., Pertuz, A., *et al.*, An overview framework of welding process selection, current state and expectative. *J. Phys. Conf. Ser.*, 2046, 012076, 2021.
9. Rivas, D., Quiza, R., *et al.*, Towards sustainability of manufacturing processes by multi objective optimization: a case study on a submerged arc welding process. *IEEE Access*, 8, 2020.
10. Jamal, J., Darras, B.M., Kishway, H.A., A study on sustainability assessment of welding processes. *J. Eng. Manuf.*, 12, 501–512, 2019.
11. Paulo Davim, J., *Sustainability Manufacturing*, pp. 1–246, Wiley, 2010.
12. Epping, K. and Zhang, H., A sustainable decision-making framework for transitioning to robotic welding for small and medium manufacturers. *Sustainability*, 10, 3651, 2018.
13. Bala, Y.G., Santhi, B., Kumar, R.D., *Sustainability in welding industries*, pp. 1–20, Wiley, 2024.
14. Castro, A.F., Silva, M.F., Silva, F.J.G., Designing a robotic welding cell for bus body frame using a sustainable way. *Procedia Manuf.*, 11, 207–214, 2017.
15. Bohórquez, O., Pertuz, A., Rincón-Ortíz, M., An overview framework of welding process selection, current state and expectatives. *J. Phys. Conf. Ser.*, 2046, 012076, 2021.

17

Mechanical and Structural Evaluation of Friction Stir Welded AA6063 and AA6069 Aluminum Alloys for Marine and Aerospace Applications

M. Arockia Jaswin* **and R. Geetha**

Department of Mechanical Engineering, Velammal Engineering College, Chennai, Tamil Nadu, India

Abstract

A comprehensive investigation of the mechanical and structural evaluation of friction stir welded (FSW) specimen of AA6063 and AA6069 aluminum alloys to explore their potential deployment in marine and aircraft manufacturing is presented in this study. This research work demonstrates the ductile strength, bending characteristics, hardness, and microstructural characteristics of the specimen synthesized by FSW process. The effect of FSW in the microstructure and phase transformation are also analyzed. From the experiments and analysis conducted, tensile strength, bending strength, and hardness of the welds exhibited remarkable enhancements compared to the base materials, which is owing to the uniform and homogeneous dissemination of precipitate particles and fine grain structure of the welds. The outcome of this analysis confirms the suitability of deploying FSW process for producing AA6063 and AA6069 alloy joints with excellent mechanical properties and microstructural properties, which can meet the growing demands in the field of aerospace and marine engineering. Also, this experimental study confirms the feasibility of applying this type of solid state welding joints in the fabrication of high-performance, compact lightweight devices and operating components in aerospace and marine engineering.

Keywords: Aluminum alloys, AA6063, AA6069, friction stir welding, tensile strength, hardness, bending strength, microstructure analysis

**Corresponding author*: winn.au@gmail.com

Sandip Kunar and Gurudas Mandal (eds.) Advanced Welding Technologies, (353–368) © 2025 Scrivener Publishing LLC

17.1 Introduction

There is always a growing demand for lightweight and high-performance materials in the field of marine, automobile, and aerospace industries [1]. This demand has led to the evolution of many materials with improved performance. With the evolution of newer materials, there is always a challenge in processing these materials. Many advanced processes have been invented in the last 20 years to improve the properties and performance of these materials [2]. Solid state welding is one such processing method, which is used to produce permanent joints with enhanced properties. The major advantage of using solid state welding is to join dissimilar metals with less distortion [3–5]. FSW, a solid state welding method, is highly emerging today as a transformative process for welding lightweight materials, which can be deployed in smart automobile industries, and the marine and aerospace industries [6]. Friction stir welding has emerged as a transformative joining technique in aircraft and marine manufacturing due to its ability to produce high-quality welds in aluminum alloys [7–10]. Among these alloys, AA6063 and AA6069 are particularly significant for their favorable mechanical properties and resistance to corrosion making them suitable for structural applications in adverse environments. This work is aimed at developing and evaluating the mechanical behavior and microstructural properties of aluminum alloys through friction stir welding for deploying them in aerospace and marine industries. These industries always rely on lightweight materials with higher performance. The outcomes of this experimental study will help in overcoming the gaps in understanding the effect of various operating factors of the friction stir welding process in the analysis of mechanical and microstructural behavior of AA6063 and AA6069 aluminum alloys [11]. The evaluation of tensile strength, micro hardness, bending strength, and microstructural analysis will confirm the weld integrity and deployment in real-time applications [12–15]. The prime objective of this work is to experimentally evaluate and compare the mechanical behavior of AA6063 and AA6069 aluminum alloys joined through friction stir welded with the properties of base metal and to examine the tensile strength, bending strength, micro hardness on base metal, and FSW AA6063 and AA6069 aluminum alloys. Also, the experimental study is extended to analyze the impact of microstructural changes in the properties of welded joints.

17.2 Materials and Methodology

17.2.1 Aluminum Alloys

The materials selected for this study are two grades of aluminum alloys, AA6063 and AA6069. In the 6000 series, alloy AA6063 is a structural alloy with medium strength. Also, this type of alloy possesses excellent surface finish and corrosion resistance making this alloy an excellent option for structural applications. Generally, the AA6063 aluminum alloy is tempered by T6 in which the metal is solutionized and aged artificially [16]. The alloy composition of AA6063 material is given in Table 17.1. Typical applications of the AA6063 alloy include the fabrication of architectural products and fabrication of pipes for irrigation systems. Table 17.2 represents the alloy composition of AA6069. It is a heat-treatable alloy, which can be used in high-pressure environments such as gas cylinders, cold impact extrusion products, and lightweight automobile components. The elements mostly alloyed with aluminum are zinc, magnesium, and manganese. Zinc is added to increase the water corrosion properties. Magnesium added with aluminum improves weldability [17].

17.2.2 Experimental Methods

The experimental evaluation starts with the selection of the aluminum alloys. The selected alloys are drawn into sheets and welded by friction stir welding. A detailed methodology followed is represented in Figure 17.1. The aluminum alloys joined by FSW and the base metal are subjected to various mechanical and microstructural studies to explore their

Table 17.1 % Weight composition of AA6063 aluminum.

Composition	Si	Mg	Fe	Mn	Cu	Cr	Zn	Ti	Al
Wt. %	0.70	0.90	0.35	0.10	0.10	0.10	0.10	0.10	Remaining

Table 17.2 Chemical composition of AA6069 aluminum alloy.

Composition	Si	Cu	Mg	Fe	Mn	V	Cr	Ti	Al
Wt. %	1.60	1.00	1.60	0.10	0.40	0.30	0.30	0.10	Remaining

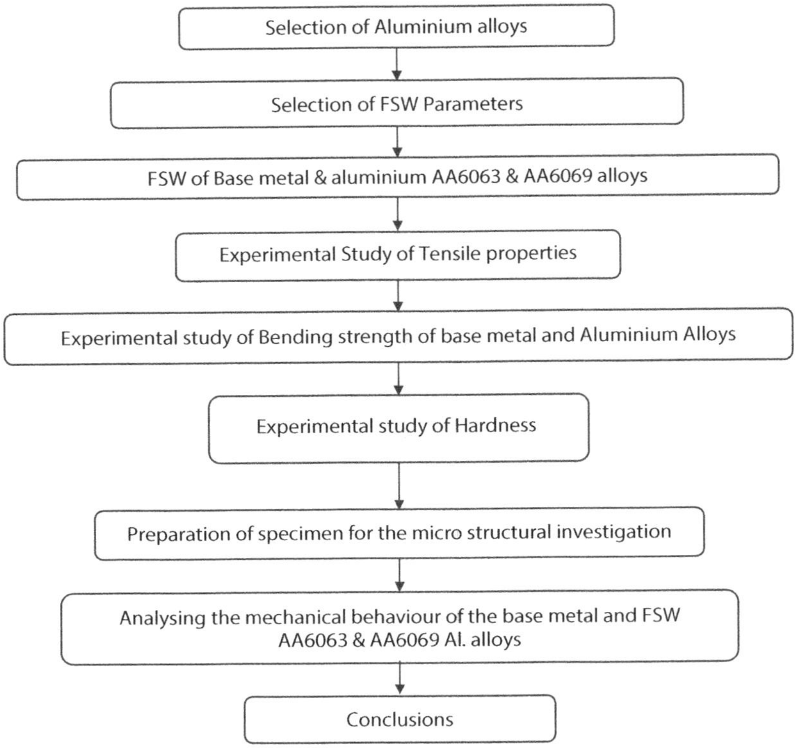

Figure 17.1 Process methodology.

compatibility to be deployed in various industries. The tool, also called Probe, used in this process is made up of titanium carbide. The titanium carbide Probe, which is cylindrical in shape, is placed in the integrating regions of the plates to be joined, and pressure is applied when the Probe starts to rotate [18]. The Probe rotates at 1,450 rpm with a feed of 45 mm/min and 10 kN force on the Probe. The offset value of the Probe is maintained in the range of 0–1.5 mm. The Probe is presented in Figure 17.2. The various parameters of the probe are represented in Table 17.3.

An FSW machine developed by Interface Design Associates Private Limited is used to do the FSW on the baseplates. The friction stir welding machine is basically a milling machine customized to meet the requirements of the friction welding process [19–21]. The machine's maximum load is 20 kN and is controlled by a four-axis control system. The machine resolution and accuracy are 5 and 10 μm, respectively. The sheets are placed on the table, and the probe is made to rotate on the fixed plates with pressure applied. The probe arrangement on the plates is shown in Figure 17.3.

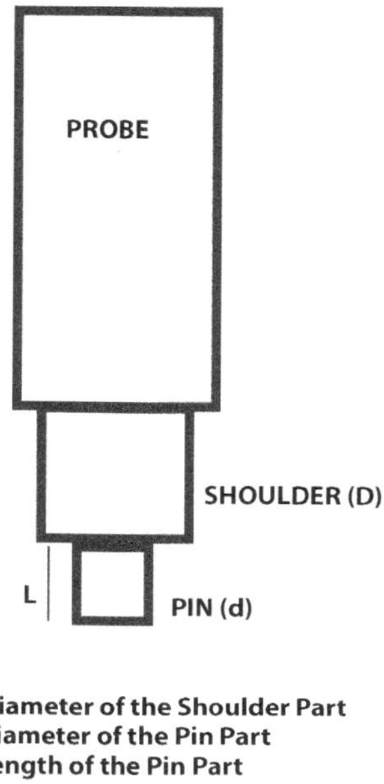

Figure 17.2 Probe used in friction stir welding machine.

Table 17.3 Probe dimensions.

Sl: no.	Parameters	Quantity (values)
1	Probe type	Cylindrical Probe with threads
2	Ratio of D/d of the Probe	3.5
3	Shoulder diameter of the Probe	20 mm
4	Diameter of the pin	5 mm

17.2.2.1 Mechanical Tests—Tensile Strength

The friction stir-welded portions are processed in a cutter and post processed to fabricate the samples for the tensile testing machine. The stress developed in the specimen during cutting is observed. The test specimen as per the ASTM standard is given in Figure 17.4. The test is carried out at a gauge length of 80 mm. The tensile test is performed in a Universal Testing Machine with reference to the ASTM Standards [22]. The test is performed

Figure 17.3 The probe and aluminum alloy sheet arrangement.

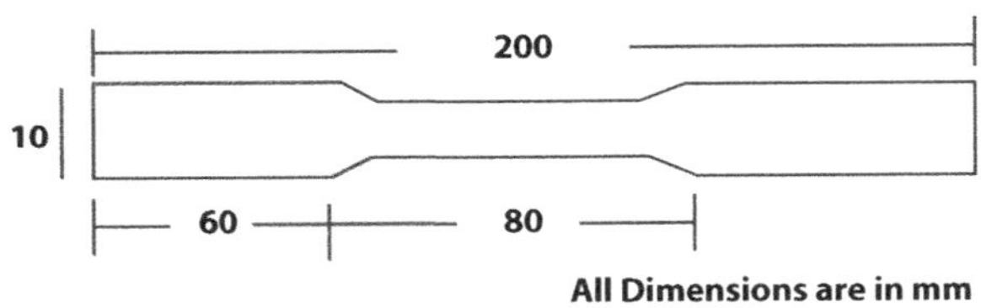

Figure 17.4 Dimensions of the tensile specimen.

by applying a load of 100 kN initially. The load is gradually increased at a rate of 2.5 kN/mm as recommended by the ASTM standards. The failure in the specimen is observed by gradual necking and final fracture of the welded material.

17.2.2.2 Bending Strength Test

The specimen for bending strength is fabricated as per the standard followed by (ASTM) Standards are given in Figure 17.5. A three-point bending test is carried out on the material, placed on the table, and supported by two jaws. The load is applied on the bending notch part exactly at the center of the center, and the readings are observed at various offset levels of 0, 0.5, 1, and 1.5 mm.

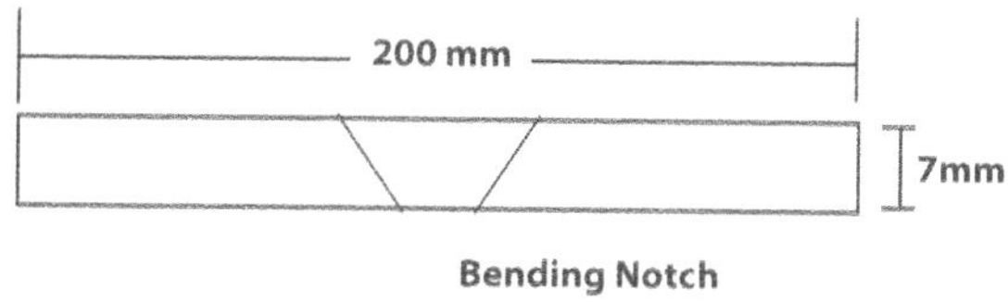

Figure 17.5 Dimensions of the bending test specimen.

17.2.2.3 *Hardness Test*

The maximum resistance developed on the welded aluminum alloys are determined by a Vickers hardness tester. A square diamond pyramid indenter is employed to study the welded aluminum alloy's load-bearing capacity [23].

17.2.2.4 *Microstructure Analysis*

The microstructural analysis is conducted on the fabricated specimen to study the Thermomechanical Affected Zone, Friction Stir Processing Zone, and Heat-Affected Zone (HAF) of aluminum alloys [24]. To effectively reveal the microstructure of the FSW alloy material, the specimen is subjected to an etching process with Kellers reagent solution. The specimen is mounted in a mold made up of phenolic powder. The specimen mounted in the mold is shown in Figure 17.6.

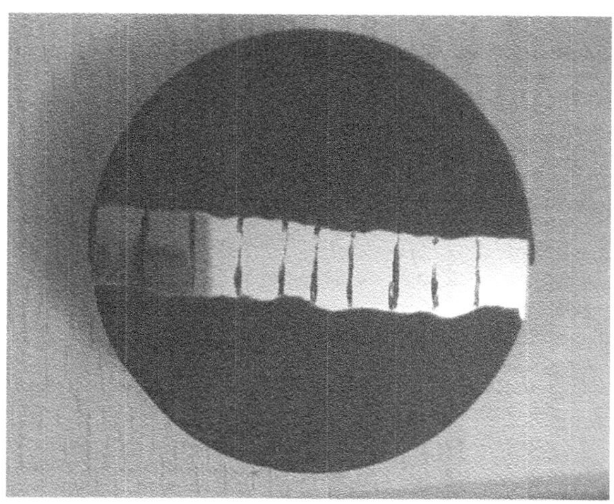

Figure 17.6 Specimen for microstructure analysis.

17.3 Results and Discussions

17.3.1 Tensile Strength

Aluminum alloy base metal is subjected to various mechanical testing, such as tensile strength, bending strength, hardness, and microstructural studies, and correlated with the test results of FSW AA6063–AA6069 alloys. Table 17.4 represents the observed values of tensile behavior of aluminum alloys individually and the behavior of friction stir welded AA6063 and AA6069 aluminum alloys. It is seen from the test values that the friction stir-welded aluminum alloys showed improved maximum tensile strength and yield strength compared to those of the AA6063 and AA6069 alloys. The percentage of elongation is also comparatively high. There is a 65% increase in yield strength in the welded aluminum alloys compared to that of AA6063 and a 54% increase in the yield strength value in the case of the AA6069 alloy.

The observed values of force during each stroke is represented in Table 17.5 for aluminum alloy AA6063. The maximum force is observed between the eighth and ninth strokes with a maximum force of 9,710 N, at which the specimen breaking happened for AA6063 Al. alloy. The graphical representation of the force applied at each stroke to aluminum AA6063 Al alloy is given in Figure 17.7. Similarly, the effect of force observed by the aluminum AA6069 alloy specimen at each stroke is presented in Table 17.6. The maximum force is observed on the 17th stroke at 12,500 N, in which the breaking of the specimen happened for AA6069 aluminum alloy.

Table 17.4 Tensile behavior of parent metal and FSW aluminum alloys.

Sl. no.	Specimen	Yield strength (N/mm²)	Maximum tensile strength (N/mm²)	% Elongation (%)
1.	Base metal AA6063	210	231	2.4
2.	Base metal AA6069	240	269	4.1
3.	AA6063–AA6069 (friction stir welded)	280	319	3.4

Table 17.5 Force observed in each stroke for aluminum AA6063 alloy.

Stroke (mm)	**1**	**2**	**3**	**4**	**5**	**6**	**7**	**8**	**9**	**10**
Force N	1,070	2,340	3,760	5,100	6,500	7,850	8,750	9,410	9,710	8,000

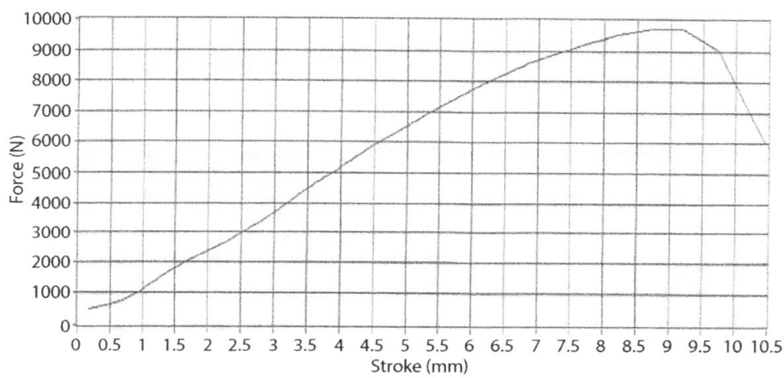

Figure 17.7 Graphical representation of force applied at each stroke for AA6063 aluminum alloys.

Table 17.6 Force observed in each stroke for aluminum AA6069 alloy.

Stroke (mm)	**1**	**3**	**5**	**7**	**9**	**11**	**13**	**15**	**17**	**18**
Force N	1,000	3,900	6,360	8,200	9,910	11,090	11,920	12,300	12,500	1,100

The graphical representation of the force applied at each stroke to aluminum AA6069 aluminum alloy is shown in Figure 17.8. The effect of force observed by the friction stir welded aluminum AA6063–AA6069 alloy at each stroke is presented in Table 17.7. The maximum force is observed at the 15th stroke with 14,600 N, which is sufficient to fracture the specimen confirming that the tensile property of FSW AA6063–AA6069 metal is significantly better when related to the tensile properties of aluminum alloys. The graphical representation of the force and the stroke is given in Figure 17.9.

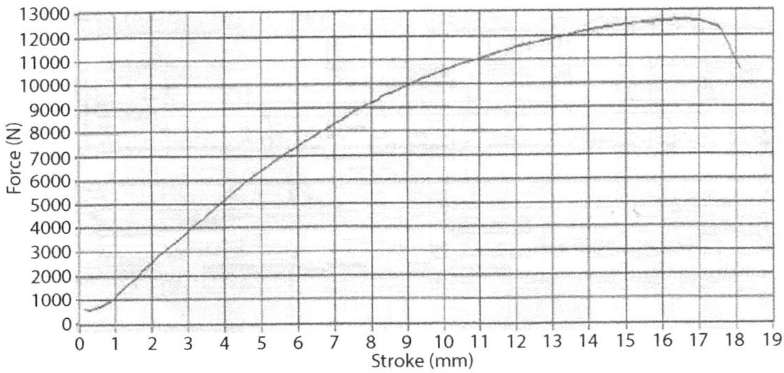

Figure 17.8 Graphical representation of force applied at each stroke for AA6069 aluminum alloy.

Table 17.7 Force observed in each stroke for friction stir-welded aluminum alloys.

Stroke (mm)	**1**	**3**	**5**	**7**	**9**	**11**	**12**	**14**	**15**	**17**
Force N	930	2,100	6,050	9,652	12,000	13,330	14,150	14,450	14,600	13,700

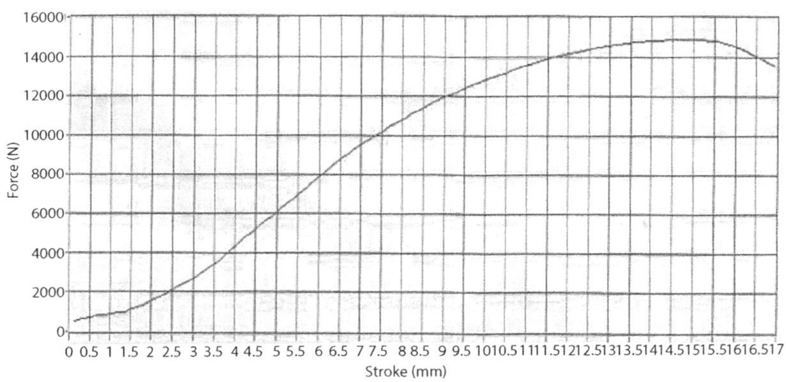

Figure 17.9 Graphical representation of force applied at each stroke for friction stir-welded aluminum alloys.

17.3.2 Bending Test

The evaluated test reports of bending strength is presented in Table 17.8. The values observed for the various specimens of aluminum AA6063, AA6069 alloys, and friction stir-welded aluminum alloys confirm that the

Table 17.8 Bending strength results of aluminum specimen and friction stir-welded specimen.

Sl. no	Test	Sample AA6063	Sample AA6069	Friction stir-welded specimen at offset			
				0 mm	0.5 mm	1 mm	1.5 mm
1	Maximum force kN	2.12	2.52	3.65	3.40	3.18	3.02

bending strength of the FSW aluminum alloys possess higher bending strength compared to the individual aluminum alloys [25–28]. The high yield strength and the improved ductile properties of the friction stir-welded joint of aluminum alloys exhibited significant resistance to crack initiation and propagation. Also, it is evident from the observed values that as the offset value increases, the force observed decreases.

17.3.3 Hardness

The load-bearing capacity of the FSW joint is evaluated on a Vicker's microhardness test machine, and the observed values are shown in Table 17.9. Hardness is measured at 0.5-kg load at various spots of the FSW joint and compared with the hardness of aluminum alloys [29]. The friction stir-welded joint exhibited maximum load-bearing capacity when compared to aluminum AA6063 alloy and AA6069 alloy. This is due to the improved ultimate tensile strength and bending strength that resulted in the improved load-bearing capacity [30].

Table 17.9 Micro–hardness test values.

Sl. no.	Specimen	Hardness at 0.5-kg load (Hv)
1.	AA6063	63
2.	AA6069	79
3.	AA6063–AA6069 (friction stir welded	112

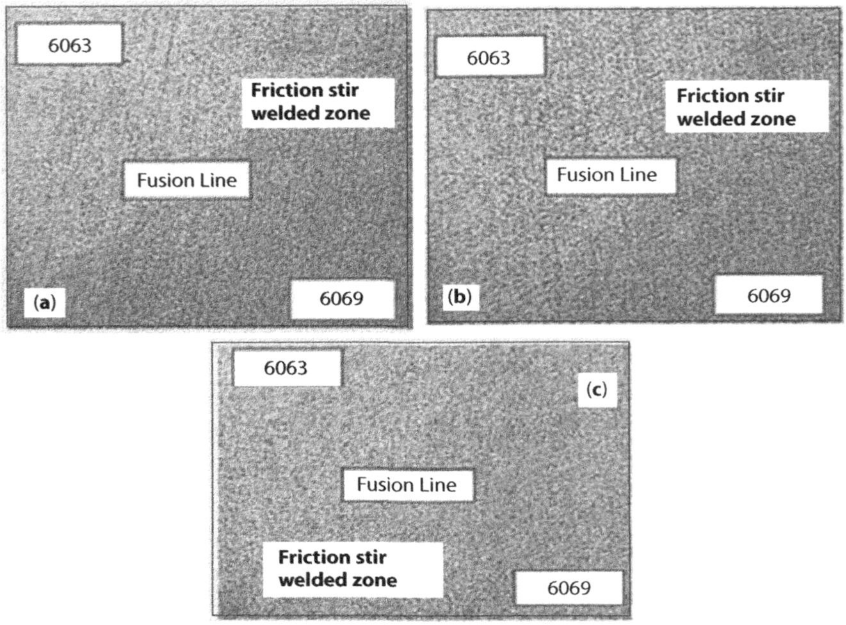

Figure 17.10 Optical microscopy images of aluminum alloys and friction stir-welded joint at (a) ×100, (b) ×200, (c) ×500.

17.3.4 Optical Microscopy Studies

The optical microscopy images of the test specimen at various magnifications (×100, 200, ×500) are shown in Figure 17.10. It is inferred from the images that there is no presence of macro-level defects such as inclusions of impurities, cracks, voids and agglomerations on the surface of the specimen [31]. This is due to the uniform variations in the heat input by the probe because of its uniform centrifugal action on the specimen. The aluminum alloy structures also exhibited a uniform structure without any defects. Also, the grain structure of the joints in the fusion line showed fine grains compared to the aluminum alloys [32].

17.4 Conclusion

The detailed experimental study conducted revealed the potential of deploying FSW process for aluminum AA6063 and AA6069 metals. The FSW exhibited significant improvements in the mechanical behavior of joints. The tensile strength, hardness, and bending strength show significant enhancements in the joints. The microstructural studies also confirmed that the surface of the joint and base metals are free from defects. Thus, the friction stir welding confirms its suitability for producing defect-free welded aluminum structures that can be used in marine and aerospace applications.

References

1. Niu, P., Li, Y., Fu, X., Li, W., Gong, J., Yang, C., Yu, D., Strengthening mechanisms of pre-stretching strain on friction stir welded 2024-to-7075 dissimilar aluminium alloys joints: Yield sequence effect. *Mater. Charact.*, 211, 113895, 2024.
2. Djili, A., Bezzazi, B., Zioui, N., Haboussi, M., Effect of natural aging on the tensile properties and the toughness of friction stir welds of 2024-T3 aluminum alloy. *J. Adv. Joining Processes*, 8, 100153, 2023.
3. Barkley, K.M., Arner, J.S., Pike, T.A., Diwakar, P., Birrenkott, C.M., Correlation of surface and interfacial temperature during differential ultrasonic spot welding. *J. Adv. Joining Processes*, 7, 100142, 2023.
4. Venugopal, V., Singh, V.P., Kuriachen, B., Underwater friction stir welding of marine grade aluminium alloys: A review. *Mater. Today Proc.*, 2023. ISSN 2214-7853, https://doi.org/10.1016/j.matpr.2023.07.182.
5. Prakash, M., Vinoth Kumar, M., Sathya Moorthi, S., Sathish Kumar, R., Muthuchari, P., Arulananth, T.S., Investigation of effect of tool offset position on mechanical properties of dissimilar aluminium alloy sheets joined by friction stir welding. *Mater. Today Proc.*, 2023. ISSN 2214-7853, https://doi.org/10.1016/j.matpr.2023.06.252.
6. Rathinasuriyan, C., Muniamuthu, S., Mystica, A., Senthil Kumar, V.S., Investigation of heat generation during submerged friction stir welding on 6061-T6 aluminum alloy. *Mater. Today Proc.*, 46, 17, 8320–8324, 2021.
7. Vysotskiy, I., Malopheyev, I., Mironov, S., Kaibyshev, R., Pre-strain rolling as an effective tool for suppression of abnormal grain growth in friction-stir welded 6061 aluminum alloy. *Mater. Sci. Eng. A*, 733, 39–42, 2018.
8. Malopheyev, S., Vysotskiy, I., Kulitskiy, V., Mironov, S., Kaibyshev, R., Optimization of processing-microstructure-properties relationship in friction-stir welded 6061-T6 aluminum alloy. *Mater. Sci. Eng. A*, 662, 136–143, 2016.
9. Li, Z., Zhang, Z., Chen, X.-G., Microstructure, elevated-temperature mechanical properties and creep resistance of dispersoid-strengthened Al-Mn-Mg 3xxx alloys with varying Mg and Si contents. *Mater. Sci. Eng. A*, 708, 383–394, 2017.

10. Komarasamy, M., Alagarsamy, K., Ely, L., Mishra, R.S., Characterization of 3" through-thickness friction stir welded 7050-T7451 Al alloy. *Mater. Sci. Eng. A*, 716, 55–62, 2018.
11. Su, Y., Li, W., Shen, J., Fu, B., dos Santos, J.F., Klusemann, B., Vairis, A., Comparing the local-global deformation mechanism in different friction stir welding sequences of Ti-4Al-0.005B titanium alloy T-joints. *Mater. Sci. Eng. A*, 823, 141698, 2021.
12. Zapata, J., Toro, M., López, D., Residual stresses in friction stir dissimilar welding of aluminum alloys. *J. Mater. Process. Technol.*, 229, 121–127, 2016.
13. Ilman, M.N., Sriwijaya, R.A., Muslih, M.R., Triwibowo, N.A., Sehono, Strength and fatigue crack growth behaviours of metal inert gas AA5083-H116 welded joints under in-process vibrational treatment. *J. Manuf. Process.*, 59, 727–738, 2020.
14. Shah, P.H. and Badheka, V., An Experimental Investigation of Temperature Distribution and Joint Properties of Al 7075 T651 Friction Stir Welded Aluminium Alloys. *Procedia Technol.*, 23, 543–550, 2016.
15. Senthil Kumar, S., Denis Ashok, S., Narayanan, S., Investigation of Friction Stir Butt Welded Aluminium Alloy Flat Plates Using Spindle Motor Current Monitoring Method. *Procedia Eng.*, 64, 915–925, 2013.
16. Galvao, I., Verdera, D., Gesto, D., Loureiro, A., Rodrigues, D.M., Influence of aluminium alloy type on dissimilar friction stir lap welding of aluminium to copper. *J. Mater. Process. Technol.*, 213, 11, 1920–1928, 2013.
17. Bisadi, H., Tavakoli, A., Tour Sangsaraki, M., Tour Sangsaraki, K., The influences of rotational and welding speeds on microstructures and mechanical properties of friction stir welded Al5083 and commercially pure copper sheets lap joints. *Mater. Des.*, 43, 80–88, 2013.
18. Abdollah-Zadeh, A., Saeid, T., Sazgari, B., Microstructural and mechanical properties of friction stir welded aluminum/copper lap joints. *J. Alloys Compds.*, 460, 1–2, 535–538, 2008.
19. Dinaharan, I., Kalaiselvan, K., Vijay, S.J., Raja, P., Effect of material location and tool rotational speed on microstructure and tensile strength of dissimilar friction stir welded aluminum alloys. *Arch. Civ. Mech. Eng.*, 12, 4, 446–454, 2012.
20. Wang, T., Zou, Y., Matsuda, K., Micro-structure and micro-textural studies of friction stir welded AA6061-T6 subjected to different rotation speeds. *Mater. Des.*, 90, 13–21, 2016.
21. RajKumar, V., Venkatesh Kannan, M., Sadeesh, P., Arivazhagan, N., Devendranath Ramkumar, K., Studies on Effect of Tool Design and Welding Parameters on the Friction Stir Welding of Dissimilar Aluminium Alloys AA 5052 – AA 6061. *Procedia Eng.*, 75, 93–97, 2014.
22. Ghetiya, N.D. and Patel, K.M., Prediction of Tensile Strength in Friction Stir Welded Aluminium Alloy Using Artificial Neural Network. *Procedia Technol.*, 14, 274–281, 2014.

23. Taher, A., Shehabeldeen, Zhou, J., Shen, X., Yin, Y., Xiaoyuan, J., Comparison of RSM with ANFIS in predicting tensile strength of dissimilar friction stir welded AA2024 -AA5083 aluminium alloys. *Procedia Manuf.*, 37, 555–562, 2019.
24. Rzaev, R., Chularis, A., Smirnov, V., Semyenova, L., The Influence of the Friction Stir Welding Parameters on the Formation of Welded Joint of Aluminum and Copper Alloys. *Mater. Today Proc.*, 11, 1, 534–542, 2019.
25. Sameer, M.D. and Birru, A.K., Mechanical and metallurgical properties of friction stir welded dissimilar joints of AZ91 magnesium alloy and AA 6082-T6 aluminium alloy. *J. Magnesium Alloys*, 7, 2, 264–271, 2019.
26. Chen, S., Zhang, H., Jiang, X., Yuan, T., Han, Y., Li, X., Mechanical properties of electric assisted friction stir welded 2219 aluminum alloy. *J. Manuf. Process.*, 44, 197–206, 2019.
27. Kalinenko, A., Kim, K., Vysotskiy, I., Zuiko, I., Malopheyev, S., Mironov, S., Kaibyshev, R., Microstructure-strength relationship in friction-stir welded 6061-T6 aluminum alloy. *Mater. Sci. Eng. A*, 793, 139858, 2020.
28. Chandrashekar, A., Reddappa, H.N., Ajaykumar B.S., A Study of Mechanical and Micro Structural Properties of Friction Stir Welded Aluminium Alloy. *Mater. Today Proc.*, 2, 4–5, 3491–3500, 2015.
29. Li, D., Yang, X., Cui, L., He, F., Zhang, X., Investigation of stationary shoulder friction stir welding of aluminum alloy 7075-T651. *J. Mater. Process. Technol.*, 222, 391–398, 2015.
30. Zhang, F., Su, X., Chen, Z., Nie, Z., Effect of welding parameters on microstructure and mechanical properties of friction stir welded joints of a super high strength Al–Zn–Mg–Cu aluminum alloy. *Mater. Des.*, 67, 483–491, 2015.
31. Wang, T., Komarasamy, M., Liu, K., Mishra, R.S., Friction stir butt welding of strain-hardened aluminum alloy with high strength steel. *Mater. Sci. Eng. A*, 737, 85–89, 2018.
32. Shengchong, M., Zhao, Y., Zou, J., Yan, K., Liu, C., The effect of laser surface melting on microstructure and corrosion behaviour of friction stir welded aluminum alloy 2219. *Opt. Laser Technol.*, 96, 299–306, 2017.

18

Experimental Analysis and Optimization of Microwave Welding Parameters for Improved Joint Strength

Kishor Jyoti Deka[1], Shubhajit Das[1]* and Sangeeta Das[2]

[1]*Department of Mechanical, School of Manufacturing and Construction, Assam Skill University, Guwahati, Assam, India*
[2]*Department of Mechanical Engineering, Girijananda Chowdhury University, Guwahati, Assam, India*

Abstract
In this study, an effort has been made to investigate the output responses of microwave welding (MW) of polymer-based material. Microwave welding parameters viz., material combination, microwave power, absorbent weight, dead weight, and infill percentage, were considered as input process parameters for analyzing the effects on lap shear strength and elongation, which were the output responses. The experiments were performed at a microwave applicator for different material combinations to study the effect of different microwave welding parameters. A total of 27 set of experiments were conducted, and further statistical analysis was done using Taguchi's design of experiments (DoE). Material combination followed by microwave power was found to have significant effect on both lap shear strength and elongation. Analysis of variance (ANOVA) was performed to find out the significant parameters during microwave welding. Response Surface Methodology (RSM) was used to develop predictive models of the output responses. Multi objective optimization of microwave welding output responses was performed using desirability analysis (DA) approach. Material combination of Acrylonitrile Butadiene Styrene (ABS)/ABS was found to have significant effect on lap shear strength and elongation.

***Keywords*:** Microwave welding, design of experiments, analysis of variance, response surface methodology, desirability analysis

**Corresponding author*: 06shubhajit@gmail.com

Sandip Kunar and Gurudas Mandal (eds.) Advanced Welding Technologies, (369–384) © 2025 Scrivener Publishing LLC

18.1 Introduction

Welding leads to an extremely crucial role in joining of metals or non-metals in day-to-day engineering works. Welding of different techniques are adopted to join different materials to create a single piece resulting in a permanent joint. Among other manufacturing processes, welding is one of the processes that find application in different categories of industries. Welding techniques are classified according to the source of energy applied for heating the metals. Microwave welding (MW) is one of the welding techniques to join mostly polymeric materials using microwave energy. MW is a non-traditional joining process, which comes under the category of environment-friendly and energy-efficient manufacturing method. Microwaves are also referred to as microwave radiation. Wavelengths of electromagnetic radiation in microwave ranges between radio and infrared waves in the electromagnetic spectrum. The frequency in microwaves ranges between 300 MHz and 300 GHz [1]. In maximum of the cases of thermoplastic materials, a frequency of radiation typically more than 2.45 GHz is required for an instant temperature rise when irradiated with radiation frequency. Lesser magnitude of frequency may not be sufficient enough to immediately increase the temperature in most of the thermoplastic materials. To attain an instant temperature rise, a material is used, shaped like a gasket, which absorbs the microwave energy at the surfaces of the joint [2]. In MW, volumetric heating occurs at the point of joint surfaces since microwave can pass into the volume of the material. Therefore, it becomes easy to attain rapid and uniform heating of materials. In volumetric heating, heat propagates from inside to outside of the material and hence produce an inverse heating profile, which is opposite to conventional heating profile [3]. Microwave heating is volume dependent and not surface area dependent. Moreover, MW is a contactless welding process. Since MW is a new development in the joining of materials and basically used for polymeric material joining, so many researchers are working on it to make it commercially impactable. It is difficult to join polymeric materials by other welding techniques, which already have a great impact in the joining of materials in the engineering field. To overcome the difficulties of joining polymeric materials, MW is one of the promising methods. The need for MW is due to the green and eco-friendly welding process, volumetric heating of materials, which leads to minimum heating time, energy efficient in relation to other welding techniques, used to join three-dimensional and complex geometries, lower manufacturing cost, low cost of equipment, better microstructure of welded joint compared to

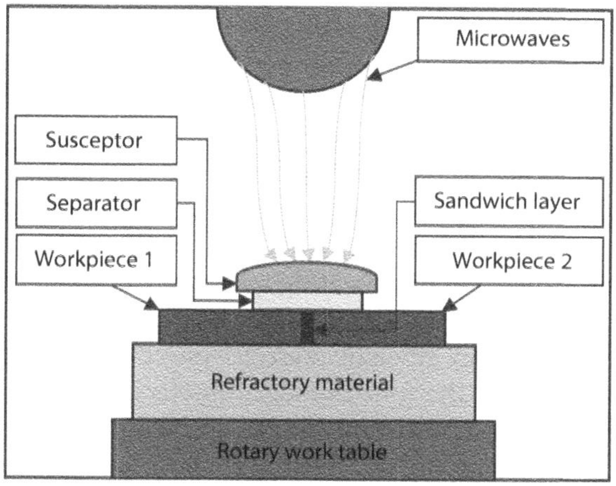

Figure 18.1 Microwave welding.

other welding methods, and improved mechanical strength of materials [2]. The input process parameters of MW play a significant role in welding quality like homogeneous joint, strength of the joint, reliability, etc. The measurable output responses of MW are shear strength of the joint, elongation, hardness, etc. Figure 18.1 depicts the MW of polymer-based materials. Susceptor material is used to absorb electromagnetic energy and later convert it to heat energy. A separator also has the capability to absorb electromagnetic energy, and it is applied to separate the susceptor material and sandwich layer. The sandwich layer is prepared like a slurry by mixing epoxy resin with metal-based nano powder. This layer acts as a heat absorber at the joint interface. Refractory materials are used to avoid reflection of microwaves from the targeted workpiece to be weld and also to place the workpiece by preparing a required slot on it [12].

This chapter aims to analyze the joint strength during MW of polymer-based material with respect to different input process parameter combinations and also optimization of the input parameters.

18.2 Review of MW for Polymer-Based Materials

Nandwani *et al.* [1] conducted a literature review on the exposure time of microwave-based welding with reference to different similar and dissimilar metals. Microwave Welding (2009), a handbook on plastics joining, discusses the microwave welding process description, advantage and disadvantage of the process, and process application with other related information [2]. Gupta *et al.* [3] reviewed the joining of metal and ceramics in the field of microwave welding. Tiwary *et al.* [4] investigated the output responses of microwave-assisted welding with different input parameters. The authors commented that the selection of input parameters have a great impact on successful welded joint. Loharkar *et al.* [5] conducted research on microwave-based heating parameters and investigated the effect of output responses and also studied the factors responsible for these parameters to achieve an efficient material processing using microwave. The result revealed that the four input parameters, such as dielectric and magnetic properties, load and applicators, and heating mechanism, have a significant role in controlling the heating process using microwave-based energy. Yarlagadda *et al.* [6] carried out an investigation using microwave energy for welding of engineering thermoplastics. The results exposed that the improvement of welding duration and bond strength of the joint interface can be attained using primers in the form of epoxy-based resin to promote material joining by means of microwave energy. Sahota *et al.* [7] studied the basic characteristics of microwave heating and its application toward joining of metallic materials. It was found that better welding result may be achieved using the nano-sized interfacing powder. Tayier *et al.* [8] conducted a research on microwave hybrid heating to join dissimilar materials and also analyzed microwave hybrid heating essential parameters and its impact on joint status. Gupta *et al.* [9] performed an experimental study to join mild steel plates using microwave radiation with nickel paste as an interfacial material. The result revealed that mild steel substrates fused efficiently with the help of microwave hybrid heating. Khandelwal *et al.* [10] conducted an investigation to weld High-Density Polyethylene (HDPE) using microwave with the help of Multiwalled Carbon Nanotubes (MWCNTs) as microwave absorber. In the results, it was shown that HDPE can be weld using microwave with MWCNTs as a susceptor. Kathirgamanathan P. used Inherently Conducting Polymers (ICP) to perform an experimental analysis of MW of thermoplastics. In the results, it was shown that MW of thermoplastics can be done using ICPs [11]. Saxena A. conducted a study on the joining of mild steel plate using microwave welding. The author also reported on the microwave joining of the same and different combinations of materials. In the result, the author stated that metallurgical bonding has been achieved in mild steel–mild steel joining using microwave hybrid heating [12].

18.3 Experimental Analysis

In this analysis, MW of polymer-based materials was explored to examine the welded joint's lap shear strength and elongation. The materials considered were Acrylonitrile Butadiene Styrene (ABS) and Polylactic Acid (PLA). As explained by various researchers, the important input factors that govern the weld strength are exposure time, quality of absorbent, power level, and welding pressure. In this study, some of the important input factors were considered to analyze the output responses of MW. The factors are Material Combination (MC), Microwave Power (MP), Absorbent Weight (AW), Dead Weight (DW), and Infill Percentage (IP). Fe_2O_3 was considered as absorbent material. During the experiments, it was observed that between MP of 700–1,100 W, AW of 0.25–0.35 g, and DW of 500–1,300 g, the weld was successful and did not cause any damage. Depending on this, three levels with five types of input parameters were considered. Table 18.1 shows the input parameters with various levels of the present study. The welded joint lap shear strength was assessed following the ASTM D3163-01 standard. This standard technique using tension loading is helpful in obtaining or analyzing the plastic lap shear joint's welded strength. This test method is important for evaluating the comparison of data for welded joint shear strength made with different thermoplastics.

The investigational matrix was formed using Taguchi's Design of Experiments (DoE) with five factors, three levels each, with L27 trail approach. Further Analysis of Variance (ANOVA) was used for statistical optimization and parameter analysis.

Table 18.1 Input parameters and various levels.

Sl. no.	Factors	Process parameters	Level (−1)	Level (0)	Level (1)
1	A	MC	ABS/ABS	ABS/PLA	PLA/PLA
2	B	MP (W)	700	900	1,100
3	C	AW (g)	0.25	0.3	0.35
4	D	DW (g)	500	900	1300
5	E	IP (%)	25	40	55

Table 18.2 Input factors and measured output responses.

Exp. no.	MC	MP (W)	AW (g)	DW (g)	IP (%)	Lap shear strength (MPa)	Elongation (mm)
1	ABS/ABS	700	0.25	500	25	8.66	3.43
2	ABS/ABS	700	0.3	900	40	7.47	2.91
3	ABS/ABS	700	0.35	1,300	55	8.39	2.82
4	ABS/ABS	900	0.25	900	55	10.55	3.52
5	ABS/ABS	900	0.3	1,300	25	9.64	3.8
6	ABS/ABS	900	0.35	500	40	10.04	4.02
7	ABS/ABS	1,100	0.25	1,300	40	7.99	3.99
8	ABS/ABS	1,100	0.3	500	55	11.72	3.74
9	ABS/ABS	1,100	0.35	900	25	9.58	5.32
10	ABS/PLA	700	0.25	500	25	5.9	2.31
11	ABS/PLA	700	0.3	900	40	3.97	2.25
12	ABS/PLA	700	0.35	1,300	55	5.23	2.06
13	ABS/PLA	900	0.25	900	55	5.72	2.24
14	ABS/PLA	900	0.3	1,300	25	2.22	1.37
15	ABS/PLA	900	0.35	500	40	9	3.28
16	ABS/PLA	1,100	0.25	1,300	40	6.98	2.83
17	ABS/PLA	1,100	0.3	500	55	7.5	2.2
18	ABS/PLA	1,100	0.35	900	25	7.06	2.51
19	PLA/PLA	700	0.25	500	25	6.31	2.95
20	PLA/PLA	700	0.3	900	40	7.14	2.34
21	PLA/PLA	700	0.35	1,300	55	2.92	1.71
22	PLA/PLA	900	0.25	900	55	8.43	3.02
23	PLA/PLA	900	0.3	1,300	25	7.36	2.99
24	PLA/PLA	900	0.35	500	40	9.97	3.8
25	PLA/PLA	1,100	0.25	1,300	40	8.96	3.8
26	PLA/PLA	1,100	0.3	500	55	7.94	2.74
27	PLA/PLA	1,100	0.35	900	25	6.53	2.95

The two workpieces are kept in such a way that one is overlapping another. In between the overlapping area, absorbent material was placed. DWs are used to squeeze tightly both the workpieces with a thermocouple to monitor the temperature rise. Then a microwave applicator was used to place the whole preparation inside the applicator. After application of microwave for a specified duration, the workpieces were removed and allowed to cure. Further, it was processed for the calculation of lap shear strength and elongation. Table 18.2 shows the measured output responses of Lap shear strength (MPa) and Elongation (mm) for the different sets of experiments, and this table was considered as the DoE data table.

18.4 Effect of Input Process Parameters on the Lap Shear Strength and Elongation

After completion of successful MW of polymer-based workpieces, it was further examined statistically for their lap shear strength and elongation. The measured output responses of lap shear strength and elongation were analyzed using means and Signal-to-Noise (SN) ratio. Table 18.3 shows the response table for means of lap shear strength, which depicts the ranking of the parameters. Quality characteristics "larger is better" criterion was selected for both lap shear strength and elongation.

As exposed in Table 18.3, MC was found as the most crucial parameter for lap shear strength. Similarly, AW shows the lowest important parameter for lap shear strength. Figures 18.2 and 18.3 represent the effect of parameters on means and S/N ratios of lap shear strength, respectively.

After investigation of a microwave welded joint, it was found that a small number of workpieces of ABS/ABS showed fracture in ductile mode, while maximum of the workpieces consisting of all PLA/PLA and PLA/ABS showed fracture in brittle mode. Evidently, MC has a great impact on lap shear strength. The maximum lap shear strength with a value of 11.72 MPa (Exp. no. 8) was observed at ABS/ABS material combination. ABS showed a positive response to microwave irradiation causing good joint strength. Table 18.4 shows the response table for means of elongation, which depicts the ranking of the parameters.

Table 18.4 reveals that, like lap shear strength, MC was found as the utmost important parameter for elongation also. Similarly, AW shows the lowest important parameter for elongation. Figures 18.4 and 18.5 represent the effect of parameters on means and S/N ratios of elongation, respectively. The highest elongation was observed at ABS/ABS material combination,

Table 18.3 Response table for means.

Level	MC	MP	AW	DW	IP
1	9.338	6.221	7.722	8.560	7.029
2	5.953	8.103	7.218	7.383	7.947
3	7.284	8.251	7.636	6.632	7.600
Delta	3.384	2.030	0.504	1.928	0.918
Rank	1	2	5	3	4

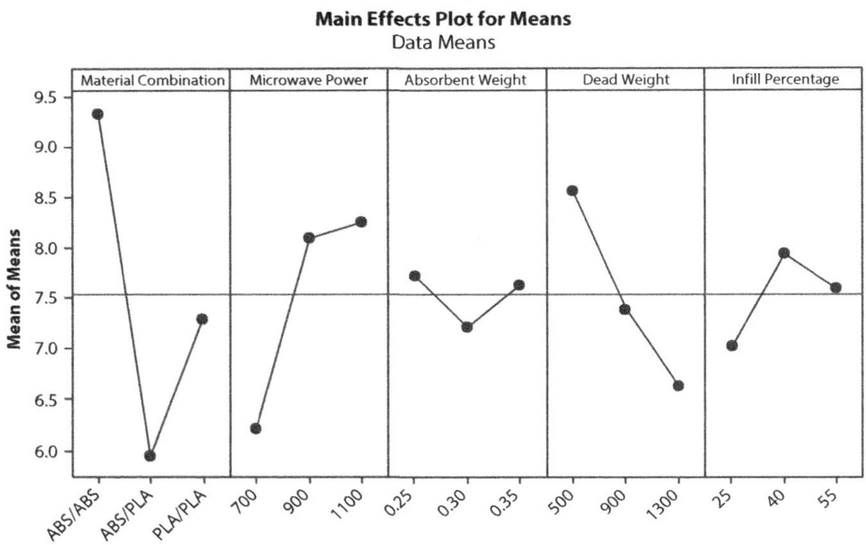

Figure 18.2 Effect of parameters on means of lap shear strength.

with a value of 5.32 mm (Exp. no. 9). It was noticed that the highest value of both lap shear strength and elongation were found at the same combination of materials, i.e., ABS/ABS.

18.4.1 Response Surface Methodology (RSM) Modeling for Lap Shear Strength and Elongation

Response surface methodology is an assemblage of both statistical and mathematical integrated approaches to examining and modelling of

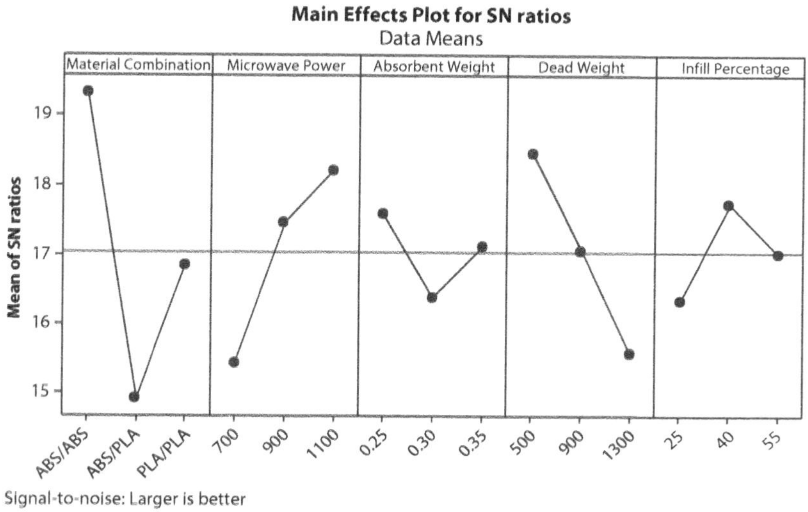

Figure 18.3 Effect of parameters on SN ratios of lap shear strength.

Table 18.4 Response table for means.

Level	MC	MP	AW	DW	IP
1	3.728	2.531	3.121	3.163	3.070
2	2.339	3.116	2.704	3.007	3.247
3	2.922	3.342	3.163	2.819	2.672
Delta	1.389	0.811	0.459	0.344	0.574
Rank	1	2	4	5	3

problems where response is obtained by several input process factors. The aim of RSM is to establish the correlation between the input and output parameters being investigated. To estimate an unidentified function, the DoE method is used. Commercially available software MINITAB 16 was used to perform the modeling of Lap Shear Strength (L_{ss}) and Elongation (E_l). Based on R^2, Pred (R^2), and Adj (R^2) values, the predictive model selection test was performed. R^2 value is the variation of output responses, and the higher the R^2 value means the model fits the data in a better way. To compare the model with numbers of predictors, the Adj (R^2) value is used. Similarly, Pred (R^2) value is used for the prediction of new observations by

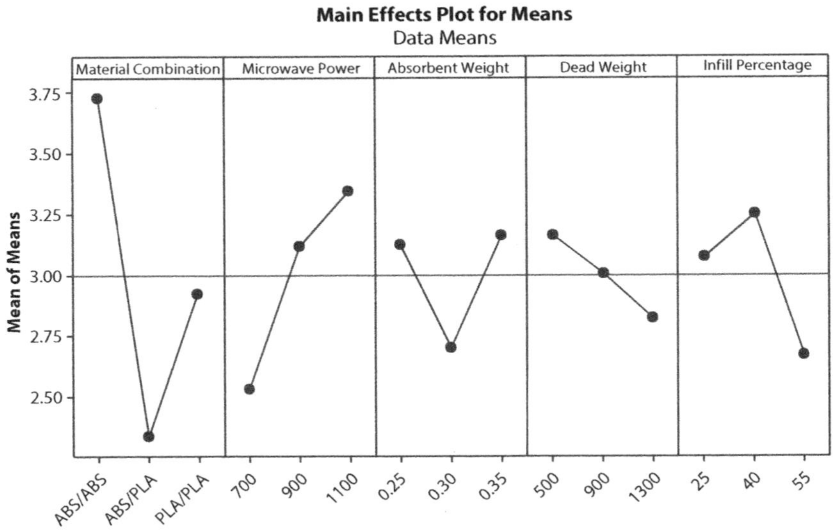

Figure 18.4 Effect of parameters on means of elongation.

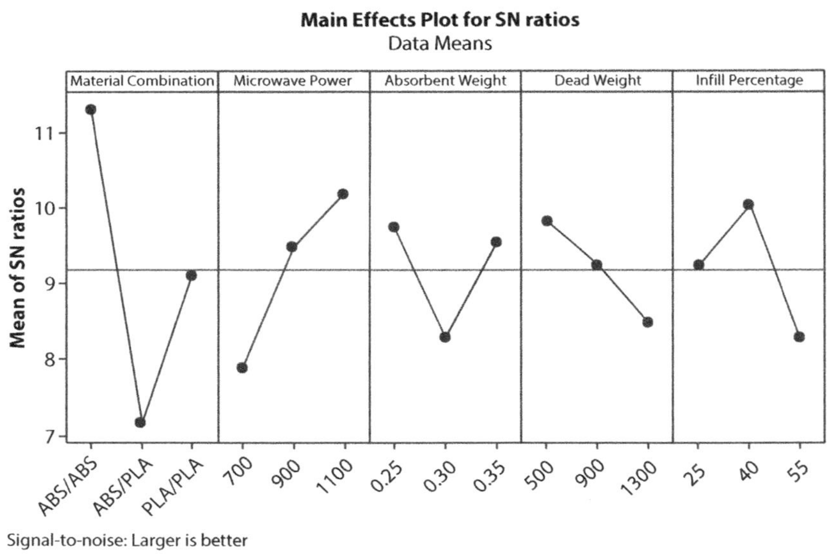

Figure 18.5 Effect of parameters on SN ratios of elongation.

the model. In this way, the model for lap shear strength and elongation has been selected, which is given by Equations 18.1 and 18.2, respectively, as shown below. A model of full quadratic was used considering all the possible interactions between the input process parameter terms.

$$\begin{aligned} L_{ss} &= -20.1577 - (5.88167A) + (0.0642542B) + (23.1000C) \\ &-(0.00670677D) + (0.181324E) + (1.86750A^2) - (3.39375\times10^{-5}B^2) \\ &-(11.6667C^2) + (2.86198\times10^{-6}D^2) - (0.00172037E^2) \\ &+(0.000954167AB) - (8.48333AC) - (1.20833\times10^{-4}AD) \\ &-(0.0205000AE) \end{aligned} \quad (18.1)$$

where "L_{SS}" denotes the lap shear strength. It contains A^2, B^2, C^2, D^2, E^2 and AB, AC, AD, AE as the square and interaction terms of the input process parameters, respectively.

$$\begin{aligned} E_l &= -39.5696 - (2.86185A) + (0.0583432B) + (158.066C) \\ &-(0.00723910D) - (0.0200484E) + (0.726154A^2) \\ &-(3.19231\times10^{-5}B^2) - (264.103C^2) + (4.41827\times10^{-6}D^2) \\ &+(0.000283974AB) - (0.747436AC) + (4.18269\times10^{-5}AD) \\ &-(0.00912393AE) \end{aligned} \quad (18.2)$$

where "E_l" denotes the elongation. It contains A^2, B^2, C^2, D^2 and AB, AC, AD, AE as the square and interaction terms of the input process parameters, respectively. These developed RSM models may be applied for predicting the lap shear strength and elongation for different possible combinations of input process parameters during MW. This will aid in minimizing the cost and time associated with actual experimentation and also helps the researchers in identifying suitable parameters and their significance in examining the suitability of the welding for different MCs.

18.4.2 Parametric Variation of Lap Shear Strength and Elongation

Figure 18.6 shows the different contour plots for lap shear strength for several combinations of MW parameters. From plots (a), (b), (c), and (d), it was seen that in case of ABS/ABS material combination, lap shear strength tends to increase with increase in MP, higher value of AW, lower value of DW, and higher value of IP, respectively. Similarly, Figure 18.7 shows the different contour plots for elongation for various combinations of MW parameters. From plots (a), (b), (c), and (d), it was seen that in case of ABS/ABS material combination, elongation tends to increase with medium value of MP, medium value of AW, lower value of DW, and lower value of IP, respectively.

18.4.3 ANOVA for Lap Shear Strength and Elongation

ANOVA is an effective way for analyzing how process variables affect responses. To evaluate the impact of each input factor on the output responses, we can obtain help using ANOVA. It also helps in identifying

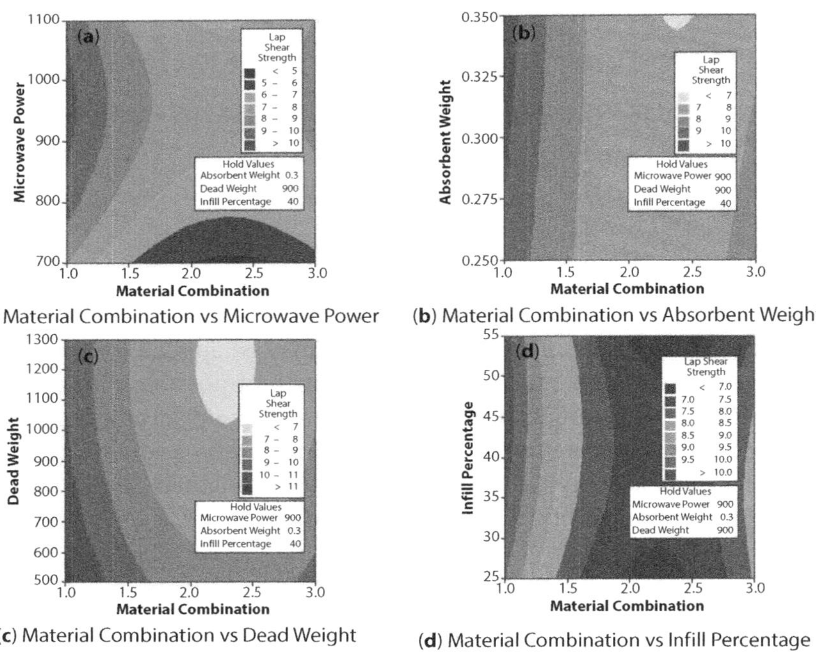

(a) Material Combination vs Microwave Power (b) Material Combination vs Absorbent Weight (c) Material Combination vs Dead Weight (d) Material Combination vs Infill Percentage

Figure 18.6 Contour plot for lap shear strength.

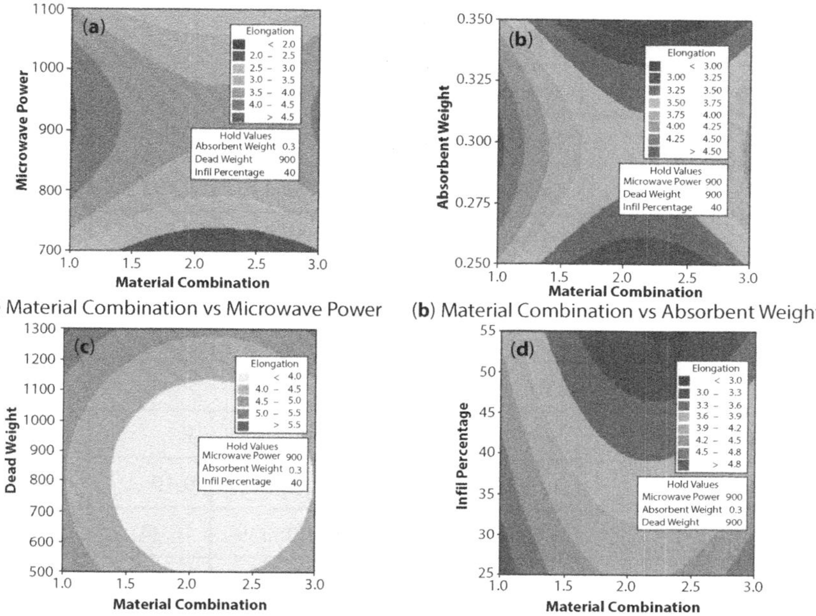

(**a**) Material Combination vs Microwave Power (**b**) Material Combination vs Absorbent Weight

(**c**) Material Combination vs Dead Weight (**d**) Material Combination vs Infill Percentage

Figure 18.7 Contour plot for elongation.

the factor interaction and estimates the significant factors. In the current study relating to both lap shear strength and elongation, ANOVA along with multi-objective optimization was used.

From Table 18.5, it was revealed that MC and MP for lap shear strength were the most substantial input parameters (F = 11.17, 10.92), followed by DW (F = 5.00). AW and IP were seen as the less significant parameters. On the other hand, IP and MP for elongation were the most substantial input parameters (F = 19.45, 16.44), followed by MC and DW (F = 9.10, 5.70). AW was found as the less significant parameter during elongation.

18.5 Multi Objective Optimization Using Desirability Analysis

To estimate the composite desirability value, the desirability analysis approach was used. Normally, composite desirability value ranges between 0 and 1. Higher composite desirability value leads to conforming to the

Table 18.5 ANOVA for both lap shear strength and elongation.

ANOVA of linear terms for lap shear strength						
Source	DF	Seq SS	Adj SS	Adj MS	F	P
A	1	18.973	18.9728	18.9728	11.17	0.007
B	1	18.544	18.5440	18.5440	10.92	0.007
C	1	0.034	0.0338	0.0338	0.02	0.890
D	1	8.266	8.5018	8.5018	5.00	0.047
E	1	0.007	0.0269	0.0269	0.02	0.902
ANOVA of linear terms for elongation						
Source	DF	Seq SS	Adj SS	Adj MS	F	P
A	1	1.09739	0.69456	0.69456	9.10	0.015
B	1	1.59782	1.25539	1.25539	16.44	0.003
C	1	0.11372	0.13347	0.13347	1.75	0.219
D	1	0.39440	0.43552	0.43552	5.70	0.041
E	1	0.27730	1.48494	1.48494	19.45	0.002

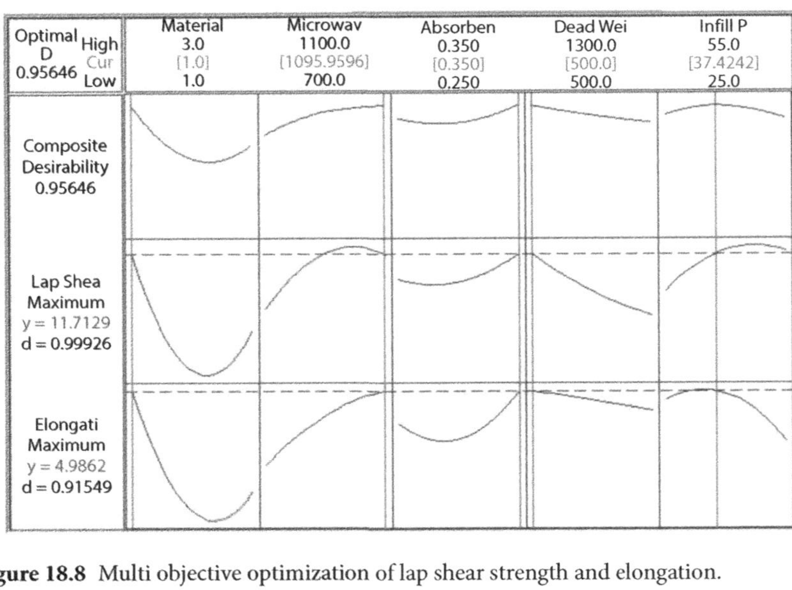

Figure 18.8 Multi objective optimization of lap shear strength and elongation.

selected optimization approach. The lap shear strength as well as elongation plot of optimization using multi objective are depicted in Figure 18.8. In multi objective optimization, the maximum lap shear strength and elongation were found to be 11.712 MPa and 4.986 mm, respectively, obtained at a composite desirability value of 0.956. The input process parameter combination was found to be MC: ABS/ABS, MP: 1095.96 W, AW: 0.35 g, DW: 500 g, and IP: 37.424%.

18.6 Conclusion

In the present study, it was found that selection of polymer-based material has a great impact on the MW technique for better output responses. In this MW investigation, ABS is responding well compared to PLA. From the analysis, it was revealed that MC was found to be the most significant parameter for lap shear strength and elongation followed by MP, respectively. From the ANOVA analysis of lap shear strength, it was found that MC and MP were the most vital input parameters followed by DW. Similarly, ANOVA analysis for elongation IP and MP were found the greatest impactable input parameters followed by MC and DW. Further simultaneous optimization of lap shear strength and elongation was found at a combination of MC: ABS/ABS, MP: 1,095.96 W, AW: 0.35 g, DW: 500 g, and IP: 37.424% at a composite desirability value of 0.956. The maximum lap shear strength and elongation predicted were found to be 11.712 MPa and 4.986 mm, respectively.

References

1. Nandwani, S., Vardhan, S., Bagha, A.K., A literature review on the exposure time of microwave based welding of different materials. *Mater. Today Proc.*, 27, 3, 2526–2528, 2020. https://doi.org/10.1016/j.matpr.2019.10.056.
2. Microwave Welding. 135–138, 2009. https://doi.org/10.1016/B978-0-8155-1581-4.50017-2.
3. Gupta, P., Kumar, S., Kumar, A., A microwave welding process. *Int. Conf. Advancements Futuristics Trends Mech. Mater. Eng.*, 2012.
4. Tiwary, V.K., Arunkumar, P., Malik, V.R., Investigations on microwave-assisted welding of MEX additive manufactured parts to overcome the bed size limitation. *J. Adv. Joining Processes*, 7, 100141, 2023. https://doi.org/10.1016/j.jajp.2023.100141.

5. Loharkar, P.K., Ingle, A., Jhavar, S., Parametric review of microwave-based materials processing and its applications. *J. Mater. Res. Tech.*, 8, 3306–3326, 2019. https://doi.org/10.1016/j.jmrt.2019.04.004.
6. Yarlagadda, P.K.D.V. and Chai, T.C., An investigation into welding of engineering thermoplastics using focused microwave energy. *J. Mater. Process. Tech.*, 74, 1-3, 199–212, 1998. https://doi.org/10.1016/S0924-0136(97)00269-0.
7. Sahota, D.S., Bansal, A., Kumar, V., Application of microwave in welding of metallic materials – A review. *Mater. Today Proc.*, 43, 1, 466–470, 2021. https://doi.org/10.1016/j.matpr.2020.11.997.
8. Tayier, W., Janasekaran, S., Tai, V.C., Microwave hybrid heating (MHH) of Ni-based alloy powder on Ni and steel-based metals - A review on fundamentals and parameters. *Int. J. Lightweight Mater. Manuf.*, 5, 1, 58–73, 2022. https://doi.org/10.1016/j.ijlmm.2021.10.002.
9. Gupta, P., Kumar, V., Chaudhary, S., Kumar, S., Singh, J.K., Chauhan, P.K., Investigation and characterization of microwave processed joint of MS1020. *Mater. Today Proc.*, 47, 13, 3877–3883, 2021. https://doi.org/10.1016/j.matpr.2021.03.523.
10. Khandelwal, P. and Gupta, V., Microwave joining of polymers. *Int. J. Sci. Res.*, 11, 7, 2319–7064, 2022. (IJSR) ISSN.
11. Kathirgamanathan, P., Microwave welding of thermoplastics using inherently conducting polymers. *POLYMER*, 34, 14, 3106, 1993.
12. Saxena, A., Review of microwave oven welding. *Proc. of 3rd IRF Int. Conf*, 2014, ISBN: 978-93-82702-62-7.

19

Metaheuristic Approaches for the Optimized Friction Stir Welding Process

Pritam Pain* **and Supriyo Roy**

Department of Mechanical Engineering, Haldia Institute of Technology, Haldia, India

Abstract

Friction stir welding (FSW) has emerged as a pivotal solid-state joining technique, especially for aluminum alloys, due to its ability to produce high-quality welds without the common drawbacks of traditional welding methods. However, optimizing the FSW process parameters to achieve superior weld quality, mechanical properties, and energy efficiency remains a complex challenge. This research focuses on employing metaheuristic approaches to optimize key parameters of the FSW, including speed of the rotation, speed of the welding, and the applied force on the axial. The study investigates the effects of several operating factors on the final tensile strength and tensile elongation of the welded joints, including pin shape, axial force, rotating speed, and welding speed. The traditional Taguchi technique was used for single-objective optimization, and the PSO algorithm was used for multi-objective optimization. The findings show that by determining parameter sets that optimize mechanical parameters, like tensile strength and elongation, PSO optimization greatly improves the FSW process. The significance of the interactions among the parameters was also studied and reported using surface plots.

***Keywords*:** FSW, aluminum alloys, ANOVA, PSO, UTS, tensile elongation

19.1 Introduction

The 2XXX, 6XXX, and 7XXX grades of age-hardened aluminum alloy are challenging to join using regular welding techniques due to their sensitivity to cracking, comparatively high strength with respect to the weight

**Corresponding author*: pritam.me.dscsdec@gmail.com

Sandip Kunar and Gurudas Mandal (eds.) Advanced Welding Technologies, (385–400) © 2025 Scrivener Publishing LLC

of the material, etc. In the production of age-hardened aluminum alloys, particularly for the aerospace and automotive industries, friction stir welding (FSW) has emerged as the practical as well as feasible alternative to conventional welding techniques. Since the FSW method eliminates the need for melting of the parent material and followed by solidification at the same time, joining the weld zone of similar as well as dissimilar materials is accomplished by varying the recrystallization by means of a spinning tool, which allows for great plastic buckles [1]. Even though friction stir welding has many advantages, there are still some flaws that can occur, like an inadequate heat that marks in inadequate material flow, defective stimulated zone during welding, less probe infiltration, residual stress inside the base metal, air gaps on the proceeding side, and rough material deposition pressure along the thickness of the parent material. Age-hardened aluminum alloys typically had to meet stringent requirements for tensile elongations (TE), tensile strengths, and fatigue performance.

A small literature survey has been done on the FSW. Kilic *et al.* [2] has revied different process parameters of FSW. Bhardwaj *et al.* [3] studied the recent development of FSW and discusses the future developments scopes. Beygi *et al.* [4] used FSW to join 5-mm-thick aluminum metal with 3-mm carbon steel metal and observed that the intermetallic compound layer delayed the crack. Bevilacqua *et al.* [5] has optimized the FSW process considering ultimate elongation and ultimate tensile stress (UTS) as the output parameters while joining AA5754 aluminum metal. Abdollahzadeh *et al.* [6] studied different types of FSW, namely, underwater FSW, vibration-assisted FSW, and conventional FSW. In their study, it shows that underwater FSW provides better microstructure of the welded materials. In the study of Abnar *et al.* [7], they used different process parameters, like tool tilt angle and plunge depth, to optimize the FSW process. Finally, regarding strain-hardened Al alloys, a few more procedures had to be taken into consideration, including surface modification, bobbin tool FSW, copper backing, and double-sided FSW. In the work of Adesina *et al.* [8], H13 FSW tools coated in AlCrN and TiAlN were used to butt weld aluminum alloy 2124, which has a thickness of 8 mm. The application of these coatings did not degrade the weld properties; rather, an improvement in the nugget zone's hardness was observed, due to the unique dynamic stirring plus refinement of the Al matrix as well as SiC particles. Das *et al.* [9] studied the surface morphology of aluminum material while machining in a CNC machine. Ahmed *et al.* [10] studied the microstructure of the mild steel material while joining using FSW process. Aonuma *et al.* [11] joined two different materials, namely, Mg–Zn–Zr with titanium alloy using FSW process and found out that the thin layer of Zn and Zr has been created at the top layer, which has a positive effect on the tensile strength of the welded material. In a few research work, it has been found that the Taguchi mythology has successfully found the best process parameters while considering the single objective function at a time [12, 13]. Sarvaiya and Singh [14] used the particle swarm optimization (PSO) to optimize the FSW process. Quarto *et al.* [15] used ANN along with PSO to optimize the FSW process and reduced the error in the welding by 5%.

Observations have shown that by choosing the right control settings, the faults of FSW could be removed. Within this study, two age-hardened aluminum alloys, AA2024-T3 as well as AA6061-T6, are taken into consideration while choosing a suitable FSW process parameter [16].

19.2 Experimental Setup

The different AA2024 and AA6061 aluminum alloys' weld connections are created using a computerized friction stir welding (FSW) process. The aluminum alloy plates AA2024 and AA606 on the FSW machine during a welding operation. Figure 19.1 displays the FSW schematic diagram. The AA2024 alloy underwent cold working, solution heat treatment, and natural aging to make it a stable condition. The compound material underwent solution heat treatment to enhance its mechanical properties, and as it aged at room temperature, its mechanical characteristics stabilized. The AA6061 alloys is solution–heat-treated and artificially aged. Table 19.1 shows the chemical elements of each alloy. The butt joints were fabricated using aluminum alloy plates measuring 150 × 75 × 5 mm^3. With respect to the longitudinal plate surface, the tool rotates perpendicularly. Based on a thorough review of the literature, the process parameters, including pin

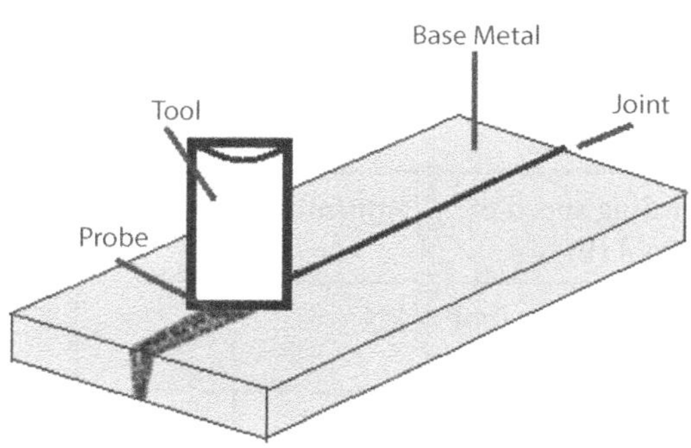

Figure 19.1 Diagrammatic representation of friction stir welding.

Table 19.1 The AA2024 and AA6061 aluminum alloys' chemical structure.

Material	Mn	Mg	Si	Fe	Cu	Cr	Zn	Ti
AA2024	0.47	1.50	0.10	0.11	4.40	0.01	0.14	0.04
AA6061	0.14	0.90	0.74	0.40	0.22	0.18	0.09	0.05

shape (PS), welding speed (F), axial force (P), and tool rotating speed (N), were chosen. Taper (tap), square (squ), and straight cylinder (cyl) were the three main pin shapes that were evaluated in this research work.

The process control parameters along with their levels are shown in Table 19.2.

After nine successful experiments, these following regression equations were formulated:

$$\text{Ultimate Tensile Strength (in Mpa)} = 112 + 0.010 *N+ 0.48 *F- 1.7 *P - 3.3 *PS- 0.00039 *N*F+ 0.0017 *N*P- 0.0019 *N*PS- 0.0078 *F*P + 0.075 *F*PS+ 0.15 *P*PS \tag{19.1}$$

$$\text{Tensile Elongation (l in \%)} = 19.2 - 0.00331 *N+ 0.024 *F - 1.30 *P+ 2.05 *PS- 0.000011 *N*F+ 0.000529 *N*P- 0.00159 *N*PS + 0.00069 *F*P- 0.0002 *F*PS + 0.048 *P*PS \tag{19.2}$$

Both of these equations are subjected to maximization problem.

Table 19.2 Process control parameter of friction stir welding.

Sl. no.	Control parameters	Unit	Different levels		
			Level 1	Level 2	Level 3
1	Rotational speed of the tool (N)	rpm	1,500	1,700	1,900
2	Forwarding speed of the tool (F)	mm/min	0.5	1	1.5
3	Axial load on the tool (P)	kN	3	6	9
4	End shapes of the tool (PS)		1 (Tap)	2 (Squ)	3 (cyl)

19.3 Taguchi Analysis

A statistical technique called Taguchi analysis tries to reduce variability in products through strong design. It studies several components and their interactions methodically and effectively using orthogonal arrays. The technique prioritizes the signal-to-noise (S/N) ratio to improve performance constancy in the face of unpredictable fluctuations. By measuring the cost of deviating from objective performance, Taguchi's loss function encourages attention to be paid to minimizing loss to society. In general, Taguchi analysis aims to produce excellent results with little need for experimentation [17].

19.3.1 Result Calculation Using ANOVA

An analysis of variance, or ANOVA, is a geometric method used to compare three or more groups' means and determine whether there are any significant differences between them. By comparing the variance within groups to the variance between groups, it evaluates the influence of one or more factors. The F-statistic, which is essential for determining if group means differ significantly, is the main result of an ANOVA. The group means differ significantly if the P-value linked to the F-statistic is less than a predetermined cutoff (usually 0.05). ANOVA makes the assumption that the variances are the same for all groups and that the data has a normal distribution [18].

19.3.1.1 Analysis to Enhance σ

The ultimate tensile strength ANOVA table is displayed in Table 19.3. The table clearly shows that, in the case of FSW, pin shapes are the most influential parameter for σ, which is followed by axial load. Tool forwarding speed is the most insignificant parameter in this case.

19.3.1.2 Analysis to Maximize δl Using ANOVA

The tensile elongation ANOVA table is displayed in Table 19.4. The table makes it clear that axial load is the most effecting parameter for δl in case of FSW, which the tool's rotating speed follows. Tool forwarding speed is the most insignificant parameter in this case.

Table 19.3 ANOVA for maximum ultimate tensile strength.

Source	DF	Adj SS	Adj MS	F value	P-value
Tool rotational speed (N)	2	52.83	26.41	0.85	13.00
Tool forwarding speed (F)	2	3.42	1.71	0.06	0.84
Axial load (P)	2	130.11	65.06	2.09	32.03
Pin shapes (PS)	2	219.87	109.94	3.54	54.12
Error	0				
Total	8	406.24			
Pooled error	6	186.36	31.06	6.54	

Table 19.4 ANOVA for maximum tensile elongation.

Source	DF	Adj SS	Adj MS	F value	P-value
Tool rotational speed (N)	2	3.28	1.64	1.82	37.05
Tool forwarding speed (F)	2	0.43	0.21	0.24	4.82
Axial load (P)	2	3.45	1.72	1.92	38.98
Pin shapes (PS)	2	1.69	0.85	0.94	19.15
Error	0				
Total	8	8.84			
Pooled error	6	5.40	0.90	4.92	

19.3.2 Result Calculation Using S/N Ratio Using ANOVA

In science and engineering, the signal-to-noise ratio, or S/N ratio, is a metric used to express the degree of signal distortion caused by noise. It is stated as the ideal signal-to-background noise power ratio, frequently represented in decibels (dB). A cleaner, clearer signal is indicated by a higher S/N ratio, whereas a more corrupted signal is indicated by a lower S/N ratio.

19.3.2.1 Analysis to Maximize σ Using S/N Ratio

From Figure 19.2, it is evident that the maximum UTS can be achieved, while N is 1,500 rpm, F is 60 mm/min, P is 9 kN, and the tool shape is taper. A square tool provides the highest UTS.

19.3.2.2 Analysis to Maximize δl Using S/N Ratio

From Figure 19.3, it is evident that the maximum TE can be achieved, while N is 1,500 rpm, F is 90 mm/min, P is 3 kN, and the tool shape is taper. When axial load increases, tensile elongation reduces gradually.

19.4 ANN (Artificial Neural Network)

Artificial Neural Networks (ANNs) are computational models inspired by the human brain, used for tasks like classification and regression. In MATLAB R2015a, ANNs are implemented using the Neural Network Toolbox. An ANN consists of neurons organized in layers: input, hidden, and output. The input layer receives data, hidden layers process it, and the output layer produces the final result. To create an ANN in MATLAB, define the network structure by specifying the number of neurons in each

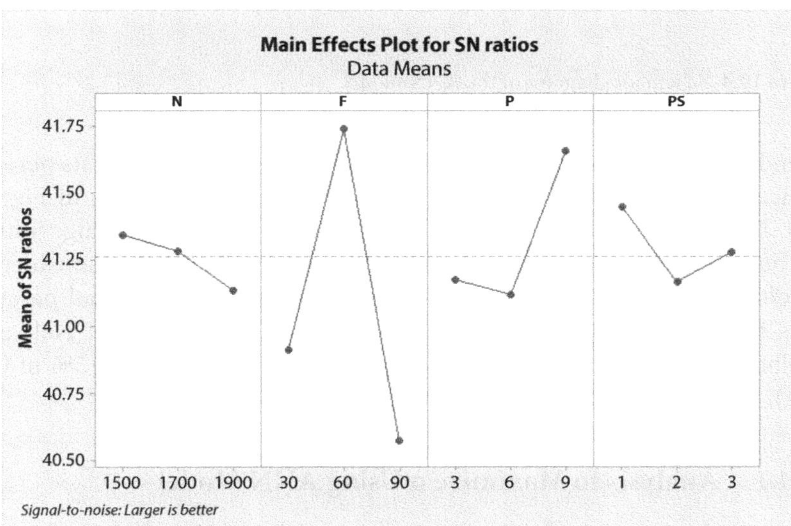

Figure 19.2 S/N ratio to maximize ultimate tensile strength.

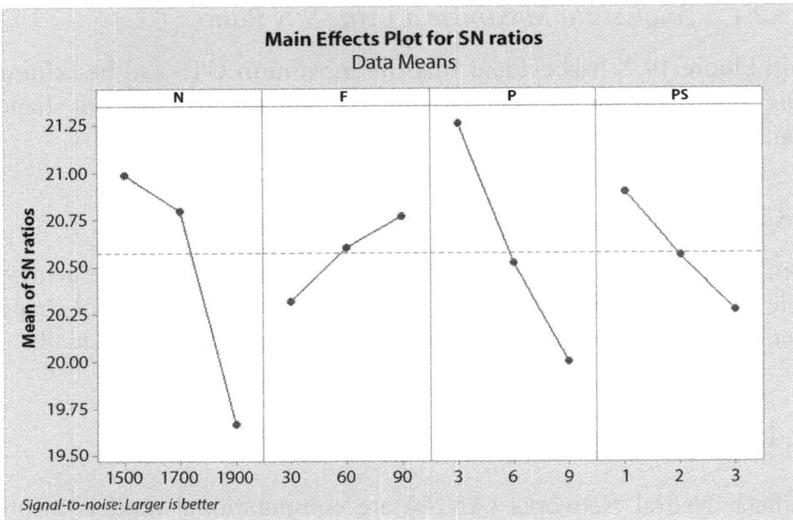

Figure 19.3 S/N ratio to maximize tensile elongation.

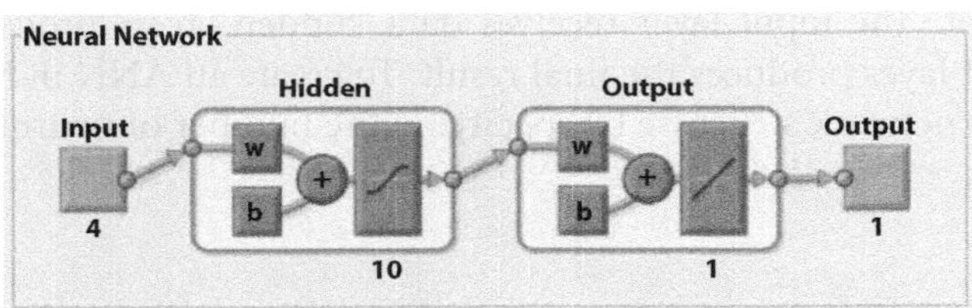

Figure 19.4 Schematic diagram of the ANN model.

layer. Initialize the network, train it using a dataset, and validate its performance using test data. The schematic model of ANN is shown in Figure 19.4. In this analysis "Levenberg–Marquardt backpropagation algorithm" has been used. In this study, four input parameters have been considered simultaneously, and in the case of output parameters, individual parameters have been considered one at a time. MATLAB's Neural Network Toolbox facilitates the training (70% of the data), validation (15% of the data), and testing (15% of the data) of the ANN model.

19.4.1 Analysis to Maximize σ Using ANN Model

Figure 19.5 represents the performance model of ANN analysis for UT. By iteratively refining the input parameters, the model helps identify the optimal settings for maximizing weld strength. Figure 19.6 represents the regression fit model for ultimate tensile strength where it is evident that for training, validation, testing, and also overall case, the regression value is more than 81%. It means the experiment model is good.

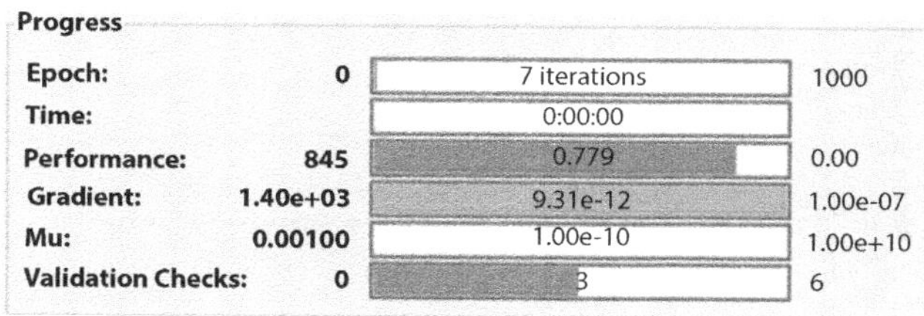

Figure 19.5 Performance model for σ.

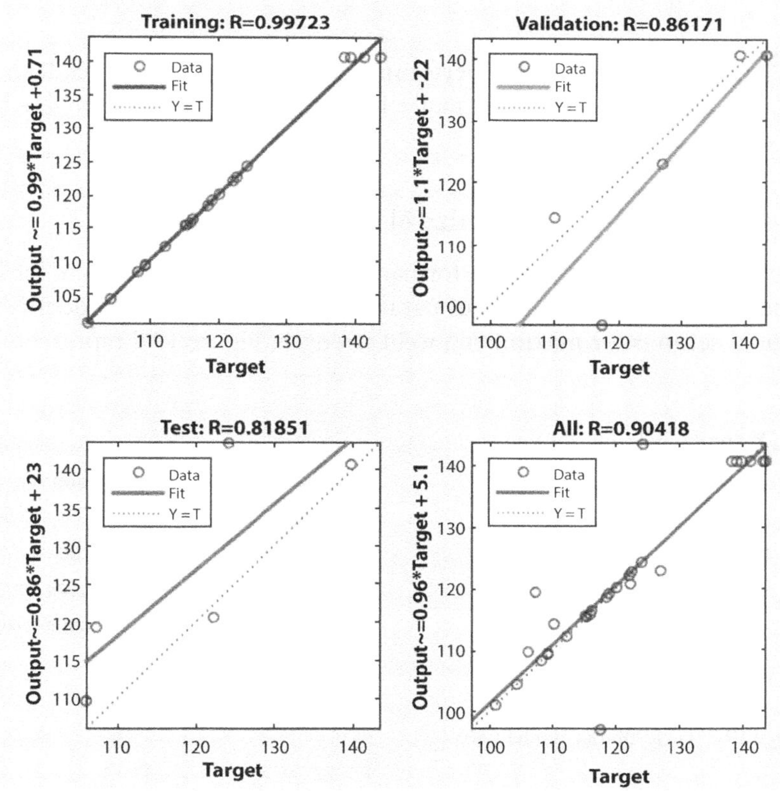

Figure 19.6 Regression model for σ.

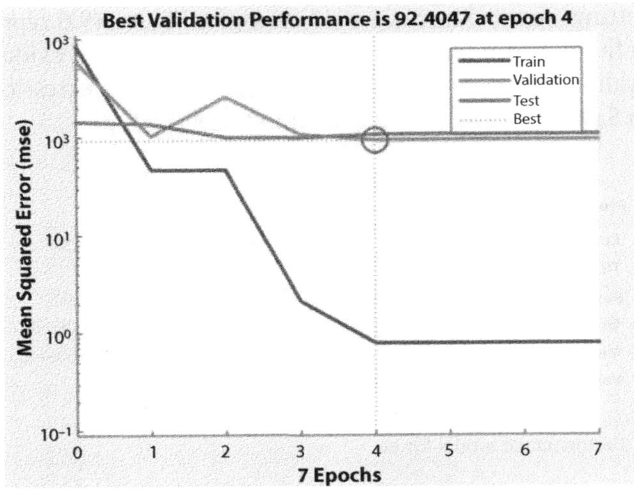

Figure 19.7 Performance plot for σ.

Figure 19.7 represents the performance plot for the UT model. It shows that the best validation performance has been observed at epoch 4, which has a value of 92.4047.

19.4.2 Analysis to Maximize δl Using ANN Model

Figure 19.8 represents the performance model of ANN analysis for UT. By iteratively refining the input parameters, the model helps identify the optimal settings for maximizing weld strength. Figure 19.9 represents the

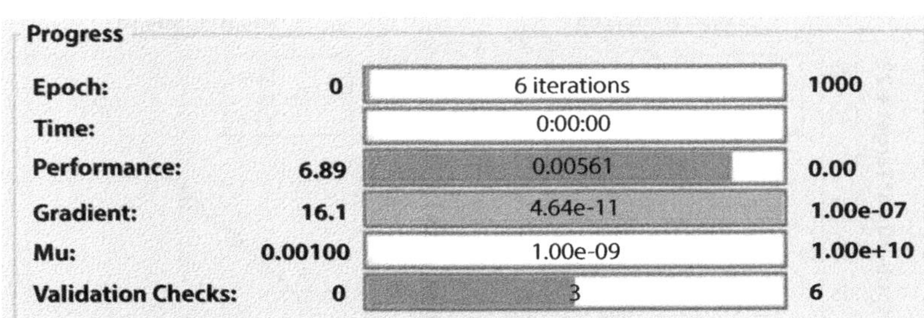

Figure 19.8 Performance model for δl.

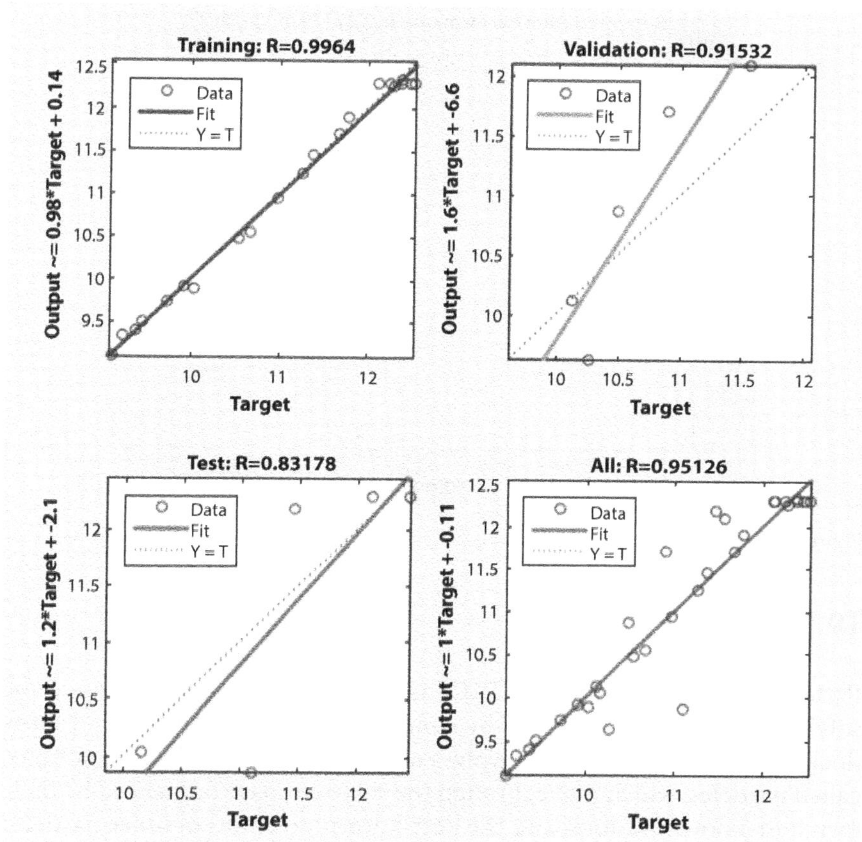

Figure 19.9 Regression model for δl.

regression fit model for ultimate tensile strength where it is evident that for training, validation, testing, and also overall case, the regression value is more than 83%. It means the experiment model is good.

Figure 19.10 represents the performance plot for the UT model. It shows that the best validation performance has been observed at epoch 3, which has a value of 0.29174.

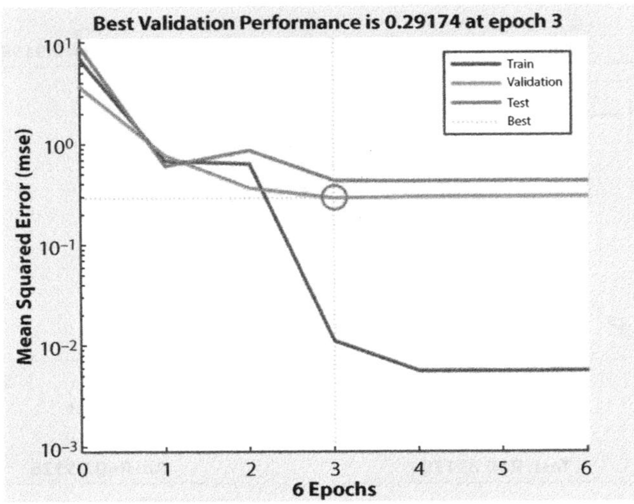

Figure 19.10 Performance plot for δl.

19.5 PSO

Particle Swarm Optimization (PSO) is a computational method used for solving optimization problems by simulating the social behavior of birds flocking or fish schooling. It involves a population of candidate solutions, called particles, which move through the solution space influenced by their own best-known position and the best-known positions of other particles [19]. Each particle adjusts its trajectory based on a combination of its current velocity, its best-known position, and the global best-known position. PSO is widely used in various fields due to its simplicity and ability to converge on optimal or near-optimal solutions efficiently. Its applications range from machine learning and artificial intelligence to engineering and economics [20].

19.5.1 Multi Objective Optimization of UTS and TE

In this particular study, both the responses, i.e., ultimate tensile strength and tensile elongation are considered better for their maximum value. Thus, this can be considered as maximization problem and the corresponding multi-objective function (MOF, shown in Equation 19.3) can be expressed as a summation of Equations 19.1 and 19.2.

MOF = Ultimate Tensile Strength (σ in Mpa) + Tensile Elongation (δl in %) (19.3)

This maximization problem is solved using PSO algorithm coded in MATLAB. The optimum combination for maximum value of multi objective function is rotational speed of 1,900 rpm, welding speed of 30 mm/min, axial force of 9 kN, and taper-shaped tool. The corresponding function value is 141.52. The improvement of multi objective function with number of iterations is shown in Figure 19.11.

However, the significance of each parameter on the multi objective function is shown Figure 19.12. During obtaining the surface plots of any two parameters, the other two parameter values were kept constant at its middle value.

Here, it is evident that the MOF value is maximum [Figure 19.12 (a)] at two conditions, higher rotational speed of the tool with lowest speed of welding and lowest speed of rotation with highest welding speed when the other two parameters were kept constant at their middle values. Figure 19.12 (b) depicts that the MOF value is maximum at the highest axial load with lowest welding speed while keeping the other two parameters constant. Figure 19.12 (c) also depicts the same trend between axial load and rotational speed. Figures 19.12 (d) to (f) illustrate that the maximum value of MOF can be achieved using taper tool.

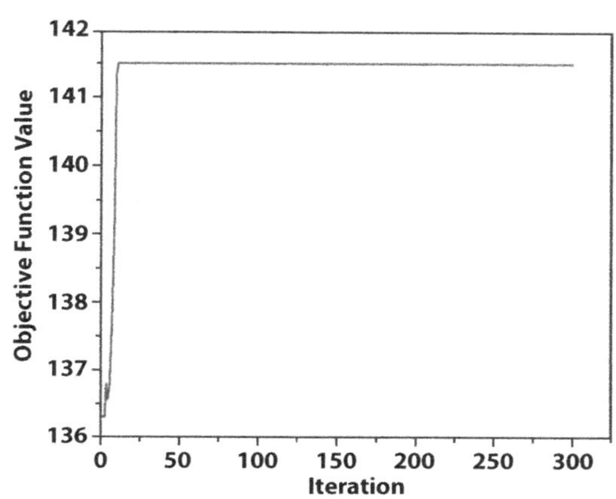

Figure 19.11 Improvement of multi objective function with number of iterations.

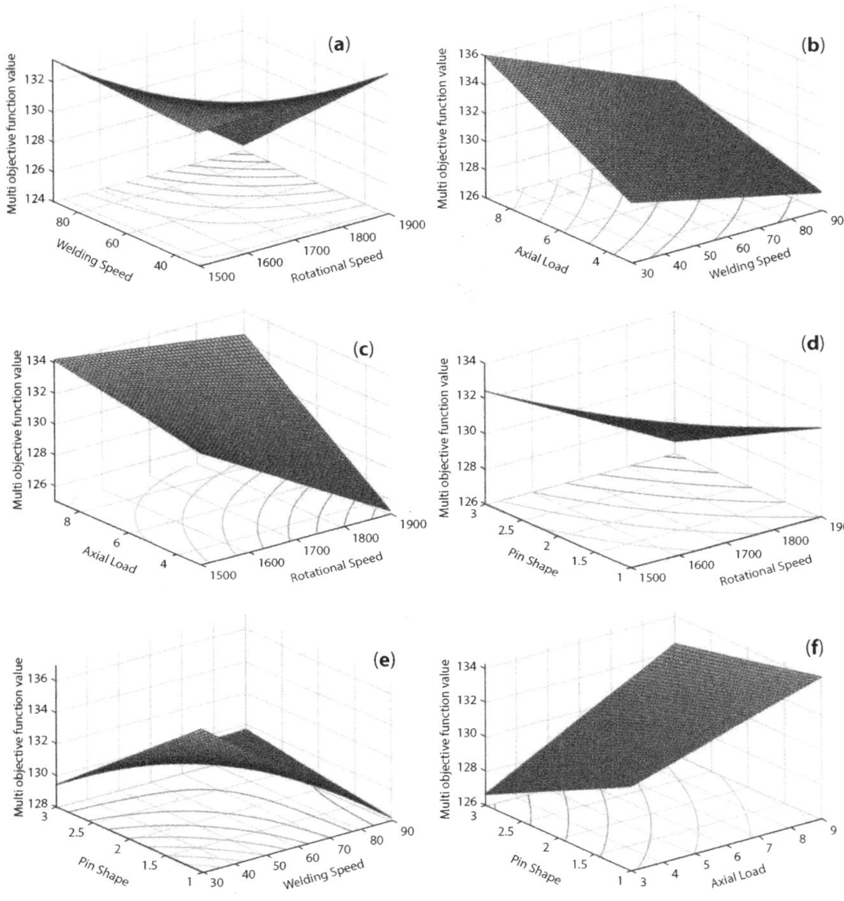

Figure 19.12 Surface plots showing the significance of parameters on the multi objective function value.

19.6 Conclusion

In this research work, adequately, the process control parameters for joining two dissimilar materials, namely, AA2024 and AA6061, has been discussed while joining using FSW process. It is observed from the ANOVA that the most effecting parameters for UTS is pin shapes (PS), and in case of TE, the most effective parameter is axial load (P). From the S/N ratio calculation, it is observed that the square tool provides the highest UTS. Also, from the same analysis, it is evident that when axial load increases, tensile elongation reduces gradually.

References

1. Mohan, D.G. and Wu, C., A Review on Friction Stir Welding of Steels. *Chin. J. Mech. Eng.*, 34, 1–29, 2021.
2. Kilic, S., Ozturk, F., Demirdogen, M.F., A comprehensive literature review on friction stir welding: Process parameters, joint integrity, and mechanical properties. *J. Eng. Res.*, 2023.
3. Bhardwaj, N., Narayanan, R.G., Dixit, U.S., Hashmi, M.S.J., Recent developments in friction stir welding and resulting industrial practices. *Adv. Mat. Proc. Tech.*, 5, 461–496, 2019.
4. Beygi, R., Mehrizi, M.Z., Akhavan-Safar, A., Safaei, S., Loureiro, A., da Silva, L.F.M., Design of friction stir welding for butt joining of aluminum to steel of dissimilar thickness: heat treatment and fracture behavior. *Int. J. Adv. Manu. Tech.*, 112, 1951–1964, 2021.
5. Bevilacqua, M., Ciarapica, F.E., D'Orazio, A., Forcellese, A., Simoncini, M., Sustainability Analysis of Friction Stir Welding of AA5754 Sheets. *Proc. CIRP*, 62, 529–534, 2017.
6. Abdollahzadeh, A., Bagheri, B., Abassi, M., Kokabi, A.H., Moghaddam, A.O., Comparison of the Weldability of AA6061-T6 Joint under Different Friction Stir Welding Conditions. *J. Mat. Eng. Perf.*, 30, 1110–1127, 2021.
7. Abnar, B., Gashtiazar, S., Javidani, M., Friction Stir Welding of Non-Heat Treatable Al Alloys: Challenges and Improvements Opportunities. *Crystals*, 13, 576, 2023.
8. Adesina, A.Y., Al-Badour, F.A., Gasem, Z.M., Wear resistance performance of AlCrN and TiAlN coated H13 tools during friction stir welding of A2124/SiC composite. *J. Man. Pro.*, 33, 111–125, 2018.
9. Das, J., Laya, D., Basu, S., Roy, S., Modelling and analysis of roughness characteristics of aluminium alloy under CNC face milling operation. *Int. J. Mach. Mach. Mat.*, 20, 299, 2018.
10. Ahmed, M.M.Z., El-Sayed Seleman, M.M., Touileb, K., Albaijan, I., Habba, M.I.A., Microstructure, Crystallographic Texture, and Mechanical Properties of Friction Stir Welded Mild Steel for Shipbuilding Applications. *Mat.*, 15, 2905, 2022.
11. Aonuma, M. and Nakata, K., Dissimilar metal joining of ZK60 magnesium alloy and titanium by friction stir welding. *Mat. Sci. Eng. B*, 177, 543–548, 2012.
12. Bayazid, S.M., Farhangi, H., Ghahramani, A., Investigation of Friction Stir Welding Parameters of 6063-7075 Aluminum Alloys by Taguchi Method. *Proc. Mat. Sci.*, 11, 6–11, 2015.
13. Anawa, E.M. and Olabi, A.G., Using Taguchi method to optimize welding pool of dissimilar laser-welded components. *Optics. Las. Tech.*, 40, 379–388, 2008.

14. Sarvaiya, J. and Singh, D., Selection of the optimal process parameters in friction stir welding/processing using particle swarm optimization algorithm. *Mat. Tod. Proc.*, 62, 896–901, 2022.
15. Quarto, M., Bocchi, S., D'Urso, G., Giardini, C., Multiresponse optimization of friction stir welding by an integrated ANN-PSO approach. *Proc. Inst. Mech. Eng. Part B J. Eng. Man.*, 237, 1858–1868, 2022.
16. Vijayan, V.S.R.D., Optimization of friction stir welding process parameters using RSM based Grey–Fuzzy approach. *Sau. J. Eng. Tech.*, 2, 12–25, 2017.
17. Sivam, S.P.S.S., Saravanan, K., Pradeep, N., Moorthy, K.S., Rajendrakumar, S., Grey Relational Analysis and Anova to Determine the Optimum Process Parameters for Friction Stir Welding of Ti and Mg Alloys. *Per. Poly. Mech. Eng.*, 62, 277–283, 2018.
18. Pain, P., Bose, G.K., Bose, D., Parametric analysis and optimization of aluminium and SS 204 material using micro-EDM system. *Int. J. Int. Des. Man.*, 17, 3025–3042, 2023.
19. Pain, P. and Bose, G.K. (Eds.), *Machine Learning Applications in Non-Conventional Machining Processes*, IGI Global, USA, 2021.
20. Hussien, R.M. and Al-Shammari, M.A., Optimization of Friction Stir Welded Aluminium Plates by the New Modified Particle Swarm Optimization. *IOP Conf. Ser. Mat. Sci. Eng.*, 1094, 012156, 2021.

20

Laser–Arc Hybrid Welding

Ravindra Nath Yadav[1]*, Basant Kumar Bhuyan[2], Sanjeev Kumar Singh Yadav[3] and Sanjay Mishra[4]

[1]*Mechanical Engineering Department, Bakhtiyarpur College of Engineering, Bakhtiyarpur, (Under Science, Technology and Technical Education Department, Bihar), Patna, India*
[2]*Mechanical Engineering Department, Manav Rachana International Institute of Research and Studies, Faridabad, India*
[3]*Mechanical Engineering Department, Harcourt Butler Technical University, Kanpur, India*
[4]*Mechanical Engineering Department, Madan Mohan Malviya University of Technology, Gorakhpur, India*

Abstract
Hybrid welding comprises the two welding processes to explore the potential of individuals with broad field of applications. In hybrid welding, the demerits of the individual process are improved into merits after combining with other as a result significant improvement in weld quality. Such welding processes provide notable solutions with significant benefits over each of constituent processes. This chapter presents a report on hybrid welding that comprises the features of laser beam welding and arc welding in a single weld pool. Such combined welding process refers to laser–arc hybrid welding process. It also covers the basic principle, merits, demerits, and applications of laser–arc-based hybrid welding. Finally, the present article is organized in such a manner in which it becomes easier for the reader to understand the crucial phenomena of such hybrid welding process.

***Keywords*:** Arc, joining, laser, fusion, hybrid, plasma, welding

**Corresponding author*: mechrny@gmail.com

Sandip Kunar and Gurudas Mandal (eds.) Advanced Welding Technologies, (401–418) © 2025 Scrivener Publishing LLC

20.1 Introduction

The modern industries focus on reducing construction weight of objects necessary for enhancement of fuel efficiency. Fuel efficiency is directly related to weight reduction. In simple terms, the reduction in weight of approximately 10% means a 3%–7% reduction in fuel consumption [1–3]. Hence, low-weight, high-strength materials, especially titanium and its alloys, high-strength steel and its alloys, heat-treated steel and super alloys, etc., are becoming the first choice in advanced manufacturing industries. In manufacturing, various operations, like cutting, welding, forming, machining, etc., are performed to get a desired product. In such manufacturing operations, welding is a common joining process, which is used to weld the structural works in the field of automobile and aerospace industries. In welding, application of heat is necessary to melt the object, and welded object can be obtained after solidification of molten metal [4]. Filler metals are applied in welding leading to an increase in the weight of the weld object. Hence, the applicability of welding is limited for structural works especially in the automobile, shipping, and aerospace industries. Instead of this, welding is found unsuitable to weld thin and fragile materials.

For welding of thin and fragile material, laser beam welding (LBW) is an effective method for joining without loss of mechanical properties. Here, laser radiation is used as heat source, which is focused at a desired location with the help of focusing lens (Figure 20.1). Laser radiation in the form of beam hits within the cavity of the joining pieces in which melting of workpiece surface of two objects occurs. Finally, the melted metal is mixed together and forms a strong weld joint after solidification [5–7]. The LBW provides weld with deep penetration, high speed, negligible heat affected zone (HAZ), and thermal damages [7, 8]. Similar and dissimilar objects with high aspect ratio (10:1) can be welded by the LBW process. Due to flexibility of the beam, narrow and tiny inaccessible locations can be easily

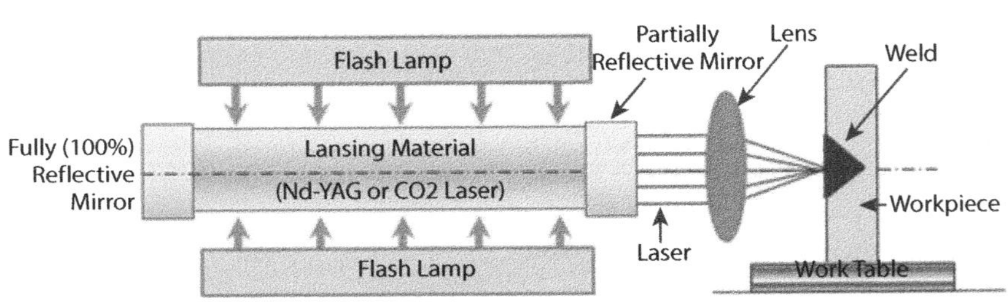

Figure 20.1 Schematic of LBW process.

welded with high accuracy. Even though, it shows several demerits like high initial cost, unsuitability for highly reflective materials, and thicker objects [2, 7–11].

The welding of thicker materials and heavy structures, especially in wind mills and shipyards, becomes of interest for manufacturing industries. The other steel-based industries, like power plants, steel producers, steel pipes, and heavy machineries, also show their interest to weld heavier sections with conventional welding processes [12]. The welding of a thick wall steel structure is encountered in the welding works in areas of wind mill and steel pipes [13]. Initially, submerged arc welding (SAW) is found suitable to meet the objectives, but numerous weld beads and seam preparation (normally double V-weld) increase the welding time [12, 13]. Recently, gas metal arc welding (GMAW) shows potential to weld such steel structures. It is a stable welding process that provides spatter [14]. However, GMAW has several shortfalls as low weld penetration, slow welding speed, and wider HAZ [15, 16]. Due to this, heavy steel industries are looking toward new technology to overcome the above problems.

To overcome the problems and compensate for the deficiency of two welding techniques, especially LBW and AW (arc welding), a new technology of hybridization of welding comes into existence, which is known as hybrid welding process. In the present case, the LBW and AW are combined in such a form that they are utilized on a single weld bead. Such a developed process is referred to laser–arc hybrid welding (LAHW) process. The LAHW process shows their potential in the field of welding due to rapid speed, deep penetration, process stability, high weld quality, and gap bridge ability [16–19].

20.2 Laser–Arc Hybrid Welding

Generally, LAHW is defined as a combination of two different types of heat sources, especially laser and electrical arc, for welding applications. The basic ideas behind the coupling of two deferent processes are to maximize the merits and minimize weakness characteristic of individual welding processes [17–20]. In hybrid welding, laser power effectively controls the penetration depth, while weld width and bridging are controlled by the arc. Hybrid welding shows several benefits over individual welding such as high welding speed, no need of edge preparation, better metallurgical control, better weld quality, and higher electrical efficiency [19, 20].

The LAHW process combines the LBW with arc welding, especially GMAW or GTAW (gas tungsten arc welding), in such a way that heat sources (laser and arc) are incident in a single weld pool [19, 21]. A typical diagram of the LAHW process and its constituents is presented in Figure 20.2. It was proposed by Steen and Eboo by combining GTAW and LBM process for cutting and welding applications [22]. They used CO_2 laser with GTAW and found unique merits such as stable arc behavior, dramatic change in welding speed for welding of thin sheet, and high penetration depth than the LBW process [22].

The LAHW shows enhanced capabilities than LBW and AW when separately applied for welding applications. The differences between arc welding, LBW, and LAHW are summarized in Table 20.1 [10, 12, 19–26]. The key benefits of LAHW are the ability to bridge gap and significantly high-speed welding with deep penetration than laser and arc welding [24, 25]. The deep penetration allows a single pass welding by LAHW process than multi pass arc welding. Such phenomenon leads to decreased welding time. The LAHW process also shows reduction in formation of brittle phases, HAZ, micro cracks ,and thermal distortion in weld object [24]. The potential of LAHW is extended to weld the advanced materials like titanium and its alloys, heat-treated steel, high-strength steel, aluminum, and super alloys [24]. The LAHW process is mostly used in automobile and shipping industries. It is a better tool for welding of different thicknesses of similar/dissimilar metal and alloys. It is also used in heavy machineries, steel industries, petroleum, and power plant sectors for welding applications. Of course, the LAHW is a unique joining process, but it needs high initial cost and setup, a skilled operator, and is only applicable for heavy equipment and machineries.

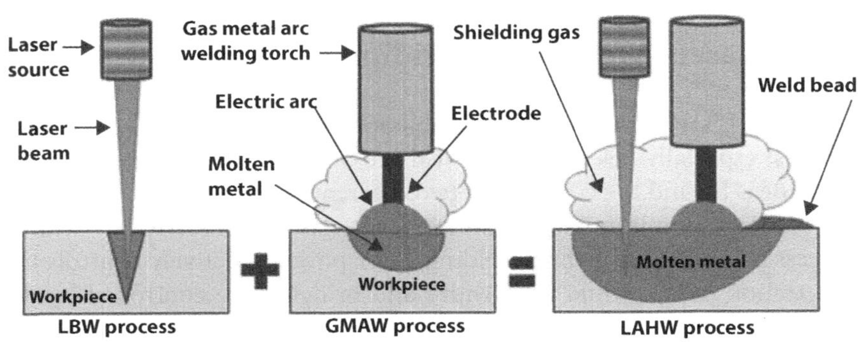

Figure 20.2 LAHW process and its constituents.

Table 20.1 Comparison between arc welding, LBW, and LAHW processes [10, 12, 19–26].

Parameters	Arc welding	LBW process	LAHW process
Power source	Electric arc	Laser radiation	Electric arc and laser radiation
Polarity	Either straight or reverse polarity based on welding requirement	No polarity is required. Laser radiation directly utilized to heat up the object	No significant effect in polarity. The process capability depends upon orientation of laser and arc at weld pool
Arc stability	Poor and unstable	No arc formation, laser leads in welding	Stable arc formation and better interaction among laser and arc
Fusion zone and penetration	Shallow fusion zone and low penetration	Narrow fusion zone and deep penetration	Wide fusion zone and deep penetration depth
Gap bridging	Good gap bridging	Poor gap bridging	Superior gap bridging
Welding speed	Slow speed welding	High speed welding	Comparatively high-speed weld
Cooling rate	Slow cooling, which reduces brittle formation	High cooling rate leads in high porosity weld	Relatively slow cooling rate, which reduces porosity in weld
Crack propensity	High propensity of cracks	Rapid cooling leads in crack formation	Reduce brittle phase formation and cracks in weld
Applications	Welding of ferrous metal and alloys	Welding of carbon steels, stainless steel, aluminum, and titanium	Welding heavy steel structure and machineries, welding Ti, Al, and advanced materials

20.3 Orientation of LAHW Process

Generally, the LAHW process can be developed in two different orientations based on application of heat source (laser/arc) in weld pool as laser leading and arc leading [23, 26]. The various orientations of the LAHW process are presented in Figure 20.3. The actual photographic view of two different configurations of LAHW is shown in Figure 20.4 [26]. In laser leading LAHW process, the laser radiation is in leading form, which is followed by an electrical arc. Such a combination of welding is applicable for deep penetration and welding of narrow sections. The laser plays a role in deep penetration, while bridging is made by arc power. The laser leading hybrid welding efficiently produces better weld profiles. Here, a low arc current supports to get a narrow bead and a stable weld pool with reduction in defects [26].

In arc leading LAHW, the electric arc is in leading role, which is followed by the laser. The application of laser after the electric arc makes a more uniform temperature throughout the weld pool, which leads to support slow cooling. Due to this, a uniform and homogeneous micro-structure can be achieved with reduction in micro crack formation and thermal defects. The arc leading hybrid welding process shows low penetration, and it requires slow welding speed for deep penetration. It suffers with the weak and unstable interaction between heat source and workpiece with increment in weld defects [26]. It is a slower welding process and prefers wider and shallow welding applications.

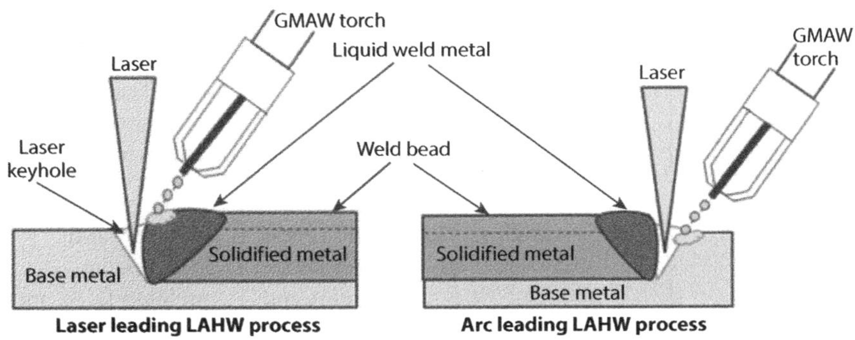

Figure 20.3 Orientation of LAHW process.

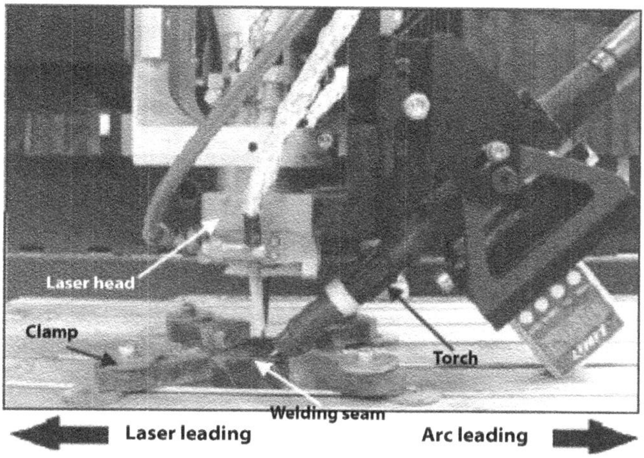

Figure 20.4 Photographic view of the LAHW process [26].

20.4 Classification of LAHW Process

In LAHW, the orientation, laser radiation, and arc energy are significantly affected the performances of the process. Hence, LAHW can be classified based on orientation and energy sources as laser leading LAHW, arc leading LAHW, laser-stabilized AW, laser plus AW, laser-GTAW, laser-GMAW, and laser plasma welding [21, 23, 27]. The classification of LAHW is presented in Figure 20.5.

Based on orientation, either the laser leading or arc leading technique can be utilized in the LAHW process (Figure 20.3). The laser leading process prefers deep penetration with high-speed welding. The process capability of laser leading is better than arc leading in terms of weld quality and bridging capacity by arc phenomenon. The heat energy-based LAHW process can be categorized into two categories as follows: laser power or arc power-based LAHW processes. Further, laser power arc-based hybrid welding can be categorized as laser-stabilized arc welding (LSAW) and laser pulse arc welding (LPAW), while arc power-based hybrid welding can be categorized into laser-GTAW and laser-GMAW processes.

The LSAW process utilizes low power laser radiation (<1 kW) approximately 500 W in which small amount of heat is generated. In this process, electric arc dominates the LAHW, and a laser keyhole is not formed [21, 28]. It provides comparatively poor weld quality with shallow penetration within the weld pool. In LPAW, the laser and arc have equal power, and one of them is in the dominating phase. It required high laser approximately

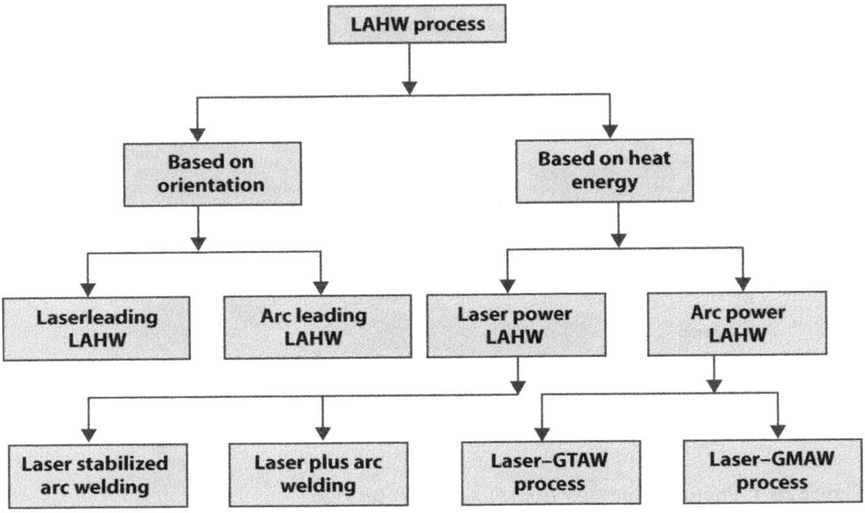

Figure 20.5 Classification of the LAHW process.

1 kW and easily forms a laser keyhole in welding. The laser radiation gives better penetration, while bridging capability is improved by arc power. Instead of this, LPAW provides better weld quality, deep penetration, and is suitable for high-speed welding applications [21, 29].

Laser-GTAW is a hybridization of LBW and GTAW to achieve a high welding rate with deep penetration and better metallurgical bonding in the presence of a protective environment of inert gas [30, 31]. The photographic arrangement and laser camera view of the laser-GTAW process is shown in Figures 20.6 (a) and (b) [30]. The schematic diagram of

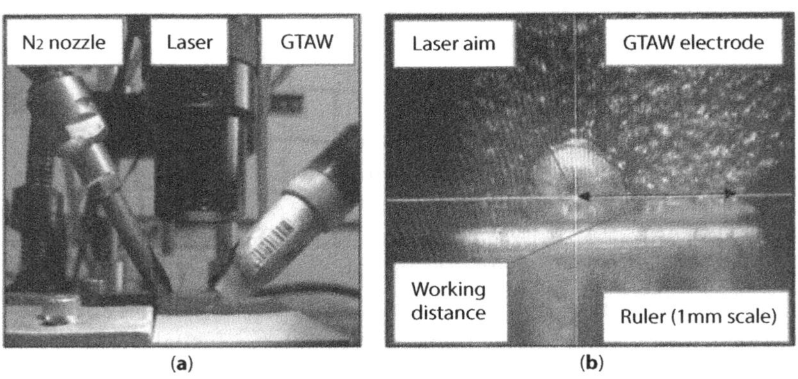

Figures 20.6 (a) Hybrid laser-GTAW process. (b) Laser camera view [30].

laser-GTAW is shown in Figure 20.7. Here, a combination of concentrated laser with low arc current is applied to weld the objects. The laser beam facilitates to achieve deep material penetration, while low current melts the base metal and filler material to enhance the fluidity of melted metal [30]. It also supports to obtain a narrow bead, stable melt pool, and reduction in defects. It is suitable for welding low-melting materials like aluminum, magnesium, copper, and its alloys, low carbon steel, stainless steel, nickel alloys, dissimilar metals and alloys, and welding of different thickness of metallic parts for industrial applications [30–33].

Laser-GMAW comprises the LBW and GMAW to achieve rapid and automated welding with high deposition rate. It is similar to the laser-GTAW process but differ in various ways like arc generation, filler metal, deposition rate, and applicability. The differences between laser-GTAW and laser-GMAW are summarized in Table 20.2 [13, 19–21, 30–37]. The schematic diagram of hybrid laser-GMAW is presented in Figure 20.8. A photographic view of the laser-GMAW is presented in Figure 20.9 [34]. In laser-GMAW, the arc is created among the wire electrode and workpiece. The wire electrode also acts as a filler material and melted due to the heat of the arc. Here, a high-intensity laser is used to create a deep keyhole, while high current is utilized to melt the wire electrode and workpiece. It is suited for deep penetration and high deposition rate [38].

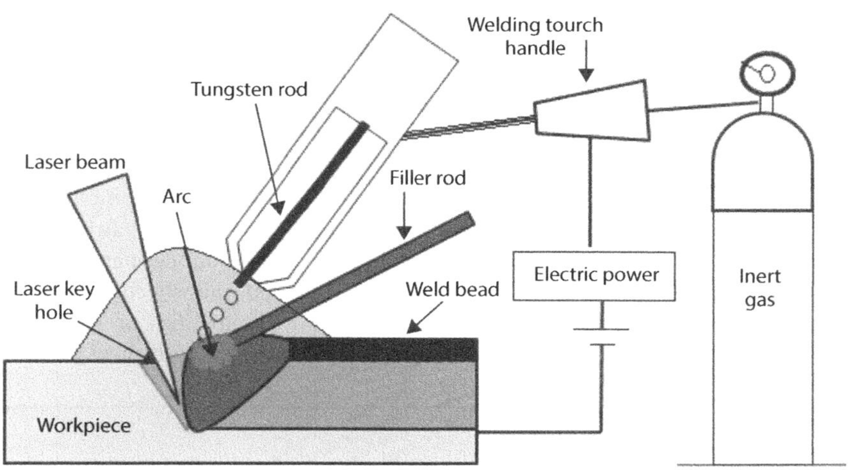

Figures 20.7 Schematic diagram of the laser-GTAW process.

Table 20.2 Differences between laser-GTAW and laser-GMAW [12, 13, 19–21, 30–37].

Parameters	Laser-GTAW process	Laser-GMAW process
Power source	AC or DC. DC is used for mild steel and steel materials, while AC is used for welding of Al	DC power is used
Arc generation	Arc is formed among tungsten electrode and workpiece	Arc is generated among wire electrode and workpiece
Filler material	Filler rod is used as filling materials by melting due to arc	No need of filler material, and electrode rod is used as filler material by melting due to arc
Inert gas	Argon is used, but higher penetration helium is applied	Mostly, argon is used
Laser beam	Deep penetration	Deep penetration and keyhole formation
Current	Low current uses for melting of workpiece and filler rod	Higher current uses for melting of electrode wire and workpiece
Deposition rate	Low deposition	Higher deposition rate
Applications	It is used for welding of steel, CI, alloy steel, stainless steel non-ferrous materials (Cu, Al, and Mg). It is suitable to weld thin objects made of titanium or steel. It is mostly used in aviation and aerospace industries for welding of sheet metal works	It is used for welding of steel, cast iron, and aluminum. It is suitable to achieve high weld and high deposition rate. It is used for repairing of automobiles parts, worn out rail wheels, welding of pipelines, pressure vessels, and sheet metal works

Laser-GMAW is a high-performance automated process that developed by combining the LBW and GMAW at a single weld platform. It provides high-speed welding with high deposition rate and narrow HAZ in the weld bead. It is suitable for thick and heavy objects. It is applicable for welding

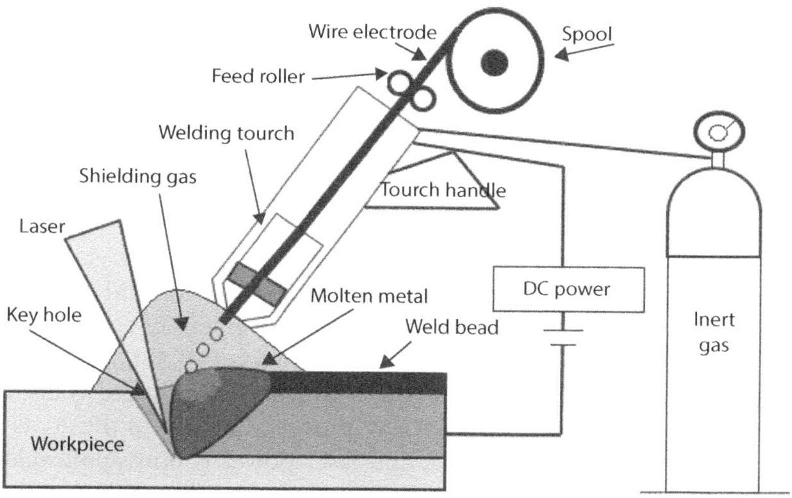

Figure 20.8 Laser-GMAW process.

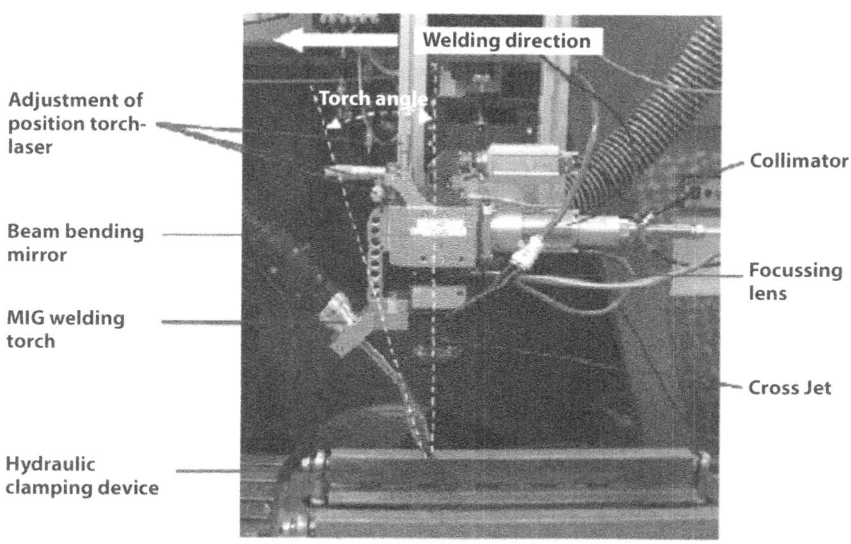

Figure 20.9 Photographic view of the laser-GMAW setup [34].

of aluminum, copper, and steel and steel [33–37]. It is a widely accepted welding method for small and medium industries for rapid production. It is used in automobile industries for repairing works, welding of worn out rail wheels, and welding pipelines and pipes. It is also used to weld sheet metal objects and pressure vessels.

20.5 Parameters of LAHW Process

The performances of the LAHW process are significantly affected by LBW and AW parameters. Generally, laser power, type of laser, focal length of beam, and spot diameter are common laser parameters. The arc welding parameters as electric power, polarity, and inert gas are significantly affected the performance of LAHW process.

In LAHW, the laser power is the most significant factor, and it speeds up the welding process with deeper penetration. It compensates the high HAZ and distortion of weld object in case of AW process [35, 36]. In LBW, laser power is utilized to melt the filler rod in which welding speed is declined approximately 20% [36]. In the LAHW process, the heat of the arc is utilized for melting of the filler rod, and laser power is only applicable for high-speed welding.

Laser beam and its type greatly affect the LAHW process and is responsible for keyhole formation. Generally, two types of lasers are used in the LAHW process as follows: Nd-YAG (wave length = 1.06 μm) and CO_2 laser (wavelength = 10.6 μm) [11, 36]. The CO_2 laser shows great electrical efficiency and lower operational cost, while the Nd-YAG laser shows great processing efficiency for the same laser power. In the LAHW process, the Nd-YAG laser is preferred for welding of aluminum and its alloys. However, higher power (8 kW) can be achieved with CO_2 laser, while Nd-YAG is limited to 4 kW. The Nd-fiber and high-power diode (HPD) lasers are also preferred for LAHW of aluminum. The HPD laser makes welding economical but only suitable for welding aluminum metal and alloys.

The focus length and spot diameter are related to lens parameters and have a lead role in keyhole formation. Most of the heat is available on the focal spot/diameter of the beam, which is utilized for melting or heating of an object during keyhole formation in hybrid welding. The focal length has a leading role in deep penetration and higher focal length suitable for a thick sheet/plate [39]. For keyhole formation, a smaller spot diameter is preferred, and the keyhole disappears with a larger spot diameter [36]. Consequently, a larger spot diameter of the beam increases the wettability of deposited metal in the LAHW process. It also provides better heat input than individual welding process, either LBW or arc welding, with a larger spot diameter.

The electrical power in LAHW is necessary for arc formation and melting of the filler metal. Either AC or DC power source is utilized in the LAHW process. A DC power source supports the stable and strong arc formation at constant polarity, while polarity changes during each cycle of the AC power. The AC power is applied in GTAW to weld aluminum and its alloys [26]. In case of GMAW of aluminum with DC power, the several difficulties, like burn through, poor gap tolerance, and creation of holes are occurred. Even though such issues of aluminum welding can be solved by the utilization of AC power in the GMAW process, but insufficient penetration at higher speed limits it [36, 40].

Polarity is another parameter that greatly affects the LAHW process. Mostly, DCEN (direct current electrode negative) or DCEP (direct current electrode positive) is employed in arc welding [41]. DCEN is preferred for GTAW, while DCEP is suitable for the GMAW process [41]. The DCEP polarity is used with AC power in case of GTAW, which requires a larger diameter of tungsten electrode and water cooling. In the welding of aluminum metal with GMAW, the DCEP polarity is set, while positive polarity effectively removes the oxide layer from aluminum [36, 41]. DCEN provides a stable arc and an easier metal transfer for welding in which a smoother weld bead can be achieved. Of course, DCEP polarity is highly efficient to weld the thicker (more than 3 mm) objects.

Inert gas is used to make a protective environment during the LAHW process. It also protects the undesired chemical reaction with weld bead and rapid cooling. Mostly, argon is used as inert gas, but sometimes, helium or nitrogen is also applied in welding [41]. Helium gives higher penetration but is much costly than argon. On the other hand, nitrogen gives poor results (unstable arc and porosity). Even though, nitrogen gas is applied as shielding gas during welding of copper and lighter works.

20.6 Process Capability of LAHW Process

The LAHW shows potential for welding of thin as well as thick similar and dissimilar materials with high speed. The welding speed is achieved by LAHW at 5 m/min for Al welding while 0.95 m/min with GMAW and 3 m/min with LBW for the same metal [36]. Instead of this, much higher welding speed (approximately 15 m/min) has been achieved for welding of sheet made of steel sheet with the hybrid laser-GTAW process [42]. The uneven thickness and dissimilar materials can be also welded by the LAHW process efficiently [13, 36, 43, 44]. The welding of unusual properties of materials, such as low–high melting point, soft–hard metal/alloys can be made possible by the LAHW process. In the same way, the different characteristic materials, like Ti–Al, Al–Mg, Al–stainless steel, Ti–stainless steel, and low strength–high strength materials are easily welded by LAHW process [44–47].

High penetration depth can be achieved by the LAHW process. It gives a higher penetration of approximately 10%–20% than LBW and 20%–50% than GMAW processes in the case of aluminum welding [36]. In hybrid welding, the heat input is lower because of the high density of the energy and high speed of the welding resulting in decreases in distortion of welded parts. The gap bridging capacity of LAHW is better than those of the LBW and GMAW/GTAW processes. It is effectively increased the gap bridging 0.14 mm of the GMAW to approximately 1.05–1.19 mm [36]. The robustness of HLAW is much better than those of the LBW and GMAW/GTAW processes. The reduction in distortion of welding objects, filler material feeding, and gap bridging ability also enhanced the robustness of such hybrid welding process. It is used to weld complex profiles also like T-section, L-section, and others. The T-joint of a thin sheet made of magnesium is effectively welded by the GTAW process and achieved mechanical properties up to 90% of the base metal [48].

The LAHW process shows significant potential in the broad field of manufacturing industries. A variety of materials, including soft/hard and thin/thick, can be easily welded by this process. The materials having unusual properties, like Sn–steel, Ti–Al, Mg–Al, Ti–steel, etc., of different thickness can be welded by the LAHW process. Even though, it suffers with several drawbacks like high initial cost and need of a skill operator and costly equipment. Here, the HAZ is higher than the LBW process, and there is difficulty in controlling the molten metal. The controlling parameters belong to the LBW and GMAW/GTAW processes in which hybrid welding becomes more complex. The deep weld for a narrow section with LAHW faces a high harness in root zone, cracks, porosity, and poor microstructure due to uneven mixing of filler material [49]. The lack of metal deposition occurs on the side wall of narrow profiles objects, which is also a major concern with LAHW process [50]. Instead of this, the LAHW process needs automation, which limits the broad applicability and suitability for small-scale industries.

20.7 Summary

Hybrid welding is a unique joining method that comprises two or more different forms of energy or processes to achieve better weld quality at rapid speed. The aim of hybrid welding is to utilize the merits of individual process and minimize the adverse effects in welding. The laser and arc energy-based hybrid welding process, such as LAHW, shows their potential to weld objects at high speed with high weld quality and negligible HAZ and thermal defects. In the LAHW, the laser plays roles in speeding up the welding process with deep penetration, while the arc is utilized to weld the filler metal and effectively bridge the gap. The current chapter focuses on presenting the theory and applications of the LAHW process including their merits and demerits. It also covers the various LAHW processes to weld the objects in different conditions to fulfill the demands of advanced industries. Of course, the entire chapter becomes beneficial for industries, professionals, and researchers who make their efforts in the same field of welding.

Acknowledgment

The authors gratefully acknowledge Elsevier Publishers and Journals for reprint/reuse permission of the figures from the published manuscript. The authors are also grateful to the copyright center for assistance to find reprint/reuse permission from various sources of journals.

References

1. Manladan, S., Yusof, F., Ramesh, S., Fadzil, M.A., review on resistance spot welding of magnesium alloys. *Int. J. Adv. Manuf. Technol.*, 86, 1805–1825, 2016.
2. Baqer, Y.M., Ramesh, S., Yusof, F., Manladan, S.M., Challenges and advances in laser welding 427 of dissimilar light alloys: Al/Mg, Al/Ti, and Mg/Ti alloys. *Int. J. Adv. Manuf. Technol.*, 5, 4353–4369, 2018.
3. Yadav, R.N., Solid-state friction welding technology for joining of lightweight metal and alloys, in: *Advances in Processing of Lightweight Metal Alloys and Composites*, R.V. Vignesh *et al.* (Eds.), pp. 147–165, Springer Nature, Singapore, 2023.
4. Little, R.L., *Welding and Welding Technology*, McGraw Hill Education, USA, 2017.
5. Parmar, R.S., *Welding Engineering and Technology*, Khanna Publishers, New Delhi, India, 2013.
6. Auwal, S.T., Ramesh, S., Yusof, F., Manladan, S.M., A review on laser beam welding of copper alloys. *Int. J. Adv. Manuf. Technol.*, 96, 475–490, 2018.

7. Lienert, T., Siewert, T., Babu, S., Acoff, V., *Welding Fundamentals and Processes*, ASM International, US, 2011.
8. Oladimeji, O.O. and Taban, E., Trend and innovations in laser beam welding of wrought aluminum alloys. *Weld. World*, 60, 415–457, 2016.
9. Nabavi, S.F., Farshidianfar, A., Dalir, H., A comprehensive review on recent laser beam welding process: geometrical, metallurgical and mechanical characteristic modeling. *Int. J. Adv. Manuf. Technol.*, 129, 4781–4828, 2023.
10. Rasmussen, D. and Dubourg, L., Hybrid laser-GMAW welding of aluminium alloys: a review. *Proc. 7th Int. Conf. Trends in Weld. Research*, Pine Mountain, USA, pp. 133–142, 2006.
11. Yadav, R.N., Mishra, S., Suryavanshi, A., Laser beam micromachining and fabrication, in: *Electro Micromachining and Microfabrication: Principles and Research Advances*, S. Kunar, G. Kibria, P. Chatterjee (Eds.), pp. 139–171, Apple Academic Press, USA, 2024.
12. Nielsen, S.E., High power laser hybrid welding-challenges and perspectives. *Phy. Proc.*, 78, 24–34, 2015.
13. Reisgen, U., Olschok, S., Engels, O., Laser beam submerged arc hybrid welding: a novel hybrid welding process. *J. Laser Appl.*, 30, 1–21, 2018.
14. Houldcroft, P.T., *Submerged-Arc Welding*, 2nd Ed., Woodhead Publishing, Cambridge, 1990.
15. Matsuyama, K., Digital control used in welding machines in Chinese Mechanical Engineering Society. *Proc. Int. Forum Weld. Techn. Auto. Ind.*, 2, 81–87, Beijing, China, 2003.
16. Xu, G.X., Wu, C.S., Qin, G.L., Wang, X.Y., Lin, S.Y., Adaptive volumetric heat source models for laser beam and laser + pulsed GMAW hybrid welding processes. *Int. J. Adv. Manuf. Technol.*, 57, 245–255, 2011.
17. Xu, G.X., Wu, C.S., Qin, G.L., Wang, X.Y., Lin, Y.S., Numerical simulation of weld formation in laser+GMAW hybrid welding II. Combined volumetric distribution mode of hybrid welding heat source. *Acta Metall. Sinica*, 44, 641–646, 2008.
18. Jokinen, T., Viherva, T., Riikonen, H., Kujanpaa, V., Welding of ship structural steel A36 using a Nd:YAG laser and gas-metal arc welding. *J. Laser Appl.*, 12, 185–188, 2000.
19. Defalco, J., Practical applications for hybrid laser welding. *Weld J.*, 86, 47–50, 2007.
20. Kah, P., Overview of the exploration status of laser-arc hybrid welding processes. *Rev. Adv. Mater. Sci.*, 30, 112–132, 2012.
21. Roepke, C., Liu, S., Kelly, S., Martukanitz, R., Hybrid laser arc welding process evaluation on DH36 and EH36 steel. *Weld. Res.*, 89, 140–150, 2010.
22. Bunaziv, I., Akselsen, O.M., Frostevarg, J., Kaplan, A.F.H., Laser-arc hybrid welding of thick HSLA steel. *J. Mater. Process. Techn.*, 259, 75–87, 2018.
23. Victor, B.M., Hybrid laser arc welding, in: *Welding Fundamentals and Processes*, T. Lienert, T. Siewert, S. Babu, V. Acoff (Eds.), ASM International, US, 2011.

24. Ribic, B., Palmer, T.A., Roy, T.D., Problems and issues in laser-arc hybrid welding. *Int. Mater. Rev.*, *45*, 223–244, 2009.
25. Acherjee, B., Hybrid laser arc welding: state-of-art review. *Opt. Laser Technol.*, 99, 60–71, 2018.
26. Casalino, G., Campanelli, S.L., Dal Maso, A.D., Ludovico, Arc leading versus laser leading in the hybrid welding of aluminium alloy using a fiber laser. *Proc. CIRP*, 12, 151–156, 2013.
27. Zhou, J. and Tsai, H.L., *Hybrid Laser-Arc Welding*, Intech. Open Sci., UK, 2012.
28. Bagger, C. and Olsen, F.O., Review of laser hybrid welding. *J. Laser Appl.*, 17, 2–14, 2005.
29. Shinn, B.W., Farson, D.F., Denney, P.E., Laser stabilisation of arc cathode spots in titanium welding. *Sci. Technol. Weld. Join.*, 10, 475–481, 2005.
30. Videira, A.M., Mendes, W.R., Ventrella, V.A., Calliari, I., Increasing the corrosion resistance in the UNS S32750 super duplex steel welded joints through hybrid GTAW-laser welding and nitrogen. *Materials*, 16, 543–12, 2023.
31. Song, G., Yu, J., Li, T., Wang, J., Liu., L., Effect of laser-GTAW hybrid welding heat input on the performance of Mg/Steel butt joint. *J. Manuf. Process.*, 13, 131–138, 2018.
32. Sierra, G., Peyre, P., Beaume, F.D., Stuart, D., Fras, G., Galvanised steel to aluminium joining by laser and GTAW processes. *Mater. Character.*, 59, 1705-1715, 2008.
33. Kaul, R., Kumar, H., Rao, B.T., Kukreja, L.M., Singh, N., Vora, H.S., Studies on characteristics of CO_2 laser-GTAW hybrid welding of austenitic stainless steel. *J. Laser Appl.*, 22, 79–85, 2010.
34. Grunenwalda, S., Seefelda, T., Vollertsena, F., Kocak, M., Solutions for joining pipe steels using laser-GMA-hybrid welding processes. *Phy. Procedia*, 5, 77–87, 2010.
35. Rao, Z.H., Liao, S.M., Tsai, H.L., Modelling of hybrid laser–GMA welding: review and challenges. *Sci. Techno. Weld. Join.*, 16, 300–305, 2011.
36. Wang, J.B., Nishimura, H., Katayama, S., Mizutani, M., Welding of aluminum alloy by using filler-added laser-arc hybrid welding process. *Weld. Int.*, 27, 98–108, 2011.
37. Han, R., Fan, X., Wang, J., Ren, D., Song, G., Enhancing all-position weldability and weld quality via an innovative laser-GMAW hybrid welding technique with copper liner. *J. Manuf. Process.*, 124, 697–707, 2024.
38. Silva, R.G.N., Rodrigues, M.B., Pereira, M., Faes, K., Evaluation of high penetration hybrid laser-GMAW welding process productivity applied in the joining of thick plates. *Int. J. Adv. Manuf. Technol.*, 121, 3271–3283, 2022.
39. Anghel, C. and Gupta, K.A., review on laser beam cutting. *Proc. 5th NA Int. Conf. Ind. Eng. Operat. Manag. Detroit*, pp. 3168–3178, 2020.
40. Tong, H., Ueyama, T., Nakata, K., Ushio, M., High speed welding of aluminium alloy sheets using laser assisted alternating current pulsed metal inert gas process. *Sci. Technol. Weld. Join.*, 8, 229–234, 2003.

41. Mandal, N.R., *Aluminum Welding*, Narosa Publishing House, New Delhi, India, 2002.
42. Arias, J.L., Romero, P., Vandewynckèle, A., Vázquez, J., Laser-TIG hybrid welding of very thin austenitic stainless steel sheets. *Laser Mater. Process. Conf. (ICALEO 2005) Congress Proceedings*, pp. 104–107, 2005.
43. Abioye, T.E., Olugbade, T.O., Ogedengbe, T.I., Welding of dissimilar metals using gas metal arc and laser welding techniques: a review. *J. Emer. Trends Eng. Appl. Sci.*, 8, 225–228, 2018.
44. Spena, P.R., Angelastro, A., Casalino, G., Hybrid laser arc welding of dissimilar TWIP and DP high strength steel weld. *J. Manuf. Process.*, 39, 233–240, 2019.
45. Gao, M., Chen, C., Gu, Y., Zeng, X., Microstructure and tensile behavior of laser arc hybrid welded dissimilar Al and Ti alloys. *Materials*, 7, 1590–1602, 2014.
46. Gao, M., Chen, C., Wang, L., Wang, Z., Zeng, X., Laser-arc hybrid welding of dissimilar titanium alloy and stainless steel using copper wire. *Metall. Mater. Trans. A*, 46, 2007–2020, 2015.
47. Gao, M., Chen, C., Mei, S., Wang, L., Zeng, X., Parameter optimization and mechanism of laser–arc hybrid welding of dissimilar Al alloy and stainless steel. *Int. J. Adv. Manuf. Technol.*, 74, 199–208, 2014.
48. Li, C. and Liu, L., Investigation on weldability of magnesium alloy thin sheet T-joints: arc welding, laser welding, and laser-arc hybrid welding. *Int. J. Adv. Manuf. Technol.*, 65, 27–34, 2013.
49. Bunaziv, I., Ren, X., Olden, V., A comparative study of laser-arc hybrid welding with arc welding for fabrication of offshore substructures. *Phys. Conf. Ser.*, 26, 1–10, 2023. DOI 10.1088/1742-6596/2626/1/012033.
50. Meng, Y., Li, G., Gao, M., Zhang, C., Zeng, X., Formation and suppression mechanism of lack of fusion in narrow gap laser-arc hybrid welding. *Int. J. Adv. Manuf. Technol.*, 100, 2299–2309, 2019.

21

Beyond Fumes and Flux: Green Welding for a Sustainable Future

M. Abdur Rahman[1*], G. Rajesh[1], S. Jeavudeen[1], R. Karunanithi[1] and N. Sri Rangarajalu[2]

[1]Department of Mechanical Engineering, B.S. Abdur Rahman Crescent Institute of Science and Technology, Chennai, India
[2]Department of Production Technology, MIT Campus, Anna University, Chennai, India

Abstract
The manufacturing sector is increasingly concerned about the environmental impact of welding, a crucial process in joining materials. As a result, there has been an upsurge in research into "green" welding technologies. These technologies aim to reduce the ecological footprint of welding while enhancing its performance. This review consists of various green welding processes, advantages concerning lower energy consumption and reduced emissions, and practical applications. Among them is Friction Stir Welding (FSW), which is regarded as a flagbearer in green technology because of its low energy requirements and minimal environmental degradation. Unlike conventional approaches for welding, it utilizes less energy and vacuum or flux materials that do away with hazardous substances. This brings not only environmental benefits but also improves worker safety by reducing exposure to toxic fumes. The review stresses the need for continuous research on novel solutions in green welding expressing their potential for considerable savings on power resources and improved quality of welded joints. In addition, toward the end there, one finds a description of how integrating approaches for greener welds fits into worldwide campaigns aimed at achieving sustainable production within the framework of Industry 4.0 and also upcoming related concepts such as Industry 5.0. By adopting energy-saving mechanisms and ecologically friendly whole work styles, like FSW technology—ultimately it serves as an advocate for environment protection—seamless touch screen wall-mounted tablets.

**Corresponding author*: abdurrahman@crescent.education

Sandip Kunar and Gurudas Mandal (eds.) Advanced Welding Technologies, (419–446) © 2025 Scrivener Publishing LLC

Keywords: Green welding, sustainable manufacturing, FSW, sustainability, Industry 4.0, Industry 5.0

21.1 Introduction

Industry is about working on products where the metal, for example, is extracted from the ground and then made into those items that define our existence. Laying these materials together with high tensile strength is crucial, and welding is regarded as one of the influential processes. While joining is done by the use of glue, welding makes use of heat to melt and join metals, and the joined portions are always stronger than the original metal. Economic design of products for manufacture entails abiding by ways and means that reduce cost and component usage. This includes setting the performance characteristic for the final welds—how strong they have to be and how they look like. Another significant aspect that arises in this process is weldability—the ability of a material to be welded. Knowing the weldability affects the selection of materials and the creation of appropriate welding procedures. It helps to guarantee that the final product is not only useful but also enduring, and was produced to high quality, which is free from defects [1]. This focuses on efficient joining methods, like welding, exemplifies the delicate balance that manufacturing strives for: these include aspects such as attaining functional attributes, demonstrating expenditures control, and material utilization. The cost of energy has increased tremendously, and this is an additional cost to industries that involve much welding. In addition, the energy employed in welding processes ranks high among key pollutants in the environment. This has created much attention to finding "eco-friendly" welding technologies that use the least energy. In layman's terms, there is a high demand for the development of a diversity weld that has an equivalent environmentally friendly and energy minimal [2]. The subject of sustainability is gradually invading industries and companies across the globe. This entails means of reducing the adverse effects on the environment and consolidating the utilization of energy. One of the most important stages of material joining in manufacturing, welding has not been left out. There is, however, a main drawback; the conventional welding techniques despite being efficient are recognized power hogs. They also emit dangerous gases within the fumes, which causes the environment around the welders to be unhealthy and dangerous for people living nearby. To overcome these problems, scientists and engineers have popularized what is known as "green" welding procedures.

These can reduce the impacts of welding on the surroundings while at the same time possibly increasing the quality and durability of the welds [3].

What was a dream a decade ago is now a reality of these newer environment-friendly technologies in welding. All these modern approaches focus on lowering energy use, minimizing the emission of toxic gases, and enhancing productivity. The focus on sustainability proves very well to fit the increase in environmental standards all over the world and ever-expanding market demands for green manufacturing [4].

With sustainability being established as one of the most crucial economic factors in industries, industries are slowly shifting to what can be referred to as "green" welding technologies. Such techniques are focused on improving efficiency and minimizing energy use and emissions to the environment. This focus relates to increased stringency of environmental standards, as well as to the trend toward environmentally friendly production. The use of green welding methods is a double-barreled benefit to industries cutting down their carbon footprint and slashing expenses. Several prominent green technologies offer distinct advantages: FSW is very economical in terms of energy usage and uses no noxious gases; hence, FSW is safe for both the workers and the environment. Laser Beam Welding (LBW) allows the best control over the process reducing the amount of materials used, and power used, and gives superior weld quality. Electron Beam Welding (EBW) is essentially used in a vacuum environment to release low gas emissions and carries out deep accurate welds using less energy. GTAW or TIG Welding produces more precise and excellent welds with less smoke and fumes and less solid waste hence using less energy. Cold Metal Transfer (CMT) works at relatively lower temperatures, this means lower energy consumption, less spatter, and negligible heat influence to the workpiece. In conclusion, technologies for green welding are transforming the manufacturing industries, thus preparing the world for a better future for the environment [5].

It is now high time that green welding technologies became an essential tool for the processes in modern industry. These innovative methods offer a multitude of benefits, such as the saving of energy, reduction of both aerial and noise pollution, the safeguarding of the health of the workers from hazardous emissions, saving of cost through low-energy usage, and meeting the requirements of more stringent environmental standards. Through green welding, industries can easily reduce the amount of emissions they contribute to the environment, increase the efficiency at which they conduct their operations, and increase their profitability without necessarily having to spend a lot of money in the process. This chapter looks deeper into the imperative concepts of green welding and briefly outlines the development of other methods including Friction Stir Welding or Laser Beam Welding. These technologies demonstrate the tremendous opportunities for the reorganization of welding processes consonantly with the principles of sustainable development of modern industry.

21.2 Green Welding: Techniques and Research Developments

21.2.1 Friction Stir Welding (FSW)

With the rising issues of sustainability, there is a change in directions especially in the industrial sector toward welding. More conventional approaches can be resource intensive expending vast amounts of energy, and emitting various pollutants that endanger worker's lives and violate environmental standards. Hence, green welding technologies are on the rise with FSW being considered as the most efficient and environment-friendly welding technique.

As earlier established, FSW's way of joining materials with metals not shared is different from other approaches that melt metals. It mechanically mixes and stirs materials at the joint and does not melt it. This drastically cuts down on power consumption and does not require shield gases or fluxes that are used in other standard welds. FSW, therefore, does not result in the formation of a fused zone; therefore, it minimizes distortion and residual stress that are typical of welding, which makes it suitable when joining aluminum, copper, and their related alloys used in aerospace, automobile, and ship industries.

FSW's applications are far-reaching. Besides, it generates high-strength joints in lightweight materials necessary for new age transportation and construction, retains accuracy, and generates weld with low heat affected zones. This is crucial in manufacturing structures that are convoluted which require the matching accuracy of the material used and ensuring that there is very little deformation that can be made.

Yang *et al.* (2024) [6] carried out a computational simulation to understand the movement of materials while welding aluminum (Al) and magnesium (Mg) alloys through a method called friction stir welding (FSW). Contrary to the traditional casting process, it does not result in fusion of the metals but heats and mixes them. While welding the researchers knew precisely how the materials were displaced, thanks to the simulations made. This, they said, pointed to low wastage of material, most of which, they noted, was the magnesium, at the bottom of the joint, while the aluminum was mostly found on top. This movement is probably caused by the transfer of material in the thickness of the joint. Within the high-velocity zone of the material movement, most of the material followed a linear kind of flow. Notably, some of the buildup was mainly spotted twirling around the welding tool.

Consequently, the present occurrences revealed can assist engineers that practice FSW of Al and Mg alloys in enhancing the design of tools, thereby guaranteeing sufficient sturdy and quality welds.

Chen *et al.* (2024) contribute to an essential area of robotic FSW that has not been well handled, namely, the stiffness of the robot affecting weld quality. Small misalignments at the end effector level discontent the welds. Their concept is a new five-factor digital twin model. This model uses real-time data as the robot's performance data and sensor information to quantify variation. Incorporating co-simulation with finite element analysis (FEA) allows for conducting virtual welding simulations to predict. This approach achieves 6. Nine percent in the form of error in lateral deviation prediction helps to improve the parameter optimization for efficiency in robotic FSW [7].

Abdollahzadeh *et al.* (2024) conducted a study to establish the optimum CFSW and SFST welding process for AMM materials while considering low temperatures and the limit of the BIMC formation. The authors employed zinc interlayer and titanium carbide nanoparticles; the latter was applied with varying degrees of rotation from 850 to 950 revolutions per minute, combined with a constant traverse rate of 45 mm per minute. The findings of this study revealed that heat input is a critical factor that influences microstructure, distribution of nanoparticles, and formation of IMCs, Al3Mg2 and Al12Mg17. It could have confirmed its efficacy at 950 rpm and 45 mm/min, which led to the improvement of microstructural refinement and mechanical properties of underwater FSW compared to the other welded joints [8].

Real-time temperature measurements in FSW were an innovative technique introduced by Abboud and colleagues (2024) in their research via tool-integrated thermocouples. It involved the use of AA5083-H111 and AA6082-T6 alloys and compared the various parameters of rotational and welding speeds. Their main finding, therefore, was that a minimum temperature had to be sustained. According to the study, it was evident that rendering 65 times the solidus temperature gave high-quality welds. The findings of this research are useful for enhancing the efficiency and effectiveness of FSW and for improving practice guidelines for using FSW in industry [9].

Yaduwanshi *et al.* (2024) [10] analyzed the heat input and thermal profile of FSW, Plasma FSW, and TIG FSW for aluminum welding. Heat generation was modeled using ABAQUS software, auxiliary arc preheating (45 amp), and the welding speed of 63 mm/min. Plasma FSW values high peak temperatures, 50%–55% of base metal melting, the efficiency of the arc, and high energy level increasing the plasticization and heat control. Discussing the outcomes of this study reinforces the development of new types of hybrid FSW opportunities for industries.

Al-Wajidi *et al.* conducted research on FSW with a focus on 5052-H32 and 6061-T651 Al alloys under MQL in the year 2024 [11]. While exploring the MQL factors, such as flow rate, nozzle orientation, and diameter, they apply the factor of variance and gray relational analysis of welding force, power consumption, and surface roughness. The probable and ideal parameters are 7. The optimization of RPO is at a rate of 5 ml/h, where surface quality is enhanced with the reduction in welding force and power used by 25% and 20%, respectively. Other factors increasing FSW performance include a nozzle orientation of 60° and a 2.5- to 3. 75-mm diameter.

Rabiezadeh *et al.* (2023) [12] also performed self-reacting friction stir welding between 3-mm-thick aluminum alloys: AA5083 and AA7039. To ensure the joints produced had minimum or no defects, they adjusted other parameters such as the transverse speed, the tool rotation speed, and the shoulder pinching gap. High junction efficiencies were demonstrated by tensile testing: former PM's statement 94.5% for AA5083–AA5083, 80.4% for AA7039, as well as 97.4% for AA7039–AA5083. These outcomes were almost to the same level and sometimes even surpassing the conventional friction stir welding. Macroscopically similar joints were seen to have a high ductile failure, while similar joints had very few numbers of brittle failures.

In their study, Kumari and Acherjee [13] also proposed a new combined approach under MCDM, namely, CARCACS for tool material selection in FSW. The abbreviation of criteria-based ranking tools is CARCACS, which is Correlation Analysis for Criterion-wise weights along with Alternative Ranking for determining the weights of the criteria and ranking the alternatives. Convenient parameters that are business-oriented signify normal CARS, excluding variations in the normalization process and weight assessment. Thus, if using CARCACS, the decision maker could address the whole process of getting the decision where H13 tool steel is defined as a proper material for FSW in the case of the need to weld the aluminum-alloy to steel material. It means that by practicing FSW, aspects such as welding performance, impacts that hit the environment, and legislation can be enhanced. Thus, FSW is positioned in the framework of industries searching for sustainable manufacturing solutions as a technology that allows for minimizing negative effects on the environment and reducing costs, as well as improving welding capabilities. It signposts a major advancement toward attaining sustainable manufacturing and following the strict environmental standards for maintaining environmental quality and enhancing workers' well-being.

21.2.2 Laser Beam Welding (LBW)

In laser beam welding, it is possible to melt the metals and join considering the high-intensity laser beam. In this method of control, there is little use of supplementary material, which in turn uses low energy. For higher levels of accuracy and productivity than PWB, LWB is preferred because it produces fewer contaminants when welding.

In contrast with the earlier models that had accuracies of 30.23% and 42.5%, Pamarthi V. V. *et al.* [14] observed that their novel monitoring technique identified the through-thickness molten pool shape with 94.3% accuracy. They observed a structure at the rear end, which looked like a tail on a pool of molten liquid, which influenced the crystallization pathways and led to early solidification, which increased the tensile strains. To minimize the shrinkage strains in the fusion zone and the tail-like feature, there was a positive change in the power ratio from 0.36 to 1.5.

Saurabh, S. K. *et al.* [15] investigated a 1 using 1,400 W of power, a speed of 1,200 mm/min, and a heat input of 70 J/mm at room temperature. The thickness of the base plate is 2 mm of Nimonic 80A super alloy suitable for CO_2 laser welding. Yield and ultimate tensile strengths are presented as 881 MPa with an average strength of 710 MPa for the tensile test. The increase in Vickers hardness started toward the base metal, and it was 280 HV near the centerline of the fusing zone. When the metallurgical studies were carried out, zones were observed to have equivalent phases at different localities, and fine-grained weld joint attributes were established using electron microscopes, optical microscopes, and X-ray diffraction methods.

Since laser beam welding processes need to be analyzed numerically, Hartwig, P. *et al.* [16] present a heat source model with a physical base. This is seen clearly by them concerning the efforts of calibrating existing models such as the Conic or Goldak models. Here, from the previous CFD calculations of Lamé curves, the melting isotherm is moving as a Dirichlet boundary condition for the FE calculation.

Heat input is directly set as body load in the new model, which calculates the melting isotherm on the base of CFD calculations. To compare with the new model's numerical data, we use the Lamé curve model and also CFD simulations with experimental data.

Some of the benefits of LBW according to Efa D. A. [17] include small HAZ, accuracy, flexibility, and automation. The selection of a setting for the process and the treatment of material that exhibits high levels of thermal conductivity together with low melting points are challenges. The following process parameters are predicted by the study utilizing Response Surface Methodology (RSM), Genetic Algorithm (GA), and backpropagation neural networks: the power is distributed through the beam ranging from 4,000 to 6,000 W, the spot size from 0.4 to 0.8 mm, welding speed from 30 to 50 mm/s, and segment from 5 to 15. From the study, it was established that the peak temperature depends on the beam power, while the residual stress depends on the segment number.

Concerning typical problems, Jiang, Y. *et al.* [18] studied the vacuum laser welding of magnesium alloy AZ31. They established that when the pressure drops from 101 to 10 kPa, the hump defect is enhanced, together with a diminished penetration depth, while the breadth of the weld lessens; however, the pressure drops to as low as 1 kPa and 0. The results of 1 kPa were equally effective. Works on using a high-speed camera proved turbulent with unsteady flow in the essential keyhole region at a pressure of 101 kPa, which became steady at 10 kPa. A newspaper article published in Practical Welding Today showed that the weld's tensile strength, elongation, as well as grain size were considerably higher in a 10-mm-thick bottom-locking weld made under vacuum due to a higher percentage of fine grains and lower porosity.

Conforming to the study of Heilmeier J. *et al.* [19], the mobility sector has transitioned toward electrified powertrains with the aim of cutting down on global emissions. Thus, for making traction battery connections and proper production of electric vehicles, laser beam welding is indispensable. The study employed the use of an optical microphone and a laser welding monitor to record the spectral and acoustic vibrations in real time in a bid to detect the weld faults. The capabilities for classifying defects with the sensors are given by statistical measures that are derived from signal correlations to systematic disturbance sources such as dirty surfaces and misalignments.

The brief of findings within the present literature related to laser beam welding has been briefly discussed in the following table. It deals with the issues concerning the heat source models, molten pool behavior, properties of the welded alloys, optimal parameters for arc welding, the influence of vacuum, and identification of defects by spectral and acoustic signals.

These studies show how laser beam welding technologies are being developed and applied in a variety of materials and industrial areas.

21.2.3 Using Electron Beam Welding (EBW)

An intense, fast-moving electron beam is used in electron beam welding to fuse materials together. Using a hoover, this technique prevents hazardous gases from escaping. EBW produces precise, deep welds with minimal energy input and has a lower negative environmental impact because of its high efficiency.

Sisodia R. *et al.* [20] used electron beam welding (EBW) to examine the effects of *in situ* localized electron beam post-weld heat treatment (LEB-PWHT) on high-strength S690QL steel welded joints. Impact, bending, tensile, and microhardness tests are used to evaluate the mechanical properties; optical microscopy is also used to examine the microstructure.

A tensile fractography SEM examination shows better fracture behavior with fewer micro-voids and dimples. The results show that weld quality can be improved using LEB-PWHT in S690QL steel applications. These features include a 5% increase in tensile strength, a decrease in HAZ hardness, and a sevenfold increase in toughness.

To increase efficiency and decrease heat input during laser beam welding, Troise, M. *et al.* [21] investigated the advantages of pulsed lasers over continuous wave lasers. These gains are especially noteworthy for workpieces with thin walls.

In aluminum alloys, pulsed techniques are known to reduce hot fractures and produce smaller grain sizes. Like pulsed lasers, pulsed electron beam welding has drawbacks, such as porosity and spikes in deep welding applications, despite its potential advantages.

Developing a pulsed electron beam welding procedure for AW 6061 sheet metal is the main goal of the study, which also discusses ideal parameters and efficiency benefits over continuous welding techniques.

Dawson H. *et al.* [22] investigated the effects of electron beam (EB) welding on PM2000 ODS steel using traverse speeds of 1,200–2,200 mm/min and heat inputs of 72–39 J/mm.

Analytical electron microscopy and microhardness measurements of the welds revealed minor differences in the fusion zone width and microstructure among the four butt welds. In the fusion zone, large columnar grains with coarsened Y–Al–O particles and a lower density than the base material were seen in a ferritic matrix.

Dutilleul T. *et al.* [23] state that electron beam welding (EBW) has a significant impact on welding thick-section components in the nuclear industry. This applies to Small Modular Reactor (SMR) projects. The process works well in one go, but operators need to be careful because it might fail. In chamber setups, like the Pro-beam K2000 system at Nuclear AMRC, workers use electron-optical image analysis to track seams. This helps to line up the beam with the joint. The equipment makes the process more reliable by boosting overall weld quality and making operations more effective. It spots problems and helps to make things better before welding starts.

Peng H. *et al.* [24] examine the electron beam welding (EBW) of L12-nanoparticle-strengthened medium-entropy alloys (MEAs), Ni43.4Co25.3Cr25.3Al3Ti3 and Ni42.4Co24.3Cr24.3Al3Ti3V3 (at.%). EBW creates strong flexible joints with large columnar grains in fusion zones (118–245 μm) bigger than those in heat-affected zones (15.6–22.3 μm) and base materials (4.6–5.6 μm). Both alloys show high yield strengths (838–858 MPa) ultimate tensile strengths (1,416–1,420 MPa), and good fracture strains (20%–21%) at 77 K. Dislocation networks stacking faults, and nanoscale deformation twins boost strain hardening and flexibility after fracture.

The advancements in post-weld heat treatment, pulsed welding techniques, microstructural effects on steel alloys, large component manufacturing capabilities, and welding of nanoparticle-reinforced alloys are the main topics found in the table that summarizes recent research about electron beam welding (EBW). This research shows how EBW technologies have been developed and applied to different materials and industries.

21.2.4 Gas Tungsten Arc Welding (GTAW) or TIG Welding

TIG welding also known as Gas Tungsten Arc welding, makes welds with a non-consumable tungsten electrode. This technique creates quality welds and has a reputation to be precise. When done right, TIG welding can save energy and create less waste and pollution than other common methods.

Zhang P. and others' research [25] looked at how well CoCrFeMnNi high-entropy alloy (HEA) and 304 stainless steel (304 SS) could be welded for use in planes and nuclear plants. The gas tungsten arc welding process made perfect butt joins. SEM/XRD tests showed an unmixed area on the 304 SS side because of quick cooling and a fusion zone with just one FCC phase. The fusion zone was a bit harder because it had carbon and smaller grains. Stress tests showed that joints with 38% stretch and about 465 MPa strength worked well for room-temperature building needs.

Xia Y. [26] examines how ultrasonic-assisted gas tungsten arc welding (UA-GTAW) improves the mechanical properties and microstructure of

Inconel 690 alloy, a key component in nuclear applications. Despite the alloy's reputation to resist corrosion and stay strong at high temperatures, large particles often show up in the material after welding. UA-GTAW boosts mechanical performance by fine-tuning solidification grains and reducing element separation. UA-GTAW joints show much higher yield strength ultimate tensile strength and elongation than regular GTAW joints. Their specific values jump from 320 MPa, 591 MPa, and 25.1% to 387 MPa, 672 MPa, and 31.6%.

Skripko S. *et al.* [27] looked into how discharge parameters affect early spatial instability in single-pulse electrode-negative gas tungsten arc welding. They examined peak arc pressures (P_{peak}), time to peak pressures (T), and final arc pressures. As the current amplitude went up from 50 to 200 A, P_{peak} (0.25 to 4.00 kPa) and P_{final} (0.16 to 1.72 kPa) increased. Bigger sharpness angles from 30° to 90° led to lower P_{peak} and P_{final} values. Changing the electrode diameter from 1.0 to 2.4 mm had an impact on these parameters. The stabilization times spanned anywhere between 5 and 35 ms, these being proportionate to what type of electrode and its properties.

21.2.5 Cold Metal Transfer (CMT)

Cold metal transfer is a sophisticated version of gas metal arc welding (GMAW). It operates at low temperatures thereby reducing heat distortion and preventing droplets from forming. It is an interesting alternative for companies that wish to reduce energy expenses while still improving weld quality because it requires less power consumption and produces clean welds. Karim *et al.* [28] analyzed the wire-arc directed energy deposition (DED) method in relation to the cold metal transfer (CMT) process to design a bimetallic structure made up of 4043 aluminum (Al) and 316L stainless steel (SS). The impact of flux heat input was studied on geometry, porosity, and microstructures with a special focus on the formation of intermetallic compounds (IMCs), particularly $FeAl_2Si$. IMC thickness varied between 5 and 18 μm under different levels of heat flux input. One of the highest tensile strengths (130 MPa) for steel/Al bimetals ever recorded was achieved almost perfectly. This implies that higher levels of energy caused worse mechanical efficiency coupled with thicker intermetallic compounds, which could also be more fragile.

Wire and arc additive manufacturing (WAAM) is acknowledged far and wide for its ability to quickly and affordably create huge, complex parts. This investigation by Wang J. *et al.* [29] used Cold Metal Transfer wire arc additive manufacturing (CMT-WAAM) to build samples from 316 austenitic stainless steel at different welding currents. Bead formation and maximum arc travel speed (MAWFS) were evaluated at 100 to 175 A currents in CMT and CMT pulse (CMT+P) modes. The EBSD study also demonstrated that CMT+P mode improved the midsection industrial applicability by reducing the average grain size from 13.426 to 9.429 μm.

Fan *et al.* [30] looked at the ZChSnSb11Cu6/Q235B composite's preparation using swing cold metal transfer (CMT) welding. The welding experiments were compared with numerical simulations through heat transport theory coupled with a double ellipsoid heat source model. It was demonstrated that the models accurately reflected changes in molten pools and heat cycles, which has been proven by experiments. A scanning electron microscope investigation found out swing CMT welding significantly strengthened interface bonding strength. Conversely, while distance from the weld centerline led to lower heating rates, peak temperatures increased at measurement points with increased welding current.

Wire arc additive manufacturing (WAAM) is a technology that has gained attention for its efficient resource utilization, particularly in large-scale production, maintenance, and structural applications. However, there are still challenges such as unsatisfactory geometry and uneven surfaces. It takes precision to determine dimensions and machining allowances, which must be adhered to, to obtain specifications of the final part. A design of experiments (DOE) and heat analysis revealed how various parameters from the cold metal transfer (CMT), including travel speed and interpass time, influence component dimensions and machining allowances. The data that were acquired in this way were used in coming up with strategies to achieve specifically set requirements [31]. Additive synthesis of components made from Al–Zn–Mg–Cu alloy was done using Metal Inert Gas Welding (MIG) and Cold Metal Transfer Advance (CMT-A). In contrast with MIG, CMT-A deposited samples had smaller grains, fewer η'-strengthened precipitates, and a supersaturated solid solution. Peak aging hardness increased by 36.8%, while hardness increased by 16.1% in comparison with MIG for CMT-A samples. This suggests that high-strength Al–Zn–Mg–Cu parts can be produced directly using CMT-A without necessarily carrying out the re-solutionizing process [32]. Recent research compiled information on different applications of Cold Metal Transfer (CMT) during welding and additive manufacturing processes which these studies support.

This allows industries to save on energy costs, reduce emissions, and improve their welding capabilities considerably. Additionally, these developments promote environmental sustainability while increasing productivity and reducing expenses in manufacturing processes.

21.3 Exploring the Frontiers of Green Welding: Latest Scientific Studies

Kuppusamy Vellingiri and co-authors (2023) focused on two different methods of extraction. They experimented with gas metal arc welding (GMAW) techniques employing filler wires of various diameters made out of SS 316 L and ER 316 L stainless steel grades. For comparison purposes, cheap welding hoods were selected, one having a square duct and another with a conical one to see whether they can capture welding fume effectively. Fumes were captured through glass fiber filters whose current intensity and gas flow rates were changed as required. It was found that in terms of removing gases, a hood with a conical duct would be more effective than one with a square duct; especially when CO_2 was added to the shielding gas combinations [33].

Heramo and Workneh (2023) optimized friction stir welding parameters for distinct aluminum alloys, namely, AA6061-T6 and AA5052-H32 [34]. They achieved the highest possible tensile strength of 183.04 MPa using square pin form, a transverse speed of 40 mm/min, and a rotational speed of 1,400 rpm. It was shown by the results obtained from the investigation that the joint characteristics as well as the overall efficiency of the processes are significantly affected by tool pin profile.

In their research on eco-friendly technology for welding operations, Rachman (2023) [35] stresses waste minimization, energy optimization, and emissions reduction. The effect of electrode modifications (E6010, E6013, E7016, and E7018) on ASTM A36 steel during shielded metal arc welding (SMAW) is studied here. Through highlighting how similar techniques affect environmental performance and welder competence employing a 4 × 3 factorial design, this article advocates for Friction Stir Welding (FSW) as a better alternative in terms of green technologies among other welding processes.

An experimental investigation was carried out by Egerland *et al.* (2023) [36] aimed at reducing exposure to fumes during arc welding, particularly while using Metal-Active-Gas (MAG) techniques. Moreover, the complete elimination of fume emissions even if their production is reduced is hard. Therefore, this study investigated various technology welder protection measures in typical manufacturing environments. It had fume emission rate measurements, chemical element analyses, and exposure assessments, which were very instrumental in improving safety precautions during welding. Further evaluations showed that fume exposure assessments beyond the welding processes were influenced by environmental conditions.

The development and state of laser and laser-arc hybrid welding methods for high-strength structural steels are covered in Siltanen (2023) [37]. Laser welding, which was once thought to be a specialized joining technique, has gained popularity as a result of lower operating costs and more readily available laser welding equipment. Smaller businesses may now afford to invest in the laser technology sector, which has grown due to lower euro/kilowatt ratios. One-micron lasers currently dominate this market. Notwithstanding the COVID pandemic's short-term difficulties, the laser technology sector is expanding and preparing for more rise, especially in fields like laser cutting.

The application of laser welding in lithium-ion prismatic cells is examined by Qu *et al.* (2023) [38], with an emphasis on non-jellyroll components such as connectors, metal cans, lids, and terminals. Even though laser welding is widely used, mechanical pressure can sometimes cause connections between these structures to fail too soon. The work replicates welding connections using mechanical testing on laser-welded sections using the Cohesive Zone Model (CZM) to construct a Finite Element Model (FEM). Based on models and experiments showing that laser-welded connections between non-jellyroll structures can break during intrusion testing, strong failure requirements are essential for prismatic cell design.

Touileb *et al.* (2023) [39] investigate the application of wire arc additive manufacturing (WAAM) to create large-volume metal components and reinforce structural features such as half-cylinder shell forms. The work focuses on minimizing thermally induced distortions through experiments and transient thermo-mechanical finite element computations. After welding beads were applied to the specimens, displacement transducers and thermocouples were used to capture heat histories and distortion data. To forecast temperature fields and distortions, the verified simulation model assessed a variety of deposition patterns and shell thicknesses. The findings highlighted the significance of weld bead orientation and sequence, which decrease with increasing shell thickness, by demonstrating a non-linear relationship between distortion and shell thickness.

Creep tests were performed on T91/Super304H welded joints by Miao *et al.* (2023) [40] at 610°C and 180–150 MPa stress. They used finite element simulations to comprehend failure mechanisms and utilized OM and SEM to assess microstructural alterations. Martensitic T91 was shown to be the primary factor controlling creep deformation; the T91 heat-affected zone was found to be susceptible. Creep life rose as stress dropped, changing the fracture type from ductile to brittle. Phases containing Nb enhanced the creep strength of S304H steel. Concentrations of stress at phase interfaces led to creep holes, which developed into cracks in accordance with the type IV cracking model.

The shielding effect of electrode frames in electrostatic precipitators (ESPs), which are essential for removing dust from flue gases, was examined by Zhao *et al.* (2023) [41]. To evaluate the properties of corona discharge and the distribution of current density on the collector plate surface, they constructed an experimental setup consisting of RS barbed electrodes and a 480 C-type dust collector. Results showed that electrode frames considerably reduced current density across them to almost zero, shielding corona discharge. To reduce dust escape pathways, a novel ESP design with a split-level frame arrangement has been proposed. This shielding effect reduces the efficiency of dust collecting.

Rahul *et al.* (2024) [42] developed a new SMAW-coated stainless-steel electrode to reduce fume and Cr(VI) during welding. They replaced nanoparticles in the flux and coated the core wire with nano-arc stabilizers. Compared to traditional electrodes, fume production was reduced by 82% and breathing zone concentration by 85%. Size-reduced synthetic rutile and nano-zirconium improved arc stability, reduced spatter, and reduced Cr(VI) in respirable and inhalable fumes by 68%.

21.4 Welding and Industry 4.0

Interoperability, or connectorship, is one of Industry 4.0's four core principles. It demonstrates how seamlessly Cyber-Physical Systems (CPS), robots, and people may communicate via networks. The aim of welding interoperability is connectivity between the welding apparatus and a central computer. This makes it possible to gather and interpret sensor data remotely in real time; this is referred to as "tele-monitoring" (or, more accurately, "tele-welding" in welding applications). This approach allows multiple users to access and centrally store data from multiple sensors at the same time without having to be physically close to the equipment. Because it allows online monitoring and management of welding operations across several machines within a workplace, machines, sensors, and human operators are all integrated into a single networked environment [43].

As the "Industry 4.0," the current industrial revolution, the same processes are convened by new products in addition to the logistics behind them and a growing consumer gang wanting ecologically friendly items. Data collection and analysis are the main focus, as digital tools are integrated into the production process. The "Internet of Things" (IoT) and

"Cyber-Physical Systems" (CPS), which are instrumental in making it feasible for networked devices to control physical objects via computers and to exchange data globally, are the most indispensable units of Industry 4.0. In the production apparatus, the sensors are utilized to gather data in real time; hence, the process management and the decision-making cycle are sped up. The progression of welding techniques is portrayed, with the beginning of early attempts to accelerate deposition and finishing with more advanced processes such as gas metal arc welding (GMAW) and resistance spot welding (RSW) [44].

The transition of manufacturing from industrial to the fourth age is almost exclusively due to globalization. Fluctuation or lack of precision products is one of the reasons why high partial differentiation is so commonplace today. The conventional means of manufacturing are fading out, which is why smart tools have been introduced, which are based on a combination of different sensing technologies. Also, these inventions help in the enhancement of the quality of the product and the reduction of the cost of the product, thus, making them a major revolution in manufacturing. The pathway from the formation of the first industry to the Industrial Revolution of 4.0 signifies the journey of the industry pointing to the advancements in welding as a crucial part of modern manufacturing.

Welding automation came into existence due to the old terms "automation" and "mechanization," which stand for the use of mechanized systems that can autonomously complete manufacturing processes. The main accomplishment is to make human participation in production a lot less or to eradicate it completely. This is done by electronic and computer components that have been custom-made for this purpose.

The rise of Industry 4.0, which is a result of the pressure exerted by globalization, was the reason that the agility and flexibility of manufacturing became a necessity. The adoption of Industry 4.0 embodies the transformation that took place, which was due to the development of high-grade products. Traditional manufacturing is no longer viable, which is why they rely on such smart systems that react to sensory input to decide what to do next. These new technologies elevate the quality of the products and, in addition, also the cost. That is, they are a definite breakthrough in production. The transition from the first industrial revolution to Industry 4.0 is the example that best characterizes this journey presenting modern welding as an essential part of the manufacturing process.

Locking technology, which is another term for welding, which began from terms, such as "automatic" and "mechanization," is the idea of a set of fully automated systems, which can run manufacturing operations independently. Its main purpose is to cut down or even entirely remove manual inputs in the process of manufacturing, and as a result, it assails ecological preservation. This custom-designed provision is made through electronic and computer-controlled devices. The well-functioning of an automated production line is the best proof thereof, which is a case where the chassis production, the assembling of parts, welding, and painting are done in an organized process of a sequence. The model, known as "fixed automation," allows the transfer system to switch between the workstations. In contrast, "programmable automation" is the programmable automation of CNC machines through numerical coordinates input via computer, which automatically places the previous tool in position during operations.

Different possibilities demonstrate the intercommunication of mechanical and electronic subsystems. In totally autonomous production lines, electronics aggregates containing electrical motors and robots make up the mechanical part, while controllers, servo motors, sensors, and automation software work as electronic components. The relation between the mechanical and electronic parts in this system is established through microcontrollers in the controllers that manage mechanical functions, process the sensor data, and produce an output [44].

The latest machines match the electronic parts with mechanical systems, thus subsequently adding flexibility and reliability, also, paving the way for a sustainable solution and adopting a green approach. The method of monitoring the equipment, with the aid of sensors installed in mechanical parts, empowers the company to run continuously and online, by tracking the machine's condition and always staying on top with the best possible equipment.

The concept of estimating the state of certain machines is indeed the only part of process automation that directly confronts human experience as it is done in the field of artificial intelligence. In a welding scenario, a process of high complexity, various parameters are more critical in the deciding stage. For instance, in arc welding, the voltage and the current; in the case of resistance spot welding (RSW), the force and the current shall be selected; in laser welding, the beam power and frequency are the critical ones with these dependencies. Besides, elements like part mismatch, variation in groove depth, the length of overlapped parts in RSW, and the reflectivity of the workpiece in laser welding will have a significant effect on the welding geometry. Adequate control of these parameters guarantees high welding quality.

If we are to monitor welding processes, we will gain the benefits of insight into the weld seam, checking the depth of penetration, finding welding defects, taking the measurements of the weld zone, and last, demonstrating power source performance. Monitoring can be done in two ways: offline or online—offline is when we install and immediately check the weld quality, and later check that we are on time. This is the more flexible method and is not dependent on immediate implementation; offline monitoring can be done at another time. It is the same as the first time but the answer is no.

Product quality is a priority in the welding industry because welding is the last step in the manufacturing process. Welding automation consists of the invention of robotic and infrastructure facilities for the maintenance of the welding process in a sustainable manner. The most important thing is to control the machine-made operations where the process itself has so many variations. One of the main issues is the manual arc welding that can create spatter because of erroneous parameter settings, which means that the whole cleaning operation will take too long. Automation is application independent; therefore, it may resolve these problems by completely ruling out human errors, by constantly keeping the specified parameters as the loop control takes its function through the feedback systems, which also maintains the goal of superior quality in the welding process using minimal reworks, and by the productivity gains it brings.

The progress of Industry 4.0 is far from the small beginnings of the concept alone, as it has evolved to the point where it has included such advanced technologies as IoT (Internet of Things), AI (Artificial Intelligence), and automation to provide efficient manufacturing. The essential items that it concentrates on the most are such things as interoperability, which provides networking connections for both Cyber-Physical Systems (CPS) and machines with humans, allowing them to interact effectively. The main advantage of using interoperable systems is that you can interconnect and get data in real time from welding equipment. This connection allows for real-time data analysis as well as remote access. Typically, this method, so-called "tele-welding," is accomplished by the connection of welding tools to the main computer, which reduces or eliminates the need for presenting one's self to the machine and thus gives a quicker processing time, causes higher productivity, and helps to make decisions on the basis of the presented data [45].

21.5 Welding and Industry 5.0

The very idea of Industry 5.0 is coming to the forefront, as the concept has been largely spread among the business circle as the next stage of development in the industry. Industry 5.0, which is quite the opposite of Industry 4.0, specifically highlighted connectivity and automation, whereas Industry 5.0 has transferred the humans' position in production to the main priority by using creativity and dexterity to operate the machines in an exceptionally efficient way. In welding, Industry 5.0 is probably making use of advanced robotics, co-bots, and AI-driven systems that improve the worker position with better techniques and capabilities. This development is directed at improving the quality of products and the sustainability of the manufacturing process, which part would entail addressing such challenges through the exploitation of the current technology and human skills [46].

Welding technologies like friction stir welding (FSW), laser welding, and electron beam welding (EBW) are expected to play a big part in the future of Industry 5.0. These methods are big supporters of what this new industrial era is all about. The main goal here is to use advanced tech to save resources, make workplaces safer for people, and cut down on harm to the environment.

Lessening the Environmental Impact: Green welding helps to save more energy than traditional arc welding because of the following reasons. FSW has a low melting point temperature, so FSW consumes less energy than arc welding. Similarly, FSW does not require any extra electrodes or gases, which cuts down further greenhouse gases.

Furthermore, reducing gases leads to a reduced footprint on the environment for FSW welds. Laser welding is also quite an efficient technique in the welding process. The laser fiber welds by direct heating of the material, which means that less heat is used. Lesser use of heat means fewer pollutants. Ultimately, lower energy leads to a lower negative impact on the environment.

Welding to Utilize Resources: There are various types of welding, which are performed to fabricate different objects. Green welding makes different types of welding techniques more efficient in a way that they waste less resources. Also, since FSW does not melt the base material, the material properties are fully preserved, and, again due to this, thinner material can be utilized, and it can also significantly contribute to the reduction of material waste in the manufacturing process, which is another core aspect of sustainability.

Enhanced Worker Safety: This is why it is logical that one of the measures linking automation of green welding applications to constructive benefits is an improvement in the safety of workers. One of the situations that might not be comfortable for a human operator is where quantitative robotic systems can be used to automate laser welding. For instance, laser welding can be mechanized or employed by intelligent machinery; this way, workers cannot be endangered by hazardous gaseous fumes and ultraviolet rays emitted in the normal welding process. Also, electron beam welding can be carried out in vacuum environment.

Integration With Industry 5.0 Technologies: Industry 5.0 technologies have also advanced the green welding process where AI and Machine learning can have a look at large data sets of welding processes and improve some of the parameters like welding speed, power setting, and material fed rate to not only stabilize the process but also would require less energy, fewer emissions and less waste. Real-time monitoring systems using sensors allow for the modification to improve the process in real-time and are one of the ways to prevent weld defects, hence increasing the efficiency of the weld process.

Collaborative Robots (or Cobots): A Cobot can be employed in the green and assembly welding zones; it takes over the tedious tasks, such as feeding and positioning of parts when working along with the human operator. This lowers intensity on the shop floor and enhances the security of manufacturing through the formation of sets of people with higher skills in carrying out the last part—finishing processes that involve judging as well as hand work.

Therefore, using green welding and optimization of Industry 5, welding technologies, even the 0 technologies, wield an opportunity to enhance the sustainability of the procedure among the manufacturers. This integration not only fosters organizational commitment toward environmental protection and meeting the set legal requirement but also increases operation capacity and viability, especially in a world where resources are limited. Similarly, industries seek sustainable manufacturing practices and green welding emerges from the pack with viable solutions that can satisfy the necessity of cleaner, safer, and efficient production for worldwide users [47–49].

21.6 Conclusion

Today, many industries pay special attention to sustainability and the minimization of negative effects on the environment, and this factor has led to the development of a great number of environmentally friendly technologies in the welding industry. These innovations are especially significant in meeting increasing energy requirements and exacting carbon reduction standards that define the current manufacturing environment. Consumers and governments continued putting pressure on the welding industry to reduce pollution and protect the health of the welders and the environment from the adverse impacts of welding. This review has therefore gone further than just the topic of Friction Stir Welding (FSW) technology to discuss other green welding technologies, the benefits as well as the uses of these technologies in sustainable manufacturing. Major technologies that have taken root and implemented in this regard include Laser Beam Welding (LBW), Electron Beam Welding (EBW), and Cold Metal Transfer (CMT). Each technology offers unique benefits: Hence, LBW and EBW have better accuracy, and the least heat-affected zone, CMT, has the least spatter and distortion-free zones, and all of them have small wastage of material and energy compared to conventional techniques. This chapter has compiled vital research that has highlighted these technologies' potential to enhance welding processes. Investigations carried out on microstructural changes, mechanical characteristics, and thermal processes illustrate their capabilities of increasing weld quality, cost effectiveness, and saving a huge amount of energy in various fields. Thus, the chosen examples highlight the role of the industry stakeholders in implementing the principles of sustainable development in the process of manufacturing products. Regarding the future perspectives, the application of green welding technologies is fully consistent with the directions of Industry 4.0 as well as the changing meaning of Industry 5.0. These paradigms focus on intelligent methods of manufacturing management with special regard to the rationalization of energy usage and material consumption, and protection of the environment. Besides compliance with environmental legislation standards, the industry also popularizes and progresses environmentally sound welding methods for enhancing competitiveness in the global market. To sum up, this chapter has aimed to showcase green welding technologies and stressed the significance of their application for improving the sustainable development of various sectors of manufacturing in the world. With these innovations, it can be possible for long-term sustainability to be observed both in lessening the impacts of environmental problems as well as in amplifying the productivity and reliability of welded structures and all other industrial applications.

21.7 Future Scope of Green Welding Technologies

The following discusses some key areas for the future scope of green welding technologies:

21.7.1 Technological Advancements

Hybrid Welding Techniques: By integrating such green welding methods as FSW with the currently prevailing technologies like arc welding, one may achieve even better efficiency as well as increase the versatility of the technology.

Automation and Robotics: Consequently, there is a need to incorporate green welding technology with automation and robotics since it increases precision, consistency, and the ability to operate remotely, hence increasing the safety and productivity of the process.

Artificial Intelligence (AI) and Machine Learning (ML): Real-time changes in the parameters of the welding process through AI and ML for green technologies' welding can lead to energy savings and enhanced weld quality.

Advanced Materials: Further and more research into new materials that can specifically be used in green welding processes will be useful in increasing the usability of the processes as well as their efficiency.

21.7.2 Sustainability Focus

Life Cycle Assessments (LCA): The rationale for carrying out LCA analyses for green welding technologies is that such a study will enable the ascertaining of the impacts of these technologies throughout their life cycle.

Standardization and Regulations: The policies and standards can, in the long run, persuade additional welders and industries to embrace environmentally sustainable practices when welding since they would be created to uphold the high standards of materials processed by the welders.

Waste Management and Recycling: However, more research can be done concerning eco-friendly approaches when it comes to the disposal of wastes produced during green welding processes.

21.7.3 Industries 4.0 and 5.0 Integration

Data Analytics and Cloud Computing: Applying data analytics and cloud computing in green welding may help to find the best solutions in resource management and energy use.

Digital Twins and Predictive Maintenance: Incorporating high-fidelity green welding systems and integrating digitized twins with predictive maintenance techniques could help save time and eliminate or reduce downtime.

Human–Machine Collaboration: Thus, technologies, such as cobots, as well as qualification upgrading, will facilitate the transition to environmentally friendly robotic welding.

21.7.4 Broader Applications

Additive Manufacturing: Stating, green technologies and welding technologies integrated with AM techniques can allow the formation of new structures of higher capacity and sustainability.

Repair and Refurbishment: Carbon-conscious welding or low carbon intensity welding can be looked into for more environmentally friendly approaches more so in repairing as well as refurbishing existing structures and equipment.

Emerging Industries: Studying and expanding green welding applications in new fields, such as aerospace, renewable energy, and EVs, may help advance their sustainable initiatives. By focusing on these areas, the future of green welding technologies holds immense promise for creating a more sustainable and efficient manufacturing landscape.

References

1. Kumar, K., Zindani, D., Davim, J.P. (Eds.), *Sustainable Manufacturing and Design*, Woodhead Publishing, Cambridge, United Kingdom, 2021.
2. Bala, Y.G., Santhi, B., Dinesh Kumar, R., Sustainability in Welding Industries, in: *Automation in Welding Industry: Incorporating Artificial Intelligence, Machine Learning and Other Technologies*, S.Q. Moinuddin, *et al.* (Eds.), pp. 215–228, Wiley-Scrivener, Hoboken, New Jersey, USA, 2024.
3. Singhal, T.S., Jain, J.K., Kumar, M., Bhojak, V., Sustainability in manufacturing: Material forming and joining processes, in: *Sustainability in Smart Manufacturing*, S. Shah, *et al.* (Eds.), pp. 118–135, CRC Press, Boca Raton, Florida, USA, 2024.
4. Torres-Treviño, L., Escamilla-Salazar, I., González-Ortíz, B., On developing a green and intelligent manufacturing system. *Expert Syst. Appl.*, 243, 122876, 2024. https://doi.org/10.1016/j.eswa.2023.122876.
5. Phillips, D.H., *Welding Engineering: An Introduction*, John Wiley & Sons, Hoboken, New Jersey, USA, 2023.
6. Yang, C., Chen, G., Qiao, J., Wu, C., Zhou, M., Zhang, G., Shi, Q., Material flow during dissimilar friction stir welding of Al/Mg alloys. *Int. J. Mech. Sci.*, 272, 109173, 2024. https://doi.org/10.1016/j.ijmecsci.2024.109173.
7. Chen, S., Zong, G., Kang, C., Jiang, X., Digital Twin Virtual Welding Approach of Robotic Friction Stir Welding Based on Co-Simulation of FEA Model and Robotic Model. *Sensors*, 24, 3, 1001, 2024. https://doi.org/10.3390/s24031001.
8. Abdollahzadeh, A., Vanani, B.B., Koohdar, H., *et al.*, Influence of Variation Ambient System on Dissimilar Friction Stir Welding of Al Alloy to Mg Alloy

by the Addition of Nanoparticles and Interlayer. *Met. Mater. Int.*, 30, 2830–2852, 2024. https://doi.org/10.1007/s12540-024-01670-4.

9. Abboud, M., Dubourg, L., Racineux, G., Kerbrat, O., Experimental Methodology to Identify Optimal Friction Stir Welding Parameters Based on Temperature Measurement. *J. Manuf. Mater. Process.*, *8*, 4, 137, 2024. https://doi.org/10.3390/jmmp8040137.
10. Yaduwanshi, D.K., Rao, C.R.M., Naidu, S.C.V.R.M., *et al.*, Thermal evaluation of aluminum welding: a comparative study of friction stir welding (FSW), plasma-fsw, and tungsten inert gas (TIG)-FSW techniques. *Int. J. Interact. Des. Manuf.*, 18, 5501–5513, 2024. https://doi.org/10.1007/s12008-024-01796-0.
11. Al-Wajidi, W. and Eltaggaz, A., Multi-objective optimization and experimental investigation of friction stir welding under Minimum quantity lubrication. *Proceedings of the Institution of Mechanical Engineers, Part E: Journal of Process Mechanical Engineering*, vol. 0, 2024, https://doi.org/10.1177/09544089241228697.
12. Rabiezadeh, A., Salafzon, A., Mostafavi, N., Dissimilar welding of AA5083/AA7039 by self-reacting friction stir welding. *J. Adhes. Sci. Technol.*, *38*, 1, 70–91, 2023. https://doi.org/10.1080/01694243.2023.2219367.
13. Kumari, A. and Acherjee, B., A novel hybrid multi-criteria decision methodology for assessing tool materials in friction stir welding process. *Int. J. Interact. Des. Manuf.*, 18, 4963–4986, 2024. https://doi.org/10.1007/s12008-024-01783-5.
14. Pamarthi, V.V., Sun, T., Das, A., Franciosa, P., A novel methodology for investigating the through-thickness molten pool shape during remote laser beam welding. *Opt. Laser Technol.*, *175*, 110794, 2024. https://doi.org/10.1016/j.optlastec.2024.110794.
15. Saurabh, S.K., Chand, P., Yadav, U.S., Study of the effect of laser beam welding on joint of Nimonic 80A superalloy: an experimental approach. *Int. J. Adv. Manuf. Technol.*, 136, 1, 1–13, 2024. https://doi.org/10.1007/s00170-024-14081-x.
16. Hartwig, P., Bakir, N., Scheunemann, L., Gumenyuk, A., Schröder, J., Rethmeier, M., Physically Motivated Heat Source Model for Laser Beam Welding. *Metals*, 14, 4, 430, 2024. https://doi.org/10.3390/met14040430.
17. Efa, D.A., Laser Beam Welding Parametric Optimization for AZ31B and 6061-T6 Alloys: Residual Stress and Temperature Analysis Using a CCD, GA and ANN. *Opt. Laser Technol.*, *175*, 110837, 2024. https://doi.org/10.1016/j.optlastec.2024.110837.
18. Jiang, Y., Jiang, M., *et al.*, Vacuum laser beam welding of AZ31 magnesium alloy: Weld formability, microstructure and mechanical properties. *Opt. Laser Technol.*, *169*, 110115, 2024. https://dx.doi.org/10.2139/ssrn.4519969.
19. Heilmeier, J., Kick, M.K., *et al.*, Inline failure detection in laser beam welding of battery cells: Acoustic and spectral emission analysis for quality monitoring. *J. Laser Appl.*, *36*, 022007, 2024. https://doi.org/10.2351/7.0001216.

20. Sisodia, R., Weglowski, M., Sliwinski, P., *In situ* localised post-weld heat treatment with electron beam welding of S690QL steel. *J. Adv. Joining Processes, 9*, 100182, 2024. https://doi.org/10.1016/j.jajp.2023.100182.
21. Troise, M., Krichel, T., Olschok, S., Reisgen, U., Investigation of the influence of pulse parameters on the resulting weld seam quality in pulsed electron beam welding of AW 6061. *J. Adv. Joining Processes, 9*, 100183, 2024. https://doi.org/10.1016/j.jajp.2023.100183.
22. Dawson, H., Hughes, J., Jimenez-Melero, E., Impact of electron beam welding on the microstructure of PM2000 ODS steel. *Fusion Eng. Des., 204*, 114528, 2024. https://doi.org/10.1016/j.fusengdes.2024.114528.
23. Dutilleul, T., Widdison, R., Kyffin, W., Development of electron optical capabilities for manufacturing of large components by electron beam welding. *Weld. World.*, 68, 4, 819–828, 2024. https://doi.org/10.1007/s40194-023-01646-9.
24. Peng, H., Huang, S., Hu, L., Luo, B., Li, L., Baker, I., Electron beam welding of L12-nanoparticle-strengthened strong and ductile medium-entropy alloys for cryogenic applications. *Mater. Sci. Eng., A, 889*, 145875, 2024. https://doi.org/10.1016/j.msea.2023.145875.
25. Zhang, P., Qi, Y., Cheng, Q., Sun, X., Welding dissimilar alloys of CoCrFeMnNi high-entropy alloy and 304 stainless steel using gas tungsten arc welding. *J. Mater. Eng. Perform., 33*, 7, 3273–3282, 2024. https://doi.org/10.1007/s11665-023-08229-1.
26. Xia, Y., Dong, B., Cai, X., Lin, S., Ultrasonic vibration assisted gas tungsten arc welding of Inconel 690 alloy: Ultrasonic effect to refine grains and improve mechanical properties. *Ultrason. Sonochem.*, 108, 106950, 2024. https://doi.org/10.1016/j.ultsonch.2024.106950.
27. Skripko, S., Gordynets, A., Kiselev, A., Slobodyan, M., Spatial stabilization of arc discharges in pulsed gas tungsten arc welding. *Weld. World,* 68, 1, 1–14, 2024. https://doi.org/10.1007/s40194-024-01781-x.
28. Karim, M.A., Jadhav, S., Kannan, R., Pierce, D., Lee, Y., Nandwana, P., Kim, D.B., Investigating stainless steel/aluminum bimetallic structures fabricated by cold metal transfer (CMT)-based wire-arc directed energy deposition. *Addit. Manuf., 81*, 104015, 2024. https://doi.org/10.1016/j.addma.2024.104015.
29. Wang, J., Zhao, Z., Du, W., Bai, P., Wang, L., Zhang, Z., Huang, Z., High-temperature compression deformation strengthening mechanism of Mg–Gd–Y–Zn–Zr alloy formed by cold metal transfer (CMT) arc additive forming. *Mater. Sci. Eng., A, 893*, 146084, 2024. https://doi.org/10.1016/j.msea.2023.145201.
30. Fan, C., Guo, H., Wang, J., Experimental and Simulation Study of Babbitt Alloy/Steel Composite in Swing Cold Metal Transfer (CMT) Welding Under Different Heat Input. *JOM*, 76, 6352–6362, 2024. https://doi.org/10.1007/s11837-024-06424-6.

31. Belhadj, M., Kromer, R., Werda, S., Darnis, P., Effect of cold metal transfer-based wire arc additive manufacturing parameters on geometry and machining allowance. *Int. J. Adv. Manuf. Technol.*, 131, 2, 739–748, 2024. https://doi.org/10.1007/s00170-023-11835-x.
32. Zou, T., Duan, S., Wu, D., Cao, Y., Matsuda, K., Guo, F., Zou, Y., Microstructure and properties of Al–Zn–Mg–Cu alloy fabricated by cold metal transfer additive manufacturing. *J. Mater. Res. Technol.*, *30*, 7816–7821, 2024. https://doi.org/10.1016/j.jmrt.2024.05.181.
33. Kuppusamy Vellingiri, S.K., Manoharan, D., Ponnusamy, S., Kettimuthu Ramadass, U., Dhanabalaselvan, V., Assessment of welding fume impacts in a confined workplace by two extraction patterns—a case study of small-scale manufacturing industries. *Environ. Sci. Pollut. Res. Int.*, 30, 4, 10037–10051, 2023, doi: 10.1007/s11356-022-22782-7.
34. Heramo, W.T. and Workneh, H.Z., Optimization of process parameters in friction stir welding of dissimilar aluminum alloys (AA6061–T6 and AA5052–H32). *Int. J. Adv. Manuf. Technol.*, 129, 9, 4473–4489, 2023. https://doi.org/10.1007/s00170-023-12546-z.
35. Rachman, T., Application of Environmentally Friendly Technology in the Welding Process. *Collaborate Eng. Daily Book Ser.*, 1, 1 Juni, 92–99, 2023.
36. Egerland, S.A., Wiesinger, M., Sharma, R., Ebert, B., Experimental study of measures preventing welders from fume exposure. *Weld. World*, *67*, 1, 1–20, 2023. https://doi.org/10.1007/s40194-022-01395-1.
37. Siltanen, J., *Laser and Hybrid Welding of High-Strength Structural Steels*, Lappeenranta-Lahti University of Technology, Lappeenranta, Finland, 2023, [Thesis].
38. Qu, Y., Sun, Z., Spettmann, C., Xing, B., Xia, Y., Zhou, Q., Model development of laser welding in lithium-ion prismatic cell for enhancing failure prediction. *Int. J. Mech. Sci.*, *259*, 108637, 2023. https://doi.org/10.1016/j.ijmecsci.2023.108637.
39. Touileb, K., Attia, E., Djoudjou, R., Benselama, A., Ibrahim, A., Boubaker, S., Ponnore, J., Ahmed, M.M.Z., Laser Weld Aspect Optimization of Thin AISI 316 SS Using RSM in Relation with Welding Parameters and Sulfur Content. *Metals*, 13, 7, 1202, 2023. https://doi.org/10.3390/met13071202.
40. Miao, K., Yin, Y.Y., Wang, T., *et al.*, Creep Deformation Mechanism and Life Prediction of T91/S304H Welded Join. *Strength Mater.*, 55, 2, 340–354, 2023. https://doi.org/10.1007/s11223-023-00529-0.
41. Zhao, H., Yang, G., Shen, J., Liu, B., Lyu, Z., Research on the shielding effect of the electrode frames in the electrostatic precipitator. *J. Air Waste Manage. Assoc.*, 73, 6, 462–470, 2023, doi: 10.1080/10962247.2023.2200402.
42. Rahul, M., Sivapirakasam, S.P., Sreejith, M., Vishnu, B.R., Vijayakumar, K., Concurrent reduction of the fume and Cr (VI) concentrations of inhalable and respirable particle using a new covered stainless-steel SMAW electrode. *J. Environ. Chem. Eng.*, 12, 1, 111632, 2024. https://doi.org/10.1016/j.jece.2023.111632.

43. Moinuddin, S.Q., Saheb, S.H., Dewangan, A.K., Cheepu, M.M., Balamurugan, S. (Eds.), *Automation in the Welding Industry: Incorporating Artificial Intelligence, Machine Learning and Other Technologies,* Wiley-Scrivener, Hoboken, New Jersey, USA, 2024.
44. Mahto, R.P. and Rout, M., Challenges and Prospects of Welding 4.0 Adoption: Implication for Emerging Economies, in: *Smart Technologies for Improved Performance of Manufacturing Systems and Services*, B.B. Behra, *et al.* (Eds.), pp. 71–93, CRC Press, Boca Raton, Florida, USA, 2024.
45. Naik, D.K., Sharma, V.P., Dinesh Kumar, R., Automation in Welding Industries, in: *Automation in Welding Industry: Incorporating Artificial Intelligence, Machine Learning and Other Technologies*, S.Q. Moinuddin, *et al.* (Eds.), pp. 37–48, Wiley-Scrivener, Hoboken, New Jersey, USA, 2024.
46. Kumar, S., Introductory Chapter: Welding in the Era of Industry 5.0, in: *Welding-Materials, Fabrication Processes, and Industry 5.0*, S. Kumar (Ed.), IntechOpen, London, United Kingdom, 2024.
47. Goka, S., Moinuddin, S.Q., Cheepu, M., Dewangan, A.K., Welding Practices in Industry 5.0: Opportunities, Challenges, and Applications, in: *Automation in Welding Industry: Incorporating Artificial Intelligence, Machine Learning and Other Technologies*, S.Q. Moinuddin, *et al.* (Eds.), pp. 263–279, Wiley-Scrivener, Hoboken, New Jersey, USA, 2024.
48. Tian, Y. and Liu, A., Responsible manufacturing toward Industry 5.0, in: *Manufacturing from Industry 4.0 to Industry 5.0*, D. Mourtzis (Ed.), pp. 231–263, Elsevier, Amsterdam, Netherlands, 2024.
49. Noeske, P.L.M., Simperler, A., Leite Cavalcanti, W., Beber, V.C., Alliott, T., Schiffels, P., Goldbeck, G., Towards Advancing Translators' Guidance for Organisations Tackling Innovation Challenges in Manufacturing within an Industry 5.0 Context. *Sustainability*, 16, 8, 3486, 2024. https://doi.org/10.3390/su16083486.

22

Toward a Greener Weld for Integrating Sustainability Into Welding Practices

Sunita Routray[1*], Ranjita Swain[2] and Rudra Narayan Mohapatro[2]

[1]*Department of Mechanical Engineering, Centurion University of Technology and Management, Jatni, Odisha, India*

[2]*Department of Chemical Engineering, C.V. Raman Global University, Bhubaneswar, Odisha, India*

Abstract

Sustainable manufacturing encompasses the consideration of environmental repercussions, energy utilization, and the economic ramifications inherent in the product creation process. Welding and processing constitute pivotal elements within the manufacturing domain. The attainment of sustainability in welding and processing has emerged as a critical area of concern in contemporary manufacturing. The present chapter expounds upon the multifaceted components of sustainability across various welding and processing methodologies. The discourse encompasses an in-depth analysis of energy conservation, material waste minimization, resource optimization, and process parameters, accentuating the environmental advantages and cost-efficiency potentials inherent in diverse welding techniques. The abovementioned processes are systematically categorized under distinct classifications of welding and processing, including fusion arc welding, gas-based welding and processing, laser-assisted welding and processing, resistance welding and processes, and solid-state welding. New technologies have also emerged, such as robotic welding, use of welding cobots, welding drone in different places of welding to increase the productivity, enhancing the quality of weld and to address welding at critical places. A career in welding is also very encouraging for welding engineers. All these factors combine to make the welding process more sustainable and acceptable in the future.

Keywords: Welding, innovation, MMR, laser beams, fabrication, ultrasonic, intermetallic, fabricating

**Corresponding author*: sunitaroutray77@gmail.com

Sandip Kunar and Gurudas Mandal (eds.) Advanced Welding Technologies, (447–476) © 2025 Scrivener Publishing LLC

22.1 Introduction

The chapter focuses on traditional as well as advanced welding processes, effective utilization of processes for industrial growth, and providing graduates with a competitive edge in the welding industry. Welding requires essential skills, including expertise in product innovation and inspection methods, which are highly sought after by employers.

According to the Maximize Market Research (MMR) report, the welding consumables market is set to surge to an impressive US$19.2 billion by 2030, up from US$13.38 billion in 2023. The global market for welding consumables is forecasted to achieve a robust CAGR of 6.2% between 2023 and 2030. Dominating the global market, the automotive sector claims the largest share, followed closely by the construction sector, as illustrated in Figure 22.1.

One of the latest advancements in welding is the use of laser technology, which can instantly melt metal and fuse two pieces with remarkable precision. Laser welding employs high-energy lasers to precisely heat and melt metals and thermoplastics creating weld pools to join them [1]. This method is incredibly fast and precise, though it cannot be performed manually. Gas tungsten arc welding (TIG) is regarded as the highest quality and most expensive arc welding process. The acronym IGBT, which stands for "Insulated Gate Bipolar Transistors," refers to high-speed switching devices used in all Weld class Inverter welding machines to regulate

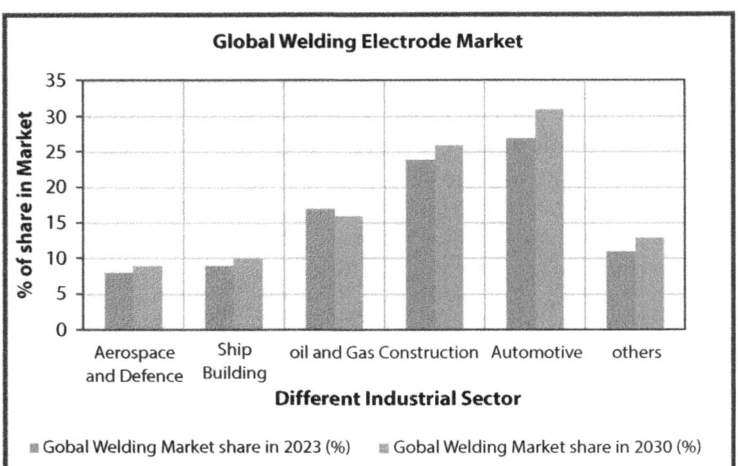

Figure 22.1 Global elector market share in 2023 and expected share % to be in 2030.

voltage. Tungsten inert gas (TIG) welding is often considered the strongest welding method due to the extreme heat it generates and the slow cooling rate resulting in high tensile strength and ductility. Metal Inert Gas (MIG) welding is also known for producing strong joints making it another excellent candidate for high-strength welds.

Various sectors employ a range of welding processes, including Arc welding, Gas welding, Resistance welding, Solid-state welding, Laser beam welding, Electron beam welding, Atomic hydrogen welding, Submerged arc welding, Plasma arc welding, Friction welding, Ultrasonic welding, and Explosive welding. However, across all sectors, four main types dominate as follows: Arc welding, Gas welding, Resistance welding, and Solid-state welding.

22.1.1 Arc Welding

This welding process includes several methods where an arc produces the heat to melt and join metals. Shielded metal arc welding (SMAW), or stick welding, uses a flux-coated electrode for welding. An electric current forms an arc between the electrode and the metals being joined. Gas metal arc welding (GMAW), commonly known as MIG welding, uses a continuous electrode and inert gas shielding to protect the weld. Flux-cored arc welding (FCAW) is similar to GMAW but can be used in a tubular wire filled with flux removing the need for external shielding gas. Gas tungsten arc welding (GTAW), or TIG welding, is generally used in a non-consumable tungsten electrode and inert gas, like argon, to shield the weld area from oxidation. The different methods have unique advantages and applications making arc welding a versatile choice for a variety of welding tasks.

22.1.2 Gas Welding

Gas welding is a process, which can be used for heat production by a gas flame to melt and join the metals. Oxy-fuel welding is the most common type of gas welding method, where oxygen and a fuel gas (usually acetylene) combine to melt the metal using flame at the welding point. Oxy-Acetylene welding is the most widely used method for gas welding process. It involves a mixture of oxygen and acetylene gas for producing flame that can reach a temperature of approximately 3,200°C. This intense heat melts the metal allowing the welder to fuse pieces together. Welding torch tool is used to mix and control the gases in the welding torch. By adjusting the composition of oxygen and acetylene flow, the welder can control the flame's size and temperature. In various gas welding applications, a filler rod is used to weld the added material ensuring a strong joint. The rod is melted by the flame temperature and deposited into the weld pool. Sometimes, a flux can be used to prevent oxidation and contamination of the weld area. The flux can be applied to the filler rod or directly to the metal surfaces for welding process. Gas welding requires strict safety precautions. Proper ventilation, flame-resistant clothing, and eye protection are essential to protect against the intense heat and bright light of the welding flame, as well as potential gas leaks. Gas welding is recognized for its versatility allowing for welding, cutting, and brazing of various metals. Gas welding process is particularly useful for repair work and widely used in industries such as automotive repair, metal fabrication, and construction.

22.1.3 Resistance Welding

Resistance welding is a process where heat generated through electrical resistance is used to join metal pieces. This method is logical and often used for high-speed, automated production. Resistance welding involves the application of pressure and passing through a high current to join through the metal piece. The electrical resistance at the contact surfaces produces heat, which causes the metal to melt and form a weld. Spot welding is the most common type, which is used to join overlapping metal sheets. Electrodes clamp the sheets together and apply current to a small spot creating a weld nugget. Seam welding is similar to spot welding but with rotating wheel electrodes that produce a continuous weld seam. It is often used for making liquid-tight containers. Projection welding involves welding projections, or raised sections, on one of the metal pieces. The projections concentrate the welding current allowing for multiple welds to be made simultaneously. In case of flash welding, the ends of the metal pieces are brought into contact, and a current is applied. The pieces are then separated slightly to create an arc (flash), which heats the ends. They are then pressed together to form a weld. Upset welding is similar to flash welding, but the pieces are pressed together before and during the heating process to ensure a strong weld.

Resistance welding is known for its speed, efficiency, and ability to be automated. It does not require filler materials or flux making it a clean process. The method is highly suitable for high-volume production environments, such as the automotive industry. Resistance welding is widely used in manufacturing processes, especially in the automotive, aerospace, and electronics industries. It is ideal for joining thin sheets of metal, wires, and small components. By leveraging the principles of electrical resistance and heat generation, resistance welding provides a reliable and efficient way to join metal parts with minimal preparation and clean-up.

22.1.4 Solid-State Welding

Solid-state welding is a group of welding processes in which joining of materials occurs without melting them. Instead, bonding is achieved through the application of pressure, sometimes with heat, but always below the melting point of the materials in use. This welding involves creating a strong bond between materials by applying pressure and, in some cases, heat. The materials do not reach their melting point. Instead, the bonding happens through diffusion, recrystallization, or plastic deformation at the interface. The different types of solid-state welding are friction welding, ultrasonic welding, explosion welding, cold welding, diffusion welding, and roll welding. In friction welding, materials are joined by being rubbed together under pressure generating heat through friction. This heat softens the materials allowing them to bond. Variants include friction stir welding and linear friction welding. Ultrasonic welding employs high-frequency ultrasonic vibrations applied to materials under pressure bonding them through localized heating and plastic deformation. Explosion welding uses an explosive charge to force materials together at high velocity, creating a bond through high-pressure impact. This method is often utilized for cladding and joining dissimilar metals. Cold welding joins materials by applying high pressure at room temperature causing them to bond through plastic deformation and atomic diffusion. Diffusion welding involves heating materials below their melting point and pressing them together for an extended period allowing atomic diffusion to form a bond. This process is typically used for high-temperature applications. Roll welding applies pressure through rollers to bond materials, often with additional heating, and is commonly used for producing metal laminates. Solid-state welding is used in various industries, including aerospace, automotive, electronics, and manufacturing. It is particularly valuable for applications requiring high precision and strength, such as turbine blades, electrical connections, and composite materials.

22.2 Efficient Welding Process Technology in Industry

By avoiding the melting of materials, solid-state welding offers unique advantages in terms of joint quality, material properties, and the ability to join dissimilar materials and making it an essential technique in modern manufacturing processes. Welding is a crucial manufacturing technique that connects two or more pieces of material by heating and connecting them, then allowing them to cool and solidify. The primary aim of welding processes in the manufacturing field is to achieve strong, durable joints that meet service requirements while minimizing costs. Advanced welding manufacturing incorporates three scientific methods to optimize the welding processes [2]. Additive manufacturing (AM) technology has evolved from traditional manufacturing methods (TM). The fundamental principle of AM involves creating objects by depositing material layer by layer. Recent studies suggest that metal components produced through AM can be joined by the various welding techniques to create larger parts. This study explores these welding methods, reviews recent research, and investigates the welding parameters and their effects on the mechanical properties and microstructures of the resulting materials [3]. The authors explored fusion welding in an Fe–Al system highlighting the challenge of excessive brittle intermetallic compound formation that degrades joints of welding. In contrast, friction welding, being a solid-state bonding process, limits intermetallic compound formation at the weld interface. However, in the Al-Fe system, where solid solubility is minimal, some intermetallic compounds still form despite of friction welding. This study utilized high-resolution transmission electron microscopy to examine the microstructure of the friction weld interface between an Al–Mg alloy and austenitic stainless steel. Additionally, the technique of friction welding was investigated [4]. The authors applied inertia friction welding methods to join 2,219 aluminum (Al) alloy bars with 304 stainless steel bars. They also conducted detailed characterization and analysis of intermetallic compounds (IMCs) at the friction interface focusing on the bonding mechanism of the welded joint. Microstructural analysis revealed two distinct IMC layers: a nanoscale Fe4Al13 layer and a larger-scale of Cu-rich layer. Notably, the Cu-rich layer was bonded to the base materials through Al_2Cu and two Fe–Al–Cu ternary IMCs [5].

A distinct investigation was conducted with an evaluation of standard gas metal arc welding (GMAW) and a high-power Tandem GMAW (TGMAW) process concerning their energy efficiency. Measurements of current, voltage, and overall equipment power were taken into account to find out energy consumption. A novel key performance indicator, Electrical Deposition Efficiency method was introduced to represent the energy efficiency of GMAW processes. Additionally, the wall plug efficiency of the equipment was determined to estimate overall energy consumption comprehensively. The results demonstrated that applying the

TGMAW process significantly intensifies energy efficiency and economic process performance. Further findings revealed that the wall-plug efficiency of the equipment remains independent of power level and material transfer mode. A 30-mm-thick structural steel plate was welded using both standard GMAW and TGMAW to illustrate the total energy savings in a practical welding scenario. The high-power TGMAW process reduced electricity consumption by more than 20%, with energy consumption and welding time reduced by 23% and 55%, respectively. In summary, the TGMAW process simultaneously facilitates superior energy efficiency and economic performance [6].

The author highlighted that welding constitutes a significant part of the processes in manufacturing social infrastructure products. Hitachi is leveraging this technology to develop optimal designs and placements for welding jigs, alongside welding procedures and other operational needs. Currently, in the research and development phase, there is also a necessity for methods to fine-tune weld designs and operational practices. This includes utilizing simulation techniques to predict and prevent different defects in the welding processes in advance [7]. The researcher developed an optimized model for electrical energy consumption in Metal Inert Gas (MIG) welding processes based on refined welding parameters. The authors orchestrate these data into a comprehensive energy map that ensures both high-quality welds and energy efficiency. The Taguchi method is used in experimental design to modify three key parameters, that is, current, voltage, and wire feed rate and to achieve an optimal balance of energy efficiency and weld quality. They meticulously calculated and documented the electrical energy used during these different experiments, then leveraged these data to calculate the most efficient electrical energy consumption for different welding parameters. These optimized parameters were subsequently integrated into a model to minimize electrical energy consumption during welding operations [8].

The investigation begins with the reasons for using various materials used in different industries highlighting the benefits such as reduced weight, enhanced fuel efficiency, and improved mechanical properties. The study then explores into an array of advanced welding techniques developed to address these challenges. This paper explores multiple fusion welding methods, including laser welding, electron beam welding, and friction stir welding. Each technique is thoroughly explained detailing its fundamental principles. The discussion emphasizes the advantages of each method, such as reducing heat-affected zones, achieving precise control over the welding process, and minimizing the formation of intermetallic compounds. The study also underscores the use of case studies and practical examples to demonstrate the successful application of advanced welding techniques on dissimilar materials [9].

The author observed that advancements in Industry 4.0 technology have led to a significant increase in data availability, production capabilities, system flexibility, and real-time information. However, they also identified a major communication gap within the manufacturing industry's IT infrastructure, which hinders seamless data flow and sharing across networks. Current solutions typically rely on offline connections resulting in a lack of real-time visibility of the entire shop floor. This limitation prevents high-level managers from effectively monitoring shop floor activities, which can delay prompt deliveries to customers. Their paper reviews current literature on data and information distribution within the manufacturing sector and proposes implementing cloud services on a CNC-based welding machine to create a smart factory [10].

The researcher aimed to replace a commercial product and expedite the manufacturing process of an axisymmetric sleeve-type workpiece. The fixture concept introduced in their study not only increased production volume but also enhanced the mechanical properties of the parts. Strength analyses enabled the company to eliminate the commercial product resulting in significant cost savings. A key advantage of the new hub solution is the efficient use of materials, such as sheets and tubes, nearly to their end, thus reducing scrap steel waste. By utilizing usable scrap material, the company minimized the amount of discarded steel. Implementing these solutions for specific operations helped eliminate errors caused by human factors. The measured times showed that the results met quantitative expectations. This proposal was successfully implemented in a company where welding is the primary manufacturing technology [11]. The experiment was performed with a comparative study on two welding techniques: friction stir welding (FSW) and gas tungsten arc welding (GTAW). The results indicated that friction welding had a superior overall sustainability score. Specifically, for an aluminum product used in the opening and closing system of an electric train door, FSW achieved a score of 0.78, while GTAW scored 0.69, representing an 11% refinement with FSW over GTAW. The research covers economic, physical, social, and environmental impacts. Furthermore, it featured an illustration using current electric vehicle (EV) components and identifies a methodology to calculate optimal welding processes for applications. In this case, the scoring indicated 0.43 for GTAW and 0.68 for FSW showing that FSW has 36% greater efficiency over conventional arc welding methods [12].

Another author introduced methodologies for assessing sustainability, which can be assessed in the welding process. They developed a comprehensive framework to evaluate the most suitable welding process for specific applications in the welding field. This approach involved welding techniques across four main domains: economic impact, environmental impact, social impact, and physical performance. Scores from each domain were gradual to derive a holistic sustainability rating. The researcher conducted case studies on three different welding processes. Among these three different welding processes, friction stir welding developed with the highest sustainability score, surpassing gas tungsten arc welding and gas metal arc welding [13]. The welding process was conducted with an investigation of bobbin friction stir welding (BFSW) process by measuring the machine's current consumption. The different phases of the BFSW welding process and its electrical demand in the welding market were analyzed. It was evident that the total energy consumption in BFSW averages approximately 130 kW across all distinguished process phases. The weld phase, which is crucial for joint formation, was accounted with the highest energy usage of 120 kWs. This total energy consumption is significantly lower compared to traditional welding technologies, even the conventional friction stir welding (CFSW) process. These findings underscore BFSW's potential sustainability benefits and its capability to promote joining technologies soon [14].

The author introduces an approach to develop welding process maps within a sustainability framework. The research used a two-stage method where meta-heuristic optimization is employed to generate a range of optimal solutions based on quality criteria on welding framework. These solutions are then calculated using a Fuzzy classifier to evaluate their sustainability across multiple dimensions of the welding process. The results shows that it is possible to derive qualitatively viable and sustainable process maps for complex engineering systems as an initial step prior to considering changes in materials, machinery, and processes [15]. Employing modeling, optimization, and classification sequentially holds potential for mitigating the inherent trade-offs between quality and sustainability across a wide range of product applications. Fusion welding is highlighted as a critical manufacturing process used for fabricating complex metal structures [16]. This method involves the use of consumables, such as filler materials, electrode coatings, shielding gases, or fluxes, which result in the production of hazardous fumes impacting the environment and human health. In contrast, friction stir welding (FSW) is recognized as a solid phase-joining process known for its energy efficiency and eco-friendly attributes earning it the designation of a green manufacturing technology due to its minimal emission of harmful gases or radiation into the environment. FSW has been extensively adopted across industries including the aerospace, locomotives, marine, and automotive sectors. The sustainability characteristics of FSW are delved into in this study emphasizing its potential benefits in various industries in the future scenario owing to its energy-efficient and environmentally friendly properties [16]. The study delves into the elements of sustainability across various welding processes and their techniques. The study highlights the aspects studied in the welding field including material efficiency, environmental advantages, energy conservation, resource utilization, and potential cost savings associated with different welding processes [17].

22.3 Importance of Welding in Industries in India and Worldwide

Welding is the most adoptable technique for metal joints in industries. The process is economical and has futuristic prospects. In India, no other process is followed like welding by the manufacturing industries. The industries that are related to the world of welding are the aerospace industry, railway industry, automotive industry, manufacturing industry, construction and infrastructure industry, and oil and gas industry. During the 20th century, forge welding, arc welding, and oxy-fuel welding were only used. With advancement in processes, shielded metal arc welding (SMAW), gas metal arc welding (MIG), and flux cored arc welding (FCAW) were developed. Presently, laser beam welding and robotic welding also emerged and are used in industries.

In India, the Indian Institute of Welding (IIW-India) is a professional body established in 1966 to develop the welding technology across and beyond the country to create awareness for career growth [18].

The "Make in India" initiative taken by the Government has set up new changes in Indian manufacturing sector including welding technology. At the same time, with advancement in technology, the use of new materials and joining of dissimilar materials in the automotive and aerospace environments has come into picture to achieve light- and high-strength products. India has new challenges in research and development, upon which the country is focusing to have greater findings.

The target set by the government of India is to reach a target of USD 5 trillion economy by 2024; the welding industry has a crucial role as a key contributor to the economy. The National Manufacturing Policy has set a 25% contribution for the welding industry to the economic growth, where it manages to contribute only 16%. Technical challenges in welding industry should be reviewed, and proper measures should be taken as welding is the most important bond that strengthens the country's economy [19].

22.4 Energy Consumption in Welding

The consumption energy mostly depends on the heat input to join the materials using any process. Twelve different welding technologies are available [20]. These welding technologies are adopted in industries for quality material. These processes are as follows:

- Welding using metal inert gas (MIG)
- Welding using tungsten inert gas (TIG)
- Stick welding is specific to shielded metal arc welding (SMAW)
- Dual-shield welding or flux cord arc welding (DSW/FCAW)
- Advanced form of TIG, i.e., welding using special torch or plasma arc welding (PAW)
- Electrode welding or electron beam welding (EW/EBW)
- Laser welding or laser beam welding (LW or LBW)
- Gas welding (GW)
- Atomic hydrogen welding (AHW)
- Electro-slag welding (ESW)

The discussion is on the sustainable welding process with reference to energy consumption and solution to reduce this consumption.

This paper introduces a methodology for selecting sustainable welding processes through weight space partitioning. A case study evaluates two advanced welding methods, manual and automatic GMAW, based on economic and environmental criteria. LCAs that include environmental evaluations significantly influence electricity and water usage, equipment expenses, and material costs, ultimately affecting the overall costs. The holistic approach offers a comprehensive view via global weight distributions mitigating biased technology selection, which is unlikely in previous studies and other criteria for weightage. To encourage confident decision-making, weight space partitions help visualize options, enabling decision-makers to assess sensitivity in their choices. The methods come under the scalability of the criteria for sustainability in view of the social aspects. As economic and environmental values diverge, vigilant tracking of technological advancements becomes crucial for informed multi-criteria assessments [21].

The environmental issues arise by emissions and utilization of inefficient energy or materials. It is often a struggle to achieve sustainable and practical outcomes simultaneously using the traditional approaches toward the conditions used in welding such as maximum penetration, minimal weld width, and low dilution. This paper introduces a process for constructing welding process maps within a sustainability framework. There are two stage processes involved in this approach as follows: (i) meta-heuristic optimization generates a range of optimum solutions of the close data that are established on the qualitative accurate value considerations, and (ii) a Fuzzy classifier evaluates these solutions across sustainability criteria. One case study on thick welds indicates that welding specifications are optimized while considering electrical energy and material usage for sustainability. The sustainability of the process maps for complex engineering systems is an initial step, preceding potential changes in machinery, materials, or processes. By sequentially applying modeling, optimization, and classification techniques, the approach aims to reconcile quality and sustainability conflicts across diverse product types derived from welding processes. This demonstration is specifically illustrated with a twin-wire welding case focusing on sustainability parts including rate of deposits, consumption of the flux, and electrical power following qualitative welding procedure requirements [15].

SMAW is extensively utilized in the joining technology in industries like marine, automotive, aerospace, construction, and petrochemicals. The increasing demand for sustainability in manufacturing requires a critical evaluation of SMAW due to its various inputs and outputs. Sustainability concerns related to SMAW include electrical energy consumption, material usage, slag formation, emissions of fumes, and risks to human health and safety. While existing literature predominantly addresses the technical and economic aspects of SMAW, environmental and social dimensions are often neglected. This study aims to assess SMAW in triple-way views incorporating the sustainability in costing, and issues on environment and societal life. Recommendations are provided to enhance the economic and sustainable performance of SMAW. The study integrates literature to define SMAW's input parameters and output streams. The key findings include the following:

- Labor and overhead costs constitute 80%–85% of overall welding operation costs suggesting potential economic gains through optimization methods to reduce resource consumption.

- Addressing fume inhalation risks through improved process design and modern power electronics offers substantial energy consumption reduction possibilities.
- Limited studies exist on the analysis in life cycle or life cycle analysis (LCA) and numerical modeling of welding operations indicating opportunities for future research to enhance sustainability assessments and predictive accuracy.

Future studies could focus on implementing LCA for welding operations and developing advanced numerical models integrating Multiphysics phenomena to improve performance predictions, particularly in metal solidification and phase-change dynamics during welding processes [22].

The aerospace and automotive industries are currently undergoing heightened demand for lightweight, high-performance components, prompting exploration into innovative welding methods capable of effectively joining diverse materials. Welding different materials presents significant challenges due to variations in physical properties, metallurgical attributes, and thermal expansion coefficients. This research is rising to explore the different beneficial criteria in this sector such as weight reduction, fuel efficiency improvement, and enhanced mechanical properties. The developed welding techniques are investigated to solve these problems. The welding processes with their principles are being examined such as LW, EBW, and friction stir welding (FSW). These techniques emphasize on heat-affected zones, achieving precise process control, and reducing intermetallic compound formation.

The successful application of advanced welding techniques on dissimilar materials studied is a practical example. In the aerospace and automotive sectors, it was demonstrated that the joining of incompatible materials, like aluminum to steel, titanium to composites, and more, is efficient and feasible. Advanced welding processes represent a transformative opportunity in the aerospace and automotive industries, which is the integration of the diverse materials used. In engineering application, material science converging with welding technology not only enables the creation of lightweight, high-performance components but also signals a paradigm shift where such characteristics are no longer mutually exclusive. Continued advancements in aerospace and automotive technologies promise to reshape future designs, manufacturing practices, and user experiences in both sectors [8].

Optimizing energy consumption while reducing carbon emissions from electricity generation at power plants is essential for sustainable manufacturing, particularly through energy-efficient methods. Optimizing the process parameters for minimizing electrical energy consumption in welding is a crucial joint technology in manufacturing. This study focuses on determining an optimized electrical energy consumption model for MIG/metal active gas (MAG) welding processes.

The Taguchi experimental design approach, using three key parameters—power (current), electrical energy (voltage), and feed rate of wire (wire feed rate)—was systematically changed to identify the efficiency of power or energy and high-quality joint conditions. The research culminated in an energy map that summarizes the electrical energy consumption across different welding scenarios ensuring both quality and efficiency in welding operations.

Reducing electrical energy consumption not only lowers operational costs for welding industries but also contributes to sustainable manufacturing practices by reducing carbon emissions associated with electricity generation. This optimized approach supports industries heavily reliant on MIG welding machines, aligning with efforts to enhance sustainability through efficient resource utilization in manufacturing processes [10].

The manufacturing industry faces significant challenges in enhancing efficiency and adaptability. Ensuring employees have access to real-time data, information, and expertise is critical for making informed decisions. The advent of Industry 4.0 technology has greatly expanded data availability, production capabilities, system flexibility, and real-time information. However, existing IT infrastructures in manufacturing often lack effective communication and seamless data sharing across networks relying instead on offline connections.

The reviews on current literature data distribution are in manufacturing and proposes the execution of cloud services for CNC-based welding machines to transform them into smart factories. The study examines the improvement and conceptual framework of Industry 4.0 highlighting current IT infrastructure challenges in manufacturing. It introduces an Industry 4.0 solution tailored for CNC-based welding machines, specifically through a cloud-founded Manufacturing Execution System (MES) planned to manufacture the qualitative product efficiently using the FSW and sheet metal forming.

The proposed cloud integration enables to monitor instantaneously and communicate inside the industries facilitating seamless data exchange between MES and ERP systems. This setup addresses current gaps by enhancing operational visibility and efficiency, thereby reducing labor and production costs. It also improves intra-factory and customer communication across different geographical locations.

Future directions could focus on extending cloud-based solutions to provide customers with real-time virtual access to production floors using closed-circuit television cameras promoting broader implementation of cloud technology in manufacturing industries [23].

The automotive and aerospace industries are currently undergoing significant transformations driven by global trends toward innovative manufacturing methods. There is a strong emphasis on developing new materials and techniques to maximize efficiency and enhance equipment performance. This includes the adoption of optimized designs using novel combinations of materials to achieve lightweight vehicle concepts. Materials, such as aluminum, fiber plastics, stronger alloys, newly developed plastics, and bi-metallic compositions, are increasingly used across vehicle components including frames, panels, structures, engine parts, and interiors.

The shift toward lightweight designs primarily relies on materials, like fiber-reinforced plastics and combinations of metal and plastic, which present challenges for welding dissimilar materials. Advanced welding technologies are therefore becoming essential, including methods for joining metal-to-metal, plastic-to-plastic, and metal-to-plastic components.

While many SMEs have upgraded to CO_2 laser welding and some have incorporated ultrasonic welding machines and other advanced technologies, further advancements are necessary to meet future challenges and quality standards. One of the critical challenges faced globally is the availability of skilled manpower, which has driven automation and the adoption of advanced welding technologies across various sectors.

Despite the initial higher investment costs associated with implementing new technologies, manufacturers are benefiting from more productive welding techniques and associated processes such as controlled welding and smart feedback systems. These advancements not only increase efficiency but also facilitate quick return on investment (ROI) and enhance adaptability in rapidly evolving industrial landscapes [7].

In the recent era, in between the 19th and 20th centuries, Industry 2.0 witnessed the advent of electrification and mass manufacturing. Innovations, such as electricity and the assembly line, significantly facilitated the large-scale production of standardized goods enhancing efficiency in manufacturing processes. The industries are continuously adopting the new innovative ideas to modernize the process in industries. Innovative methods play a vital role in improving production efficiency, weld quality, and ensuring worker safety. Moreover, achieving sustainability goals involves implementing resource optimization and energy-efficient practices. Utilizing AI and ML technologies can optimize the utilization of resources, including energy and other consumables, thereby enhancing the economic viability, sustainability, and environmental friendliness of welding processes.

The welding technology is attached with automation and robotics to make it an automation and robotic welding. To execute welding operations with increased productivity, higher quality, and reduced costs, the robotic arms are utilized. Robotic systems efficiently manage complex or repetitive welding tasks, making them applicable in various industries, including manufacturing, construction, and automotive. Welding is playing an important role in the social infrastructure products, and with increased global procurement and production practices, ensuring weld quality assurance and improving welding efficiency have become paramount in manufacturing industries. To enhance reliability and shorten production times by predicting welding process behaviors, Hitachi has developed a technology for simulating a large structure for welding. Optimization of this technology is by designing and placing welding jigs with the determination of welding procedures and operational requirements [24].

The impact on social life and environment of this process was evaluated with the help of assessment of the life cycle as well as of the social life. Both assessments are used for the evaluation of the welding operation impact on the life cycle and social life cycle. A sustainable joint operation is developed and selected on focusing on the triple bottom line approach for both the industry and research community, and the solution is achieved. Different categories were evaluated, such as possibly in acidification, eutrophication, global warming (100 years), and photochemical ozone creation, for the assessment of the impact of the environment. The factors affecting the social life are fair salary, healthiness, and protection. The results help industry efforts in adopting a sustainable joint operation. The LCA results indicate that manual metal arc welding (MMAW) exhibits higher environmental impacts compared to gas metal arc welding (GMAW) and laser–arc hybrid welding (LAHW). MMAW's greater material and electricity consumption per meter of weld seam, particularly due to titanium dioxide used in electrode coatings, significantly contributes to acidification and eutrophication. GMAW's impact is influenced the consumption of filling material, which can be minimized with techniques like modified spray arcs reducing flange angles from 60° to 30°. Within the study scope, LAHW emerges as the preferred technology. The social LCA reveals adequate wages for welders and potential health risks associated with different welding processes. LAHW demonstrates the lowest health risks, whereas MMAW poses the highest risks due to fume formation. These findings highlight the importance of minimizing manual welding technologies to safeguard welder health [25].

Today's manufacturers face a significant challenge: managing production lines that deliver high-quality products on time and within budget, while also meeting demands for energy efficiency and cost savings. Welding operations, a crucial component of manufacturing, are no exception. In fact, welding alone consumes approximately $15 million worth of electricity annually in the United States and approximately $99 million globally.

To minimize the operating costs and electricity consumption, various processes are used in welding for achieving a quality weld. Modern manufacturing environments need an energy-efficient and reliable welding program to handle these challenges. Three steps may be implemented to enhance overall efficiency and reduce energy requirements in your welding operations as follows: optimizing power output, calculating input power requirements, and evaluating operating costs [25].

Automated arc welding of gas metal (A-GMAW) process and manual arc welding using gas metal (M-GMAW) operation are compared for the evaluation of the best one. Manual welding using a conventional arc with lower power input and more consumption of sources have limitations, but in the automatic process, robotic movement is utilized to achieve more power input and employs a modified spray arc, which reduces resource consumption and flange angles.

High productivity, flexibility, and ability are operated both manually and automatically using gas metal arc welding (GMAW). In steel construction, M-GMAW is used due to its adaptability to varying geometries. However, manual welding imposes constraints on welding speeds and deposition rates due to human limitations in process power and precision compared to robotic welding systems. For longer processing times, it is crucial to have a larger root gap and groove angles to accommodate increased material consumption.

Conversely, A-GMAW with its more operational power achieves a faster rate of material deposit and rate of welding. Modern arc processes developed by welding power source manufacturers have enhanced A-GMAW capabilities significantly. Energy consumption in A-GMAW includes both the electricity used for robot movements along defined trajectories and that consumed by the welding source, as quantified in the calculated results and process inputs provided in Table 22.1.

Environmental impacts of A-GMAW encompass material consumption and electricity usage for robot manufacturing. However, conveyance of the robot to the operational area was not considered in the Life Cycle Assessment (LCA) of the input energy of welding. Detailed inventory tables shown in Table 22.2 indicate the raw material costs.

Table 22.1 LCA results with cost analysis of manual welding.

	$\mathbf{n}_{cp}$	$\mathbf{m}_{cp}$
EP in PO_4 equipment	$3.3*10^{-2}$	$7.5*10^{-4}$ kg m^{-1}
Costs	20,000 €	13€ m^{-1}
GWP in CO_2 equivalent	1,260 kg	4kg m^{-1}

Table 22.2 LCA results with cost analysis of automatic welding.

	$\mathbf{n}_{cp}$	$\mathbf{m}_{cp}$
EP in PO_4 equipment	11.77 kg	$3.8*10^{-4}$ kg m^{-1}
Costs	80,000 €	5.3 € m^{-1}
GWP in CO_2 equivalent	7,939 kg	2 kg m^{-1}

The cost of the process can be determined by evaluating the total welding costs encompassing initial equipment investments and operational expenses. In this study, equipment costs are represented by the intercept (ncp) indicating the cost of the power using GMAW, with additional investments required for the welding robot in A-GMAW. Operational costs, including daily wages, filling material, input power, and sheltering gas, are measured using the operating time period per meter of joint length of welding and rate of consumed power. Average labor costs are sourced from German federal statistics, while electricity prices are based on Eurostat database figures for medium-sized companies in Germany. Prices for protecting gas and filling raw material are delivered by the vendors.

The intercepts (ncp) and slopes (mcp) for the evaluation of costing and impact on the environment are tabulated in Table 22.1 for M-GMAW and Table 22.2 for A-GMAW illustrating comprehensive insights into the comparative costs and environmental impacts of each welding process.

Robot welding systems are known for their high duty cycles and substantial power consumption. Given the current high-energy costs, enhancing energy efficiency has become a critical focus for fabricators directly impacting customer sustainability goals.

Böhler Welding has made significant advancements in supporting its customers through several initiatives as follows: a) Introducing more energy-efficient welding machines. b) Reducing the carbon footprint in the production of steel used for welding wires. c) Developing and supplying copper-free welding wires. d) Enhancing Mig/Mag welding processes. e) Implementing environmentally friendly pickling and passivating chemicals.

The development of Böhler Welding's patented greenWave technology exemplifies their commitment to energy efficiency. This technology optimizes power draw from the network, thereby enhancing the overall efficiency of their power sources.

Böhler Welding is also investing in sustainable welding consumables leveraging voestalpine's initiative to produce green steel for their welding products. This supports customers in meeting sustainability targets, particularly in automotive applications.

Their copper-free welding wires feature advanced surface engineering that reduces harmful chemicals and lowers energy consumption during production. These wires operate at higher welding speeds, improving kilowatts per kilogram of weld metal deposited and generating fewer fumes, thereby reducing health risks and lowering energy demands in the fume removal processes. Improvements in Mig/Mag welding processes focus on boosting productivity and reducing costs. Böhler Welding continuously refines processes like RapiDeep and QuickPulse to achieve high welding travel speeds in robotic applications. Their synergic welding programs ensure optimal electric arc performance by harmonizing Böhler Welding filler metals with power sources ensuring superior quality and efficiency [26]. It is also found that the welding processes and the additive manufacturing method are the well-known methods [3].

The problems in energy consumption can be reduced by applying different techniques by different industries. The solutions are discussed below.

22.4.1 Reduction of Energy Consumption During Welding

Energy saving can be done in the welding process by reducing the heat input and arc energy has been supplied to the workpiece for the welding process. Some of the literature are discussed below.

Research indicates that the application of the TGMAW (tandem gas metal arc welding) process can significantly enhance both energy efficiency and economic performance. Findings also suggest that the equipment's wall-plug efficiency remains consistent across different power levels and material transfer modes. In an experiment, a 30-mm-thick structural steel plate was joined using both standard GMAW (gas metal arc welding) and TGMAW demonstrating that the high-power TGMAW process reduces electricity consumption by more than 20% compared to that of the standard GMAW [5].

It has been observed that 80%–85% of the overall cost in welding operations is attributed to labor and overheads. To maximize economic benefits, various optimization methods can be employed to reduce resource consumption and costs in the SMAW (shielded metal arc welding) process [22].

Data were used to optimize electrical energy consumption based on different welding parameters. These optimized parameters helped develop a model that reduces electrical energy consumption in the welding process, which, in turn, lowers the overall cost for industries [11].

Modern manufacturing environments require an energy-efficient and reliable welding program that not only produces high-quality welds using multiple processes but also minimizes operating and electricity costs. To achieve this, consider the following steps:

1. Assess the efficiency of your existing equipment.
2. Consider switching to inverter technology.
3. Closely track shop floor production and efficiency.

Revamping welding operations and equipment to achieve energy savings might seem challenging, but even minor changes can have a significant impact. Each manufacturing facility has unique energy usage patterns in their welding shops making a thorough assessment of your plant's needs and power usage the first crucial step toward generating savings. From there, you can decide how best to upgrade, streamline, and monitor your welding equipment to suit your shop's specific needs and savings goals.

There are six ways to save energy in the welding industry [27].

i. Reduce after work, reduce energy consumption: The time saving in the process technology reduces the energy consumption. Hence, reduced after work means reduced energy consumption.
ii. Choose smart solutions to save energy: Reduced Gap Technology (RGT) is an intelligent welding solution that enables reliable narrow gap welding of materials with a thicknesses of up to 30 mm, without the need for special equipment or accessories. Together with other special processes, RGT increases welding efficiency and reduces overall energy consumption. Investing in smart solutions saves time, energy, and a welder's nerves.
iii. Always choose the optimized set-up for your welding work:

The extra length of welding cable and the minimum size alone, i.e., the cross-sectional area of the welding cables will ultimately lead to energy loss. Hence, good quality copper cables are more reliable and economical in the welding.

iv. Use the latest technology to save energy:
This work consists of careful material choices, shaping industry standards, and improving the existing technologies (while also adapting new ones). Energy efficiency is one part of our work. With Kemppi, your welding is energy-conscious welding.

v. Invest in the future:
Kemppi participated in Power 2 Power, a European co-funded innovation project to enable sustainably competitive European technologies and a strong innovation and production capacity.

vi. Do not settle for anything below the standards: As one of our own initiatives and following the practice of continuous development, Kemppi has already complied with these standards many years before the Eco-Design Directive took effect.

22.5 Environmental Impact of Welding

The high temperature generated during welding vaporizes the substances present in an arc, which get oxidized by encountering atmospheric air. The pollutants are very minute and lie in the nano-size range. It also releases several toxic gases like carbon monoxide, ozone, and nitrogen oxides. Hence, welding causes significant air pollution, hampering human health and the ecosystem. Welding is essential for economic growth, and at the same time, human health is also a primary concern. It is important to follow and adopt the paths to reduce the environmental impact and ensure the sustainability of the welding industries.

22.5.1 Environmentally Friendly Welding Techniques

One of the techniques to reduce the environmental impact of welding is to adopt welding processes that produce less waste material or gases. Friction stir welding (FSW) and magnetic pulse welding (MPW) are the processes that produce less pollution because it does not require fillers or flux. Diffusive welding is also an eco-friendly process that creates a strong bond without using flux. A non-contact welding process called laser welding uses a high-powered laser light for welding of parts. Some other welding processes, such as laser arc hybrid welding and gas metal arc welding, are also eco-friendly.

Even though the solid-state welding techniques are eco-friendly in nature, these are not ideal for every joining process and require more development [28].

Using eco-friendly materials for welding can significantly lessen environmental impact. Traditional welding materials, like nitrate-based fluxes, emit harmful fumes and create toxic waste products that damage the environment, whereas using water-based fluxes or electrode coatings can reduce environmental impact to a greater extent. The high energy consumed during welding operation can be reduced considerably by different ways to reduce the environmental effect. The use of energy-efficient equipment, which requires less power, using compressed air for welding, and switching to LED lighting in the welding area, can reduce the energy consumption a lot.

During welding, pollutants are released into the air, which are very harmful for the environment and affect the health of human beings adversely. The pollutants include nitrogen oxide, carbon monoxide, and volatile organic compounds. Industries must adopt a variety of ways to reduce the detrimental effects. These may include the use of environmental friendly welding materials, selecting the right kind of welding processes, practicing proper waste management, etc. Usually, industries use nitrate-based fluxes, which generate harmful fumes, which can be replaced with eco-friendly water-based fluxes. These materials generate less harmful fumes and waste during welding. The welding processes, like flux-cored arc welding and TIG welding, generate a lot of fumes and splatter, which can be replaced with less fume- and splatter-generating processes like laser or electron beam welding. Handling of waste generated during welding is also crucial to reduce the environmental impact. Proper disposal of welding waste products, such as spent welding electrodes, empty cylinders, and other by-products, is essential. Correct labeling and storage of this waste can minimize the risk of spills and accidents.

Welding is essential in industries, and at the same time, the environmental impact should also be reduced for its sustainability. The use of eco-friendly welding materials, adopting the right welding process, reduction in energy consumption, and following proper waste management techniques can ensure a sustainable and healthy process in industries for economic growth.

22.6 Industry 4.0 in Welding

With the implementation of Industry 4.0 to develop a smart factory, which works autonomously with less human interaction, the welding process is also expected to boom in all technological advancement. A computer must be fed with all technical details of a process so that it can take decisions like an expert welding personnel. The welding equipment must be provided with the latest technology and sensors to assess all production-relevant information. A powerful network system, high data storage, data and communication security should be provided with the system to make it more sustainable. Until complete digitization has not taken place in the welding sector, humans will play a key role in accomplishing successful welding processes. The communication bridge between the human and welding machines is one of the most successful factors. To digitalize the welding knowledge, a proper networked modular cell must be equipped with all welding parameters, consumables depending upon the task assigned. Currently, the welding parameters are selected by the welding technologist. If these tasks are to be supported by high-end software systems, it is essential to digitize existing knowledge and make it automatically retrievable. This is a very difficult task to accomplish as quantification of all existing knowledge is tedious [29].

To implement Industry 4.0 in welding, some criteria must be followed:

- The technical details should be digitized and fed to the system in such a way that a computer must be able to take a similar decision as a welding technologist.
- Welding equipment must have high-end information receiver, communication system, and high-tech sensors to digitize information and present a real-time behavior of the system.
- Robust infrastructures and sufficient storage system should be installed to transfer and store data. Proper security system must be provided to protect the data system.

22.7 New Welding Techniques to Merge in the Future

Due to the advancement in technology, many new welding processes are invented. Now, welding technology is not only confined to manual welding but also has jumped to many advanced welding processes. The advanced welding processes are discussed in detail:

The metal inert gas (MIG) welding, gas tungsten arc welding (GTAW), and stick welding are still in use by industries, but the use of advanced metals is also increasing day by day. Advanced metals include titanium aluminide intermetallic, nickel aluminide intermetallic alloys, iron aluminide intermetallic alloys, oxide dispersion strengthened (ODS) alloys, metal matrix composites (MMCS), nickel-based super alloys, Al–Li alloys, Mg alloys, super martensitic stainless steel, etc. These alloys are primarily used in aerospace industry. Previously, welding was not used widely in this sector, but with advancement in welding techniques, this sector has also gained momentum in adopting these techniques.

The advanced welding processes are discussed below, which are employed to provide the best results in different industrial sectors.

22.7.1 Magnetic Arc Welding Process

The magnetic arc welding technique uses a high-power magnetic field near the weld zone, which controls the arc. The arc creates heat and moves along a predetermined path. The heat melts the filler material and base metal to form a strong weld. This welding process is adaptable for thick plate welding and where space is a constraint. This method produces accurate welding as the flame is stable and useful for a wide range of materials [30].

22.7.2 Explosive Welding

As the name shows, the explosive welding process uses explosive energy to join the metals. This process allows the metals to be kept close, and a small explosive is exploded. The explosion produces a shock that forces one metal to join the other metal with a tremendous force. Joining of dissimilar metals can be done by explosive welding, which is difficult to be done by the traditional process. The joint produced by this welding is smooth and of superior quality.

22.7.3 Ultrasonic Welding

This process is carried out in solid state. Ultrasonic vibration is used for this process to produce frictional heat between the parts. Dissimilar metals can be joined by this method without using any filler material or flux.

22.7.4 Friction Stir Welding

Friction stir welding uses heat generated by applying mechanical force between the joining parts. The parts are rotated at a high speed and generate heat. Then softening of metals takes place resulting in joining of the parts. This welding process produces a flawless joint and is used for dissimilar metals.

22.7.5 Electron Beam Welding

Electron beam welding is a highly effective welding process, which uses a high-velocity electron beam on the parts for welding. The beam is generated by an electron beam and focused on the working piece by means of electromagnetic lenses. The beam penetrates the parts, melt them, and fusion takes place. This process is carried out in a vacuum medium to nullify the effect of surrounding air.

22.7.6 Laser Welding

Laser beam welding (LBW) is a very effective method of welding, which uses a beam of concentrated laser to fall on the welding parts. The heat generated by this beam helps the parts to melt and get fused with each other. The movement of the concentrated beam is easy to control and produces a clean smooth weld bead.

22.7.7 Hybrid Welding

Hybrid welding is a combination of two or more welding processes to have better weld joints. This process incorporates the benefits of all combined welding processes to produce a smooth high-strength weld joint.

For example, laser welding and GTAW welding processes are combined to produce high-quality welds with minimal distortion and porosity. Laser welding gives heat control due to a concentrated beam, and GTAW provides additional filler material. The benefits of both combine to produce a stable weld.

22.7.8 Robotic Welding

Robotic welding is introduced in many industries to increase productivity and high-quality consistent weld. This is an automated production, which needs high initial investment in equipment and training of personnel to operate the robot. The process is highly accepted due to the following advantages:

- Process automation

- Initial installation cost is high, which is balanced with high productivity and no labor cost
- High-quality product
- Human beings are not exposed to harmful fumes or radiation
- Improvement in efficiency that helps in mass production

22.7.9 Welding Cobots

In some industries, cobots are also used for the welding process. A welding cobot is designed to assist human welders. Equipped with sensors and computer programming, cobots can work without requiring safety barriers or cages. They can handle repetitive or physically demanding tasks and are programmed to warn operators of any dangerous situations in the workplace. In contrast, traditional robotic welding machines are preprogrammed and can pose a risk if someone comes into their path.

22.7.10 Welding Drones

Welding drones are aerial vehicles facilitated with cameras, sensors, and robotic arms to execute inspections and welding operations. These drones are useful for operation in environments where human beings cannot reach. Operated remotely, welding drones supervise the welding process and can adjust operations as needed. Dangerous locations, like high buildings, tanks containing chemicals, oil platforms, etc., are handled by this welding process.

22.8 Welding as a Career

Seeing all the prospects of welding processes in different industries, it can be concluded that welding has a prosperous career for the young welding engineers. In this aspect, various points can be discussed to see the importance of welding as follows:

- The industries, such as construction, manufacturing, automotive, aerospace, and shipbuilding, are difficult to sustain without the welding process.
- With advancement in technology, new materials, tools, and methods are coming up day by day. This offers new challenges as well as opportunities for the young generation.

- In the building of new infrastructures and bridges, industries have created the need for a higher numbers of welders.
- Skilled welders can get opportunities worldwide, can work with national and international companies.
- Skilled welders for robotic welding or underwater welding have high demand. These areas provide job stability and higher packages.
- Once experienced, the welders have the opportunity to change from one industry to another. By learning different eco-friendly practices and materials, welders are more flexible to change from one industry to another.

The above discussions ensure a sustainable welding career for young aspirants. Practicing and learning new methods of welding will ensure a sustainable welding career.

22.9 Conclusions

The welding technology can be more sustainable by reducing harmful effects on the environment, using the resources properly, and creating a worker-friendly workplace. New and advanced technologies can be adopted to ensure energy saving and reduce greenhouse gas emissions. Efficient handling of new materials with specific techniques can reduce waste generation. Advanced welding processes are more acceptable for their precision, good product quality, and low waste generation. Pollution produced from the welding process is a big challenge for its sustainability. These challenges can be handled by improving ventilation, using cleaner methods, handling waste, and prioritizing the health of people working in these industries. Safety concerns in the working environment are crucial and can be addressed by providing training in the workplace. By using smart technologies, the welding technology can be more sustainable and will enhance the economic growth of the nation.

References

1. Kenji, Oi. and Masatosh, M., Recent Trend of Welding Technology Development and Applications. *JFE GIHO*, 34, 1–7, 2014.

2. Zhang, Y.M., Yang, Y.P., Zhang, W., Na, S.J., Advanced welding manufacturing: a brief analysis and review of challenges and solutions. *J. Manuf. Sci. Eng.*, 142, 11, 110816, 33, 2020. https://doi.org/10.1115/1.4047947.
3. Karayel, E. and Bozkurt, Y., Additive manufacturing method and different welding applications. *J. Mater. Res. Technology*, 9, 5, 11424–11438, 2020. https://doi.org/10.1016/j.jmrt.2020.08.039.
4. Sivasankari, R., Balusamy, V., Venkateswaran, P.R., Buvanashekaran, G., Kumar, K.G., Characterization of magnetically impelled arc butt welded T11 tubes for high pressure applications. *Def. Tech.*, 11, 3, 244–254, 2015. http://dx.doi.org/10.1016/j.dt.2015.03.005.
5. Dang, Z., Qin, G., Wang, J., Bonding mechanism and fracture behavior of inertia friction welded joint of 2219 aluminum alloy to 304 stainless steel. *Mater. Sci. Eng. A*, 866, 144641, 2023. https://doi.org/10.1016/j.msea.2023.144641.
6. Sproesser, G., Pittner, A., Rethmeier, M., Increasing performance and energy efficiency of Gas Metal Arc Welding by a high power tandem process. *Procedia CIRP*, 40, 642–647, 2016. https://doi.org/10.1016/j.procir.2016.01.148.
7. Zhang, X., Tsukamoto, T., Tarasawa, S., Miura, Y., Welding Simulation Technologies for Highly Efficient Production of High-quality Social Infrastructure Products. *Hitachi Rev.*, 61, 6, 249–253, 2012.
8. Rajemi, M.F. and Seman, M.T.A., Sustainable Energy Efficiency for MIG Welding Process. *Inter. Jour. Innovative Techno. Exploring Eng.*, 8, 5S, 287–291, 2019. ISSN: 2278-3075.
9. Kumari, S., Patro, E.K.R., Singh, S., Chandra, P.K., Kareem, S.A., Kansal, L., Advanced Welding of Dissimilar Materials for Aerospace and Automotive Applications. *E3S Web Conferences*, 430, 01108, 2023. https://doi.org/10.1051/e3sconf/202343001108.
10. Adetunla, A. and Madonsela, N., Harnessing Industry 4.0 Technology to Improve Productivity in Manufacturing Industries: a Case of CNC Based Welding Factory. *IEEE 12th International Conference on Mechanical and Intelligent Manufacturing Technologies*, pp. 180–185, 2021, https://doi.org/10.1109/ICMIMT52186.2021.9476203.
11. Lonkwic, P. and Tofil, A., Supporting Welding Work in the Aspect of Increasing Production Process Efficiency. *Adva. Sci. Techno. Res. Jour.*, 17, 1, 8–14, 2023. https://doi.org/10.12913/22998624/157418ISSN2299-8624.
12. Hoyos, E., Maria, M.C., Backer, J.D., Martin, J., Sustainability Score Comparison of Welding Strategies for the Manufacturing of Electric Transportation Components. *Sustainability*, 158650, 11, 1–17, 2023. https://doi.org/10.3390/su15118650.
13. Jamal, J., Darras, B., Kishawy, H., A study on sustainability assessment of welding processes. *Proc. Inst. Mech. Engineers J. Eng. Manuf.*, 234, 3, 2019. https://doi.org/10.1177/ 0954405419875355.
14. Sued, M.K., Samsuri, S.S.M., Kassim, M.K.A.M., Nasir, S.N.N.M., Sustainability of Welding Process through Bobbin Friction Stir Welding.

IOP Conf. Series Mater. Sci. Eng., 318, 012068, 1–7, 2018. https://doi.org/10.1088/1757-899X/318/1/012068.

15. Sharma, A., A fundamental study on qualitatively viable sustainable welding process maps, Jour. *Manuf. Sys.*, 46, 221–230, 2018. https://doi.org/10.1016/j.jmsy.2018.01.002.
16. Majeed, T., Wahid Mohd, A., Nadeem Alam, M., Mehta, Y., Siddiquee Arshad, N., Friction stir welding: A sustainable manufacturing process. *Mater. Today Proc.*, 46, 15, 6558–6563, 2021. https://doi.org/10.1016/j.matpr.2021.04.025.
17. Mehta, K., Sustainability in Welding and Processing, in: *Materials forming, Machining and Tribology*, pp. 125–145, 2018, Corpus ID: 117324792https://doi.org/10.1007/978-3-030-03276-06.
18. (https://iiwindia.com/about-us/reports/).
19. Girotra, C.C., Role of the Welding Industry in the Growth of Indian Economy. *Indian Weld. J.*, 53, 2, 46-51, 2020.
20. *Saif 2023 Annual Report*, https://www.saif.com/aboutsaif/publications/annual-report-2023.html.
21. Sproesser, G., Schenker, S., Pittner, A., Borndorferb, R., Rethmeier, M., Chang, Y.J., Finkbeiner, M., Sustainable welding process selection Based on weight space partitions. *Procedia CIRP*, 40, 127–132, 2016.
22. Alkahla, I. and Pervaiz, S., Sustainability assessment of shielded metal arc welding (SMAW) process. *IOP Conference Series: Materials, Science and Engineering*, vol. 244, p. 012001, 2017, https://doi.org/10.1088/1757-899X/244/1/012001.
23. Sproesser, G., Chang, Y.J., Pittner, A., Finkbeiner, M., Rethmeier, M., Sustainable Technologies for Thick Metal, Plate Welding, in: *Sustainable Manufacturing, Sustainable Production, Life Cycle Engineering and Management*, pp. 71–84, 2017, https://doi.org/10.1007/978-3-319-48514-0_5.
24. Kumar, S., Introductory Chapter: Welding in the Era of Industry 5.0, in: *Welding Materials, Fabrication Processes and Industry 5.0*, 2024, https://doi.org/10.5772/intechopen.1003918.
25. Albright, M., Product Manager, The Lincoln Electric Company, Three Simple Ways to Reduce Energy Consumption in Your Welding Operations, https://www.lincolnelectric.com/en/.
26. Voestalpine Böhler Welding, 2024, https://www.voestalpine.com/welding/global-en/company/news-and-events/boehler-welding-increases-the-efficiency-of-its-customers-with-perfect-solutions-in-terms-of-automation/.
27. Rousku, A., Welding values by Kemppi, Welding industry – 6 ways to save energy by Energy-saving, Industrial welding, Welding process, welding production, 2023, https://weldingvalue.com/2023/01/6-ways-to-save-energy/#2827d24b.
28. Wahid, M.A., Majeed, T., Alam, M.N., Friction stir welding: A sustainable manufacturing process. *Mater. Today Proc.*, 46, 1–2, 1–6, 2021.
29. Ryalat, M., El Moaqet, H., AlFaour, M., Design of a smart factory based on cyber-physical systems and internet of things towards Industry 4.0. *Appl. Sci.*, 13, 2156, 2023. https://doi.org/10.3390/app13042156.

30. Fukumoto, S., Tsubakino, H., Okita, K., Aritoshi, M., Tomita, T., Amorphization by friction welding between 5052 aluminum alloy and 304 stainless steel. *Scr. Mater.*, 42, 8, 807–8123, 2000. https://doi.org/10.1016/S1359-6462(00)00299-2.

23

Advancements in Robotic Welding Sensing Technology: A Review

S. Ajithkumar*, B. Arulmurugan, V. Manoj Mohan Prasath, P. Utchimahali Muthu Raja and M. Aravindh

Department of Mechanical Engineering, KPR Institute of Engineering and Technology, Coimbatore, Tamil Nadu, India

Abstract

Robotic welding enhances product quality by making the welding process easier and more accurate in its application. Sensing technology developments are powerful means to increase the functionality of robotic welding systems. This paper provides insight into the current breakthroughs in robotic welding sensing techniques covering vision-based sensing, weld pool monitoring, seam tracking, defect detection, and quality inspection. Vision systems enable real-time observation of the welding process and provide accurate control over weld quality and supervision. Integrating temperature and force sensors into the welding head allows for monitoring the weld pool and penetration providing complete control over the welding process to optimize welding conditions for stronger connections. Seam tracking and positioning techniques increase welding accuracy by properly aligning the welding torch in relation to the workpiece. The procedures for defect detection and quality inspection provide identification and classification of weld faults ensuring that high-quality welds can be attained. Sensing technologies fused with artificial intelligence provide better analytics of data, hence process optimization and adaptive control. The Industrial Internet of Things (IIoT) enables real-time data interchange, remote monitoring, and centralized management of robotic welding systems. In this research, several applications of sensing technology in intelligent robotic arc welding processes are discussed. The results show their potential for higher productivity and quality. The scope of sensing technology in the future in robotic welding processes is explored with new trends and opportunities for further improvement.

**Corresponding author*: ersajithkumarbe@gmail.com

Sandip Kunar and Gurudas Mandal (eds.) Advanced Welding Technologies, (477–496) © 2025 Scrivener Publishing LLC

***Keywords*:** Robotic welding, sensing technology, Artificial Intelligence, Internet of Things, patent landscape

23.1 Introduction

Robotic welding has modernized the industry with respect to accuracy, reliability, and effectiveness in welding procedures. Only a few years ago, sensing technologies influenced this area of welding procedures by introducing their use in robotic welding systems. Sensing technologies enable a robot to feel its environment and adapt, enhancing weld quality and improving process control while increasing productivity. In robotic welding, the application of sensing technology implies the utilization of various types of sensors associated with monitoring systems for the continual collection of accurate information during the welding operation. Multiple types of sensors collect critical information about the welding parameters, the geometry of the joint, material properties, surroundings, etc. The data collected are then reviewed and processed to make appropriate and prompt decisions, followed by changes to ensure the maximum quality of the welds [1, 2].

The welding arc sensors pick up voltage, current, and length information about the welding arc. Then, the information is put into the proper control of the welding process, ensuring maximum possible stability of the electric arc, amount of heat input, and penetration depth. The system allows early detection of anomalies because it constantly monitors the arc while providing real-time data to make swift corrections to maintain constant weld quality. Apart from arc monitoring, sensors are vital in retrieving joint geometry and fit-up data. Among other techniques, temperature data from infrared thermography monitors and modulates the weld pool. The size, shape, and temperature of the weld pool are critical factors that determine perfect fusion with no defects; these factors are drastically affected by the precision of the welding process [3]. Further, it reads environmental variables, like temperature and humidity, and provides real-time adjustment of welding parameters under such conditions, most notably gas flow rate and welding speed.

Advanced computer algorithms and control systems have been designed to analyze these data to improve the welding process. By the continuous feedback loop, the performance of the robotic welding system is always at its optimum level. It is capable of producing high-quality welds every time. Visual inspection systems, ultrasonic testing, and radiographic inspection techniques are used to detect surface defects, internal irregularities, and inconsistencies in the weld structure. Early defect detection allows for the correct action to be taken in time, thus minimizing scrap and rework and ensuring that the product conforms to the required quality standards [4, 5]. Sensing in robotic arc welding has even further implications than controlling the weld pool and detecting saleable defects. Improvements include enhanced weld seam monitoring and accurate torch positioning, especially with complex or irregular joint trajectories. Improving continuously, sensor technology—in its most recent developments—is characterized by increasing resolution and faster response times, thus enabling accurate tracking even in highly dynamic welding processes. Artificial intelligence and machine learning will allow knowledge acquisition from previous data, adjustment of tracking methods to different joint configurations, and continued improvement in productivity and precision over time [6].

Data analytics needs to be incorporated into sensor-gathered data for weld seam tracking. Real-time analysis discovers patterns, trends, and anomalies, explaining the welding process. It fine-tunes the variables of welding, ensuring optimum quality, and shortens the cycle time by working out the most efficient positioning of the torch and welding speed for various joint shapes. Moreover, welding equipment goes along with the IIoT to increase the possibilities of sensing technology. Regarding connectivity to a network via IIoT, the welding systems can communicate without any problem and make provisions for remote monitoring and real-time adjustments [7, 8]. This type of interface will help collect data from several systems at the corporate level to provide resource management, predictive maintenance, and permanent improvement for enhanced productivity and effectiveness of the process under consideration. The overall purpose of the book chapter is to provide an in-depth review of the recent advances in sensing technology related to robotic welding.

It investigates new developments in weld pool monitoring, defect detection, and quality evaluation, which have established that sensors play a critical role in enhancing welding quality and process control. This chapter will give the patent landscape analysis to investigate new research and development directions and emerging trends in robotic welding sensing technology. The results are expected to help further enhance the subject area.

23.2 Sensing Technology in Robotic Welding

The collaborative application of many sensing techniques, such as vision systems, force/torque, temperature, gas, acoustic/vibration, and seam tracking, is essential for accurate and effective robotic welding. Owing to these integrated sensors, robots can adjust to changes in the welding environment, guaranteeing precise welds and excellent outcomes. Because of this integration, robotic welding has shown to be a stable and effective method that can meet the needs of contemporary welding.

23.2.1 Vision-Based Sensing

Robotic welding uses cameras, laser scanners, and other visual sensors to record and analyze visual data. With this system, the robots will be able to see and properly perceive their environment of welding, trace weld seams, and facilitate adjustments in real-time torch positioning and quality control. The reason this technology has been a breakthrough for robotic welding is that it gives them the ability to "see" and understand their surroundings [9]. These find major use in the form of 2D and 3D vision sensors. Cameras are the most common 2D vision sensors providing visual feedback regarding the welding process. They can cover the welding area strategically. The laser line scanner can project a laser line onto the welding surface and analyze its deformation to obtain information about joint position and weld pool shape. Light travel time measurement allowed for accurate depth perception, as in time-of-flight techniques, such as 3D mapping with ToF. Structured light project patterns were captured by a camera, and their analysis regarding deforming enabled a 3D representation of the welding area. Stereo vision uses two cameras and picture disparity analysis to calculate depth and reconstruct a 3D welding environment [10].

The vision sensors continuously take pictures of the joint area and, together with image processing algorithms, track the weld seam. This feature provides correct monitoring of weld seams, even in complex joints, and constantly produces accurate welds due to the maintained correct torch direction and positioning. The system provides information on the exact location and orientation of both the joint and the workpiece. It uses visual data analysis to correctly orient the torch, thus providing positioning accuracy, which reduces errors for better weld quality and integrity. Vision-based sensing, on the other hand, further provides in-process control and inspection of weld quality by photographing the process of welding and detecting defects such as porosity, lack of fusion, and excessive spatter through image analysis algorithms. Fault detection and evaluation improve the quality of welds and reduce post-weld examination through the robotic system. The data regarding the size and shape of the weld pool can be captured with the help of vision sensors. As shown in Figure 23.1, the monitoring of GTAW is designed on the lines of ensuring that ideal

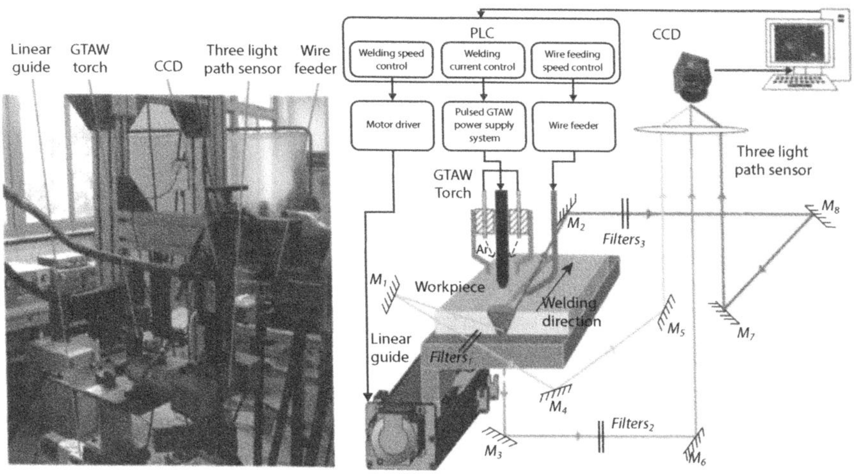

Figure 23.1 Schematic of arc welding monitoring system [11]. This figure is available under an open-access license.

welding conditions are maintained. Vision sensors can be used to track the size and shape of the weld pool in real time enabling appropriate adjustments in welding settings for consistent penetration and fusion quality in every weld [11].

Advanced techniques of image processing play an important role in the use of vision-based sensors in welding. It retrieves vital information from pictures used in immediate surveillance and decision making. Image enhancement techniques, such as noise reduction or contrast improvement, increase the quality of the image itself and bring details relevant to the weld pool and the geometry of the joint into a sharper view, thereby enhancing visibility for more accurate analysis and interpretation. Image segmentation is the process of dividing photographs into different and distinct regions. This technique in image segmentation has been used in applications, such as robotic welding, where it may be applied in separating the weld pool from the background in ways of extracting relevant information accordingly [12, 13]. Feature extraction involves the determination of visual components, like edges and contours, which would play an important role in assessing the welding characteristics. Robotic welding follows a machine learning and pattern recognition technique to identify patterns, such as joint types and flaws, to recognize objects. Calibration is the aligning of the virtual path with reality or, basically, the position of a joint in real life for accuracy. Compensation is making adjustments at runtime for correct weld placement and the parameters needed. This procedure improves the trajectories of welding, reduces deformation, and allows real-time modifications that ensure constant weld quality and stability.

23.2.2 Force Torque Sensing

Force and torque sensors give important feedback on the welding tool's interaction with the workpiece in the course of the welding process. Such technologies allow the system to monitor and control applied stresses and torques to make high-quality welds with very accurate process control. The force sensors measure the welding forces and are normally mounted on the welding tool or at the robot's end effector. They provide instant feedback regarding the force applied while welding. Force sensing in robotic welding makes use of varied sensors, which include, among others, strain gauges, piezoelectric sensors, and capacitive sensors. Strain gauge sensors have an important role in a wide range of applications, from welding to the integrity assessment of different structures, load distribution, and even weld quality assessment. They do this by tracing changes in electrical resistance, which they pick up during the process of welding. They are able to come up with very precise measurements of what is occurring and are usually placed in critical areas such as joints and weld beads [14, 15]. They provide continuous monitoring of the robotic welding process for performance assessment, excessive distortion detection, and knowledge acquisition about the nature of load distribution. This will allow the optimum settings for welding by placing many strain gauges at strategic positions, hence assuring improved quality control and process efficiency.

Piezoelectric sensors work very well in this area of welding because they can measure the force and vibrations with good accuracy. They further integrate within systems to generate electrical charges at the occurrence of force, which are then used for immediate monitoring and control. They are capable of measuring vibrations that arise from robot movements and other variables, all of which the robotic system can use to determine the stability and precision of the welding process. With improved sensitivity and response time, these devices offer an excellent opportunity for providing accurate monitoring and making quick adjustments. Typical applications include seam tracking, gap identification, and proximity sensing in robotic welding. Capacitive sensors work by capacitance; this allows them to precisely measure changes in electrical capacitance. During weld seam monitoring, the sensors track changes in capacitance as the welding torch travels, thus ensuring that the weld seam is correctly positioned [16]. In gap detection, changes in capacitance are measured because of material changes, thus enabling the system to recognize these gaps and adapt to them by a real-time response. Moreover, they act as proximity switches because they detect objects without actual contact, ensuring the correct alignment of the parts to be welded, hence triggering the welding process. The fact that they do not require physical contact makes them reduce the possibility of wear and tear on the workpiece [17].

In addition, the robotic welding machines are fitted with force sensors that offer different ranges and can be configured to provide accurate management and monitoring of the welding forces. These are devices used in the maintenance of ideal contact on the workpiece; this prevents underpenetration and excessive force. The torque sensor has been designed to monitor rotational forces with precision for measuring torque. Modern development encompasses minimized size, sophisticated data-processing capabilities, and real-time monitoring by means of wireless connectivity [18]. Multi-axis torque sensors measure various directions of torque; thus, accurate control over welding parameters is ensured enhancing weld quality and creating opportunities for optimization and predictive maintenance recommendations. It is noted that equipping these gadgets with Industry 4.0 technologies in robotic welding systems makes them efficient and productive.

23.2.3 Thermal Sensing

Due to the development in thermal sensor technology, high-resolution temperature measurement capabilities have been developed lately. More generally, new sensors provide more accurate and precise temperature data that allow better control of the welding process and, hence, better weld quality. It measures the temperature around the welding zone. They realize the capture of heat data and provide feedback in real time, thus allowing exact control of the welding conditions. Applying a number of heat sensors, such as pyrometers, IR sensors, and thermocouples, into robotic welding has been used. Thermocouples, used for welding purposes, are based on the Seebeck effect. The phenomenon appears at the point of contact of two different metals in a temperature gradient generating an electromotive force [19]. In principle, it is this electromotive force (EMF) that shows a potential difference across the connection relative to a reference point—proportional to the temperature difference between the connection and a reference point. In the welding environment, they use metal wires (thermoelements), such as Type K, Type T, and Type J, to establish a measuring connection. It is inexpensive, easy to set up, and of compact dimension for use in robot welding cells. It provides a wide temperature measurement range, suitable for many welding processes and fast real-time measurement [20]. While they are rugged, their accuracy is lower, and due to oxidation, aging, and thermal drifting, calibration and maintenance may be required recurrently to guarantee an effective temperature reading.

Besides, robotic welding uses infrared sensors for temperature measurement and control. They are also suitable for non-contact temperature measurements in highly dynamic welding processes because the thermal radiation issued by objects is detected without direct contact. They offer a fast response time enabling measurement within very short times and very accurate temperature readings, a critical function in real-time monitoring for creating ideal welding conditions. These sensors can profile temperature through the weld joint and thus provide very accurate control of the welding conditions [21, 22]. These non-contact devices will allow temperature measurement remotely; therefore, their positioning can vary, which in turn will help reduce interference with the welding process, especially from angles or other difficult locations.

The pyrometers are complex infrared sensors that enable non-contact temperature measurement during high-temperature processes by making use of thermal radiation emitted by a hot object, hence keeping away from the welder or the industrial environment. Recent developments ensure high resolution for proper control of sophisticated welding processes. These pyrometers create very accurate temperature readings due to filtering and signal processing algorithms, which remove noise and consider ambient conditions. Robotic welding multipoint temperature sensing systems are equipped with a variety of sensors that detect the temperature and thus allow for adjustments to be made in real time ensuring excellent welds time and again. Wireless heat sensors enhance mobility and flexibility. Data analysis provides process optimization and quality control and may contribute to documentation [23].

23.2.4 Gas Sensing

Gas detection ensures the highest level of safety during welding while maintaining the quality of the welds. Gas sensors detect and monitor shielding gases, like carbon dioxide and argon, which prevent weld pool contamination. It ascertains ideal gas concentrations and flow rates for the exclusion of errors and guarantees excellent welding. During the process of welding, gas sensors detect dangerous vapors and fumes such as volatile organic compounds (VOCs), nitrogen oxides, and ozone. Robotic systems turn on safety features to protect the environment and the health of the welder, including ventilation and alarms. In gas sensing for robotic welding, a variety of sensors detect and measure gas concentrations [24, 25].

Electrochemical sensors, which are used in welding for certain gas identification, use electrochemistry concepts along with sensing, counter, and reference electrodes. When the sensor electrode comes in contact with the target gas, a current proportional to the gas concentration is created. These sensors offer high sensitivity and accuracy at very low gas concentrations, thus offering real-time monitoring and safety control. Their fast response allows them to make quick changes in hazardous situations, thus ensuring improved safety. They also support process management through constant observation of gas concentrations and optimize weld quality by optimizing welding parameters such as gas flow rates and composition [26]. Infrared (IR) sensors can identify a gas, say methane (CH_4), through an analysis of how certain infrared wavelengths are absorbed. They emit light in the infrared portion of the spectrum, measure the quantity of light absorbed, and calculate with high accuracy the amount of gas being considered [27, 28]. Their fast, precise, and selective measurements facilitate the detection of gas leakage without contact and in almost no time.

Photoionization detection (PID) sensors measure volatile organic compounds using the processes of ionization. It involves the ionization of volatile organic compound molecules in the environment through UV light. As a result, this generates an electric current directly proportional to the concentration of the gas. It will provide a very accurate reading of the level of VOCs and thus can be effectively used in monitoring VOC concentrations in robotic welding. Robotic welding systems use metal oxide gas sensors for the detection of gases such as carbon monoxide and hydrogen [29, 30]. These devices utilize changes in electrical conductivity in the presence of gasses. It usually consists of a metal oxide substance, usually tin dioxide. On account of their conductivity change, such sensors are able to effectively and accurately measure and detect the presence and level of the gas in robotic welding applications.

Catalytic bead sensors detect flammable gases during welding. The device contains two beads: one exposed to the gas and one that acts as a reference. The principle of operation involves the reaction of the gas on the catalytic bead. Temperature change and resistance depend on the concentration of the gas upon contact with the catalytic bead. This method is highly reliable and sensitive, thus enabling the identification to be very fast, hence improving workplace safety. Besides, gas sensors offer a real-time response for the effective control of gases, hence offering protection continuously [31]. These devices improve efficiency and quality when integrated into welding control systems because they provide smooth communication with other devices. This technology gives an assurance of safety in welding operations, regulations, welfare of the workers, and the environment because of the continual monitoring of gas composition and concentration.

23.2.5 Acoustic and Vibration Sensing

Acoustic sensors identify the sound waves produced during welding and analyze them to provide information regarding weld quality and defect detection, as well as process monitoring. These record the sound waves that arise during the process of welding, which are affected by gas flow, electrode contact, and torch movement. After doing this, such sensors will transform the sound waves into electrical signals for analysis. They are strategically located near the welding area, normally on the welding torch, robot arm, or welding fixture—almost always based on the type of welding application and the characteristics of the acoustic signal to be acquired [32]. It can identify defects with reasonable proficiency; by analyzing different acoustic signals, it identifies issues like cracks, porosity, lack of fusion, and spatter.

The weld quality is characterized by amplitude, frequency content, and duration in the acoustic signal analysis. They identify the degree of weld quality by comparing these data with the reference criteria [33]. Acoustic data handled by the sensors is efficiently done, hence providing an accurate evaluation of the weld quality and process stability. The following are some other approaches: frequency domain analysis, time-domain analysis, statistical analysis, and pattern recognition. For example, Lv *et al.* [34] investigated the use of audio-sensing technology for the automatic control of welding penetration. With an arc sound signal of GTAW in different workpiece forms, they utilized a backpropagation artificial neural network (BPANN), piecewise controller to provide real-time control of welding penetration. The results of their study indicated that, as shown in Figure 23.2, this controller changed the welding current quickly and effectively due to variations in the arc sound signal.

The acoustic sensor is combined with additional technologies, such as vision, force, and temperature sensors, resulting in a comprehensive perspective of welding processes. This integration improves the ability to detect, monitor, and regulate defects. System accuracy and dependability are enhanced through the integration of data from many sensors. Recent developments in acoustic sensor technology focus on enhancing sensitivity, precision, and functionality, such as integrating high-frequency sensors,

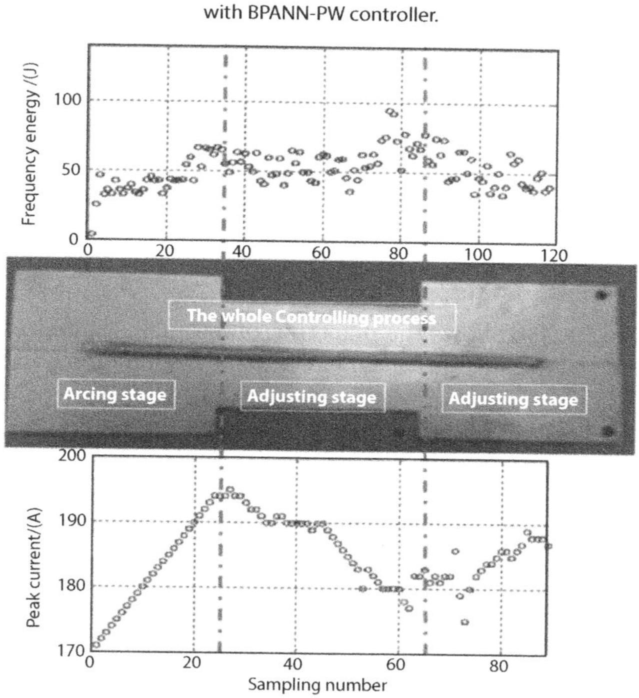

Figure 23.2 Peak current rectification and arc sound energy in welding via the BPANN-PW controller [34]. This is reproduced with permission from Elsevier.

advanced signal processing techniques, artificial intelligence, machine learning, and real-time wireless data transfer. Robotic welding allows for non-destructive testing, detection of defects in difficult regions, immediate feedback, and optimization of the process assuring constant quality and increased productivity [35]. In addition, vibration sensor systems transform mechanical vibrations into electrical signals, quantifying attributes, such as amplitude and frequency, in the context of welding. It is capable of detecting vibrations of both low and high frequencies. Typical categories comprise accelerometers, velocity sensors, and displacement sensors [36, 37]. These sensors evaluate the quality of welds and identify any abnormalities that could affect them. Furthermore, it enables the ongoing surveillance and regulation of welding procedures by consistently monitoring for sudden alterations that could indicate issues. Advanced signal-processing techniques are utilized to carefully analyze the data helping to identify trends and deviations from normal operations.

23.3 Integration of Sensing Technology and Artificial Intelligence in Robotic Welding

The combination of AI technologies with sensor technology has upgraded the efficiency and effectiveness of robotic welding to a high degree. It merges advanced algorithms with sophisticated sensors for efficiency and accuracy in robotic welding. It gives a robot welding system the capability to obtain real-time data, which helps in making informed decision making and the capability to adapt to any changing welding conditions. Furthermore, it also allows the detection of anomalies and provides real-time feedback so that adaptive control schemes can be implemented [30]. Sensing technology is imperative for measuring temperature, force, torque, gas, vibration, and auditory signals, among others, which, in turn, provide meaningful input to the AI system. Preprocessing ensures that data are of good quality, homogeneous, and aligned before being fed to AI algorithms. Techniques for feature extraction are applied to transform raw data into relevant features. In machine learning, the Support Vector Machine, random forests, and neural networks are algorithms trained to recognize patterns or perform regression. On the other hand, deep learning has the capability to do complex analyses by acquiring knowledge straight from sensor data. Real-time use of trained AI models can enable the welding process to monitor the data continuously for abnormality detection for immediate feedback into adaptive control strategies [38, 39]. Artificial intelligence, through sensor technology, enables knowledge acquisition in robotic welding and learns to modify its behavior in a continuous way.

Such AI models are able to work in real time, changing and updating themselves in different welding conditions, thus providing additional resiliency and flexibility to the models. Several key benefits can be realized in this area of robotic welding integration. With the sensor data analyzed by artificial intelligence algorithms, abnormalities can be found, and equipment malfunctioning can be forecasted in advance, thus facilitating planning proactive maintenance and minimizing downtime, which is one of the main benefits. The system achieves optimized welding characteristics through dynamic changes in parameters, such as weld speed, voltage, current, and gas flow rates, by analyzing sensors in real-time data [40]. This ensures a consistent quality of welding performance. Further, the AI system is trained with safety measures to prevent accidents from happening since sensors detect potentially dangerous conditions. Sensing technologies, coupled with artificial intelligence, could provide data used in driving informed decisions, besides offering critical knowledge on how to improve operations for quality control and welding performance improvement [41]. Thus, this would allow continuous improvement of the welding process to be afforded and enabled with well-informed choices.

23.4 Industrial Internet of Things in Robotic Welding

IoT technologies provide interconnectivity, communication, and data sharing of robotic welding systems with other devices, sensors, and activities within the industrial ecosystem. It involves many sensors that collect data on temperature, force, vibration, gas concentration, and weld quality. Due to interlinkage within the IoT network, these sensors collect and monitor data instantaneously. This provides immense intelligence with respect to the welding process. After that, this information is sent to the cloud platforms, where these data, aided by sophisticated analytics—including machine learning and AI algorithms—derive practical insights [42]. These can facilitate real-time predictive maintenance, process optimization, and quality control. IIoT allows remote supervision and regulation of the robotic welding process guaranteeing real-time data availability from any geographical location. It enhances the current situation understanding, enables timely decision making, and helps in maintaining equipment by forecasting faults and optimizing maintenance resources [43].

Moreover, the IIoT allows for the immediate enhancement of welding settings by utilizing AI algorithms that assess sensor data and past performance. These algorithms determine the optimal welding parameters, such as voltage, current, speed, and motion trajectories, that can be automatically implemented by the robotic welding system. Furthermore, IIoT enables smooth integration between robotic welding systems and enterprise systems, such as enterprise resource planning (ERP) and manufacturing execution systems (MES), thereby optimizing production planning, inventory management, and resource allocation. This integration enhances the overall manufacturing workflow and productivity. Additionally, it improves safety and risk management by continuously monitoring safety-critical parameters and promptly issuing real-time alerts to guarantee a secure working environment for operators [44, 45]. Furthermore, it provides complete visibility and traceability throughout the process of robotic welding, which leads to improved management of the supply chain, reduced time between order and delivery, and increased operational effectiveness.

Ensuring data encryption and implementing authentication mechanisms are of utmost importance for security. It offers the capacity to scale and adapt to accommodate increasing data volumes and additional sensors as demands rise.

23.5 Future Scope

Prediction of future sensing technology development relies on several sources, which include industrial bulletins and journal articles. However, this article emphasizes that the patent landscape holds significance in its most updated nature. The insights in relation to heat source models and artificial intelligence in sensing can be understood from the analysis of patents. Effective keyword selection is important for the purpose of navigation through relevant patents. In this work, "English ALL" with the "single-family member" limitation has been chosen for a more in-depth patent landscape analysis [46]. This approach makes it possible to recognize areas of research and sensing technology's orientation.

Table 23.1 gives an overview of the patent landscape for sensing technology in robotic welding. Searching for the terms "robotic welding" and "sensor" returned a whopping 19,873 patents in this highly specialized area, which directly speaks to the great interest and distinctly different research done to this day in the betterment of sensing technology in robotic welding. Data on the interest and investment in the development of sensing technology for robotic welding has increased significantly, therefore acknowledging the role of sensors in bringing efficiency, precision, and generally improving the performance in the process of robotic welding. The United States of America dominates the use of robotic welding sensing technology, with a total number of 10,544 filed patent applications, thereby indicating the dominance of the position and efforts put toward developing the sector. The second position is held by the Patent Cooperation Treaty, which reflects the interest and cooperation of the globe in technology. China is third, indicating full participation and contribution to further developments. B26J and B23K are the most prominent IPC codes, which show the areas of influence of artificial intelligence and advanced sensing mechanisms to make the welding process more efficient. Yearwise analysis has shown that the number of filed patents has been on a constant rise during the last 10 years.

Table 23.1 Patent landscape analysis for robotic welding sensing technology.

Country	Patent count	Major contenders	Patent count	IPC code	Patent count	Year	Patent count
United States of America	105,44	Ethicon LLC	610	B26J	3,264	2014	781
Patent Cooperation Treaty	6,508	Gilag Gmbh International	386	B23K	3,238	2015	814
China	918	Illinois Tool Works INC	268	A61B	2,821	2016	906
European Patent Office	876	Applied Materials INC	213	G05B	1,476	2017	1,091
India	304	Fanuc co	210	G06F	1,150	2018	1,420
Japan	175	The Boeing Company	204	G01N	1,035	2019	1,543
Canada	142	X Development LLC	192	B29C	816	2020	1,606
United Kingdom	129	General Electric Company	160	H01L	798	2021	1,806
Republic of Korea	123	Ethicon Endo-Surgery Company	174	G01B	499	2022	2,106
Australia	48	Covidien LP	170	G06T	498	2023	964

23.6 Conclusion

This chapter summarizes significant developments in sensing technology applied to robotic welding. The integration of a wide range of sensors into robotic welding systems, including vision-based sensing, force and torque sensors, heat sensing, gas sensors, acoustic sensors, and vibration sensors, enables enhanced process control and quality assurance and strengthened safety measures. It ensures that vision-based sensing will give accurate determination of weld path planning and identification of defects. Force and torque sensors allow the manipulation of the welding parameters with a high degree of accuracy. Thermal sensing enables the identification of possible irregularities in heat distribution with a view to control. Gas sensors regulate the gas levels and identify noxious smells. Acoustic sensors enhance the identification of defects. Vibration sensors are used for supervision, regulation, and safety assurance. Such sensing technologies improve the quality of welds, increase production, and enhance workers' safety by providing availability of control with precision, identification of defects at an early stage, and continuous refinement of welding parameters.

Sensing technology integrated with AI and the IIoT makes very accurate adjustments based on real-time data and sensor feedback to improve weld quality, reduce faults, and increase production. This proves the potential of integration and its flexibility for adapting and optimizing based on sensor inputs. IIoT optimizes control and workflow performance through remote monitoring and real-time sharing of data. Moreover, predictive maintenance and strong data analytics are possible, which ensure fast repairability and reduce downtime. The results of the patent landscape analysis represent the industry's commitment to the further development of sensing technology in robotic welding and to creating innovation in that domain.

References

1. Zheng, C., An, Y., Wang, Z., *et al.*, Hybrid offline programming method for robotic welding systems. *Robot Comput. Integr. Manuf.*, 73, 102238, 2022.
2. Rout, A., Deepak, B.B.V.L., Biswal, B.B., Advances in weld seam tracking techniques for robotic welding: A review. *Robot Comput. Integr. Manuf.*, 56, 12–37, 2019.
3. Lee, K., Hwang, I., Kim, Y.M., Lee, H., Kang, M., Yu, J., Real-time weld quality prediction using a laser vision sensor in a lap fillet joint during gas metal arc welding. *Sens. (Switzerland)*, 20, 6, 2020.
4. Mishra, D., Roy, R.B., Dutta, S., Pal, S.K., Chakravarty, D., A review on sensor based monitoring and control of friction stir welding process and a roadmap to Industry 4.0. *J. Manuf. Process*, 36, 373–397, 2018.
5. Mathivanan, K. and Plapper, P., Artificial neural network to predict the weld status in laser welding of copper to aluminum. *Procedia CIRP*, 103, 61–66, 2021.
6. Wu, D., Huang, Y., Zhang, P., Yu, Z., Chen, H., Chen, S., Visual-Acoustic Penetration Recognition in Variable Polarity Plasma Arc Welding Process Using Hybrid Deep Learning Approach. *IEEE Access*, 8, 120417–120428, 2020.

7. Liu, Q., Chen, C., Chen, S., Key Technology of Intelligentized Welding Manufacturing and Systems Based on the Internet of Things and Multi-Agent. *J. Manuf. Mater. Process.*, 6, 6, 2022.
8. Yusof, E.M.M., Yusof, M.I., Ali, R., Harjimi, I.H., Bahrin, Q.K., Welding station monitoring system using internet of thing (IoT). *Indones. J. Electr. Eng. Comput. Sci.*, 18, 3, 2020.
9. Muhammad, J., Altun, H., Abo-Serie, E., Welding seam profiling techniques based on active vision sensing for intelligent robotic welding. *Int. J. Adv. Manuf. Technol.*, 88, 1–4, 2017.
10. Xu, Y., Fang, G., Chen, S., Zou, J.J., Ye, Z., Real-time image processing for vision-based weld seam tracking in robotic GMAW. *Int. J. Adv. Manuf. Technol.*, 73, 9–12, 2014.
11. Chen, C., Lv, N., Chen, S., Welding penetration monitoring for pulsed GTAW using visual sensor based on AAM and random forests. *J. Manuf. Process.*, 63, 152–162, 2021.
12. Huang, W. and Kovacevic, R., Development of a real-time laser-based machine vision system to monitor and control welding processes. *Int. J. Adv. Manuf. Technol.*, 63, 1–4, 2012.
13. Huang, Y., Xiao, Y., Wang, P., Li, M., A seam-tracking laser welding platform with 3D and 2D visual information fusion vision sensor system. *Int. J. Adv. Manuf. Technol.*, 67, 1–4, 2013.
14. Ding, C., Han, Y., Du, W., Wu, J., Xiong, Z., *In Situ* Calibration of Six-Axis Force-Torque Sensors for Industrial Robots with Tilting Base. *IEEE Trans. Rob.*, 38, 4, 2022.
15. Gan, Y., Duan, J., Chen, M., Dai, X., Multi-robot trajectory planning and position/force coordination control in complex welding tasks. *Appl. Sci. (Switzerland)*, 9, 5, 2019.
16. Mishra, D., Gupta, A., Raj, P., *et al.*, Real time monitoring and control of friction stir welding process using multiple sensors. *CIRP J. Manuf. Sci. Technol.*, 30, 1–11, 2020.
17. dos Santos Paes, L.E., da Silveira, C.A., Pereira, M., dos Santos Magalhães, E., Vilarinho, L.O., Thermocouple positioning through capacitive discharge for temperature monitoring in laser welding. *J. Braz. Soc. Mech. Sci. Eng.*, 43, 10, 2021.
18. Kenda, M., Klobčar, D., Bračun, D., Condition based maintenance of the two-beam laser welding in high volume manufacturing of piezoelectric pressure sensor. *J. Manuf. Syst.*, 59, 117–126, 2021.
19. Arora, H., Singh, R., Brar, G.S., Thermal and structural modelling of arc welding processes: A literature review. *Meas. Control (United Kingdom)*, 52, 7–8, 2019.
20. Nayak, L.J. and Roy, G.G., Thermocouple temperature measurement during high speed electron beam welding of SS 304. *Opt. (Stuttg)*, 201, 163538, 2020.

21. Yu, P., Xu, G., Gu, X., Zhou, G., Tian, Y., A low-cost infrared sensing system for monitoring the MIG welding process. *Int. J. Adv. Manuf. Technol.*, 92, 9–12, 2017.
22. Yu, R., Huang, Y., Peng, Y., Wang, K., Monitoring of butt weld penetration based on infrared sensing and improved histograms of oriented gradients. *J. Mater. Res. Technol.*, 22, 3280–3293, 2023.
23. Schmailzl, A., Käsbauer, J., Martan, J., *et al.*, Measurement of core temperature through semi-transparent polyamide 6 using scanner-integrated pyrometer in laser welding. *Int. J. Heat Mass Transf.*, 146, 118814, 2020.
24. Mvola, B. and Kah, P., Effects of shielding gas control: welded joint properties in GMAW process optimization. *Int. J. Adv. Manuf. Technol.*, 88, 9–12, 2017.
25. Feng, S., Farha, F., Li, Q., *et al.*, Review on smart gas sensing technology. *Sens. (Switzerland)*, 19, 17, 2019.
26. Hamzeh, R., Thomas, L., Polzer, J., Xu, X.W., Heinzel, H., A sensor based monitoring system for real-time quality control: Semi-automatic arc welding case study. *Procedia Manuf.*, 51, 201–206, 2020.
27. Caio, L.B.A., Silva, A.M.A., Bestard, G.A., Vieira, L.S., de Carvalho, G.C., Alfaro, S.C.A., Mild steel GMA welds microstructural analysis and estimation using sensor fusion and neural network modeling. *Sensors*, 21, 16, 2021.
28. Jung, J.J. and Song, S.J., Development of a Flame Detector using a 4-wavelength Infrared Sensor. *Fire Sci. Eng.*, 35, 4, 2021.
29. Vasilev, M., Macleod, C.N., Loukas, C., *et al.*, Sensor-enabled multi-robot system for automated welding and in-process ultrasonic nde. *Sensors*, 21, 15, 2021.
30. Wang, B., Hu, S.J., Sun, L., Freiheit, T., Intelligent welding system technologies: State-of-the-art review and perspectives. *J. Manuf. Syst.*, 56, 373–391, 2020.
31. Samotaev, N., Pisliakov, A., Filipchuk, D., *et al.*, SOI Based Micro-Bead Catalytic Gas Sensor, in: *Springer Proceedings in Physics*, vol. 255, 2021.
32. Sadeghi, S., Karimi, N.Z., Fotouhi, M., Hasani, M., Najafabadi, M.A., Pavlovic, A., Residual stress evaluation in friction stir welding of aluminum plates by means of acoustic emission and ultrasonic waves. *FME Trans.*, 46, 2, 2018.
33. Zhu, T., Shi, Y., Cui, S., Cui, Y., Recognition of Weld Penetration During K-TIG Welding Based on Acoustic and Visual Sensing. *Sens. Imaging*, 20, 1, 2019.
34. Lv, N., Xu, Y., Li, S., Yu, X., Chen, S., Automated control of welding penetration based on audio sensing technology. *J. Mater. Process. Technol.*, 250, 81–98, 2017.
35. Asif, K., Zhang, L., Derrible, S., Indacochea, J.E., Ozevin, D., Ziebart, B., Machine learning model to predict welding quality using air-coupled acoustic emission and weld inputs. *J. Intell. Manuf.*, 33, 3, 2022.
36. Schwarz, E.B., Bleier, F., Guenter, F., Mikut, R., Bergmann, J.P., Improving process monitoring of ultrasonic metal welding using classical machine

learning methods and process-informed time series evaluation. *J. Manuf. Process*, 77, 54–62, 2022.
37. Balachandar, K., Jegadeeshwaran, R., Gandhikumar, D., Condition monitoring of FSW tool using vibration analysis-A machine learning approach. In. *Mater. Today Proc.*, 27, 2970–2974, 2019.
38. Wang, J., Mu, C., Mu, S., Zhu, R., Yu, H., Welding seam detection and location: Deep learning network-based approach. *Int. J. Press. Vessels Pip.*, 202, 104893, 2023.
39. Huo, W., Bakir, N., Gumenyuk, A., Rethmeier, M., Wolter, K., Strain Prediction Using Deep Learning during Solidification Crack Initiation and Growth in Laser Beam Welding of Thin Metal Sheets. *Appl. Sci. (Switzerland)*, 13, 5, 2023.
40. Shi, C., Sun, H., Liu, C., Tang, Z., Research on Weld Seam Bead Recognition Based on Convolution Neural Network. *Sci. Program.*, 2022, 1626384, 2022.
41. Kershaw, J., Yu, R., Zhang, Y.M., Wang, P., Hybrid machine learning-enabled adaptive welding speed control. *J. Manuf. Process.*, 71, 374–383, 2021.
42. Barot, R.S. and Patel, V.J., Development and Investigations of IoT Enabled Wire Feed and Weld Speed Measurement for Submerged Arc Welding. *Mapan J. Metrol. Soc. India*, 37, 4, 2022.
43. Posch, G., Jürgen, B., Krissanaphusit, A., Internet of things/industry 4.0 and its impact on welding. *Yosetsu Gakkai Shi/Journal Jpn. Weld. Soc.*, 86, 4, 2017.
44. Chen, C., Lv, N., Chen, S., Data-Driven Welding Expert System Structure Based on Internet of Things, in: *Transactions on Intelligent Welding Manufacturing*, 2018.
45. Ke, Q. and Liu, X., Internet-of-Things monitoring system of robot welding based on software defined networking, in: *2016 1st IEEE International Conference on Computer Communication and the Internet, ICCCI 2016*, 2016.
46. WIPO - Search International and National Patent Collections, 2023, https://patentscope.wipo.int/search/en/search.jsf.

24

Robotic Welding and Automation: Cutting-Edge Technology for Industry

S. Ajithkumar*, B. Arulmurugan and S. Idhayaraja

Department of Mechanical Engineering, KPR Institute of Engineering and Technology, Coimbatore, Tamil Nadu, India

Abstract

State-of-the-art robotic welding today is one of the most critical innovations in industrial operations that can offer improved profitability, reliability, and flexibility against other traditional methods of welding. This demand for individual and flexible solutions raises the need for more advanced and automated welding technologies, which cover variability in parts, conditions, and limitations of the process. It provides the current status on intelligence and automation in robotic welding focusing on recent advancements and challenges in this area. Furthermore, it gives the role that sensor feedback plays in a robotic welding process, including vision-based sensing, force, and heat sensing. It also takes into consideration the notion of supervised learning, whereby the system uses labeled data in predicting the performance of the welds and adjusts the process variables to realize intelligent robotic welding. The chapter reviews adaptive robotic welding systems capable of self-adjusting the route and parameters of welding by analyzing the sensor input and real-time data obtained in the process of welding. It further discusses the concept of path correction robotic welding, wherein the welding path is corrected on a real-time basis for errors in position and environmental disturbances. This review presents an important view on current developments and future directions regarding intelligence and automation in robotic welding for opening up avenues toward productivity enhancement and quality improvement in the manufacturing industry.

***Keywords*:** Robotic welding, sensing system, signal monitoring, adaptive robotic welding system

**Corresponding author*: ersajithkumarbe@gmail.com

Sandip Kunar and Gurudas Mandal (eds.) Advanced Welding Technologies, (497–512) © 2025 Scrivener Publishing LLC

24.1 Introduction

Robotic welding represents state-of-the-art technology in the welding industry by automating and thus optimizing the welding process. It gives manufacturers the potential to produce high-quality welds and enables them to be repetitive in their results while getting a productivity bonus from accuracy and efficiency. This normally includes industrial robots having special tools and sensors for welding applications. Robotic arms, integrated with welding tools, like welding torches, make it possible to position and place the welding process with accuracy. These robots come with sensors providing information on variables, such as arc voltage, wire feed speed, and travel speed, to provide real-time monitoring and adjustments in the welding operation [1]. It does this by maintaining the most suitable welding parameters through continual observation and adjusting to the welding conditions, hence producing welds of high quality at very high productivity levels.

The programming and control systems of a robotic welding system are the heartbeat, which aids in automating and optimizing the welding process. Normally, sophisticated software along with algorithms is used to program the movements of the robots, welding paths, and other parameters. In this way, it can automate highly complex welding tasks that turn into a challenge or are very time consuming for human welders, like multi-pass welds or complex joint configurations [2]. Using manipulators, the exact movements of the robot ensure quality through accurate and repetitive welds increasing productivity.

Programming and control systems also ensure good communication and integration within the whole production line. This ensures perfect coordination and synchronization between the robotic welding system and other production processes of manufacturing. For instance, it takes instructions for part loading and unloading and sends data to the upstream and downstream processes. As such, this would streamline communication and integration in the manufacturing process and reduce cycle times while maximizing efficiency. Figure 24.1 shows the process planning of robotic welding with multiple assignment techniques. The process planning of robotic welding refers to a well-organized method that allows finding the optimum welding parameters, sequences, and strategies for performing a wide range of welding tasks using robots efficiently [3, 4]. The technique of process planning used depends upon the specific characteristics of the welding task involved.

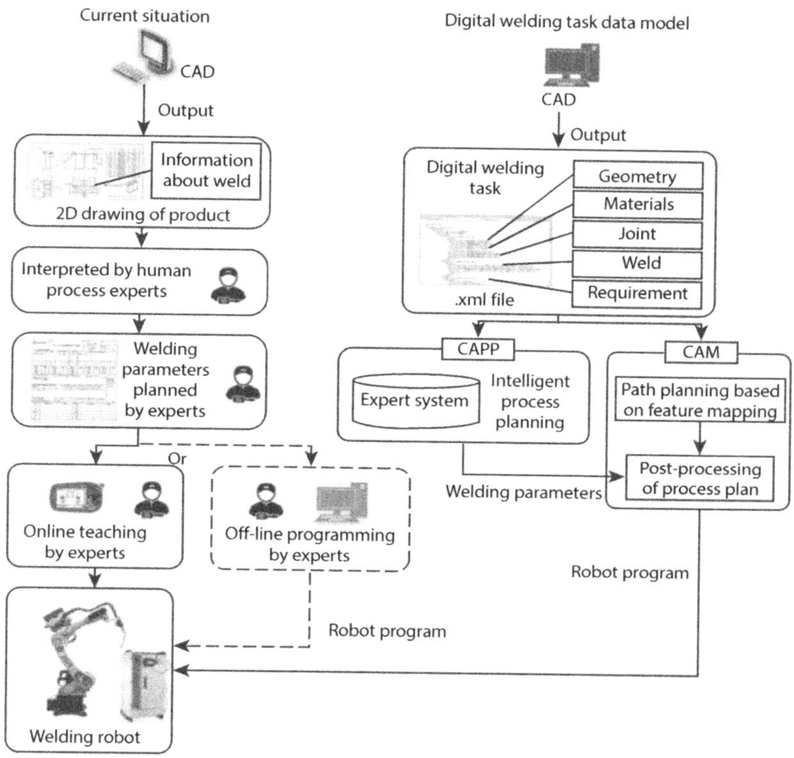

Figure 24.1 Robotic welding process planning with multiple assignment techniques [3]. Reproduced with permission from Elsevier.

Robotic systems for welding improve workplace safety by reducing the exposure of humans to welding processes' hazards, such as heat, fumes, radiation from the arc, and related risks. The robots bear such rough conditions while performing with consistency and accuracy. Skilled welders can supervise and program the robotic systems while utilizing their knowledge to fine-tune the parameters of the welding process and solve complicated challenges in welding. It is due to this collaboration of robots with human experts that creates interaction for maximum efficiency and quality of welds [5].

Besides the inherent capabilities of the robotic welding systems, incorporating artificial intelligence further enhances the performance in welding. The AI algorithms make such a system execute complex tasks more precisely and adaptively. It adjusts the welding parameters on real-time feedback gained through data analysis and machine learning techniques that achieve high productivity [6]. It also analyzes the data from sensors to detect defects, such as cracks and porosity, while welding is in progress so that action can be taken immediately. Moreover, AI makes possible adaptive welding, the possibility of adjusting the welding parameters due to workpiece geometry changes and environmental variations. This adaptiveness guarantees constant quality, even in difficult situations.

AI-based vision systems have a huge role in detecting defects in robotic welding. By utilizing AI and deep learning, it analyzes images or videos of the weld in real time to identify porosity and cracks. This technology not only means it allows for continuous monitoring of the welding process but also facilitates data analytics at an advanced level for process optimization by extracting valuable insights. AI also allows seamless integration of robotic welding systems with Industry 4.0 technologies, such as cloud computing and IIoT, for real-time data sharing, remote monitoring, and control through AI-driven predictive maintenance, a vital link for improving productivity, quality, and safety in robotic welding. Given the ever-improving level of technology, it means that AI-driven innovations in robotic welding would open wide avenues for even more sophisticated and able systems in the subject area [7, 8].

This chapter explains the sensing systems, signal monitoring, supervised learning, adaptive robotic welding, and path planning. Herein, it explains the role of sensors in perceiving the environment and interacting with it and their importance in quality control and process monitoring. Finally, it addresses issues on the integration of supervised learning techniques and the concept of adaptive robotic welding systems to enhance capabilities. This paper further discusses planning paths and trajectory optimization for correct motion planning. In addition, it deals with prospects of artificial intelligence, sensing, learning algorithms, collaborative robotics, augmented reality, and safety features, which are enablers for improved productivity, quality, and flexibility in robotic welding and automation.

24.2 Sensing System

Sensing systems are central to the robotic welding process in providing feedback that is used in the control loop. For example, changes in the workpiece, changes in joint geometries, and other environmental perturbations can be monitored in the outputs. All such integrations are a source of accuracy and efficiency in the quality of welding. Examples of such sensing systems include vision systems, laser sensors, force/torque sensors, seam tracking sensors, gas and arc sensors, and temperature sensors. The authors work with the control algorithms to make it feasible for the robot to learn from the current mission; through this, self-adaptation is given achieving the goal of movement and welding parameter optimization along with overall optimization of welding execution in full [9].

Robotic welding incorporates vision systems, including cameras and lighting systems, while advanced image processing ensures correct perception. Some of these cameras are of different types, mounting on the robot arm or tool, or in the work area, and capturing clear images. Effective illumination is a necessary element where LED, fiber-optic, or laser illumination finds a place. Image processing extracts very useful information, such as weld seams and reference points allowing the tracking to be precise and seam characteristic identification. Anomalies detected could be cracks or porosity enabling immediate feedback for adjustment [10]. Enhanced images help in real-time analysis that enables fast and accurate correction while welding. They are integral systems in identifying and extracting features such as weld seam position, shape, and dimensions. They use edge detection and contour analysis algorithms. More significantly, among the above vision-guided robots, is weld seam tracking for correct welding. Calibration compensates for each image distortion, and alignment ensures correct positioning [11]. Real-time feedback provided to the robot controller may adjust welding parameters to correct placement and adapt to variations on the workpiece.

Real-time and very accurate feedback is provided by laser sensors in robotic welding. These sensors apply laser technology, utilizing techniques of time-of-flight and triangulation to evaluate the reflections from a workpiece. By analyzing the reflected laser light, these sensors determine the characteristics of weld joints so that accurate positioning and alignment can be assured. They detect gap sizes for gap bridging as well. These sensors, when integrated into the robotic system, provide facilitation for on-the-fly adjustments, hence improving efficiency in welding and developing a sensor-to-robot relationship [12].

Force/torque sensors in robotic welding are mounted on the arm or tool of the robot for very accurate control and adjustments to produce high-quality welds. They come in a variety of different types, from simple to complex: six-axis sensors provide complete feedback on all six degrees of freedom. Three-axis sensors measure translational forces, while single-axis sensors measure forces along one axis. Measurement principles applied in these sensors include strain gauge-based, piezoelectric, and capacitive sensors. The sensors detect contact with the workpiece and, on that basis, start the welding process precisely. They maintain consistent contact forces during the whole welding [13, 14]. Continuously measured values of forces and torques permit a real-time path adjustment for any geometry irregularity. In sophisticated applications, force-based welding dynamically reacts to changing joint conditions depending on the measured forces instead of predefined parameters.

Gas and arc sensors play a very important role in robotic welding systems, as they give valuable feedback on the processes involved. The control and parametric optimization of gas flow and arc characteristics are done by these sensors. The gas sensor, including O_2 sensors, monitors shielding gases with respect to both composition and flow. This guarantees an appropriate flow rate for the maintenance of a protective atmosphere inside preventing contamination and ensuring the integrity of the weld. Gas flow sensors regulate the rate of shielding gas flow to strike the middle ground for adequacy in protection yet efficient to minimize waste. The result of continuous monitoring ensures constant and efficient gas flow that greatly improves welding quality and efficiency [15]. Shielding gas composition sensors, on the other hand, are devices used to continuously monitor shielding gas makeup to determine percentages of argon, and carbon dioxide, among others. This means that the makeup of the gas mix should be ensured to meet predefined goals to achieve optimum weld quality and intrinsic shielding properties [16]. Gas pressure sensors will be fitted for measuring shielding gas pressure, a very important feature to ensure reliable and constant gas flow during the process of welding.

The arc sensors in robotic welding play a critical role in facilitating control over important variables, such as the length of the arc, voltage, and current, as a function of expected quality at the weld. The voltage measured across the arc by an arc voltage sensor determines the penetration and heat input. This continual monitoring allows for adjustments to the parameters so that quality can be assured and problems averted, whether it is from excessive heat or poor penetration. The arc current sensors measure the arc current, which affects the size of the bead, its penetration, and the energy input. This control prevents the part from being under- or overwelded. Arc-length sensors measure the length between the electrode and the workpiece to maintain stability for results of similar quality [17]. They provide feedback for robot movement adjustments that aim at optimizing heat transfer and welding quality.

Temperature sensors are applied to robotic welding because these sensors are used for the control of the temperature of the workpiece and hence the welding parameters. It provides an opportunity for heat control with accuracy that is quite high for all such heat-sensitive materials. The two main types include contact-based thermocouples and resistance temperature detectors, and the others are non-contact, such as infrared thermometers and pyrometers [18]. These sensors provide information about temperature to the robotic control system, which makes dynamic changes to those parameters to maintain the required temperature range. This is crucial for proper fusion and distortion control of sensitive materials. Continuous temperature control ensures the quality of the weld and the safety of the workpiece. All these sensors, correctly associated with control algorithms and corresponding software, improve performance, productivity, and weld quality.

24.3 Signal Monitoring

The robotic welding signal monitoring system is a technology and methodology by which different signals are controlled during the process of welding. The signals control welding parameters, such as welding current, voltage, wire feed speed, or any other important variable, through continuous measurement, analysis, and control. This is very important in terms of ensuring precision control, quality, and performance optimization in robotic welding [19]. A signal monitoring system is composed of sensors, data acquisition equipment, and software algorithms specifically developed to gather and analyze real-time data from the welding operation.

24.3.1 Welding Current Monitoring

The system for monitoring arc current is an integral part of the overall robotic welding process since it has a direct influence on weld performance. This is caused by the direct relationship between weld current and weld heat input. Enough heat is provided by the welding current to result in the proper fusion of the base metal by melting the infill material and creating a molten pool. The control has to be precise to avoid distortion because it is necessary to control the current to allow correct fusion without overheating the workpiece. Effective current control ensures proper penetration depth, bead shape, and base metal and filler material fusion in a constant, high-quality weld. Monitoring of welding current is very important for the quality of the weld. Low values of current represent incomplete fusions and defects, while high values are connected with overheating [20]. The anomalies it can detect are bad electrode contact, which generates an unstable arc that originates an irregular weld bead. During welding, current variations give manufacturers a chance to solve problems of electrode contact for optimum results. Directly linked to the wire feed rate, the welding current determines the heat available for melting the filler wire. Such wire feed problems as jamming and speed variation are surely going to occur with current variations [21, 22]. Current monitoring detects problems quickly to adjust for constant wire feed and stable arc provision of uniform welds. It is also useful in establishing power source irregularities for a constant power supply and therefore constant weld performance.

24.3.2 Voltage Monitoring

Voltage monitoring in robotic welding is very necessary for optimum control and quality in the welding process. With voltage monitoring, manufacturers can have stable arcs, regulate penetration depth, and achieve consistent weld quality. The voltage in welding is of great importance in maintaining a stable welding arc since it controls the amount of energy supplied to the welding process and regulates further the length and characteristics of the arc. The voltage monitoring will enable the system to detect unstable arc deviation, the formation of inconsistent weld beads, and even the extinguishment of an arc. Voltage monitoring, besides penetration depth control main factors, ensures that weld penetration is neither too shallow nor excessive [23]. Change in voltage may lead to low-quality welds since it will affect the shape of the bead, the fusion of base metals with the filler material, and the presence of defects. This enables voltage monitoring by the manufacturer for any irregularity that may affect weld quality and allows corrective action to be taken in time. The monitoring of voltage ensures stable, consistent voltage output, of vital importance in applications where uniform quality and repeatability are required [24]. Besides, problems such as faulty components or electrical disturbances in the power source can also be detected allowing timely solutions to enable consistent welding performance.

24.3.3 Wire Feed Speed and Temperature Monitoring

Robotic welding requires wire feed speed monitoring, which is very important for the entry of the filler wire into the welding arc at a controlled rate. This provides for the generation of constant weld quality. It directly influences the deposition rate, further affecting the shape, size, and the control of heat input of the bead. Monitoring wire feed speed provides stable weld pool maintenance, which can ensure fusion between base metals and filler material [25]. It also aids in consumable management facilitating the reduction of wastage of filler wires. Temperature control by means of thermocouples and infrared sensors prevents overheating and distortion of material. The signal monitoring system provides real-time feedback that controls the welding parameters and keeps the temperature at the desired level [26].

24.3.4 Data Acquisition and Analysis

The robotic welding signal monitoring system is made of data acquisition and analysis. The sensors capture data regarding the process of welding, which involves a process of transduction of physical quantities into electrical signals. Devices like analog-to-digital converters (ADCs) digitize the analog signals. High sampling rates indicate precise data taken for every instance, including the transients occurring in the course of welding. This is a very careful process that helps to make informed decisions. Signal processing manipulates acquired data and makes an analysis that may include filtering, signal conditioning, and feature extraction. Data analysis quantifies and summarizes information through statistical methods [27]. A number of more advanced techniques, related to signal processing algorithms or machine learning and identifying patterns and anomalies, are used to detect problems with the welding process. In this regard, the results of the analysis give real-time feedback and thus trigger correction actions within a robotic welding system.

The results of data analyses are subsequently presented in plots, graphs, and charts for ease of interpretation. Real-time monitoring provides instant feedback regarding the welding process, while detailed reports summarize the analysis findings, trends, and areas for improvement. Such information may be used for supplementary analysis, documentation, and quality control. This signal monitoring system can easily be integrated into an automatic welding set-up with a closed-loop feedback system that permits optimization and adjustment of the welding process for further improvement in weld quality and process efficiency, hence the performance enhancement of robotic welding [28].

24.4 Supervised Learning Intelligent Robotic Welding

In intelligent robotic welding, it is impossible to avoid supervised learning because it aims to train a model on labeled data. In this way, it teaches robots how to adjust their behavior about welding according to predefined instructions. For that, first of all, a dataset must be prepared which contains input features and corresponding output labels. This consists of information related to sensor data and desired welding behavior. These processes are either real-world welding experiments or virtually simulated processes, followed by pre-processing for quality assurance. During model training, patterns between the input and output are learned from this pre-processed dataset by modifying model parameters so that prediction differences can be reduced with actual labels. What optimization algorithm may be used will depend on the applied approach: machine learning through linear regression, SVM, decision trees, or neural networks [29]. Training goes on iteratively until the model reaches a satisfactory performance level. The evaluation is based on accuracy, precision, recall, mean squared error, and so on. After training and testing, the model is deployed in a robotic welding system. While the robot is welding, it acquires sensor data; the model instructs it on various parameters such as position, speed, power, and the like, tuning them for the quality and efficiency desired. This process is iterative, where the robot learns from every new welding data and keeps on increasing its performance by continuous improvement and retraining itself [30].

24.5 Adaptive Robotic Welding System

An adaptive robotic welding system combines robotic automation with state-of-the-art sensing, data analytics, and control to carry out optimum real-time welding processes. The technology automatically self-adjusts to the conditions of welding, the material at hand, and the results desired. The adaptive robotic welding system is developed by making use of advanced sensors through the real-time acquisition of data about the welding process. Analyzing data will help in understanding the current welding conditions, detecting deviations or problems, and extracting insights for process optimization. Herein, adaptive control algorithms come as one of the most important parts of the creation of an adaptive robotic welding system. That is, they hold the modification and optimization potential for the dynamic control parameters of the welding process with respect to real-time sensor data [30, 31].

In model-based adaptive control of robotic welding, real-time adaptation is achieved with the help of a mathematical process model that describes the basic relationships between control variables and the intended effects of the welding process. Control changes are calculated by the adaptive control algorithm from differences between sensor feedback and model predictions. Joseph Kershaw *et al.* [32] applied a hybrid machine-learning framework for real-time bead width prediction and adaptive welding speed adjustment in robotic GTAW welding. In other words, the weld pool image processing would include a Convolutional Neural Network (CNN) framework to estimate the bead width, a multilayer perceptron to correlate welding speed and time with bead width, and a gradient-descent controller using a Multilayer Perceptron (MLP) to adaptively change the speed. Figure 24.2 shows a schematic of hybrid machine learning for adaptive welding speed adjustment. The findings of this study have established considerable agreement between the actual and predicted bead width values using MLP and CNN models.

Gain-scheduling and self-tuning algorithms are adaptive control techniques that optimize performance on the basis of real-time sensor data. They adapt control inputs in such a way that they adequately react to feedback. Model Reference Adaptive Control minimizes the error when actual welding behavior is compared to some reference model. Reinforcement learning maximizes rewards from the welding process. Neural network-based control applies artificial neural networks to adjust control parameters to sensor data in such a way that the welding process is optimized online. The dynamic adjustments improve weld quality, productivity, and process efficiency. Due to the adaptive concept of the control mechanisms,

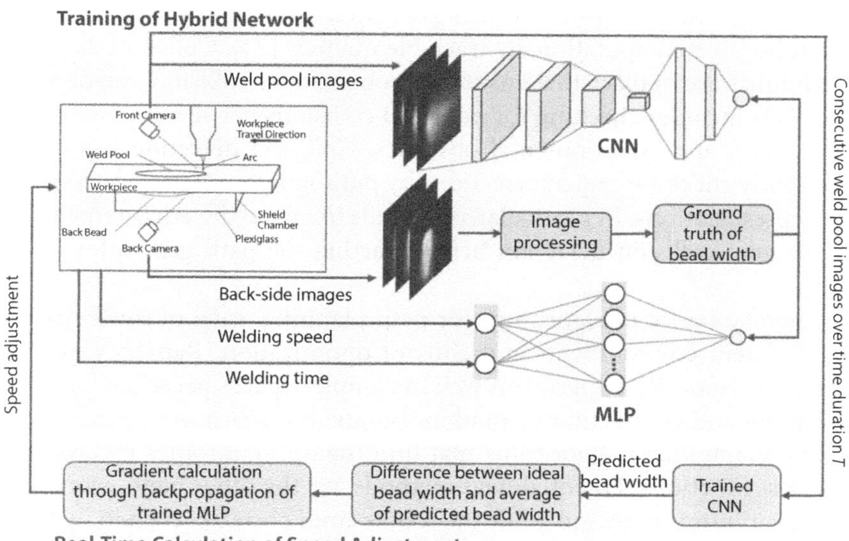

Figure 24.2 Schematic of hybrid machine learning for adaptive welding speed adjustment [32]. Reproduced with permission from Elsevier.

state-of-the-art in-line robotic welding systems are able to permanently readjust welding parameters from sensor signals to keep weld quality constant. Machine learning refines performance over time by determining optimal settings from historical data. It integrates systems for varied activities, materials, and welding methods. This uses GMAW, GTAW, and laser welding for efficient, high-quality outcomes in various applications. Tight tolerances, accurate control, and adaptability to changing conditions make it excel in complex welding procedures [33].

24.6 Path Planning and Trajectory Optimization

Path planning is a process for determining a collision-free path from a starting point to reach the target welding position. The algorithm of path planning takes into consideration the kinematics, workspace, and environmental constraints of the robot. This generates a series of intermediate positions the robot should follow to reach the target position. The overall goals are to reduce traveling distance, welding time, and other relevant factors that include energy consumption and avoiding collision. The very initial step of path planning involves developing a model of the environment in which the robot operates. The continuous workspace and workpiece geometry are discretized into a grid to simplify the planning process. This discretization enables modeling of the environment and the motion of the robot in a computationally tractable manner [34]. Collision detection techniques are in place that ensure the robot eludes collision during welding. This includes checking for potential collisions between the robot, the workpiece, and environmental obstacles. Collision detection algorithms efficiently check for any intersections by putting into use model geometrical representations. In particular, one needs to model the environment and implement collision detection before starting the path generation phase [35].

There are a lot of algorithms for path planning, each of them suitable for different environments and aims of optimization. Rapidly exploring Random Trees (RRT) performs well in complex spaces because of efficient sampling and connection of random points in a space. A* Search is one of the graph search algorithms that find the optimal routes by changing the cost functions. Potential field methods, on the other hand, work with energy minimization and guide the robots amid obstacles even in dynamic environments. Finally, trajectory optimization ensures smooth, jerk-free robotic motion after choosing the proper path. These include speed, acceleration, and joint limits, among others considered in numerical optimization methods [36]. Moreover, combined with the path-planning algorithms, it allows for precise and efficient welding without any danger or welding quality loss.

24.7 Future Scope

The future of robotic welding and automation thus becomes very promising. Advanced sensing, such as vision systems and tactile feedback, will pave the way to improving welding accuracy. Collaborative robots are to become even safer and more intuitive to use allowing people to collaborate seamlessly, thereby increasing productivity and flexibility in manufacturing. Breakthroughs in AI and ML will further upend the robotic welding landscape with autonomous adaptation and optimization. Reinforcement learning will then fine-tune the skills of robots, Internet of Things integration will improve monitoring and maintenance, and the combination of robotic welding with additive manufacturing will open up exciting perspectives by easing production and extending design freedom [37, 38].

Moreover, both augmented reality (AR) and virtual reality (VR) show further disruptive potential in the training and programming of robotic welding systems. It will be possible for operators, through AR/VR interfaces, to view and simulate the welding process in a virtual environment making the programming, training, and troubleshooting easier. It also allows for operator skill enhancement and better overall system performance [39]. Safety is going to become quite imperative in the future of robotic welding. This includes advanced safety sensors, improved human presence detection, and collaborative safety features that ensure safe interaction between humans and robots during welding.

24.8 Conclusion

This chapter discussed an overview of robotic welding and similar automation. This was detailed extensively regarding prime aspects associated with it. This accurate environmental perception is facilitated by sensing technologies, including vision systems, force/torque sensors, and thermal sensors. Adaptive learning ensured by supervised intelligent robotic welding integrated with AI assures continuous performance and increases performance over time. These are the adaptive robotic welding systems able to adjust these parameters according to the welding conditions, achieving performance optimization, delivering consistent quality in weld conditions, and handling variations effectively. Path planning and trajectory optimization are done to ensure efficient and accurate movement of the tool. Algorithms based on RRTs, A* Search, and Potential Field Methods are used to generate collision-free path, accounting for the distance traveled, energy used, and weld quality associated with it. All the way to further paths, trajectory optimization techniques address velocities, accelerations, joint limits, weld quality, and dynamic constraints. Further improvements in robotic welding and automation are expected to continue such innovations in this field. These developments will increase the levels of automation and welding quality improving productivity in a myriad of industries.

References

1. Zheng, C., An, Y., Wang, Z., *et al.*, Hybrid offline programming method for robotic welding systems. *Robot Comput. Integr. Manuf.*, 73, 102238, 2022.
2. Loukas, C., Williams, V., Jones, R., *et al.*, A cost-function driven adaptive welding framework for multi-pass robotic welding. *J. Manuf. Process.*, 67, 545–561, 2021.
3. Shen, W., Hu, T., Zhang, C., Ye, Y., Li, Z., A welding task data model for intelligent process planning of robotic welding. *Robot Comput. Integr. Manuf.*, 64, 101934, 2020.
4. Kovarikova, Z., Duchon, F., Trebula, M., *et al.*, Prototyping an intelligent robotic welding workplace by a cyber-physic tool. *Int. J. Adv. Manuf. Technol.*, 125, 9, 4855–4882, 2023.
5. Kah, P., Shrestha, M., Hiltunen, E., Martikainen, J., Robotic arc welding sensors and programming in industrial applications. *Int. J. Mech. Mater. Eng.*, 10, 1, 2015.
6. Kesse, M.A., Buah, E., Handroos, H., Ayetor, G.K., Development of an artificial intelligence powered tig welding algorithm for the prediction of bead geometry for tig welding processes using hybrid deep learning. *Met. (Basel)*, 10, 4, 2020.
7. Li, Y., Yu, B., Wang, B., Lee, T.H., Banu, M., Online quality inspection of ultrasonic composite welding by combining artificial intelligence technologies with welding process signatures. *Mater. Des.*, 194, 108912, 2020.
8. Eren, B., Guvenc, M.A., Mistikoglu, S., Artificial Intelligence Applications for Friction Stir Welding: A Review. *Met. Mater. Int.*, 27, 2, 2021.
9. Xu, F., Xu, Y., Zhang, H., Chen, S., Application of sensing technology in intelligent robotic arc welding: A review. *J. Manuf. Process.*, 79, 854–880, 2022.
10. Rout, A., Deepak, B.B.V.L., Biswal, B.B., Mahanta, G.B., Weld Seam Detection, Finding, and Setting of Process Parameters for Varying Weld Gap by the Utilization of Laser and Vision Sensor in Robotic Arc Welding. *IEEE Trans. Ind. Electron.*, 69, 1, 2022.

11. Lee, K., Hwang, I., Kim, Y.M., Lee, H., Kang, M., Yu, J., Real-time weld quality prediction using a laser vision sensor in a lap fillet joint during gas metal arc welding. *Sens. (Switzerland)*, 20, 6, 2020.
12. Li, G., Hong, Y., Gao, J., Hong, B., Li, X., Welding seam trajectory recognition for automated skip welding guidance of a spatially intermittent welding seam based on laser vision sensor. *Sens. (Switzerland)*, 20, 13, 2020.
13. Ding, C., Han, Y., Du, W., Wu, J., Xiong, Z., *In Situ* Calibration of Six-Axis Force-Torque Sensors for Industrial Robots with Tilting Base. *IEEE Trans. Rob.*, 38, 4, 2022.
14. Gan, Y., Duan, J., Chen, M., Dai, X., Multi-robot trajectory planning and position/force coordination control in complex welding tasks. *Appl. Sci. (Switzerland)*, 9, 5, 2019.
15. Lee, Y., Cheon, J., Min, B.K., Kim, C., Optimization of gas shielding for the vacuum laser beam welding of Ti–6Al–4 V titanium alloy. *Int. J. Adv. Manuf. Technol.*, 123, 3–4, 2022.
16. Charalampos Loukas, N., Warner, V., Jones, R., *et al.*, A sensor enabled robotic strategy for automated Defect-Free Multi-Pass High-Integrity welding. *Mater. Des.*, 224, 2022.
17. Hamzeh, R., Thomas, L., Polzer, J., Xu, X.W., Heinzel, H., A sensor based monitoring system for real-time quality control: Semi-automatic arc welding case study. In. *Procedia Manuf.*, 51, 201–206, 2020.
18. Guo, S., Wang, D., Chen, J., Feng, Z., Guo, W., Predicting Nugget Size of Resistance Spot Welds Using Infrared Thermal Videos With Image Segmentation and Convolutional Neural Network. *J. Manuf. Sci. Eng. Trans. ASME*, 144, 2, 201–206, 2022.
19. Zhao, D., Wang, Y., Liang, D., Ivanov, M., Performances of regression model and artificial neural network in monitoring welding quality based on power signal. *J. Mater. Res. Technol.*, 9, 2, 2020.
20. Mazlan, A., Daniyal, H., Mohamed, A.I., Ishak, M., Hadi, A.A., Monitoring the quality of welding based on welding current and ste analysis. In. *IOP Conf. Ser.: Mater. Sci. Eng.*, 257, 012043, 2017.
21. Lv, N., Xu, Y., Li, S., Yu, X., Chen, S., Automated control of welding penetration based on audio sensing technology. *J. Mater. Process. Technol.*, 250, 81–98, 2017.
22. Uliasz, M., Ornat, A., Burghardt, A., Muszyńska, M., Szybicki, D., Kurc, K., Automatic Evaluation of the Robotic Production Process for an Aircraft Jet Engine Casing. *Appl. Sci. (Switzerland)*, 12, 13, 2022.
23. Sumesh, A., Nair, B.B., Rameshkumar, K., Santhakumari, A., Raja, A., Mohandas, K., Decision tree based weld defect classification using current and voltage signatures in GMAW process. In. *Mater. Today Proc.*, 5, 8354–8363, 2018.
24. Ali, S., Agrawal, A.P., Ahamad, N., Singh, T., Wahid, A., Robotic MIG welding process parameter optimization of steel EN24T. *Mater. Today Proc.*, 62, 239–244, 2022.

25. Li, R., Dong, M., Gao, H., Prediction of bead geometry with changing welding speed using artificial neural network. *Materials*, 14, 6, 2021.
26. Javadi, Y., Mohseni, E., MacLeod, C.N., *et al.*, High-temperature in-process inspection followed by 96-h robotic inspection of intentionally manufactured hydrogen crack in multi-pass robotic welding. *Int. J. Press. Vessels Pip.*, 189, 104288, 2021.
27. Kumar, V., Parida, M.K., Albert, S.K., The state-of-the-art methodologies for quality analysis of arc welding process using weld data acquisition and analysis techniques. *Int. J. Syst. Assur. Eng. Manage.*, 13, 1, 2022.
28. Kumar, V., Albert, S.K., Chandrasekhar, N., Jayapandian, J., Performance Evaluation of Arc Welding Process Using Weld Data Analysis. *Trans. Indian Inst. Met.*, 71, 12, 2018.
29. Liu, Y.K. and Zhang, Y.M., Supervised Learning of Human Welder Behaviors for Intelligent Robotic Welding. *IEEE Trans. Autom. Sci. Eng.*, 14, 3, 2017.
30. Wang, P., Kershaw, J., Russell, M., Zhang, J., Zhang, Y., Gao, R.X., Data-driven process characterization and adaptive control in robotic arc welding. *CIRP Ann.*, 71, 1, 2022.
31. Soufi Enayati, A.M., Zhang, Z., Najjaran, H., A methodical interpretation of adaptive robotics: Study and reformulation. *Neurocomputing*, 512, 381–397, 2022.
32. Kershaw, J., Yu, R., Zhang, Y.M., Wang, P., Hybrid machine learning-enabled adaptive welding speed control. *J. Manuf. Process.*, 71, 374–383, 2021.
33. Zhai, A., Zhang, H., Wang, J., Lu, G., Li, J., Chen, S., Adaptive neural synchronized impedance control for cooperative manipulators processing under uncertain environments. *Robot Comput. Integr. Manuf.*, 75, 102291, 2022.
34. Fang, H., Ong, S., Nee, A., Robot path planning optimization for welding complex joints. *Int. J. Adv. Manuf. Technol.*, 90, 9–12, 2017.
35. Rout, A., Bbvl, D., Biswal, B.B., Optimal trajectory generation of an industrial welding robot with kinematic and dynamic constraints. *Ind. Robot*, 47, 1, 2020.
36. Singh, G. and Banga, V.K., Kinematics and trajectory planning analysis based on hybrid optimization algorithms for an industrial robotic manipulators. *Soft Comput.*, 26, 21, 11339–11372, 2022.
37. Hameed, A., Ordys, A., Możaryn, J., Sibilska-Mroziewicz, A., Control System Design and Methods for Collaborative Robots: Review. *Appl. Sci. (Switzerland)*, 13, 1, 2023.
38. Borboni, A., Reddy, K.V.V., Elamvazuthi, I., AL-Quraishi, M.S., Natarajan, E., Azhar Ali, S.S., The Expanding Role of Artificial Intelligence in Collaborative Robots for Industrial Applications: A Systematic Review of Recent Works. *Machines*, 11, 1, 2023.
39. Li, X., Yi, W., Chi, H.L., Wang, X., Chan, A.P.C., A critical review of virtual and augmented reality (VR/AR) applications in construction safety. *Autom. Constr.*, 86, 150–162, 2018.

25

Additive Manufacturing Integration with Welding: A Focus on Wire Arc Additive Manufacturing (WAAM)

M. Sivakumar[1*], C.T. Justus Panicker[1], N.S. Balaji[2], Marxim Rahula Bharathi B.[3] and G. Suresh[1]

[1]*Department of Production Engineering, National Institute of Technology Tiruchirappalli, Tiruchirappalli, Tamil Nadu, India*
[2]*Department of Mechanical Engineering, SRM Institute of Science and Technology, Tiruchirappalli Campus, Tiruchirappalli, Tamil Nadu, India*
[3]*Department of Mechanical Engineering, Aditya College of Engineering, Surampalem, Andhra Pradesh, India*

Abstract
A real revolution of modern manufacturing is Additive Manufacturing (AM), which makes it possible to produce intricate shapes without material losses. Traditional welding processes combined with AM especially with WAAM has brought this potential to a higher level of deposition rate and cost benefits. At WAAM, the various welding techniques used include GMAW, GTAW, SAW, Spin Arc Welding, and CMT together with PAW that makes it possible for the company to manufacture large metal structures with complicated geometries. Quality and consistency in WAAM production parts are ensured by employing tools such as vision systems, spectral analysis, and digital twin technology, facilitating precise, high-speed, and optimized manufacturing processes. This enhances material and resource utilization, product quality and safety, due to real-time/ongoing data collection, as well as modeling and forecasting. By encompassing data management and optimized process controls, the idea of digital twin forms the basis of the fourth industrial revolution for manufacturing. This chapter provides an overview of WAAM and its relation to welding, with a particular focus on the technological advancements it brings to advanced manufacturing through automated systems, real-time monitoring, and sensing.

**Corresponding author*: siva.sky06@gmail.com

Sandip Kunar and Gurudas Mandal (eds.) Advanced Welding Technologies, (513–546) © 2025 Scrivener Publishing LLC

***Keywords*:** Additive manufacturing, welding techniques, wire arc additive manufacturing, automation, digital twin technology

25.1 Introduction

AM has revolutionized modern manufacturing by allowing for structures of any geometry to be made with little to no excess material. Combining the conventional welding with AM, particularly when producing the metallic parts, has also developed the prospective application of this technology [1, 2]. Another chapter is dedicated to the types of welding process used in AM with emphasis on WAAM and interaction on automation together with sensors to enhance the productivity of the process.

WAAM merges the concept of arc welding with the concept of AM with the aim of developing large metal components with improved mechanical characteristics. The examples inclusive of WAAM involve the use of a wire feedstock, which is melted by an electric arc to form a weld pool, which is used in building the required geometry by the use of an additive layer by layer [3]. It is superior to some of the other AM technologies, such as laser melting and sintering, in terms of depositions rates and cost effectiveness. This is so because WAAM utilizes conventional welding equipment and filler wire and, hence, yields higher deposition rates and lower material costs.

The methods of welding adapted for WAAM are GMAW, GTAW SAW, spin arc welding, cold metal transfer, and plasma arc welding [4]. There is a further advantage of each technique and a range of applications for which it is best suited, which lends WAAM its flexibility in the production of large and complex metal components.

MIG, also known as GMAW, welding is often utilized in WAAM because of its efficiency and low material consumption. This activity requires the constant passing of a metal wire through a welding torch in which, through an electric arc, the wire is fused and accumulated in a layered manner. TIG, also known as GTAW, is highly adaptable to fine control of heat input and is used in applications that need fine parts. SAW, because of high rate of material deposition and high penetrating power, can be effectively used for building large components, which have high thickness rates. Spin arc welding also incorporates a rotational motion of the filler wire compared to the base and backup metals; this improves the material deposition, though minimizing heat input. CMT has fine control overheat input and material deposition; hence, it gives very clean builds with very little thermal distortion. Plasma arc welding, short for PAW, is a variation of TIG welding in that the welding zone has a narrow opening to form a plasma state around the arc, which is useful in the formation of high temperature melting and deposition.

Quality control is an essential aspect in WAAM to get uniform quality of the final parts. Vision sensing, spectral sensing, acoustic sensing, electric signal analysis, and thermal sensing systems are used to give back constant feedback, while manufacturing is ongoing [5]. These technologies assist in identifying imperfections, such as porosity, voids, and cracks, for corrective action to be taken to retain the process parameters [6].

Automation in WAAM, likewise the use of robotic systems, as well as control algorithms, adds up efficiency and accuracy to the process [7]. Automation helps to minimize the levels of manual interferences, enhance on the rate of production, and provide standardization on the mass productions. In addition to that, through the use of the integration of digital twin technology, it is possible to have a virtual mirror of the manufacturing process that provides for real-time monitoring and simulation of the flow.

This chapter discusses in detail the applications and use of welding in synonym with AM with special reference to the latest process known as WAAM. It also analyses the technique of welding, how control mechanism is applied, and how automation and digital twins advance the performance of manufacturing.

25.2 Wire Arc Additive Manufacturing

Years of experience in welding processes makes WAAM a revolutionary enhancement in additive manufacturing. Rosli *et al.* [8] give a detailed description of various welding used in the WAAM such as cold metal transfer (CMT), GTAW, GMAW, and PAW with or without the pulsed mode.

WAAM is also shown to provide much higher rates of deposition and cost effectiveness compared to other forms of AM, namely, LBM and LBS. In particular, a deposition rate of WAAM per unit of time is, in fact, 2.5–3.4 times higher than a laser-based one. This efficiency is mostly due to the approach of what is used in WAAM, which involves using standard welding equipment and the filler wire and both of which are relatively cheaper than the resources used in laser based additive manufacturing. Also, the powder materials used in laser sintering and melting are difficult for bulk use because of the requirements of their transportation and storage.

In WAAM, the welding electrode is connected to one pole of the power supply, while the substrate on which the process is carried out is connected to another pole. When the power is connected and the electrode encounters the base material, there is an arc production. This first contact is used to ionize the air gap between the electrode and the substrate reducing electrical resistance and causing an electrical discharge in the form of an arc as shown in the Figure 25.1.

The arc as such has multiple zones in it [9]. On one end, the so-called cathode spot, from where the electrons are emitted, is situated. It can be a mobile plant, a thermionic plant or a normal mode plant. During the mobile mode, cathode spots are seen to be flitting across the material surface at a very fast rate. The thermoionic mode is characterized by the presence of bright spot at the electrode tip, whereas a normal mode is when the spot matures enveloping the molten electrode tip. After the cathode spot, there is the cathode drop zone, which is followed by a voltage drop as illustrated in the Figure 25.2. Electrons then pass through the arc column, which is a region of a higher degree of brightness and temperature before getting to an anode where they release heat of condensation. The arc column can rise to the temperatures of 6,000°C to 20,000°C [10] that breaks molecular gases into ions and electrons.

In the WAAM process, the electrode is consumed since it is molten by the heat produced by the arc. It is an electrode, which has been molten and placed over the substrate where a feeding system feeds the electrode wire regularly. This make it possible to go on depositing and, at the same time, forming the structure that is required. For a long time now, WAAM has

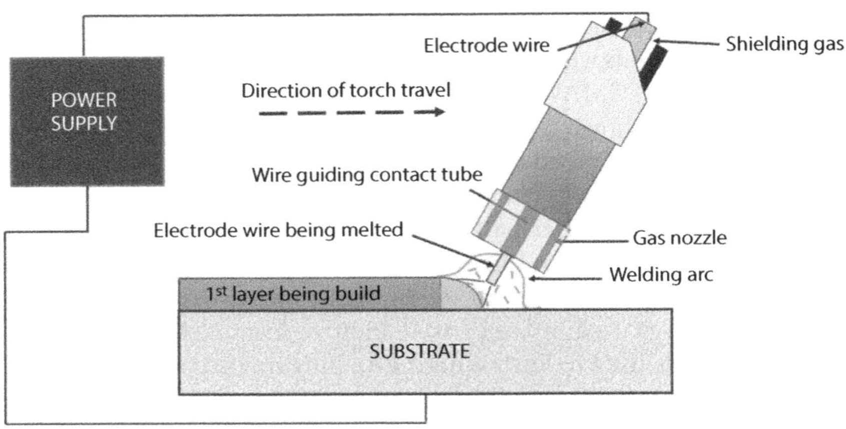

Figure 25.1 Schematic setup of wire arc additive manufacturing.

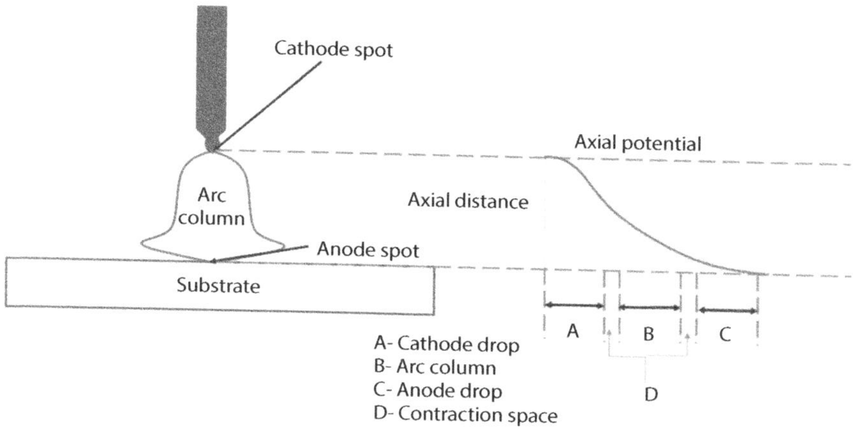

Figure 25.2 Arc column in welding.

been called several other names including 3D welding, SFF, SM, SMD, and SW [11]. These terms stress the versality of the technique and the use of various welding methods to create given shapes and dimensions.

25.2.1 Various Welding Techniques Used in WAAM

The welding techniques that have been identified in WAAM by Rosli *et al.* [8] are cold metal transfer (CMT), gas metal arc welding (GMAW), gas tungsten arc welding (GTAW), and plasma arc welding (PAW). Further, there are some variants of these processes including spin arc GMAW, pulsed PAW, super-GTAW, tandem-GMAW, and variable polarity-CMT.

25.2.1.1 Gas Metal Arc Welding (GMAW)

To understand these processes and whether they are effective in WAAM, it is necessary to discuss their functioning in detail. Gas metal arc welding (GMAW) or metal inert gas (MIG) welding is widely used in AM for making of metal parts [12]. This process, also known as WAAM, involves the use of a wire-fed metallic material that is supplied and fashioned into a bead, which is molten and added in a layer-by-layer method to create the desirable part.

In GMAW, a metal wire is fed through a shielded welding torch, which performs the dual role of the source of heat and the feed rod as shown in Figure 25.3. Owing to the direct contact between the bare face of the wire and the workpiece, an electric arc is produced, and the resulting heat

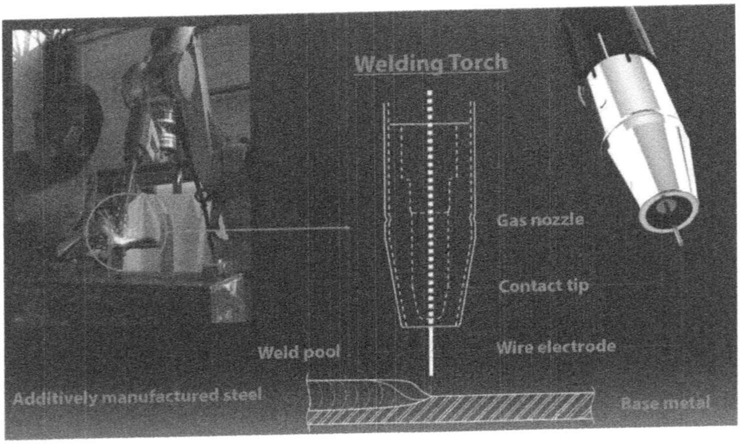

Figure 25.3 Setup for gas metal arc welding [7].

melts the wire. The removable gas, for example argon, helium or mixture of protective gases with reactive ones, like CO_2 or O_2, is enabled to circulate around the weld area shielding the molten material from other gaseous materials that come with it from reacting out with the molten material which may otherwise lower the quality of the weld. The molten wire produces small droplets, and these droplets are laid on the work piece. The arc is stroked again and again and when it strikes, parts of the workpiece surface as well as the previously deposited material are re-melted to ensure good layer to layer bonding [13]. The material is built up in layers using a welding torch to follow a path prescribed by computer numerical control (CNC) or robotic systems and thus produces intricate shapes of metals, which are used to come up with metal parts that correspond to the customer's requirements.

GMAW in additive manufacturing has high efficiency compared to the subtractive manufacturing and low wastage of the material. The process utilizes easily obtainable wire feedstock and conventional welding equipment and therefore represents an economical method of manufacturing big metal sections. Depending on the wire and shielding gas used, the material can be applied to steels, aluminum, and even titanium. This is because the various layers that are deposited are remelted to provide metallurgical bonding and minimize on formation of defects. GMAW is widely used in AM to create complex and lightweight structural parts for aerospace vehicles, custom high-performance parts for automobiles, large corrosion resistant structures for marine vessels, and to repair or replace damaged or worn-out component in repair and maintenance.

25.2.2 Gas Tungsten Arc Welding (GTAW)

Tungsten inert gas welding (TIG) or gas tungsten arc welding (GTAW) is an exacting and highly flexible welding process applied in wire arc additive manufacturing (WAAM). GTAW is a process, in which a non-consumable tungsten electrode makes an electric arc through the base material or substrate. There are two primary methods for striking the arc as follows: the touching method whereby the electrode comes into contact with the substrate to make a short circuit and the high-frequency method whereby an AC current of several kHz is superimposed on the welding current to induce ionization, hence stabilizing the arc as the electrode comes close to the substrate [14]. An inactive gas, commonly ARGON or HELIUM, passes through the welding zone to protect the weld pool from the surrounding environment causing fluctuations in arc current.

A filler wire is continuously inserted into the arc zone, and due to heat that results from the arc, the filler wire melts and puts down the material in layers on the substrate (Figure 25.4). Thus, a filler wire is continuously fed into the arc zone, the high temperature in the arc melts the wire and deposits the material on the substrate in layers (refer to Figure 25.4). A welding torch is moved either manually or through directions that are programmed into a computer and moved along paths that create the desired component geometry. Mentioned variables, including filler wire deposition rate, welding current, voltage, as well as torch speed, are regulated to meet appropriate mechanical properties and dimensional tolerance [15].

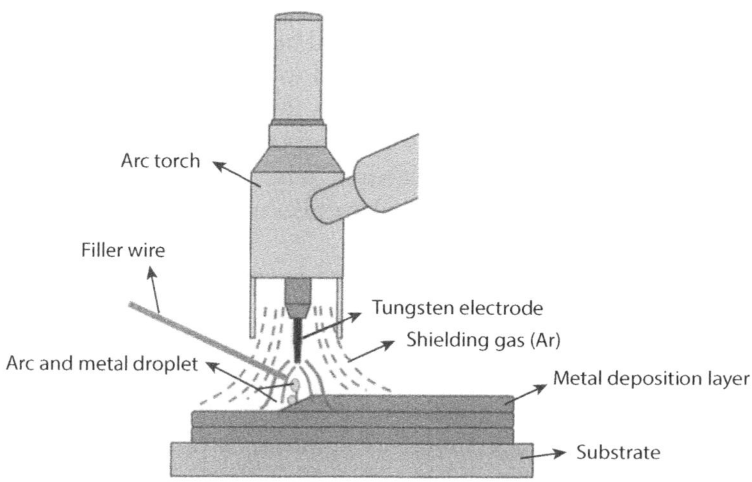

Figure 25.4 Schematics of the setup of GTAW.

GTAW has certain benefits when utilized in the process of additive manufacturing as follows: fine accuracy, desirable weld bead appearance, and achievable high degree of weld deposit metals versatility. It also affords accurate control of the heat input to minimize the chances of distortion and residual stresses in the fabricated structures. This technique is more useful for applications in aerospace industries, automobile, tool and die making, and industries needing prototypes. The equipment required for GTAW is a power source having a tungsten electrode, a filler wire, shielding gas supply, substrate material, and the movement system in the torch [16].

25.2.3 Submerged Arc Welding

Submerged arc welding (SAW) in liberty metallurgy, also known as submerged arc additive manufacturing (SAAM), uses a granular flux and a consumable electrode to build up metallic deposit in a layer-by-layer manner [17]. The procedure involves substrate cleaning and preliminary preparation, after which it is fixed to a workbench that is located within a container that can keep extra flux and slag as illustrated in Figures 25.5(a) and (b).

Tightly packed flux in the form of granules is fed through a holding hopper and is sprayed through a flux delivery tube immediately before the occurrence of an arc. The flux delivery system itself is designed is such a manner that the flux falls in line with the consumable electrode, thereby giving a complete coverage to the molten pool and arc plasma. This kind of configuration makes it possible to achieve an alternating scan pattern, and this leads to the straight wall with standard bead images. By the process of welding, the heat that is produced is used to melt the electrode and part of the substrate to form a weld pool. This flux lies on the weld pool and shields it from the adverse effects of incorporation of atmospheric gases and enhances the metallurgical reaction. While depositing each layer, the movement of the torch in the x–y plane also is in accordance with some set paths.

Whenever one layer is completed, the burner then moves upward to remove the solidified slag before starting the next layer. Between the layers, it regulates and maintains the right temperature to allow for correct deposition of layers. In this case, interlayer temperatures of approximately 300°C are used to achieve good wetting and proper fusion of the materials involved. The present technique is significantly different from welding and direct additive manufacturing where usually the interlayer temperature is comparatively lesser. Signs for temperature tracking points relate to interlayer temperatures, control points where heating element is used and has small tolerance, with-points for heating and measuring temperatures TC1, TC2, TC3, and TC4, to provide the right condition or substrate for every consequent layer as is shown in Figure 25.5(c). In this manner, going through the various steps, one on top of the other, the 3D part of a certain component is accomplished. It is laminated in that the layers are deposited individually, and each layer is scanned in a different path on the x–y plane [Figure 25.5(d)]. Since the SAAM process enables one to achieve very near or even equal to the desired size and mechanical requirements, this process is suitable for application in industries.

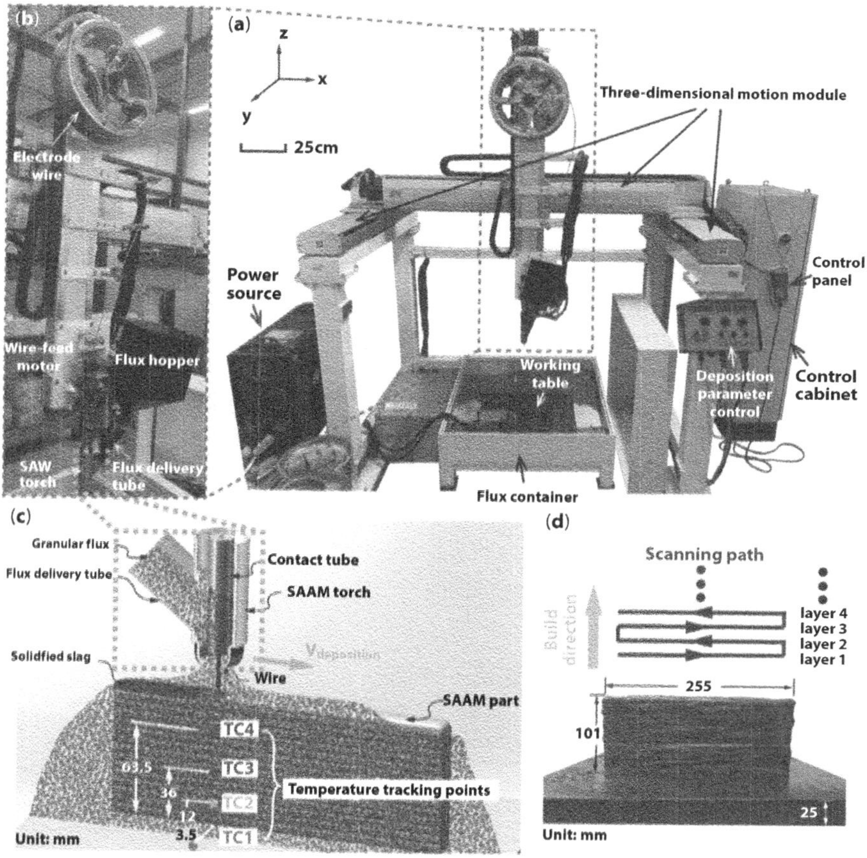

Figure 25.5 (a) Submerged arc additive manufacturing (SAAM) setup, (b) SAAM torch and flux delivery, (c) SAAM process with flux and temperature tracking, (d) scanning path and layer buildup [17].

25.2.4 Spin Arc Welding

Spin/rotational arc welding (RA) is a variation of conventional gas metal arc welding (GMAW), which was developed specifically for use in wire arc additive manufacturing (WAAM). This technique uses a special torch and control strategy for achieving circular motion to the filler wire at the time of welding [18, 19]. The operation principle of RA is shown in Figure 25.6, where the motor is used for the rotational motion of the whole assembly, while the spin/rotational arc torch is the torch to contain and control the direction of the filler wire, the eccentric sleeve that enables the rotation of the filler wire by accommodating the motor and bearing, and bearing that enables smooth rotation of the filler arc wire [20]. The filler wire, usually consumable, melts to form the weld bead and rotates around the weld axis and is driven by the motor via an eccentric sleeve. The layer, which is being constructed, is the substrate upon which the material is placed to create the layers of the additive build.

In the process, the arc chatters over the substrate and the filler wire at a definite RPM of rotation. This rotation makes the filler wire rotate in an arc form over a set diameter and therefore deposits material uniformly. The process on which the RA welding run is based is the centrifugal droplet transfer, which means that both centrifugal and gravity force is used to create very good welds. In this process, the centrifugal force from the electrode that is rotating assists in the formation of molten drops from the wire, which is then directed angularly. Another novelty that RA welding offers is the reduced net heat density into the system because of the

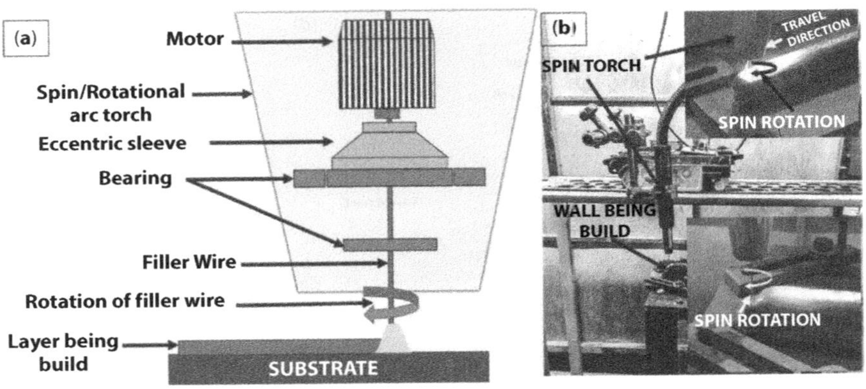

Figure 25.6 (a) Schematic representation of the spin/rotational arc welding process in WAAM. (b) Experimental setup [21].

rotation, thus faster cooling of the build and improved mechanical properties on the welds.

25.2.5 Cold Metal Transfer (CMT)

Cold metal transfer (CMT) is an advance welding technology that has been incorporated into AM that afford control overheat input and material layering. In the CMT process, the feeding of the filler wire is adjusted with the release of molten droplets (Figure 25.7), thus controlling the heat input to the depositing material [22]. In each of the short-circuit phases, the power is switched off and the wire pulled back—it helps in droplet detachment without spatter formation [23]. Something done here leads to a much neater and more precise process than that outlined in claim 3. The heat input is obtained only during the arc phase and is sustained for a short time only, thus reducing the amounts of thermal distortions and residual stresses [24]. The torch allows the positional control of wire feed rate at a very fast rate to build layers successively, and the arc length is mechanically governed to maintain a steady arc for material deposition regardless of the rate at which it is done. In fact, the whole process is absolutely controlled by electrical signals, thus increasing reliability and decreasing sensitivity to operator skills.

CMT has the following advantages in additive manufacturing technology: high precision and accuracy attributed to electronic control of heat input and wire motion, low heat distortion and spatter, and thus produces cleaner parts [26]. Restricted heat input reduces thermal displacement and stress, which remains in the final product, hence enhancing the mechanical properties of the steel. The sludge-free quality of CMT lowers post CMT processing in terms of time and energy used, thus improving general efficiency. The automation and electronic control in CMT eliminate or also

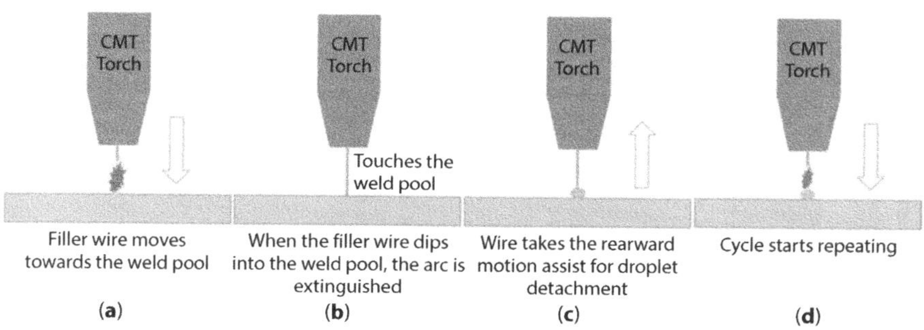

Figure 25.7 Mechanism of cold metal transfer [25].

make the load operator skills, a less factor that affects the ability and consistency of the additive manufacturing process.

25.2.6 Plasma Arc Welding (PAW)

There is a modification of TIG welding known as plasma arc welding (PAW), which involves the provision of a narrow channel around the arc into which a limited volume of inert gas is injected. There are different types of PAW-based AM known, but the most widely used one is known as the transferred arc type [27] shown in Figure 25.8. In the transferred type of PAW, the constricted arc is developed between the non-consumable electrode and the substrate plate. Here, in the non-transferred type, the arc is drawn between the electrode and the nozzle [28].

The narrow arc developed in PAW results in ionization of the input gases into negative and positive ions as in a plasma. When these gases get away from this area, they are reformed from their ionized state, release transformed energy in the form of heat with much higher temperature of 20,000°C. Severe heat is used here to cause the filler material, which is supplied to the plasma in a continuous manner, to melt and deposit on the substrate. Some geometries may be made by navigating the torch in the specified direction.

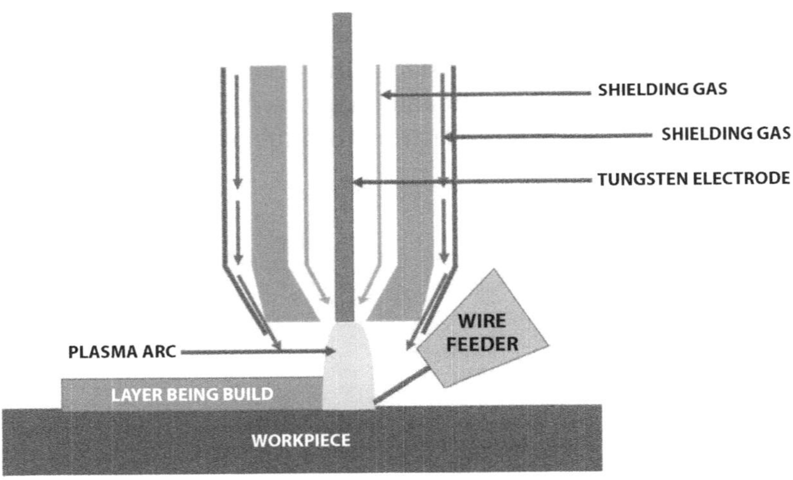

Figure 25.8 Plasma arc-based WAAM.

25.3 Integration of Automation in WAAM

Automation in wire arc in additive manufacturing enhances the effectiveness in the manufacture of large-scale metal structures by greatness. This adaptation of automated technologies reconstructs the conventional form of welding by applying robotic systems and modern computational software for perfection and efficiency [3].

25.3.1 Transformation Through Automation

At WAAM, the metal is deposited in a layer-by-layer process with the help of robotic arms and computer numerical control (CNC) machines. Such systems enable the control of the speed and path of the welding torch to an extraordinary degree and uniformity throughout the manufacture. An example of how such technologies can be implemented works in an automated system employing the use of robots is illustrated in Figure 25.9.

25.3.2 Enhancements in Productivity and Efficiency

Welding automation would also like to introduce that each welding task could be continuous, which dramatically enhances the throughput if the

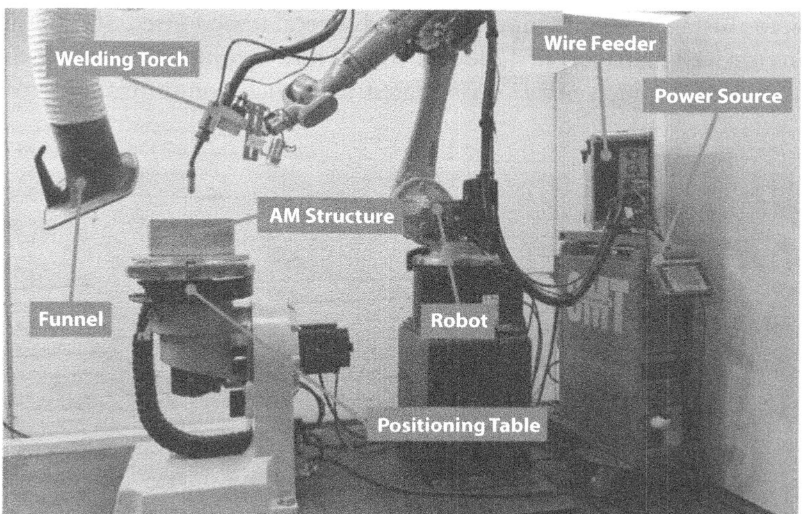

Figure 25.9 WAAM process automated by a robotic arm [29].

process was fully automated, as it is the case in WAAM. This is especially helpful in industries that require quantity, for example, automobile and aircraft manufacturing. Full automated systems have standard speed in operation and working parameters; hence, they do not stop as often as non-automated systems, they also make efficient use of resources, and this all results in decrease in production costs.

25.3.3 Quality and Consistency

First, it is possible to state that the application of automated WAAM significantly increases the quality of the builds as well as their uniformity. In manual processes, because of the elements of human factors, such as fatigue and distractions, the probability of each layer to have the same specific characteristics as defined in the automated system would not be realizable. This kind of application achieves essential standards when it comes to quality and dimensions and is used daily in products like aerospace and critical infrastructure. The WAAM technique can be used to build quality structures, such as steel bridges, as depicted in Figure 25.10.

25.3.4 Safety Enhancements

Automation when incorporated into WAAM also increases safety since weld operators are protected from unfavorable conditions typical of welding, for instance, high temperatures and several toxic fumes. These conditions are well manageable if they are forced into automated systems, which will eliminate much of risks associated with such conditions and enable

Figure 25.10 Steel bridge made by WAAM process [7].

human workers to take supervisory and maintenance roles. This shift not only serve the interest of workers but also the general working environment and satisfaction of the workers.

25.3.5 Material and Resource Efficiency

Computer-controlled WAAM systems have the ability of effectiveness of deposition material due to the accurate and consistent setting of the factors like heat input and travel speed of the machine. This capability helps to minimize the cost of materials, and consumption of welding consumables, which include electrodes and shielding gases. Real-time evaluation and control equally augment the manufacturing process productivity meaning sustainability and even cost reduction in the industries that employs a lot of materials.

25.4 Monitoring and Control Methods in WAAM

25.4.1 Quality Assurance in Additive Manufacturing

During the WAAM manufacturing, several defects can be realized in the structure and appearance of the manufactured parts. These defects are characteristic for almost all methods of fabricating parts using WAAM, laser AM, and tradition arc welding because many of these methods have similar physical foundation. Some of the common defects include porosity, void formation, and cracking, which are because of contaminated raw materials, poor setting of process parameters, extraneous disturbances, and instability of the process. These defects worsen mechanical characteristics and result in variation in signals from the process such as visible, infrared, and electrical [30].

With reference to the case of WAAM, arc welding, and laser AM, the following is a figure showing both the similarities and the differences of the variety of defects as shown in Figure 25.11. Some of the general central defects present in all the three processes are porosity, oxidation, and existence of cracks. There are some defects specific to WAAM, such as deformation, lack of fusion, and undercut, and there are arc welding defects such as penetration and depressions displayed in one sector and the other, respectively. Laser AM specific defects, such as lamination and balling, are presented in the third segment.

To handle such defects, it becomes crucial to employ enhanced sensing mechanisms that, in essence, include monitoring and feedback mechanisms

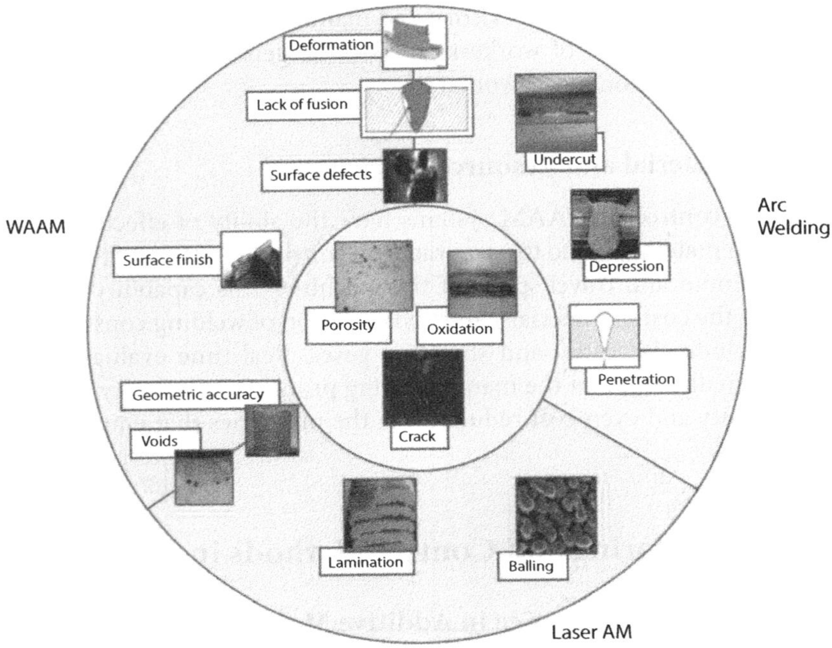

Figure 25.11 Common defects in additive manufacturing processes [30].

employed to identify deviations from normal signals in a process and correct for the changes. Major sensing techniques include visual inspection by surface detection and geometric irregularities, spectral examination to detect cases, like oxidation and contaminations, electrical response sensing for cases of fusion, as well as penetration. The use of these sensing techniques is critical to quality control since they help establish that parts are of acceptable quality, process control since they assist to stabilize process parameters, and economic control because they prevent wastage of materials and need for subsequent rework.

25.4.2 Sensing Techniques for Defect Detection in WAAM

In the case of WAAM, the process is accompanied by the use of the following: It is with these methods that one is able to identify the defects and quality of the manufactured components. This paper also presents separate works on the different types of sensing as follows: vision, spectral, acoustic, electric signals evaluation, and thermal. Table 25.1 below gives an overview of the diverse sensing techniques that apply to WAAM for different kinds of flaws, stating the application, likely tools, and advantages [30].

Table 25.1 Sensing methods for defect detection in WAAM [30].

Defect type	Sensing method	Description	Common tools	Key benefits
Porosity & Void	Vision	Visual monitoring of the process using cameras.	High-speed cameras, CCD cameras, Line lasers	-Real-time analysis and feedback -Detection of surface defects such as pores and voids -Helps in adjusting process parameters for optimal results
	Spectrum	Analyzes spectral signals emitted during the process to detect composition and phase changes	Optical emission spectroscopy, spectrometers	-Identifies elemental concentrations -Diagnoses welding perturbations -Effective for porosity detection
	Acoustic	Records and analyzes sound waves generated during the process to detect internal defects	Acoustic emission (AE) sensors	-Non-contact and non-destructive -Real-time monitoring and defect classification -Uses advanced signal processing techniques (e.g., PCA, clustering)

(*Continued*)

Table 25.1 Sensing methods for defect detection in WAAM [30]. *(Continued)*

Defect type	Sensing method	Description	Common tools	Key benefits
	Electric	Monitors electrical signals like current and voltage during the WAAM process	Current and voltage sensors	-Detects variations indicating defects or unstable process conditions -Provides comprehensive understanding of process dynamics -Real-time data monitoring
	Thermal	Uses thermal imaging to monitor temperature changes and detect defects	Infrared thermography, thermal cameras, pyrometers	-Rapid and non-contact method -Real-time monitoring and feedback -Effective for detecting cracks and thermal variations
Crack	Vision	Visual detection of cracks using high-speed imaging	High-speed cameras	-Real-time feedback -Helps in process optimization -Detects surface and internal cracks
	Spectrum	Spectral signal analysis to detect phase changes that could indicate cracks	Spectrometers	-Monitors phase transformations -Identifies crack-related spectral signals -Effective defect diagnosis

(Continued)

Table 25.1 Sensing methods for defect detection in WAAM [30]. (*Continued*)

Defect type	Sensing method	Description	Common tools	Key benefits
	Acoustic	Uses sound waves to detect cracks and other internal defects	Acoustic emission (AE) sensors	-Non-destructive -Real-time crack detection -Advanced analysis techniques
	Electric	Monitors changes in electrical characteristics to detect cracks	Current and voltage sensors	-Identifies variations indicating cracks -Comprehensive monitoring -Real-time data
	Thermal	Thermal imaging to detect cracks by monitoring temperature profiles	Infrared thermography, thermal cameras	-Non-contact -Real-time crack detection -Effective for detecting thermal profile changes
Surface defect	Vision	Visual inspection of surface defects using cameras	High-speed cameras, CCD cameras, line lasers	-Detects surface flaws -Real-time monitoring -Helps in process optimization
	Spectrum	Analyzes spectral signals for surface defect detection	Optical emission spectroscopy, spectrometers	-Identifies elemental anomalies -Effective for detecting surface defects

(*Continued*)

Table 25.1 Sensing methods for defect detection in WAAM [30]. (*Continued*)

Defect type	Sensing method	Description	Common tools	Key benefits
	Electric	Monitors electrical signals to identify surface defects	Current and voltage sensors	-Detects electrical anomalies -Real-time monitoring -Helps in process adjustment
	Thermal	Thermal imaging to monitor surface temperature and detect defects	Infrared thermography, thermal cameras	-Rapid, non-contact method -Effective for surface defect detection
Melt pool dynamics	Vision	Real-time visual monitoring of melt pool behavior	High-speed cameras, CCD cameras, line lasers	-Analyzes melt pool dynamics -Real-time feedback -Helps optimize process parameters
	Spectrum	Spectral analysis to monitor melt pool composition and behavior	Optical emission spectroscopy, spectrometers	-Identifies elemental composition -Monitors phase changes -Effective for real-time adjustments
	Acoustic	Uses acoustic signals to monitor melt pool dynamics	Acoustic emission (AE) sensors	-Non-contact -Real-time monitoring -Detects internal melt pool changes

(*Continued*)

Table 25.1 Sensing methods for defect detection in WAAM [30]. (*Continued*)

Defect type	Sensing method	Description	Common tools	Key benefits
	Electric	Monitors electrical signals to analyze melt pool behavior	Current and voltage sensors	-Detects electrical variations -Real-time monitoring -Helps in process optimization
	Thermal	Thermal imaging to observe melt pool temperature changes and behavior	Infrared thermography, thermal cameras	-Non-contact -Real-time feedback -Effective for detecting thermal behavior

25.4.3 Vision Sensing

Vision sensing is a direct sensing technique in which camera and vision processing strategies are employed for observing the WAAM process. It is employed in characterization of the melt pool, layers and surface, and beads geometries. The high-speed cameras, the CCD cameras, and the line lasers are image-receiving instruments that are processed in real time. Several defects, including pores, humps, and undercuts, can be classified by the deep learning algorithms focusing on visual information. Also, this method gives an enhanced feedback of the outcome to correct the process parameters for the best results.

25.4.4 Spectral Sensing

Spectral sensing uses optical emission spectroscopy to identify the chemical content and phase transformations that occur during WAAM. In this method, a spectrometer is used to monitor the spectral signals produced when the process is ongoing. These signals can be suggestive of defects. What is more, they can help in the understanding of the physical processes, which happen. Spectral sensing has emerged as a useful technique in estimating the elemental concentrations as well as diagnosing welding perturbations.

25.4.5 Acoustic Sensing

Acoustic sensing is one where AE is captured and processed and used as precursor for the identification of flaws such as crack and porosity. AE sensors record the sound signals produced in the welding process and pass useful information about material properties. This is a non-contact and non-destructive technique, by which the defects could be identified as a result of amplitude and frequency of the signals produced. The classification of the defects is performed with the aid of clustering and other signal processing methods, including the PCA, for monitoring the process in real time [30].

25.4.6 Electric Signal Analysis

Electrical signal analysis involves checking the change in current and voltage during the WAAM process. Fluctuations of these signals can point out to defects or variability of the process conditions. Based on the electrical characteristics, it is also possible to identify porosity, voids, and incorrect arc length. It can be applied in cooperation with other sensing methods to get information regarding the studied processes.

25.4.7 Thermal Sensing

Thermal monitoring involves the use of infrared thermography to monitor variations in the thermal field during WAAM process. In this method, thermal images are taken because of an object's surface temperature, and data are processed to determine cracks or inhomogeneous material characteristics.

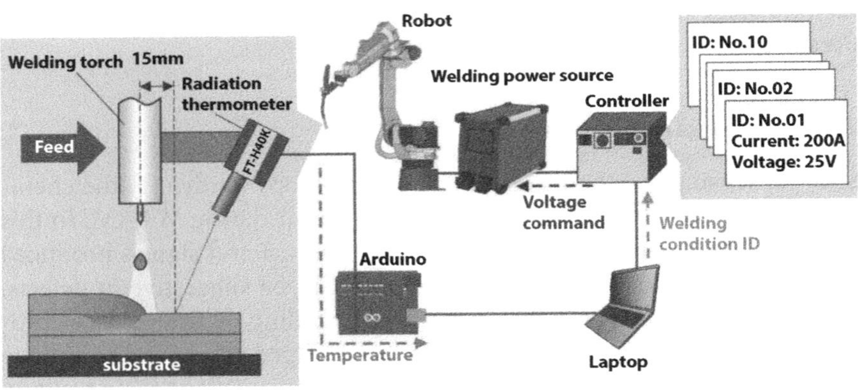

Figure 25.12 Schematic of the thermal sensing and voltage control system in WAAM [31].

There are other thermal imaging techniques, which the scanning process can be done rapidly without causing any contact with the object to be imaged. This is especially helpful where some form of thermal contrast is identified as having the potential of causing defects in the product [30].

The thermal sensing and voltage control system in WAAM employs an ARC Mate 100 iC robot for wielding the torch and having a Power Wave R350 as the power source (Figure 25.12). The temperature before the torch is measured with a spot radiation thermometer. Heat input is regulated by means of voltage, determined from the temperature, and is supplied to the welding machine. The robot controller can store numerous welding condition data set, and each data set is assigned with an ID. At the time of deposition, the temperature information of the welded section is transmitted to a laptop, which then picks the welding condition ID corresponding to the desired deposited metal shape and then transmits this ID to the controller to regulate the power source.

25.5 Digital Twin in WAAM

Digital twin, which was first defined by Professor Grieve at the University of Michigan in 2002, has acquired the necessary importance to create effective feedback between the physical and digital worlds. Remarkably, the application of the DT concept is in various industries, but the definition of digital twins remains conceptual and ambiguous. Originally employed in the context of product life cycle management, digital twins are intended to monitor the life cycle of the assets in a company and create value from the life cycle perspective based on efficient information flows. Digital twins are not uniform in either definition or implementation because of the large number of existing solutions and ideas, which is why the industry is called dynamic. Some of the applications include diagnostics, prognosis, real-time performance enhancement, connected value creation, and constant performance monitoring [32].

In the area of WAAM, the definition of digital twins given by Glaessgen and Stargel is known broadly as "integrated multi-physics, multi-scale, probabilistic simulations." The WAAM process is multi-stage, with refinement, and online monitoring and, as such, necessitates an elaborate and systemized computer program [33]. An efficient model that links the physical and easily understandable tangible world with the WAAM process can make gentle enhancements as the speed of manufacturing can be increased, accuracy can be improved, and the occurrence of defects and cost can be reduced. Some of the pre-requisites that would be required for developing and using WAAM digital twins include enhanced software features and comprehensiveness of knowledge of different metal materials.

25.5.1 Digital Twin System Model for WAAM

A new reference model for a digital twin system that focuses on generic capabilities and functionalities for WAAM is presented here. This model encompasses various key activities, including simulation of the WAAM processes, data gathering, monitoring in real time, accurate prediction, and visualization using physical models for which real-time sensor feeds form part of continuously updated models. In this model, every "concept" frame of the WAAM system is an abstraction that represents an active object or element that can potentially interact and share data with other frames. This structure supports the integration and coordination of the manufacturing process improving the flow on the production line.

The WAAM digital twin model comprises three essential components as follows: the physical twin, the digital twin, and the cloud digital twin. The physical twin oversees the real business of making the part, including acquisition of raw metal material all the way to the finished part. Real-time is the primary data acquisition through the network of the sensors that carefully observe every stage of the process. These data are then transferred to the cloud digital twin, which is mainly concerned with data and analysis server and manages all project-related data. The use of cloud digital twin is therefore important in management since it incorporates current and past data in the management of the manufacturing process.

The digital twin itself is a virtual copy of the actual manufacturing process facilitating complex modeling and other improvements with reference to the data inputs fed in real time. This environment enables one to anticipate any challenges that may come about and can design different experiments that can be tried out without affecting real manufacturing processes. This is so because of the feedback loop that is provided by having a physical twin and a digital twin ensuring that the existence of the system is always dynamic and sensitive to changes. Through continuously enhancing this developed and proposed digital twin system, substantial step changes within the manufacturing process in terms of efficiency, precision, and flexibility can be achieved when incorporating this digital twin technology in advanced manufacturing system.

Figure 25.13 represents the digital twin model employed in the current research and introducing the digital twin into the mainstream of WAAM. This model contains an updated identical copy of the totality of the physical layout of manufacture making it easy to monitor, forecast, and manage

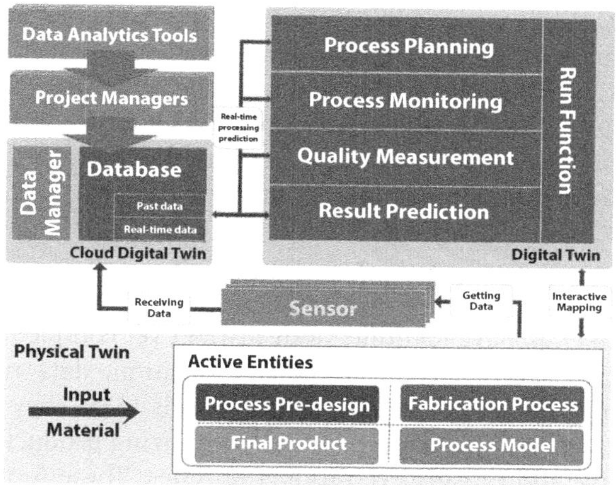

Figure 25.13 WAAM system using digital twin [32].

the physical fabrication procedure. The digital twin is divided into three core components as follows: the physical twin, the dynamic twin, and the cloud dynamic twin. The cloud digital twin is mainly used to handle the data collected in the process. The physical twin represents the actual WAAM processing unit that applies its stages of pre-designing, processing, and post-processing metal materials to produce parts that characterize finished products. The actual physical twin that one would see is a platform of multiple sensors that enables real-time data capture.

The collected data are constantly sent to the cloud digital twin for a detailed examination of the information and efficient organization of the project. The cloud digital twin processes this data because it is typically repeatable and referential; it may even keep a database of prior data at the same time. The interactive digital twin perfectly reflects the physical twin and constantly imports and exports data with the cloud using sensors. It is used in the online monitoring, analysis, and even prediction of the fabrication process. What makes it possible to perform process planning, process monitoring, quality management, and result prediction within the context of a digital twin is this interactive mapping.

25.5.2 External Monitoring and Sensing

External monitoring and sensing are primary components of DT framework for WAAM. Some of the components of a standard WAAM setup are computer, robotic arm, robot controller, welding torch, welding unit, and shielding gas. While it is possible to realize the WAAM system of accumulation without sensors, it is impossible to create a digital twin with their help without their constant feedback during the fabrication process. These systems are per se not incorporated into the digital twin but are required to supply the necessary knowledge regarding the processing status, thus increasing the realism and capabilities of the digital twin.

25.5.3 WAAM Digital Twin Functions

Jakob Bönsch *et al.* have identified four distinct yet complementary functions of digital twins as follows: data acquisition, data manipulation, modeling, and the use of decision-making tools [33]. They are essential in acquiring, archiving, assessing, as well as utilizing product information for both active and prospective production lines. These functions, when combined, work synergistically since digital twins create a link that allows efficient and effective mechanisms of flow of information in manufacturing and other industrial processes.

25.5.3.1 Data Collection

Various forms of data, which can be used to develop digital twin in the WAAM process, include the systematic gathering of data regarding specific elements related to fabrication. This entails the properties, features, states, and pattern of conduct and behavior noted at the time of the process. This task requires significant involvement of the Internet of Things (IoT) gadgets, particularly sensors and actuators, for the purpose of perceiving the concrete adjustments and for changing the signals at that. The data generated include factors, like flow and temperature distribution of the molten pool, which is very important for the quality of the parts being produced as well as the improvement of the fidelity of the digital twin.

25.5.3.2 Data Management

WAAM processes generate a large amount of data at each phase at the design phase of the product, process planning, manufacturing, post processing, and quality control phases. Due to this, proper management and processing of historical and real-time data are essential for efficient purposes. Cloud computing and edge computing are also instrumental in handling these data, and hence, the digital twin functionalities are implemented successfully. These technologies also allow for efficient storage, collection, and utilization of all the information that may be necessary in enhancing the operation of the WAAM process.

25.5.3.3 Data Processing

Data processing is the stage that is all about making the data easier to understand and to work with after it has been gathered. In this stage, functions like CNN, SVM, and DL are used to process acquired data and generate models. These models are essential in the quality inspection and other operational procedures such as assembly and testing. The results of this data processing serve as the base for the operations, control, and forecast of the functions of the concept of the digital twin.

25.5.3.4 Simulation

Simulation process, in context of the digital twin for WAAM, is based on the optimization of the parameters of the process and detailed examination of temperature and stress fields, which appear during the deposition step. To do that, several modeling approaches are used, with the finite element models being the most common. These models are useful in determining the material properties of the part and residual stresses that are likely to be experienced. Through its role as the main part of the digital twin system, simulation guarantees the safety of equipment operation at various conditions. This approach enables one to avoid problems that might be likely to occur with the manufacturing process and make minute adjustments to the manufacturing process for the purpose of improving the efficiency and quality of the manufactured products.

25.5.3.5 Prediction

Prediction is an important component of improvement of the welding process since it involves the ability to predict future behavior and make corresponding changes toward achievement of a set goals. As for the use of a digital twin, it allows the system to predict changes in the physical state and adjust the parameters thereto. It also provides for constant enhancement of the wire arc additive manufacturing process so that its efficiency and effectiveness is optimized.

25.5.3.6 Automatic Decision Making

At the core of the WAAM digital twin design, automatic decision making is a challenging problem. For high-speed WAAM processes, the experienced technicians can make decisions in split second; however, the decision making for the digital twin should be made more precise and must not depend on the technicians for reliable operations. The function of prediction is crucial for technical maintenance and optimization of the system in the absence of an expert. More work is required to get digital twins to the state where automatic decision making can be as accurate as is desired.

25.6 Improving WAAM Outcomes with Digital Twin and Augmented Reality Innovations

For WAAM, the relationship between the material properties, the product design, the process parameters, post-processing, and the quality of the final product is still not fully understood and remains rather challenging. The ability to reproduce experimental results under similar circumstances often poses difficulties, and issues relating to metal part fabrication indicate that there is a requirement to urge for improvements in the WAAM process with regard to the stability of the process. Nonetheless, each phase of a WAAM product's life cycle produces vast amounts of data. Proper utilization of these data can improve the quality of the manufactured metal parts and reduce on the possibilities of having faulty parts, which compromises the reliability of the components. To effectively solve the challenges in WAAM, good data management as well as data analysis skills are required. That is why, the establishment of a strong, efficient, and integrated data platform is the key for further advancement in the understanding of the WAAM processes as well as for the reduction of the time and the costs required for the development of the parts and the improvement of their quality. Müller *et al.* highlighted three critical requirements for an effective metal additive manufacturing product lifecycle data management system: This is because the knowledge that needs to be created includes the following: (1) quick and convenient access to comprehensive product knowledge, (2) clear record of where specific product knowledge is located, and (3) ability to automate the flow of information [34]. All these criteria focus on the aspects of consistent dissemination and automation of data in the context of the entire product's life cycle.

Digital twin technology surfaces as a sturdy platform for integrating the digital world with the real world, effectiveness in the management of digital assets, and providing complex data analysis to improve the control over processes. Regarding the use of value-added technologies, such as digital twins, data management in WAAM can be enhanced considerably.

There is a need to broaden the scope of digital twins in terms of the range of possibilities, the ease of use, compatibility, versatility, and possibility to scale, as well as the possibility of their application. The proper acquisition and handling, as well as the revision of data, could greatly enhance the reliability of process parameters and the manufacturability of parts, as well as the quality control. In addition, the digital twin is used to control and optimize the process parameters to enhance the quality of the final parts. Also, the WAAM environment can benefit from the integration of augmented reality tools, such as Hololens 2, in terms of visualization and interaction. Hololens 2 offers real-time information and enables the optimization of the manufacturing process by placing digital data on the actual environment.

The configuration of the digital twin and the employed visualization process at WAAM are illustrated in Figure 25.14. This environment employs one or more sensors for real-time data transfer and mimics such processes as microstructure formation and residual stress accumulation. Data management is based on physics principles for the enhanced capability in adaptive and automated control decision making in the WAAM process. The diagram shows a detailed process from data acquisition from

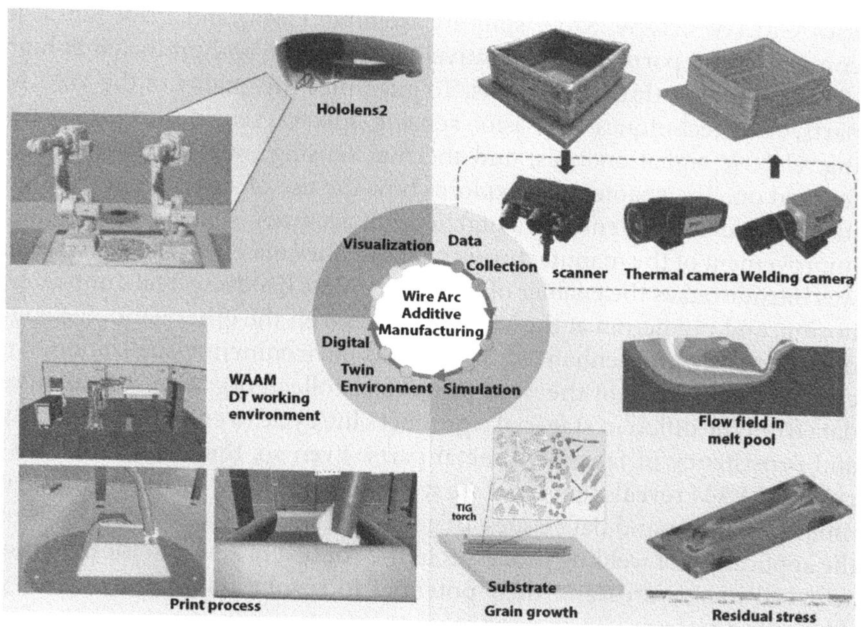

Figure 25.14 Digital twin in WAAM illustrating data collection, simulation, and process optimization [32].

scanners, thermal imaging camera, and welding-specialized camera right up to the visualization of the WAAM procedure, digital twin interface, and the simulation of the grain growth pattern and the residual stresses.

However, the WAAM process is still not free from challenges and ambiguities even to this date. Three primary challenges remain. They are as follows: (1) repetitive data analysis due to multi-disciplinary modeling and simulation problems that have an impact on material properties and structure stability; (2) the WAAM parameters where data samples, such as molten pool thermal cycles, increase exponentially; and (3) problems in data collection such as unusable data and difficulties in data storage. As data usage rises, dealing with over data and addressing issues in real time is rather difficult. Last, digital twin technology is open in a certain sense, and thus, the selection of data usage and negotiation is critical.

25.7 Summary

The welding process integration in additive manufacturing chapter of the book describes the process of combining the conventional welding methods with AM, with emphasis on the wire arc AM. It discusses how WAAM uses GMAW, GTAW, SAW, spin arc welding, CMT, and PAW fabricate complex metal parts that are relatively large in size and produced at high deposition rates, thus at low costs. To guarantee the quality of the WAAM parts, other techniques, like vision sensing, spectral sensing, acoustic sensing, electric signal analysis, and thermal sensing, were considered and focused on. The chapter also explores how the use of automation and digital twin technology ensures a high level of accuracy, speed, and real-time improvement of the manufacturing process. The concept of the digital twin was introduced as the enabler of the holistic data handling, real-time monitoring, and prediction at the WAAM process; on the other hand, the augmented reality tools enhanced the WAAM environment visualization. An emphasis was made of the need to initiate, collect, organize, and analyze data from the different stages of a product's life cycle to enhance the quality and consistency of fabricated metal parts. Even so, the chapter that discusses WAAM reveals that there are still some problems, including those of multi-parameter and data issues. In general, it provides a broad analysis of the application of welding processes in the context of AM and the progress achieved in this area and their potential to revolutionize contemporary production systems.

References

1. Munusamy, S. and Jerald, J., Effect of *in-situ* intrinsic heat treatment in metal additive manufacturing: a comprehensive review. *Met. Mater. Int.*, 29, 3423, 2023.
2. Balaji, N.S., Velmurugan, C., Kumar, M.S., Sivakumar, M., Asokan, P., Experimental investigation on mechanical properties of FDM-based nylon carbon parts using ANN approach. *Surf. Rev. Lett.*, 30, 2350028, 2023.
3. Munusamy, S. and Jerald, J., Effect of build orientation on the microstructure and mechanical properties of wire arc additive manufactured Grade 91 steel/Monel 400 bimetallic components. *Vacuum*, 227, 113429, 2024.
4. Evans, S.I., Wang, J., Qin, J., He, Y., Shepherd, P., Ding, J., A review of WAAM for steel construction – manufacturing, material and geometric properties, design, and future directions. *Structures*, 44, 1506, 2022.
5. Jin, L., Zhai, X., Wang, K., Zhang, K., Wu, D., Nazir, A., Jiang, J., Liao, W.H., Big data, machine learning, and digital twin assisted additive manufacturing: a review. *Mater. Des.*, 244, 113086, 2024.
6. Cunningham, C.R., Flynn, J.M., Shokrani, A., Dhokia, V., Newman, S.T., Invited review article: strategies and processes for high quality wire arc additive manufacturing. *Addit. Manuf.*, 21, 672, 2018.
7. Feucht, T., Lange, J., Erven, M., Costanzi, C.B., Knaack, U., Waldschmitt, B., Additive manufacturing by means of parametric robot programming. *Construct. Rob.*, 4, 31, 2020.
8. Rosli, N.A., Alkahari, M.R., Abdollah, M.F.B., Maidin, S., Ramli, F.R., Herawan, S.G., Review on effect of heat input for wire arc additive manufacturing process. *J. Mater. Res. Technol.*, 11, 2127, 2021.
9. Weman, K., *Welding Processes Handbook*, pp. 41–60, CRC Press, New York, 2012.
10. Jüttner, B., Cathode spots of electric arcs. *J. Phys. D Appl. Phys.*, 34, 103, 2001.
11. Rodrigues, T.A., Duarte, V., Miranda, R.M., Santos, T.G., Oliveira, J.P., Current status and perspectives on wire and arc additive manufacturing (WAAM). *Materials*, 12, 1121, 2019.
12. Munusamy, S. and Jerald, J., Impact of laser shock peening on mechanical properties of wire arc additive manufactured Grade 91 steel and Monel-400 bimetallic components. *Met. Mater. Int.*, 1, 1, 2024.
13. Hu, Q., Miao, J., Wang, X., Li, C., Fang, K., Microstructure and properties of ER50-6 steel fabricated by wire arc additive manufacturing. *Scanning*, 1, 7846116, 2021.
14. Zhong, H., Qi, B., Cong, B., Qi, Z., Sun, H., Microstructure and mechanical properties of wire + arc additively manufactured 2050 AlLi alloy wall deposits. *Chin. J. Mech. Eng.*, 32, 1, 2019.
15. Ungethüm, T., Spaniol, E., Hertel, M., Füssel, U., Analysis of metal transfer and weld geometry in hot-wire GTAW with indirect resistive heating. *Weld. World*, 64, 2109, 2020.

16. Benakis, M., Costanzo, D., Patran, A., Current mode effects on weld bead geometry and heat affected zone in pulsed wire arc additive manufacturing of Ti-6-4 and Inconel 718. *J. Manuf. Process.*, 60, 61, 2020.
17. Li, Y., Wu, S., Li, H., Dong, Y., Submerged arc additive manufacturing (SAAM) of low-carbon steel: effect of *in-situ* intrinsic heat treatment (IHT) on microstructure and mechanical properties. *J. Manuf. Process.*, 46, 102124, 2021.
18. Justus Panicker, C.T. and Senthilkumar, V., Rotational arc based dual wire arc additive manufacturing of low alloy steel with nichrome inclusions for improved mechanical properties. *Mater. Lett.*, 332, 133498, 2022.
19. Kannan, G.R., Umar, M., Bharathi Kannan, T.D., Sathiya, P., Sivakumar, M., Role of arc rotational speed and post-weld heat treatment on the microstructure and mechanical characteristics of 15CDV6 HSLA steel weld joints made by spin arc GMAW. *Mater. Today Commun.*, 40, 109861, 2024.
20. Panicker, C.T.J. and Senthilkumar, V., Effect of arc rotation on the enhancement of mechanical properties of ER70S6 in WAAM. *Vacuum*, 220, 112837, 2024.
21. Selvi, S., Vishvaksenan, A., Rajasekar, E., Cold metal transfer (CMT) technology - an overview. *Def. Technol.*, 14, 28, 2018.
22. Lee, S.H., Optimization of cold metal transfer-based wire arc process regression. *Metals*, 32, 1, 2020.
23. Raut, L.P. and Taiwade, R.V., Wire arc additive manufacturing: a comprehensive review and research directions. *J. Mater. Eng. Perform.*, 30, 4768, 2021.
24. Srinivasan, D., Sevvel, P., Solomon, I.J., Tanushkumaar, P., A review on cold metal transfer (CMT) technology of welding. *Mater. Today Proc.*, 64, 108, 2022.
25. Tomar, B. and Shiva, S., Cold metal transfer-based wire arc additive manufacturing. *J. Braz. Soc. Mech. Sci. Eng.*, 45, 1, 2023.
26. Artaza, T., Suárez, A., Veiga, F., Braceras, I., Tabernero, I., Larrañaga, O., Lamikiz, A., Wire arc additive manufacturing Ti6Al4V aeronautical parts using plasma arc welding: analysis of heat-treatment processes in different atmospheres. *J. Mater. Res. Technol.*, 9, 15454, 2020.
27. Zhou, J., Jia, C., Guo, M., Chen, M., Gao, J., Wu, C., Investigation of the WAAM processes features based on an indirect arc between two non-consumable electrodes. *Vacuum*, 183, 109851, 2021.
28. Köhler, M., Fiebig, S., Hensel, J., Dilger, K., Wire and arc additive manufacturing of aluminum components. *Metals*, 9, 608, 2019.
29. Xia, C., Pan, Z., Polden, J., Li, H., Xu, Y., Chen, S., Zhang, Y., A review on wire arc additive manufacturing: monitoring, control and a framework of automated system. *J. Manuf. Syst.*, 57, 31, 2020.
30. Abe, T., Kaneko, J., Sasahara, H., Thermal sensing and heat input control for thin-walled structure building based on numerical simulation for wire and arc additive manufacturing. *Addit. Manuf.*, 35, 101357, 2020.

31. Li, H., Shi, X., Wu, B., Corradi, D.R., Pan, Z., Li, H., Wire arc additive manufacturing: a review on digital twinning and visualization process. *J. Manuf. Process.*, 116, 293, 2024.
32. Glaessgen, E.H. and Stargel, D.S., The digital twin paradigm for future NASA, and U.S. Air Force vehicles. *Collection of Technical Papers - AIAA/ASME/ASCE/AHS/ASC Structures, Structural Dynamics and Materials Conference*, 2012, doi: 10.2514/6.2012-1818.
33. Bönsch, J., Elstermann, M., Kimmig, A., Ovtcharova, J., A subject-oriented reference model for digital twins. *Comput. Ind. Eng.*, 172, 108556, 2022.
34. Müller, J.R., Panarotto, M., Malmqvist, J., Isaksson, O., Lifecycle design and management of additive manufacturing technologies. *Procedia Manuf.*, 19, 135, 2018.

26

Current Scenario, Future Scope, and Challenges in Welding

N.S. Balaji[1]*, M. Sivakumar[2], Marxim Rahula Bharathi B.[3] and Suresh G.[4]

[1]*Department of Mechanical Engineering, SRM Institute of Science and Technology, Tiruchirappalli Campus, Tiruchirappalli, Tamil Nadu, India*
[2]*Department of Production Engineering, National Institute of Technology Tiruchirappalli, Tamil Nadu, India*
[3]*Department of Mechanical Engineering, Aditya College of Engineering, Surampalem, Andhra Pradesh, India*
[4]*Department of Mechanical Engineering, Mount Zion College of Engineering and Technology, Pudukkottai, Tamil Nadu, India*

Abstract
Welding is one of the oldest technological inventions; ever since the Bronze and Iron Ages, welding has been performed manually, and it has advanced to the level of automatic and intelligent in the current world. Integration of robotics and computer technologies has a positive impact, mostly in welding, by increasing its working rate and accuracy as well as in safety measures. It is due to such technologies as sensors, data analytics, or machine learning that one can speak about the creation of intelligent welding systems that can perform such complicated tasks as welding without human intervention, revolutionizing traditional practices. Further, Welding High Entropy Alloys (HEAs), alloys that possess high entropy, have features of a large number of phases and a variety of mechanical properties. The families of these alloys call for appropriate welding parameters and new applications of the filler materials to enhance joint performance. To further improve welding technology, it is important to overcome the presented challenges. Further evolution is assumed to be caused by Industry 4.0 stressing on more automation and necessary flexibility to satisfy the complex requirements of the new manufacturing context. This comprises self-optimizing, self-organizing, and intelligent welding solutions that throw the field more application areas and challenges. This summary provides the major developments and future trends in

**Corresponding author*: nsbalajicad@gmail.com

Sandip Kunar and Gurudas Mandal (eds.) Advanced Welding Technologies, (547–572) © 2025 Scrivener Publishing LLC

welding technology and indicates how knowledge of smart manufacturing is being used to develop how this will sustain welded component advanced materials such as welding high entropy alloys.

***Keywords*:** Welding scopes, welding challenges, high entropy alloy welding, machine learning in welding, copper welding

26.1 Introduction

Welding, being an important aspect in the construction industry, can be traced back to thousands of years, or more accurately to the Stone Age period of approximately 6000 BC. Some of the earliest techniques were using hammers to force heated metal to flow between two surfaces or objects—quite primitive compared to the current welding technologies [1–3]. At first, throughout the 1880s, welding was largely contained to a blacksmith's forge and had little other uses. Humphry Davy and Edmund Davy were British chemists identified as piggybackers of the arc-welding method in the Industrial Revolution in the 18th and 19th centuries. It had been used by the French scientist De Mertens to adhere parts of railroad tracks together. These innovations were proceeding and influenced by the demands of the two world wars and caused invention of early methods of welding as follows: resistance welding, gas-shielded welding, and friction welding, and after that, ultrasonic and laser welding [4–8].

Today, in view of the modern trends in manufacturing, welding finds its importance in a wide range of fields such as in aerospace manufacturing, shipbuilding industry, construction engineering and railways, automotive vehicle manufacturing, as well as in microsystems. SME is a relatively new concept and can be said to have emerged from conventional welding techniques that employed a lot of human power; it involved highly skilled welders and was subject to their talents and experience. These manual methods also put the workers at risk in case of heat dissipation, toxic gases, dust, and more often yielded low-quality welds [9–12].

Computer technology, laser technology, and other sophisticated information technologies have given traditional welding the boot. However, the advancement especially in the use of robotics has especially given the welding industry a new dimension of automation. At present, almost over 50% of industrial robots are welding robots. These robots that evolved in the 1970 century can be divided into three main generations [13–15]. The Progress of Welding Technology is highlighted as follows in Figure 26.1.

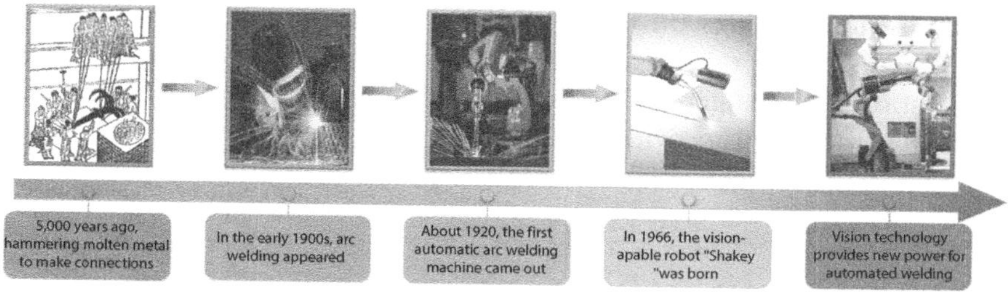

Figure 26.1 Step-by-step growth of welding technology [1].

The first generation of welding robots was of a "teaching-reproduction" type that was capable of doing routine weldings that were done manually earlier. These robots were easy to manipulate, but programming had to be done in-house by qualified personnel and did not possess versatility, and as such could not come up with corrections on their own in case of an error.

The second generation of industrial robots called "off-line programming" robots used computer graphics, welding environment data, and CAM/CAD data in their operation. These robots were somewhat flexible, as improved by sensing systems. Hence, while the present generations' capabilities are even more developed, they are still used in the welding industry in lesser extent in comparison with the first generation. The latest development in welding technology happens to be the third generation of so-called "intelligent welding robots." These are fully equipped with sensing and information processing systems that make it possible to accomplish welding operations on their own. They can collect information about the workpiece, decide on how the welding is to be done, perform the welding task, and interface other equipment for the so-called "flexible" production. While the intelligent welding robots are as yet entirely experimental, they represent the pinnacle of automated welding production and have aroused considerable interest among researchers internationally.

These examples show how far science has evolved and progressed from manual handling methods to more advanced automated systems and the great emphasis that is being placed on improving efficiency, safety, and accuracy of the welding techniques used in a number of industries [16–21].

26.2 Current Trends in Intelligent Welding Systems

26.2.1 Overview of Intelligent Welding

Defining intelligent welding accurately is a problem because the literature provides various descriptions of the term. This chapter focuses on welding methods and explores on the application of intelligence in manufacturing systems. Based on the technologies of the past decades that allow the integration of intelligence into welding systems, a general definition for intelligent welding is provided [20].

26.2.1.1 Progress of Welding Systems

The welding systems' evolution started with the first arc welding machines that were invented in the 19th century. The first of these machines demanded a human operator to make a weld, a person who would then also sense, analyze, decide, operate, control, cognize, and learn. This was an entirely manual process that required much human effort: efficiency, uniformity of production, and quality of the outcome depended heavily on human intellect and dexterity, while suffering from the human factor's inability to handle the complexities of the composite task.

Over the years, there has been a shift from completely manual welding to less of it because of many factors, some of which include the following: The evolution of welding can be divided into four phases:

Phase I: Manual welding is most reckless and lack efficiency and consistency as opposed to the automated technology. Indeed, in this phase of welding, the elegance of the weld depends on the welder, which causes variation in the quality of the weld and rate of production.
Phase II: Automation of production includes the use of robotics, although the automation of the process was still complex and difficult to simulate. Computer controls came to replace human operators but did not get rid of inconsistencies typical for welded joints made by hand.
Phase III: Application of "teach and playback" robots to manage automation but without feedback regarding issues and variation. This phase marked advances involving robotic welding of work pieces with the kind that could repeat programmed welding paths. But these systems were still defunct of real-time adaptation change in the welding environment.
Phase IV: Automation through intelligence in the sophisticated welding systems for managing and regulating the performance as well as the quality of welding. This transition signal change from Human-Physical System (HPS) to Human-Cyber Physical System (HCPS). Autonomous welding systems use sensors, analytical tools, and machine learning patterns to modify and optimize welding processes in real time, thus increasing the quality of the welds.

An intelligent welding system can be regarded as the enhancement of welding technology, which is worthwhile to develop and apply in the industries. These systems hence possess versatility and sophistication through integration of artificial intelligence and minimize human enterprise to produce high-quality welded products. It was found that this constant progression of welding technology is still advanced in enhancing mass production with high acuteness [22–25].

26.2.1.2 Fundamentals of Intelligent Welding

Smart welding combines innovation in methodologies and artificial intelligence in mimicking, as well as supplementing, and sometimes even substituting, mechanical intelligence for human intelligence in the welding setting. By converting the best of human and physical systems into intelligent cyber systems in the welding processes, they allow computational evaluation, precise command, and better sensing.

Its integration definitely facilitates the overall process of knowledge management, transfer, and utilization in the welding system. In so doing, it leads to increase in the quality of work done, increase in production rate, and increase in the welding process. Thus, human experience and know-how are unobtrusively and systematically introduced into the cyber-physical system by means of advanced software and detailed knowledge resources. This transfer eliminates unnecessary fluctuations in the welding process while improving the quality of those welds produced.

The process is illustrated in Figure 26.2 with the combined human and computer intelligence concept illustrating how decisions by cyber physical systems are able to regulate complex welding processes is displayed. Intelligent welding mentioned above is another initiation of integration of

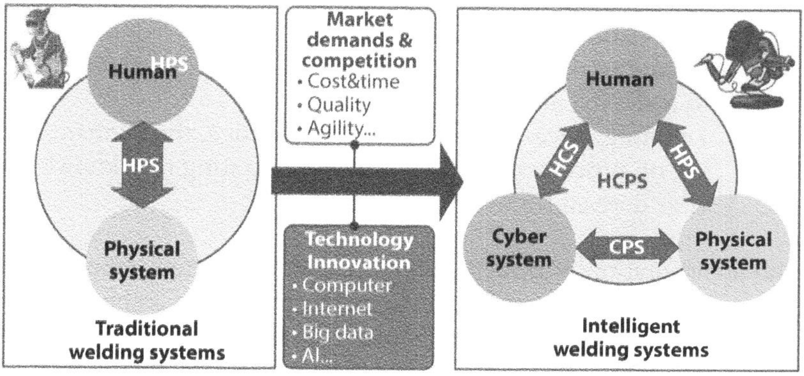

Figure 26.2 Evolution of traditional welding system from to intelligent welding system [21].

human intelligence and machine intelligence in modern manufacturing process that has led to massive development in productivity and quality [25–27].

26.2.1.3 Fundamental Intelligence Traits

Present day computing capability has provided for use of a number of tools of artificial intelligence like search techniques, optimization methods, geometric reasoning methods, and artificial intelligence modeling cognitive functions like learning. While generalized machine intelligence, capable of performing the full range of human cognitive abilities, is still a distant goal, certain human intelligence attributes can be effectively duplicated or transferred to machines within intelligent welding systems:

(i). Context-dependent dynamic perception: Selected manufacturing resources, in a state of constant flux, shifting due to changing states of the resources.
(ii). Identification: Global identification of different states, conditions, and elements incorporated into the components or sub-assemblies.
(iii). Categorization: Categorizing objects or circumstances in view of certain characteristics, for classroom, prediction, inference, decision, and interaction.
(iv). Goal setting and prioritization: Coordinating and direction giving and ensuring that the activities are goal directed.
(v). Decision making: Choosing the activities for achieving targets in accordance with the impact of the gains and losses.
(vi). Prediction and planning: Predicting the future conditions and deciding what action to take.
(vii). Negotiation: Balancing different and possibly conflicting intentions of many organizations within one global system to reach the most optimal state.
(viii). Learning: Several, gaining new knowledge, thus enhancing on processes through time by integrating new knowledge derived from different sources.

The incorporation of these intelligence traits into the welding systems give a new breakthrough in manufacturing promising improved, standardized, and quality welding production. Intelligent welding systems are changing with advancement in design and technology research carried out to finally realize an intelligent and flexible welding environment [21, 28].

26.3 Advances in Welding of High Entropy Alloys (HEAs)

Arc welding of high entropy alloys (HEAs) is a challenge since comparing HEAs to the conventional alloys, they have very different composition, structure, and properties. These different composition of metals create weldability problems, which have not been mainly addressed at the research level. The welding of HEAs is a relatively new subject; the first publication on this topic of HEAs was done in 2016 by Zhang *et al.*, where they used FeCoNiCrCu as the filler metal for dissimilar brazing joints. Sometime later, the weldability of CrMnFeCoNi using electron beam welding was carried out by Wu *et al.* Only a few articles were published on welding of HEAs with less than 10,000 citations up to 2016, but the number and citations of articles related to the welding of HEAs shot up tremendously after 2016 [29–31]. This has risen especially in the year 2020 showing that welding HEAs is gradually emerging as a prominent area of research as the concepts of HEA-related ideas evolve and as the number of publications and citations from Welding of HEA's Materials increased and as shown by Figure 26.3.

It is very viable to understand different metal joining processes to apply high entropy alloy (HEA)-based material in real-time scenarios. Up to the present time, many welding methods have been considered as earlier defined, which can be grouped basically into three classifications. The first

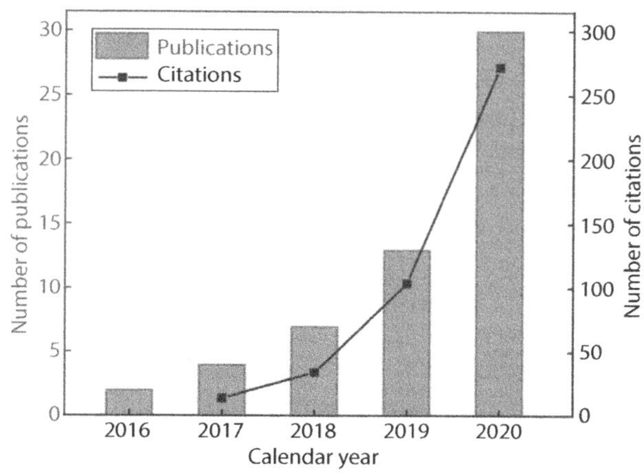

Figure 26.3 Number of publication and citations from Welding of HEA's Materials [30].

group includes the well-known fusion-based welding techniques including the GTAW or TIG welding, the EBW, as well as the LBW. The second group is of the welding techniques that are done on solid state and are few as follows: friction stir welding (FSW), explosive welding, and resistance seam welding (RESW). The third type of welding comprises the diffusion reaction welding, which includes brazing, soldering, and diffusion bonding [32–34].

Distribution of research papers over the different welding processes differs, and more than half of the publications are related to brazing, laser welding, and FSW. This is succeeded by articles on GTAW, diffusion bonding, RESW, EBW, rotary friction welding (RFW), and EXPW. The mere fact that research in these areas is so frequent demonstrates the feasibility and effectiveness of these methods in the joining of HEA materials.

Gas tungsten arc welding or GTAW is a versatile fusion-based welding technique that is capable of welding the HEA-related materials irrespective of their constitution, initial conditions, and mechanical strength. However, the strength coefficient of GTAW joints is rather lower than those of the other welding processes, although it is a multipurpose welding process. This limitation shows the importance of future works and improvement to further improve the mechanical properties of GTAW joints.

More research on welding HEAs is thus important as it helps to realize the potential that these superior materials may offer in several performance-intensive applications. Future development will continue to refine welding methods and improve knowledge of how the compositions of HEAs affect weldability and therefore improve realism in applications. Such sustained endeavor will open the way for HEAs to be employed in ways that appropriately fit the utilization of the material that is necessitated in highly demanding industries [35–39].

Figure 26.4 presents the strength coefficients of the joints containing HEAs by different welding techniques based on the tensile strength of base metal as reported in the Web of Science publications till October 2020. Most importantly, LBW exhibits good performance universally in different strength HEAs; however, the strength coefficient of LBW joints decreased slightly with higher BM strengths in HEAs. Thus, the results of tensile tests carried out on samples containing joints made through EBW are higher and have more favorable values of the strength coefficient compared to GTAW and LBW welding processes. For that reason, the control of the interfacial diffusion and reaction is significant in diffusion bonding, such as resulting in the extensive variation in the strength coefficients of the bond or joint, which is also applicable for the soldering and brazing processes.

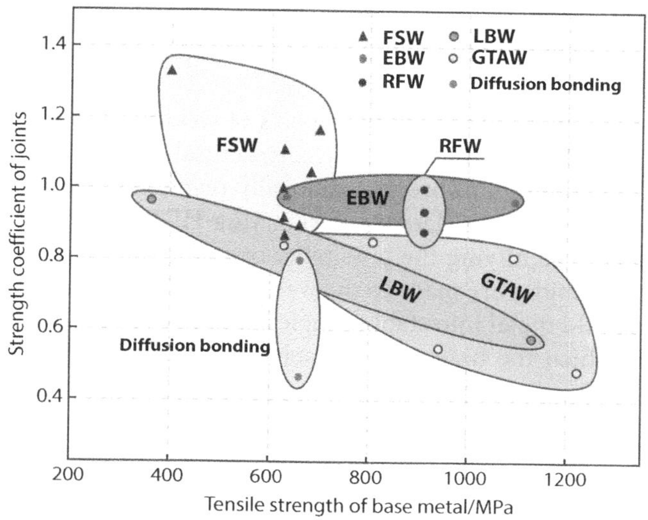

Figure 26.4 HEA-related joints under different welding techniques as a function of the tensile strength of BM and strength coefficient [30].

The strength coefficients derived for the solid-state welding techniques are given below; rotary friction welded (RFW) and friction stir welded (FSW) joints offer coefficients of strength between 0.8 and 1.4, which is considerably higher than by other methods. However, FSW application is limited to area, where strength is comparatively low, flow characteristics are appropriate, and few difficulties, like severe tool wear, in FSW exist.

In addition, there is a distinction as to whether the joints are configured using discontinuous or continuous welds, and additional research on butt and similar joints is double that of lap and dissimilar joints. In the dissimilar HEA joints, it is common to join the HEAs with super-alloys, aluminum alloy steels, magnesium alloys, and even different types of HEAs [40, 41].

However, the chemical composition of both the base metals of the HEAs and the welding procedures and the types of joints used are different, and aspects such as the proportion of the elements of the welding parts of the joints significantly influence the joint performance. Typically, the compositions of HEAs are established on a transition element set of cobalt, chrome, iron, and nickels or refractory material of niobium, tungsten, tantalum, and molybdenum with further constituents in bulk HEAs being aluminum, manganese, titanium, copper, silicon, vanadium, and zirconium. In the welding parts, the most frequently adopted HEAs are single-phase

FCC equi-atomic CrMnFeCoNi type referred to as “Cantor alloys” (48%), AlCoCrFeNi alloys (24%), CoCrFeNi alloys (20%), and the rest 8% of other HEAs.

Overall, in the chapter, various impacts of welding parameters and the initial states of HEA parts on the microstructural change and mechanical behavior of welded joints are systematically reviewed in the state-of-the-art HEA-related welding. Despite designating HEAs more as filler materials, it ends by signifying the possibility of enhancing joint performance whenever alterations are made to the welding process. These are the ability to alter the weld metal into stable simple solid solution phases, decreasing the rate at which the base metal dissolves, avoiding the bridging of the intermetallic brittle phases, and increasing solid solution strengthening in welds. Further, HEA's joints do possess the best strength and ductility at room and subzero temperature as well as fine cryogenic toughness, and they offer sound anti corrosive and anti fatigue properties [42, 43].

26.4 Advanced Techniques in Welding Defect Identification and Categorization

26.4.1 Weld Defect Detection and Characterization

The imperfections that are always associated with welding impurities, since the process is complicated and prone to interferences, in addition to lessening the aesthetic appeal of the welded joint, are also dangerous to the overall stability of the structure and may pose some risks. It is therefore important to establish methods for evaluating welding quality since this influences the quality of the final product, more specifically, methods for detecting and categorizing welding imperfections are significant. In addition to conventional image processing and pattern recognition-based approaches for detecting defects in weld images, the state of the art ML methods facilitates better accuracy and flexibility in handling similar problems [44, 45].

26.4.2 Machine Learning-Based Approaches

In the field of image classification, ML approaches can effectively work with large amounts of data and organize contextual information, so the rock-solid reliability of classifiers can be improved. What these models do is that these models learn numerous samples, and through the iterative learning, they gain interpretable and traceable ability to detect images, which these models have not seen before, making these models incredibly flexible and precise. Yu *et al.* came up with a methodology for diagnosing variable groove penetration where temperature characteristics during welding were used to create a database; the data were then used to train an SVR model to determine penetration states. Kumar *et al.* rectified the problem of feature overlap of primary and secondary surface by gray out image, which increased the accuracy for detecting defects to 95% using feature match by Backpropagation Neural Network (BPNN).

In a study by Zhang *et al.* they proposed the use of CNN to recognize welding defects whereby the arc light disturbances were incorporated in the image; the model was able to attain over 90% recognition capability. In research by Liu *et al.*, a bilinear feature extraction strategy was proposed for identifying small-scale laser welding defects: the proposed technique was verified through three image datasets with diverse granularities, with the test achieving a classification efficacy of 97%, 45% for nine different categories of defects [46–49].

26.4.3 Employing Image-Processing Techniques

Elaborate image-processing mechanisms of the conventional style are less expansive in terms of expected workflow and are useful in tasks less complex than those of ML and smaller dataset measurements. Two of the authors of this paper, Chu and Wang, used laser triangulation to acquire weld information, K & C Information import 3D data into MATLAB for modeling purposes, and the edge and displacement defects were easily recognized based on width threshold.

In addition, Yang *et al.* used curvature information from 3D images to evaluate the weld seam quality, the shadow shape for reconstructive analysis, and a single image for the shadow shape to correlate seam shape and image brightness. Vasudevan *et al.* employed an infrared camera during welding to oversee the thermal evolution and came up with a way of associating weld defects to temperatures.

Such techniques, irrespective of whether they are based on ML or traditional IPP, create a marked improvement in the ability to identify, categorize, and interpret the potential consequences of weld imperfections. This assists in the developmental of product quality and safety in the various industrial applications [50–53].

26.5 Innovations in Vision Technology for Welding Automation

26.5.1 Enhancements in Hardware Capabilities

Advancements in hardware technology, crucial for supporting sophisticated vision systems in welding, can be categorized into two main types. The two main tech drivers are (1) the improvements in the computer hardware and (2) the improvements in the technology of visual sensing. ENIAC, which was the first program, could only perform up to 5,000 operations per second, while modern processors manage more than billion operations in each second and therefore shows advancement in computing ability. This increase is important in deployment of advanced image processing algorithms that support advanced machine vision systems. Modern cameras, integrated as data collectors in these systems, have become more than tools with multi-tasking capabilities to perform what hitherto was the exclusive preserve of computers [54–56]. Similarly, Yamashita *et al.* developed a multi-sensor camera, which is capable of not only recording temperature data of actual welding sites at a particular instant but also display in the form of two-dimensional temperature profiles and thereby improving the inspection process [57]. To the best of the authors' knowledge, the observation of ultrasonic weld formation and defect formation using high-speed imaging technologies was made by Lee *et al.* and recently used by Huang *et al.* They used a high-speed CCD camera in conjunction with electrical signal acquisition. Recent technological advancements in the availability of million frame level high-speed cameras provides better understanding and documentation of the dynamic welding actions in addition to providing better insights into the weld mechanisms and phenomena [58, 59].

26.5.2 Advancements in Composite Sensing Systems

The composite sensing strategies are proven to have helped in enhancing the reliability of vision systems in welding robotics. The studies conducted by Lee *et al.* and Huang *et al.* pointed out that the use of the data of the preceding frame as a reference improves the accuracy of the current image analysis [58]. Some of the data inputs include arc spectra, sound, and voltage all of which feed the composite system to give a complete display of the welding environment and enhance the accuracy of the defect readings. Zhang *et al.* as well as Gao *et al.* have developed initial framework that combines high frame-rate vision sensors with other sensory equipment that makes the results of the analyses more deep and elementary [60, 61]. As an example of this approach, Deng *et al.* deepened this idea by creating a system that records plasma intensity, light intensity, and temperature dynamics throughout the experiment, which helps to get a full data set for defect definition [62].

26.5.3 Integration of Image Processing and Machine Learning

The most central concept of machine learning (ML) is data orientation. Conventional classifiers, including SVM and KNN, as well as learning based on artificial neural networks, including ANN and CNN, have been applied notably in different diverse industries including welding. Such models are adept at learning useful features based on massive noisy inputs and are most successful in the image-based defect identification and segregation. Therefore, Ireri *et al.* obtained near-perfect accuracy rates for color and texture classes using RBF-SVM classifiers. These application highlights the revolutionary role of ML in the enhancement of conventional techniques of analyzing image and promising a better future of complex, efficient, and accurate systems [63–65].

26.5.4 Multi-Intelligent Group Control Technology

Multi-robot systems, especially for welding, are a cutting-edge application in the field of robotic automation, which are patterned after biological swarms, along the lines of ant colonies and beehives. These systems, which by definition self-organize through interactions within networks, can span larger distances in space and also deal with integrated and interdependent tasks. In this area of study, the aspects of task division, route determination, balance, and coordination among various robots have been researched on to enhance motion parameters and the layout of the whole system to achieve superior performance and accuracy. Apart from improving weld seam tracking and identification of defects, the application of vision technologies helps to solve some challenging tasks, for instance, trajectory planning and workpiece identification. This coupling improves the capacity of dealing with complex or complex geographical parts and guarantees thin or uneven welded products indicating a trend toward a fully automatic welding system [66, 67]. This section provides an outline of the expected future development of technological improvements in robotic welding stressing on the use of improved hardware and robust algorithms to transform welding practices illustrated in Figure 26.5.

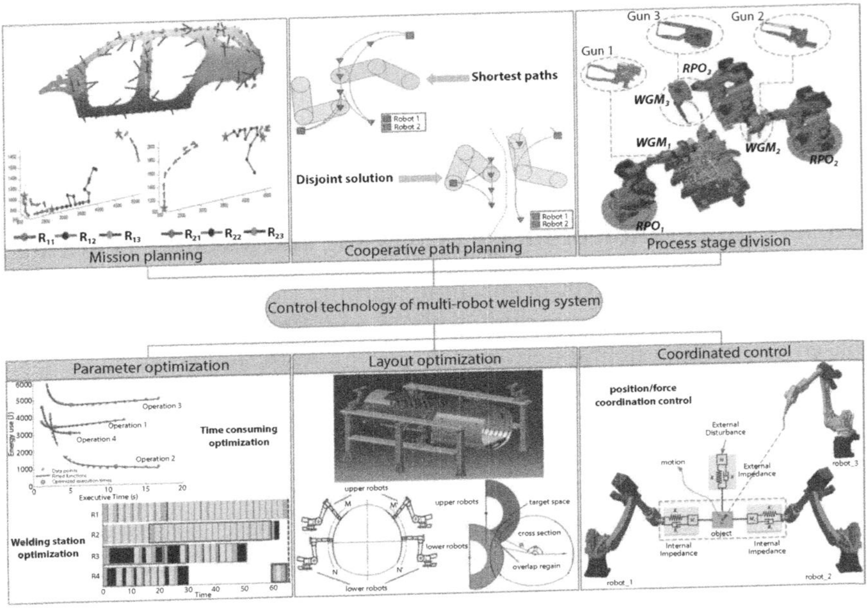

Figure 26.5 Multi-robot welding system [1].

26.6 Integration of Industry 4.0 and Cyber Physical Systems in Welding Technology

26.6.1 Industry 4.0 and Cyber Physical Systems

Industry 4.0, which began as a strategic initiative in Germany, signifies the complete integration of traditional manufacturing processes with modern information technology systems. It covers horizontal integration by value networks, vertical integration, manufacturing with networks, and integration of digital engineering systems all along the value delivery network. It hints that we can seldom segregate it from smart manufacturing, cyber physical system or CPS, enterprise architecture or EA, or information and communications Technology or ICT.

In the aspect of welding technology, Industry 4.0 is observable in the manner in which intelligent parts and interfaces interface to eliminate conventional procedures. Reisgen *et al.* discussed several applications of the Industry 4.0 framework in the case of shielded GMAW process. They talked about a system that is networked and self-sufficient and at the same time backed by CPS making them integrated products of assurance. This approach can be best explained by Industry 4.0, which can reduce the time and amount of error in welding by SMART technologies and improve the welding environment by making a connected manufacturing system [68–73].

26.7 Complexities in Copper Welding: Emphasis on Laser Beam Welding

26.7.1 Copper Welding Difficulties

The welding of copper is somewhat problematic because of the inherent properties of this material. It is also very challenging to realize 100% joint efficiency in copper welds actually when using high power density processes, like LBW for instance, because concerns, like weld bead appearance and stability of weld pool, come into the frame. Not even such innovative techniques, such as friction stir welding (FSW), help to avoid these complexities. The welding methods that are practically used in the industrial sector, like GMAW and GTAW, can rarely produce the weld efficiency as of FSW. The main problem for welding copper is its high thermal diffusivity of 401 W/mK. It causes such problems as distortion because the coefficient of linear thermal expansion is 16. Also, at high temperatures, the surface of the copper forms a cuprous oxide film, which becomes a menace to the quality of welds when fusion welding techniques are employed. This oxide formation can cause porosity or even cracks in the weld metal, and a higher percent of oxygen also decreases the toughness of the weld metal [74].

26.7.2 Challenges in Laser Beam Welding (LBW) of Copper

LBW falls under the category of high power density welding processes owing to the employment of a concentrated beam. The major benefit of LBW is the low heat input at the weld zone, which compares well with other arc welding processes favored for copper welding. Nonetheless, a major problem arising with LBW of copper is that the material is highly reflective to lasers, and so, most of the laser energy is reflected instead of being absorbed. This problem is compounded by copper's reflectivity dependence on the temperature and the wavelength of the laser source; for example, at wavelengths greater than 0.8 μm, iron (Fe) absorbs approximately 45% of laser energy, while for copper (Cu), only approximately 5% laser energy is absorbed by it. This absorptivity can triple once the melting point of the material is reached, and during the welding process, its control becomes difficult, and so, maintaining the stability of the weld pool is affected. Such instability can lead to defects, such as porosity, hence readily affecting the appearance of the weld bead. In LBW, the geometry of the weld pool is deep, and the keyhole is narrow, which makes the process difficult. The turbulence of the keyhole makes the weld pool unstable and may cause ejection of liquid metal because of the circulating molten vapor at the keyhole's region, which gives surface defects and porosities.

However, because LBW has such a small heat input into the material, the HAZ and particularly the FZ usually have a microstructure of which coarse columnar grains having reduced tensile strength, which are a characteristic. Fractures, according to tensile examination, initiate at the weld and the HAZ regions as well as even with the addition of alloying elements using filler wires. Attaining cent per cent joint efficiency continues to remain unachievable in LBW of copper. This detailed analysis reveals that the welding of copper is inherently challenging, much more so under high power density conditions, for example, in LBW, where the pipe's inherent characteristics unambiguously determine the effectiveness of the process [75, 76].

26.7.3 Challenges in Gas Metal Arc Welding (GMAW)

GMAW is one of the most popular and widely used processes of the conventional fusion welding methods characterized by its large and dispersed burning zone and higher heat input in comparison with, for example, LBW, due to less power density involved. Compared to other welding processes, such as the GTAW, the GMAW process has high heat input because of the low welding speed, though it has relatively high deposition rates. The following problems accompany this process: spatter, poor fusion from parameter selection with view to reducing heat input, and considerable distortion due to high heat input when deposition rate is to be enhanced. The high level of remelting inherent in GMAW, because of the copper's fluidity when melted, causes this mode to be most complex. Commonly, it is possible to use GMAW to weld only in pure copper considerably in limited ways.

For instance, TIG joining of 4-mm-thick copper plates may need two passes of MIG (a kind of GMAW) but one for the combination of laser-MIG. In the traditional MIG weld ,the material undergoes a double pass, and it heats up and cools-off repeatedly—which is not a healthy scenario for the material. In general, in MIG, the welding current is slightly higher and speed is much lower than in most of the other hybrid welding processes, and thereby, the heat input is considerably different. The need to do several passes, even when using a high power density source, can worsen the material properties, thus resulting in a slow cooling rate and the formation of a coarse grain structure. This structure is generally found to show lower strength, MIG weld joints of tensile strength nearly reaching up to 69% of the base metal tensile strength [77, 78].

26.7.4 Challenges in Gas Tungsten Arc Welding (GTAW)

GTAW, also called shield metal arc welding, where an inert gas is used for shielding, is greatly valued for the ability to produce a number of high purity welds and is versatile regarding the type of joints that can be welded and therefore can be applied in most engineering disciplines. However, GTAW is a conventional combination of two fundamentally different methods of arc welding—gas and electric arc—with moderate heat input but relatively low power density that has some peculiarities concerning copper welding. Because of high thermal diffusivity in copper, preheating and the control of interpass temperatures are critical because the mechanical properties of the weldment depend on the rate of cooling where low cooling rates reduce the mechanical properties of the weldment.

The effect of welding parameters on the bead geometry and depth-to-width (D/W) ratio is critical; achieving a high D/W ratio with GTAW is particularly challenging due to copper's high thermal conductivity. Typically, the weld bead is wider and shallower due to the spread of isotherms across the weld area. Achieving 100% joint efficiency is highly improbable, as the high heat input and necessary preheating lead to slower cooling rates resulting in lower tensile strength, reduced hardness, and a coarse grain structure due to softening annealing effects.

Furthermore, the use of helium as a shielding gas, although it helps manage copper's high thermal conductivity, does not significantly enhance joint efficiency, which typically remains approximately 64% post-welding. Additionally, copper welding with GTAW can lead to micro-cracking, particularly noted within a critical crystalline temperature range of 1,070°C to 1,058°C during cooling. This issue is exacerbated when helium is used as the shielding gas maintaining the base metal within this critical temperature range and leading to potential cracking.

Figure 26.4 illustrates the strength coefficients of joints involving high entropy alloys (HEAs) categorized by various welding methods, benchmarked against the tensile strength of the base metal (BM) as recorded until October 2020, using data from Web of Science publications. Notably, laser beam welding (LBW) demonstrates broad effectiveness across HEAs of varying strengths, although the strength coefficient for LBW joints declines significantly with HEAs exhibiting higher BM strengths.

In contrast, joints formed through electron beam welding (EBW) show a higher and more consistent strength coefficient compared to those created using gas tungsten arc welding (GTAW) and LBW. The control over interfacial diffusion and reactions plays a crucial role in diffusion bonding affecting the broad variability in the strength coefficients of these joints, which similarly applies to soldering and brazing [75, 80–82].

26.7.5 Challenges in Friction Stir Welding of Copper

Friction stir welding (FSW), developed by The Welding Institute in 1991, falls under the category of solid-state welding processes and is celebrated for largely circumventing the metallurgical issues common in traditional welding. Key advantages of FSW include minimal distortion, preservation of alloying elements, and the creation of a refined microstructure. Most notably, FSW eliminates the need for melting and subsequent solidification of the metal, which are typical stages prone to defects in conventional welding methods.

Despite these benefits, FSW of copper presents significant challenges, primarily due to its impact on the microstructure of the weld joint. Factors, such as tool rotation speed, welding speed, and the geometry of the tool pin significantly influence the outcome. The ratio of tool shoulder diameter to pin diameter (D/d ratio) and the intricacies of the pin and shoulder geometries are crucial in determining the mechanical properties of the weld. For instance, using a simple cylindrical pin often results in welds with substantially lower tensile strength.

The most critical parameters affecting the mechanical properties in FSW are the tool rotational speed and the welding speed, often expressed as the Rs/Ws ratio (tool rotational speed to welding speed), which also indicates the level of heat input during the welding process. FSW operates by two principal phenomena: heating the material to the recrystallization temperature and mechanically deforming the grains of the plates to be joined. Achieving an optimal balance between these effects is essential; insufficient deformation can lead to an inadequate microstructure, whereas excessive heat can cause softening annealing making it challenging to produce a defect-free joint.

Furthermore, the total heat input in FSW is directly proportional to the tool rotation speed. Excessively high rotation speeds can lead to a slow cooling rate resulting in the formation of a coarse grain structure. This coarse grain structure, in turn, leads to a reduction in tensile strength. Numerous researchers have identified softening annealing as a significant obstacle in achieving desirable mechanical properties in FSW of copper.

This section highlights the complex interplay of process parameters in FSW that must be meticulously managed to overcome the inherent challenges posed by copper's physical properties [83–85].

26.8 Summary

The book chapter "Current Scenario, Future Scope, and Challenges in Welding" delves into the evolution of welding technology from its primitive beginnings during the Bronze and Iron Ages to the advanced, automated, and intelligent systems utilized in modern industries like aerospace and automotive. It highlights the transformative impact of robotics and computer integration on the efficiency, safety, and precision of welding operations. Focusing on intelligent welding systems, the chapter discusses how these incorporate advanced sensors, data analytics, and machine learning to enable robots to perform complex tasks autonomously and adaptively. Additionally, it addresses the challenges in welding high entropy alloys (HEAs), which have complex microstructures and properties, proposing research into optimized welding parameters and the use of HEAs as filler materials to improve joint performance. The chapter concludes by projecting future advancements in welding technology, emphasizing the shift toward more autonomous systems driven by Industry 4.0 innovations, which are set to further revolutionize welding in line with the evolving demands of modern manufacturing environments.

References

1. Guo, Q., Yang, Z., Xu, J., Jiang, Y., Wang, W., Liu, Z., Sun, Y., Progress, challenges and trends on vision sensing technologies in automatic/intelligent robotic welding: State-of-the-art review. *Rob. Comput. Integr. Manuf.*, *89*, 102767, 2024.
2. Xu, Y. and Wang, Z., Visual sensing technologies in robotic welding: recent research developments and future interests. *Sens. Actuators, A*, 320, 112551, 2021.
3. Dong, H., *Robot Weld Trajectory Planning and Dynamic Weld Pool Control Method Based on Welder's Intelligence*, Dalian University of Technology, Dalian, [Chinese], 2019.
4. Xiong, X.H., Wang, D.S., Wei, J., Zhao, P., Ren, R., Dong, J.P., Cui, X., Resistance welding technology of fiber reinforced polymer composites: a review. *J. Adhes. Sci. Technol.*, 35, 1593–1619, 2021.

5. Zhang, X.Y., Zha, X.Q., Gao, L.Q., Hei, P.H., Ren, Y.F., Influence of shielding gas on microstructure and properties of GMAW DSS2205 welded joints. *Materials, 14,* 2671, 2021.
6. Sathishkumar, G.B., Sethuraman, P., Chanakyan, C., Sundaraselvan, S., Arockiam, A.J., Alagarsamy, S.V., Elmariung, A., Meignanamoorthy, M., Ravichandran, M., Jayasathyakawin, S., Friction welding of similar and dissimilar materials: a review. *Mater. Today, 81,* 208–211, 2023.
7. Ajay, V., Babu, N.K., Ashfaq, M., Kumar, T.M., Krishna, K.V., A review on rotary and linear friction welding of inconel alloys. *Trans. Indian Inst. Met., 74,* 2583–2598, 2021.
8. Liu, J., Cao, B., Yang, J., Texture and intermetallic compounds of the Cu/Al dissimilar joints by high power ultrasonic welding. *J. Manuf. Process.,* 76, 34–45, 2022.
9. Leon, M.D. and Shin, H.S., Review of the advancements in aluminum and copper ultrasonic welding in electric vehicles and superconductor applications. *J. Mater. Process. Tech.,* 307, 117691, 2022.
10. Oliveira, J.P., Shen, J., Zeng, Z., Park, J.M., Choi, Y.T., Schell, N., Maawad, E., Zhou, N., Kim, H.S., Dissimilar laser welding of a CoCrFeMnNi high entropy alloy to 316 stainless steel. *Scr. Mater.,* 206, 114219, 2022.
11. Chen, C., Zhou, H., Wang, C., Liu, L., Zhang, Y., Zhang, K., Laser welding of ultra-high strength steel with different oscillating modes. *J. Manuf. Process.,* 68, 761–769, 2021.
12. Hong, T.S., Ghobakhloo, M., Khaksar, W., Robotic welding technology. *Compr. Mater. Process.,* 6, 77–99, 2014.
13. Wahidi, S.I., Oterkus, S., Oterkus, E., Robotic welding techniques in marine structures and production processes: A systematic literature review. *Mar. Struct., 95,* 103608, 2024.
14. Medatovna, K.Z. and Igorevich, D.D., Welding equipment modernization. *Int. J. Hum.-Comput. Stud.,* 3, 10–13, 2021.
15. Zhang, L.X., Shi, J.M., Li, H.W., Tian, X.Y., Feng, J.C., Interfacial microstructure and mechanical properties of ZrB2SiC C ceramic and GH99 superalloy joints brazed with a Ti-modified FeCoNiCrCu high-entropy alloy. *Mater. Des.,* 97, 230–238, 2016.
16. Xu, F.J., Xu, Y.L., Zhang, H.J., Chen, S.B., Application of sensing technology in intelligent robotic arc welding: a review. *J. Manuf. Process.,* 79, 854–880, 2022.
17. Wang, X.W., Wei, J.B., Zhou, X., Xia, Z.L., Gu, X.S., Dual-objective collision-free path optimization of arc welding robot. *IEEE Rob. Autom. Lett.,* 6, 6353–6360, 2021.
18. Bedaka, A.K. and Lin, C.Y., CAD-based offline programming platform for welding applications using 6-DOF and 2-DOF robots, in: *International Conference on Advanced Robotics and Intelligent Systems (ARIS),* pp. 1–4, 2020.

19. Lund, H., Penttilä, S., Skriko, T., A knowledge-based multipass welding distortion estimation method for a multi-robot welding off-line programming and simulation software. *Procedia Manuf.*, 51, 302–308, 2020.
20. Chen, S.B. and Lv, N., Research evolution on intelligentized technologies for arc welding process. *J. Manuf. Process.*, 16, 109–122, 2014.
21. Wang, B., Hu, S.J., Sun, L., Freiheit, T., Intelligent welding system technologies: state-of-the-art review and perspectives. *J. Manuf. Syst.*, 56, 373–391, 2020.
22. Pan, Y., Heading toward artificial intelligence 2.0. *Engineering, 2*, 409–413, 2016.
23. Kusiak, A., Smart manufacturing must embrace big data. *Nat. News, 544*, 23, 2017.
24. Kusiak, A., Smart manufacturing. *Int. J. Prod. Res.*, 56, 508–517, 2018.
25. Shao, X., Li, X., Gao, L., Zhang, C., Integration of process planning and scheduling—a modified genetic algorithm-based approach. *Comput. Oper. Res.*, 36, 2082–2096, 2009.
26. Shin, M., Ryu, K., Jung, M., Reinforcement learning approach to goal-regulation in a self-evolutionary manufacturing system. *Expert Syst. Appl.*, 39, 8736–8743, 2012.
27. Ji, Z., Yanhong, Z., Baicun, W., Jiyuan, Z., Human–cyber–physical systems (HCPSs) in the context of new-generation intelligent manufacturing. *Engineering, 4*, 624–636, 2019.
28. Zhou, J., Li, P., Zhou, Y., Wang, B., Zang, J., Meng, T., *Identifying product and process state drivers in manufacturing systems using supervised machine learning*, Springer International Publishing, Switzerland, 2015.
29. Wu, Z., David, S.A., Feng, Z., Bei, H., Weldability of a high entropy CrMnFeCoNi alloy. *Scr. Mater.*, 124, 81–85, 2016.
30. Li, J., Meng, X., Wan, L., Huang, Y., Welding of high entropy alloys: progresses, challenges and perspectives. *J. Manuf. Process.*, 68, 293–331, 2021.
31. Sharma, A.S., Yadav, S., Biswas, K., Basu, B., High-entropy alloys and metallic nanocomposites: processing challenges, microstructure development and property enhancement. *Mater. Sci. Eng. R Rep.*, 131, 1–42, 2018.
32. Sokkalingam, R., Mishra, S., Cheethirala, S.R., Muthupandi, V., Sivaprasad, K., Enhanced relative slip distance in gas-tungsten-arc-welded Al0.5CoCrFeNi high-entropy alloy. *Metall. Mater. Trans. A, 48*, 3630–3634, 2017.
33. Sokkalingam, R., Mastanaiah, P., Muthupandi, V., Sivaprasad, K., Prashanth, K.G., Electron-beam welding of high-entropy alloy and stainless steel: microstructure and mechanical properties. *Mater. Manuf. Process.*, 35, 16, 1–10, 2020.
34. Karthik, G.M., Panikar, S., Ram, G.D.J., Kottada, R.S., Additive manufacturing of an aluminum matrix composite reinforced with nanocrystalline high-entropy alloy particles. *Mater. Sci. Eng., A, 679*, 193–203, 2017.
35. Mishra, R.S. and Ma, Z.Y., Friction stir welding and processing. *Mater. Sci. Eng. R Rep.*, 50, 1–2, 1–78, 2005.

36. Praveen, P., Yarlagadda, P.K.D.V., Kang, M.J., Advancements in pulse gas metal arc welding. *J. Mater. Process. Tech.*, 164, 1113–1119, 2005.
37. Munusamy, S. and Jerald, J., Effect of *in-situ* intrinsic heat treatment in metal additive manufacturing: a comprehensive review. *Met. Mater. Int., 29*, 12, 3423–3441, 2023.
38. Kannan, G.R., Umar, M., Kannan, T.D.B., Sathiya, P., Sivakumar, M., Role of Arc Rotational Speed and Post-Weld Heat Treatment on the Microstructure and Mechanical characteristics of 15CDV6 HSLA steel weld joints made by Spin Arc GMAW. *Mater. Today Commun.*, 40, 109861, 2024.
39. Munusamy, S. and Jerald, J., Impact of Laser Shock Peening on Mechanical Properties of Wire Arc Additive Manufactured Grade 91 Steel and Monel-400 Bimetallic Components. *Met. Mater. Int.*, 1–13, 2024.
40. Garcia Filho, F.C. and Monteiro, S.N., Welding joints in high entropy alloys: a short-review on recent trends. *Materials, 13*, 1411, 2020.
41. Lopes, J.G. and Oliveira, J.P., A short review on welding and joining of high entropy alloys. *Metals (Basel), 10*, 212, 2020.
42. Lu, Y., Chen, J., Zhang, Y., Predictive control based double-electrode submerged arc welding for filet joints. *J. Manuf. Process.*, 16, 415–426, 2014.
43. Huang, J., Pan, W., Chen, J., Shao, Y., Yang, M., Zhang, Y., The transient behaviours of free surface in a fully penetrated weld pool in gas tungsten arc welding. *J. Manuf. Process.*, 36, 405–416, 2018.
44. Chandrasekhar, N., Vasudevan, M., Bhaduri, A.K., Jayakumar, T., Intelligent modeling for estimating weld bead width and depth of penetration from infra-red thermal images of the weld pool. *J. Intell. Manuf.*, 26, 59–71, 2015.
45. Yu, R.W., Guo, S., Huang, Y., Dai, H.J., Qiu, S.B., Peng, Y., Wang, K.H., Prediction of variable-groove weld penetration using texture features of infrared thermal images and machine learning methods. *J. Mater. Res. Technol.*, 23, 1039–1051, 2023.
46. Yu, R.W., Huang, Y., Peng, Y., Wang, K.H., Monitoring of butt weld penetration based on infrared sensing and improved histograms of oriented gradients. *J. Mater. Res. Technol.*, 22, 3280–3293, 2023.
47. Balaji, N.S., Velmurugan, C., Kumar, M.S., Sivakumar, M., Asokan, P., Experimental Investigation on Mechanical Properties of Fdm-Based Nylon Carbon Parts Using Ann Approach. *Surf. Rev. Lett., 30*, 05, 2350028, 2023.
48. Balaji, N.S. and Jayabal, S., Artificial neural network modeling of mechanical behaviors of zea fiber–polyester composites. *Proc. Inst. Mech. Eng., Part E: J. Process Mech. Eng., 230*, 1, 45–55, 2016.
49. Balaji, N.S., Jayabal, S., Sundaram, S.K., A neural network based prediction modeling for machinability characteristics of zea fiber-polyester composites. *Trans. Indian Inst. Met.*, 69, 4, 881–889, 2016.
50. Kumar, G.S., Natarajan, U., Ananthan, S.S., Vision inspection system for the identification and classification of defects in MIG welding joints. *Int. J. Adv. Manuf. Tech.*, 61, 923–933, 2012.

51. Zhang, Z., Wen, G., Chen, S., Weld image deep learning-based on-line defects detection using convolutional neural networks for Al alloy in robotic arc welding. *J. Manuf. Process.*, 45, 208–216, 2019.
52. Zhang, B., Shi, Y., Cui, Y., Wang, Z., Chen, X., A high-dynamic-range visual sensing method for feature extraction of welding pool based on adaptive image fusion. *Int. J. Adv. Manuf. Tech.*, 117, 1675–1687, 2021.
53. Liu, T., Zheng, P., Chen, H., Zhang, L., An attention-based bilinear feature extraction mechanism for fine-grained laser welding molten pool/keyhole defect recognition. *J. Manuf. Process.*, 87, 150–159, 2023.
54. Yang, L., Li, E., Long, T., Fan, J., Mao, Y., Fang, Z., Liang, Z., A welding quality detection method for arc welding robot based on 3D reconstruction with SFS algorithm. *Int. J. Adv. Manuf. Tech.*, 94, 1209–1220, 2018.
55. Sreedhar, U., Krishnamurthy, C.V., Balasubramaniam, K., Raghupathy, V.D., Ravisankar, S., Automatic defect identification using thermal image analysis for online weld quality monitoring. *J. Mater. Process. Tech.*, 212, 1557–1566, 2012.
56. Vasudevan, M., Chandrasekhar, N., Maduraimuthu, V., Bhaduri, A.K., Raj, B., Real-time monitoring of weld pool during GTAW using Infra-Red thermography and analysis of Infra-Red thermal images. *Weld. World*, 55, 83–89, 2011.
57. Yamashita, S., Yamamoto, M., Shinozaki, K., Kadoi, K., Mitsui, K., Usui, H., *In-situ* temperature measurement using a multi-sensor camera during laser welding. *Q. J. Jpn. Weld. Soc.*, 33, 93–97, 2015.
58. Lee, S.S., Kim, T.H., Hu, S.J., Cai, W.W., Abell, J.A., Analysis of weld formation in multilayer ultrasonic metal welding using high-speed images. *ASME J. Manuf. Sci. Eng.*, 137, 031016, 2015.
59. Huang, Y., Xu, S., Yang, L., Zhao, S., Liu, Y., Shi, Y., Defect detection during laser welding using electrical signals and high-speed photography. *J. Mater. Process. Tech.*, 271, 394–403, 2019.
60. Zhang, Z., Chen, H., Xu, Y., Zhong, J., Lv, N., Chen, S., Multisensor-based real-time quality monitoring by means of feature extraction, selection and modeling for Al alloy in arc welding. *Mech. Syst. Signal. Pr.*, 60–61, 151–165, 2015.
61. Gao, X., Sun, Y., You, D., Xiao, Z., Chen, X., Multi-sensor information fusion for monitoring disk laser welding. *Int. J. Adv. Manuf. Tech.*, 85, 1167–1175, 2016.
62. Deng, F., Huang, Y., Lu, S., Chen, Y., Chen, J., Feng, H., Zhang, J., Yang, Y., Hu, J., Lam, T.L., Xia, F., A multi-sensor data fusion system for laser welding process monitoring. *IEEE Access, 8*, 147349–14735, 2020.
63. Ablad, E., Spensieri, D., Bohlin, R., Carlson, J.S., Stromberg, A., Spatial-Temporal Load Balancing and Coordination of Multi-Robot Stations. *IEEE Trans. Autom. Sci. Eng.*, 20, 4, 1–12, 2022.

64. Gan, Y., Duan, J., Chen, M., Dai, X., Multi-robot trajectory planning and position/force coordination control in complex welding tasks. *Appl. Sci.*, 9, 924, 2019.
65. Tan, S., Yang, J., Ding, H., A prediction and compensation method of robot tracking error considering pose-dependent load decomposition. *Rob. Comput. Integr. Manuf.*, 80, 102476, 2023.
66. Li, D., Yang, J., Zhao, H., Ding, H., Contact force plan and control of robotic grinding towards ensuring contour accuracy of curved surfaces. *Int. J. Mech. Sci.*, 227, 107449, 2022.
67. Ireri, D., Belal, E., Okinda, C., Makange, N., Ji, C., A computer vision system for defect discrimination and grading in tomatoes using machine learning and image processing. *Artif. Intell. Agric.*, 2, 28–37, 2019.
68. Reisgen, U., Mann, S., Middeldorf, K., Sharma, R., Buchholz, G., Willms, K., Connected, digitalized welding production—Industrie 4.0 in gas metal arc welding. *Weld. World*, *63*, 4, 1121–1131, 2019.
69. Reisgen, U., Olschok, S., Krichel, T., Gach, S., Determination of the influence of welding parameters on the efficiency of electron beam welding by measurement of backscattered electrons. *Vacuum*, *159*, 182–185, 2019.
70. Zhou, J., Li, P., Zhou, Y., Wang, B., Zang, J., Meng, T., Toward new-generation intelligent manufacturing. *Engineering*, *4*, 11–20, 2018.
71. Kagermann, H., Helbig, J., Hellinger, A., Wahlster, W., Recommendations for implementing the strategic initiative INDUSTRIE 4.0: securing the future of German manufacturing industry; final report of the Industrie 4.0 Working Group, Forschungsunion, 2013.
72. Wang, L., Törngren, M., Onori, M., Current status and advancement of cyber-physical systems in manufacturing. *J. Manuf. Syst.*, 37, 517–527, 2015.
73. Lu, Y., Industry 4.0: a survey on technologies, applications and open research issues. *J. Ind. Inf. Integr.*, 6, 1–10, 2017.
74. Doshi, S.J., Gohil, A.V., Mehta, N.D., Vaghasiya, S.R., Challenges in fusion welding of Al alloy for body in white. *Mater. Today Proc.*, 5, 2, 6370–6375, 2018.
75. Zhang, Y.M., Yang, Y.P., Zhang, W., Na, S.J., Advanced welding manufacturing: a brief analysis and review of challenges and solutions. *J. Manuf. Sci. Eng.*, 142, 11, 110816, 2020.
76. Darji, R., Badheka, V., Mehta, K., Copper welding: Challenges and future scope.
77. Zhang, L.J., *et al.*, A comparative study on the microstructure and properties of copper joint between MIG welding and laser-MIG hybrid welding. *Mater. Des.*, 110, 35–50, 2016.
78. Raju, L.S., Kumar, A., Indreswaraiah, P., Effect of Welding Speed on Mechanical Properties of Friction Stir Welded Copper. *Int. J. Adv. Manuf. Technol.*, 4, 2, 68–71, 2014.
79. Reisgen, U., Olschok, S., Jakobs, S., Turner, C., Sound welding of copper: Laser beam welding in Vacuum. *Phys. Procedia*, *83*, 447–454, 2016.

80. Lin, J.W., Chang, H.C., Wu, M.H., Comparison of mechanical properties of pure copper welded using friction stir welding and tungsten inert gas welding. *J. Manuf. Process.*, 16, 2, 296–304, 2014.
81. Praveen, P. and Yarlagadda, P.K.D.V., Meeting challenges in welding of aluminum alloys through pulse gas metal arc welding. *J. Mater. Process. Tech.*, 164, 1106–1112, 2005.
82. Posinasetti, P. and Yarlagadda, P., Meeting challenges in welding of aluminium alloys through pulse gas metal arc welding. *J. Mater. Process. Tech.*, 164, 1106–1112, 2005.
83. Galvao, I., Verdera, D., Gesto, D., Loureiro, A., Rodrigues, D.M., Analysing the challenge of aluminum to copper FSW, in: *Proceedings of 9th international symposium on friction stir welding*, Huntsville, Alabama, US, 2012, May.
84. Meng, X., Huang, Y., Cao, J., Shen, J., dos Santos, J.F., Recent progress on control strategies for inherent issues in friction stir welding. *Prog. Mater Sci.*, 115, 100706, 2021.
85. Murr, L.E., Li, Y., Trillo, E., McClure, J.C., Fundamental issues and industrial applications of friction-stir welding. *Mater. Technol.*, 15, 1, 37–48, 2000.

Index

Printed and bound by CPI Group (UK) Ltd, Croydon, CR0 4YY
17/09/2026
14964264-0007